Health and Safety—The Modern Legal Framework

Second edition

Ian Smith MA LLB (Cantab)
of Gray's Inn, Barrister, Devereux Chambers;
Clifford Chance Professor of Employment Law, University of East Anglia;
member of the ACAS panel of industrial arbitrators

Christopher Goddard LLB (Hons)
of the Middle Temple, Barrister, Devereux Chambers

Stephen Killalea LLB (Hons)
of the Middle Temple, Barrister, Devereux Chambers

Nicholas Randall LLB (Hons)
of the Middle Temple, Barrister, Devereux Chambers

Butterworths
London, Dublin and Edinburgh
2000

United Kingdom	Butterworths, a Division of Reed Elsevier (UK) Ltd, Halsbury House, 35 Chancery Lane, LONDON WC2A 1EL and 4 Hill Street, EDINBURGH EH2 3JZ
Australia	Butterworths, a Division of Reed International Books Australia Pty Ltd, CHATSWOOD, New South Wales
Canada	Butterworths Canada Ltd, MARKHAM, Ontario
Hong Kong	Butterworths Hong Kong, a Division of Reed Elsevier (Greater China) Ltd, HONG KONG
India	Butterworths India, NEW DELHI
Ireland	Butterworth (Ireland) Ltd, DUBLIN
Malaysia	Malayan Law Journal Sdn Bhd, KUALA LUMPUR
New Zealand	Butterworths of New Zealand Ltd, WELLINGTON
Singapore	Butterworths Asia, SINGAPORE
South Africa	Butterworths Publishers (Pty) Ltd, DURBAN
USA	Lexis Law Publishing, CHARLOTTESVILLE, Virginia

A CIP Catalogue record for this book is available from the British Library.

ISBN 0 406 93144 5

Typeset by B & J Whitcombe, Nr Diss, Norfolk, IP22 2LP
Printed and bound in Great Britain by William Clowes Limited, Beccles and London

In memory of Georgia Giles

Foreword

Not many years ago (or so it now seems to me) a Lord Justice, hearing an appeal, and upon being presented with a list of authorities, remarked, 'Why have we been provided with authorities? This is a personal injuries case'. In those days, the personal injuries practitioner needed little more than a couple of quite thin textbooks, a working knowledge of about six leading cases, and a way with the judge.

How times change! The law is ever changing, ever expanding, and (dare one say it?) becoming ever more complex. It becomes ever more difficult to keep abreast of the changes. New cases throw up new principles or modify old ones. New rules of procedure need to be mastered.

In no area has there been more wholesale change than in the area of workplace safety regulation. There are multifarious publications (some may say too many) to which the practitioner may turn for guidance about case law and procedure. There is, however, very little to be found on the law library bookshelves to help with the many Regulations, often couched in obscure language, and which regulate the many working activities in which people are employed or otherwise engaged. This book was published to fill the gap. It succeeds in its objective. It provides a straightforward and useful guide through the various intricacies of the Regulations.

This is the second edition. Where appropriate, sections have been rewritten. New material has been inserted. Account has been taken of court decisions, both at home and in Europe, which affect interpretation of regulatory provisions. I commend this book to all who need to come to grips with this important part of our law.

Alistair MacDuff

District Court Judge at Birmingham

Preface

It is now seven years since the publication of *Health and Safety—the New Legal Framework* and although several developments have occurred during the intervening years we hope that this book will meet the same objectives which were set for the first edition. The introduction of the so called 'six-pack' of Regulations in 1992 represented a revolutionary change in the structure of health and safety law in this country and the first edition of this work was designed to be a useful, thought-provoking and informative guide to the new framework. Our central purpose was to produce a book which provided a short introduction to the Regulations and which placed them in their historical and European dimensions. We also sought to provide some insight into their likely impact on health and safety litigation and some relevant precedents were included to assist the interested practitioner. However, the over-riding object was to provide a distillation of the major principles in a concise and digestable form, along with the full text of the relevant Directives, the Regulations themselves and the accompanying Codes of Practice and Guidance Notes. In so doing we hoped to have provided a good introduction to the new changes not only for the interested practitioner but also for employers, workers, trade unions and all interested parties.

We have been flattered by the responses received with respect to the first edition and the validity of producing a second edition has been driven in no small part by the helpful and encouraging comments received from colleagues and others in the past. Once again our purpose has been to produce a book which can provide the reader with a concise introduction to the law whilst also providing the full text of all of the relevant material, including the Codes of Practice and Guidance Notes which are often neglected by the authors of legal textbooks. In the circumstances we hope to have produced an able companion to the other principal health and safety works in the Butterworths stable, namely *Munkman on Employer's Liability* and *Redgrave's Health and Safety*.

One important change from the first edition has been the inclusion of a chapter dealing with criminal prosecutions. Of all the changes introduced since the first edition, the increasing willingness of the relevant authorities to resort to criminal sanctions in the area of health and safety is probably the most significant. This trend of resorting to the criminal law represents an important shift in cultural attitudes to safety in general, and is likely to gather pace with the increasing demands from the public that those who hold responsibility for safety should be properly answerable for it, both within an individual and corporate capacity. The high profile disasters on our national rail network with the terrible accompanying loss of life are likely to fuel the public debate in this area.

Once again we are indebted to the staff at Butterworths for their assistance and support. We remain responsible for any errors, omissions or false prophecies, although we trust that our risk assessments have reduced the possibility of such errors to a minimum.

The law is stated as at 1 August 2000.

Ian Smith
Christopher Goddard
Stephen Killalea
Nicholas Randall

Devereux Chambers, London

Contents

Table of Statutes

Table of Statutory Instruments

Table of European Legislation

Table of Cases

PARA

PARA

M

N

O

P

R

PARA

W

Chapter 1

Introduction

1.1 On their introduction in 1992, the health and safety Regulations which form the subject matter of this book represented a wholesale reform in this important area of the law. As a direct result of their implementation a substantial body of the pre-existing law was repealed or revoked and much of what was left was amended or qualified. The changes introduced in 1992 had a substantial impact on the way in which personal injury litigation was conducted and heralded a new approach from practitioners, employers, workers and unions. These changes were further accentuated by the civil procedure reforms introduced in the Civil Procedure Rules after the Woolf Report.

CONCENTRATING ON THE RISK

1.2 The Regulations represented a major shift away from the old system of looking to the type of workplace or premises in which the accident occurred in order to establish which, if any, statutory provisions applied. This is because, as their titles suggest, the 1992 Regulations and their successors concentrate on combating specific categories of risk as opposed to looking to the premises in which that risk has occurred. The Regulations were important for introducing universality of coverage and this change represented a substantial extension of health and safety protection for large numbers of workers who were previously unprotected.

THE EUROPEAN DIMENSION

1.3 The 1992 Regulations were also of particular significance because of their European origin. In many instances they represented the first major exposure for courts and practitioners to the influence of the European Union, with potentially far-reaching consequences. It is well established that when considering the impact and coverage of the Regulations it is necessary to look first at the wording and overall scheme of the Directive from which they are taken. In addition it is no longer sufficient for the state or its emanations to defend its actions on the basis of the domestic Regulations if they fail to meet the standards of the Directive from which they are derived. These factors are likely to increase in importance as the European Court of Justice provides its interpretations of the Directives in the future.

IMPACT ON CIVIL LIABILITY

1.4 As predicted in the first edition of this work the influence of the Regulations has also had an impact on the law of negligence in the general field of civil liability. The Directives can be seen as not only setting specific legal standards, but also as setting acceptable standards for health and safety practice throughout the European Union. It is certainly now arguable that a failure to meet the standards of a Directive will represent a breach of duty in negligence even if the risk in question is not directly covered by the provision. In addition, the general duty on employers to carry out risk assessments and consult with their workforce will be of considerable importance when considering questions of foreseeability at common law.

1.5 It is hoped that this book will provide a useful introduction to the legal framework for all those who work in the area of health and safety. Wherever possible the authors have attempted to highlight the areas in which difficulties may arise as well as providing a general commentary on the substance of the Regulations. In addition, chapters deal in some detail with the basic structures of European law, the ever increasing importance of criminal prosecutions and the practical impact of the Regulations on personal injury litigation.

Chapter 2

Interrelationship with existing provisions

2.1 It is important to realise that the Regulations which form the subject matter of this book do not exist in isolation; in spite of the fact that they constitute the single most important development in health and safety law in nearly two decades (and to a large extent stand together under the umbrella of the Management of Health and Safety at Work Regulations 1992 and 1999, enacting the 'Framework' Directive[1]), they are primarily meant to fit into much of the existing legislation.

Large parts of the old statute law were, of course, repealed and replaced (subject to transitional provisions), in particular the old factory legislation; on the other hand, the Health and Safety Commission (HSC) were concerned that the enactment of the Regulations should not compromise the more modern legislation (for example, the Control of Substances Hazardous to Health Regulations 1988) which had been produced as a result of domestic initiatives.[2] Thus, although efforts were made in some contexts to rationalise the coverage of old and new,[3] as a general rule it was accepted that there may well be overlaps and that this in turn was no bad thing, given that the overall aim was to produce a matrix of modern laws applying to all industrial undertakings. This was made clear in the first paragraph of the first Approved Code of Practice (ACOP), 'Management of Health and Safety at Work':

> 'The duties of the Management of Health and Safety at Work Regulations, because of their wide-ranging general nature, overlap with many existing regulations. Where duties overlap, compliance with the duty in the more specific regulation will normally be sufficient to comply with the corresponding duty in the Management of Health and Safety at Work Regulations. However, where the duties in the Management of Health and Safety at Work Regulations go beyond those in the more specific regulations, additional measures will be needed to comply fully with [the Regulations].'

Similarly, there are areas of overlap within the six sets of Regulations themselves, particularly in relation to the very general requirements of the 'Framework' Regulations.[4]

The most important area of overlap, however, was with the Health and Safety at Work etc Act 1974 (HSWA 1974) itself. This Act, the product of the deliberations of the Robens Committee,[5] began the process of revolutionising our industrial safety laws, and continues to act as the basis and framework for all subsequent Regulations. The relationship between the Act and the Regulations may conveniently be explained in three stages: (1) what the Act had already achieved; (2) how the Regulations took these achievements further; and (3) how the Regulations advanced the law beyond the Act.

1 Council Directive 89/391/EEC; see Chapter 5.
2 Ironically, one of the Commission's fears about qualified majority voting under art 118a of the Treaty of Rome was that it could result in a setting of *lower* health and safety standards than those already applying in the UK, usually recognised as having the most highly developed system in the EC: '... our large body of recently achieved law is at risk to the rapid counterplay of argument and compromise, with the risk that less carefully constructed and industrially acceptable solutions will emerge' (HSC, 'Plan of work for 1990–91 and beyond', para 95). Of course, the desire for higher Community-wide standards is only partly altruistic or social in its nature; it is also based on economic ideas of the 'level playing field', ie that firms in one EC country should not be able to undercut firms in other EC countries through having lower health and safety obligations and hence lower unit labour costs. On this latter basis, the Regulations can be seen as part of the 'single market' philosophy, without the need to stray into the 'social dimension' minefield.
3 See eg the Personal Protective Equipment at Work Regulations 1992, reg 3(3) (para **8.6**) which provides that the Regulations do not apply to six specified areas where existing Regulations already require the provision of specified items of protective equipment.
4 See eg the Provision and Use of Work Equipment Regulations 1998 Guidance Notes, paras 20 *ff* (see p 430).
5 *Report of the Committee on Safety and Health at Work* (Cmnd 5034 (1972)). For fuller consideration of the report and the resultant Act, see Smith & Wood *Industrial Law* (7th edn, 2000), ch 13.

WHAT THE ACT HAD ALREADY ACHIEVED

2.2 The Health and Safety at Work etc Act 1974 was based on a philosophy of active accident prevention, rather than the traditional role of the law of tort in largely reacting to accidents that have occurred, by concentrating on the awarding of compensation. The following are the main practical applications of the philosophy.

Comprehensive coverage of workplaces and processes

2.3 One of the first criticisms by the Robens report was of the vast complexity of the existing system,[1] which had grown up piecemeal, with the result that certain types of workplace or process were well covered but certain other areas were left uncovered, possibly because no major disaster had yet occurred in that area prompting Parliament to cover it in yet another piece of specific, reactive legislation.[2] The Health and Safety at Work etc Act 1974, by contrast, applies generally to all workplaces, employers and employees, and extends to many self-employed persons, and also to others such as manufacturers, designers and importers of articles to be used at work. Moreover, the domestic Regulations previously enacted prior to 1992 reflect this approach by covering certain hazards or processes *as such*, wherever they arise or are carried out. Good examples were the Regulations covering the protection of eyes,[3] noise,[4] the use of lead,[5] the use of asbestos,[6] the control of industrial major hazards[7] and the control of substances hazardous to health.[8] Likewise, provisions relating to first aid[9] and the reporting of accidents, diseases and dangerous occurrences[10] are in common, *not* reliant on the detailed nature of the workplace in question.

1 It was contained in 11 major statutes and nearly 500 statutory instruments.
2 For example, the Factories Act 1961 contained exhaustive provisions, but their application depended on the premises in question qualifying as a 'factory' within s 175(1); that definition, with voluminous case law on the meaning of phrases such as 'manual labour', 'for trade or gain' and 'close, curtilage or precinct', took seven pages of previous editions of Smith & Wood's *Industrial Law* to explain in outline only. On the other hand, before the HSWA 1974, universities were covered by no protective legislation, in spite of the fact that science faculties frequently had the potential ability to devastate the surrounding countryside for miles around.
3 Protection of Eyes Regulations 1974, SI 1974/1681.
4 Noise at Work Regulations 1989, SI 1989/1990.
5 Control of Lead at Work Regulations 1980, SI 1980/1248.
6 Control of Asbestos at Work Regulations 1987, SI 1987/2115.
7 Control of Industrial Major Accident Hazards Regulations 1984, SI 1984/1902.
8 Control of Substances Hazardous to Health Regulations 1988, SI 1988/1657.
9 Health and Safety (First Aid) Regulations 1981, SI 1981/917.
10 Reporting of Injuries, Diseases and Dangerous Occurrences Regulations 1985, SI 1985/2023.

Comprehensive duties on employers and self-employed

2.4 Sections 2–6 of the HSWA 1974 place general duties on employers, the self-employed, persons otherwise in control of premises and designers, manufacturers, importers or suppliers of articles for use at work. These duties are owed not just to employees, but also to others who may be affected by work activities. They are cast in terms of reasonable practicability for the most part, and the principal provision, the duty in s 2(1) on 'every employer to ensure, as far as is reasonably practicable, the health, safety and welfare of all of his employees', is then filled out in s 2(2) to cover the main areas traditionally covered by the common law duty of employer to employee, namely safe plant, systems of work, storage and transport of substances, training and supervision, place of work, access and working environment. What has always marked this off however, from its common law background is that it is specifically provided[1] that these duties do *not* support civil liability. Instead, they are the basis for criminal prosecution and administrative enforcement by the Health and Safety Executive.

1 HSWA 1974, s 47(1).

Self-regulation

2.5 Section 16 of the HSWA 1974 provides for the HSC to approve codes of practice giving guidance on the practical application of the general duties in the Act. This element of self-regulation and practical guidance was an important element in seeking to maximise safety, rather than relying on an 'attitude of minimum compliance' engendered by the older, more technical approach to the drafting of industrial safety legislation. Any such codes of practice are specifically made admissible in criminal proceedings, and may, of course, be relevant also in civil proceedings, subject to the usual caveat that they are not in themselves law, and so a failure to follow a suggested practice does not in itself render the employer liable.[1]

1 HSWA 1974, s 17. In criminal proceedings where a provision of an ACOP is not complied with there is a reversal of the burden of proof on to the employer to show that he has complied with the statutory obligation in question in some other way: s 17(2).

Joint regulation

2.6 The involvement of employees through recognised trade unions (where appropriate) was another key idea behind the HSWA 1974. This is advanced by means of safety representatives and safety committees, with statutory rights to be consulted and to have time off work for the necessary training and exercise of their functions.[1] The government originally took the view that these existing provisions largely complied with the new EC initiatives and so made only minor amendments. However, this approach was rejected by the ECJ and the government were forced to amend the law in 1996.[2]

1 Safety Representative and Safety Committees Regulations 1977, SI 1977/500.
2 See Chapter 12, below.

Duties on employees

2.7 Sections 7 and 8 of the HSWA 1974 place statutory duties on employees: (1) to take reasonable care for their own and other people's safety; (2) to co-operate with the employers to the extent necessary to allow those employers to comply with their statutory obligations; and (3) not intentionally or recklessly to interfere with or misuse 'anything provided in the interests of health, safety or welfare in pursuance of any of the relevant statutory provisions'. As with the general duties on the employer, breach of these provisions does not give rise to civil liability, but could lead to criminal liability.

Enforcement agencies and powers

2.8 A priority under the HSWA 1974 was the unification of the inspectorates, and the modernisation of their administrative powers. The Health and Safety Commission was established as the governing body, with the Health and Safety Executive as the executive arm. Wide powers are given to inspectors;[1] although in most cases they will proceed by way of education and suggestion, legal teeth are given in the form of the inspector's powers to issue an improvement notice[2] or (where there is a risk of serious personal injury) a prohibition notice[3] or physically to seize and render harmless any article or substance which is a cause of imminent danger of serious personal injury.[4]

1 HSWA 1974, s 20. Subsection (2) specifies remarkably wide powers of entry, investigation, sampling, etc in heads (a) to (l), followed by (m): 'any other power which is necessary for the purpose mentioned in subsection (1) above'.
2 HSWA 1974, s 21; such a notice specifies the contravention in question and requires the person to remedy it within a set period.
3 HSWA 1974, s 22; such a notice actually requires the cessation of the activity in question unless and until the situation is remedied. Appeal against both kinds of notice lies to an industrial tribunal, but in the case of a prohibition notice it remains in force pending the appeal: s 24(3). Such appeals are relatively rare. In 1999/2000 inspectors issued 6,954 improvement notices and 4,350 prohibition notices.
4 HSWA 1974, s 25.

Criminal penalties

2.9 Part of the Robens philosophy was that, given the aim of encouraging the maximisation of safety, criminal enforcement should be relegated to a longstop *but* at the same time the available penalties should be made realistic and effective in case of need.[1] A long series of offences is set out in s 33 of the HSWA 1974, including breach of the general provisions in ss 2–8, contravention of any health and safety Regulations,[2] obstruction of an inspector and breach of an improvement or prohibition notice. The desire to increase the 'ultimate deterrent' effect was achieved in two ways: (1) although some offences are summary only, the principal offences can now be tried on indictment, in which case the penalty is an unlimited fine (particularly important in the prosecution of a large company in a high-profile case) and/or a maximum of two years' imprisonment;[3] (2) the *level* of the responsibility within an organisation is kept deliberately vague and, although the policy is invariably to prosecute the organisation not an individual, there is nothing in law to *prevent* an individual being prosecuted and, ultimately, imprisoned.

1 For many years the maximum penalty under the factories legislation was £40.
2 Thus the six sets of Regulations do *not* contain enforcement provision, but instead are enforceable under these established provisions. Section 40 contains another reversal of the burden of proof: where the allegation is of failure to do something as far as practicable or reasonably practicable, it is for the accused to prove that it was *not* practicable or reasonably practicable to do more, or that there was no better practicable means than was in fact used.
3 The normal penalty maximum in summary proceedings is a fine up to level 5 on the standard scale, but as from March 1992 certain stiffer penalties were introduced; thus, breach of ss 2–6 of the HSWA 1974 (the general duties on employers) carries a maximum of £20,000 or six months' imprisonment or both. In 1999/2000 there were 2,253 prosecutions, leading to 1,602 convictions; the average fine was £6,744.

Health and safety Regulations

2.10 Finally, the HSWA 1974 laid the basis for the progressive replacement of all the existing (largely outmoded[1]) legislation by modern health and safety Regulations. The procedure is laid down[2] and, once they are enacted, not only are they criminally enforceable in the normal way (above) but also it is specifically provided that they *do* support civil liability unless they provide otherwise.[3]

1 The principal safety provisions in the Factories Act 1961, ss 12–14 still reflected a nineteenth-century cotton mill, ie s 12 'prime movers' (steam boilers), s 13 'transmission machinery' (overhead belts) and s 14 'dangerous parts of machinery

other than prime movers or transmission machinery' (the actual machines powered by the belts).

2 HSWA 1974, ss 15, 82.

3 HSWA 1974, s 47(2). The Management of Health and Safety at Work Regulations 1999 do exclude civil liability for their breach (reg 22: see para **5.20**), but the other five sets of Regulations do not, and so will support civil liability in an action for breach of statutory duty. Moreover, it is the contention in this book that, despite reg 22, breach of the Framework Regulations *will* be relevant in establishing common law negligence.

HOW THE REGULATIONS TOOK THESE ACHIEVEMENTS FURTHER

Replacement of existing legislation

2.11 The original aim at the time of the HSWA 1974 was for the repeal and replacement of all of the old legislation within two decades. Although a great deal of work was done in producing health and safety Regulations, these tended to be concerned with new areas, or only repealed and replaced specific parts of the old legislation, so that by the mid-1980s we still had with us most of the original factory legislation, etc, and it seemed that the original aim would not be realised. However, this most recent involvement of the EC through art 118a (now art 137) of the Treaty of Rome revitalised that aim and produced, with effect from 1 January 1993, the new overall framework that had been long awaited.

Use of Codes of Practice

2.12 As already seen, the device of the Code of Practice was an essential element of the Robens philosophy. One original idea was that such codes could be used *instead* of Regulations, but that did not come to fruition. Instead, Codes have been used in the more traditional way to flesh out certain sets of Regulations. However, it is arguable that the dependence on ACOPs or Guidance Notes in the new structure is quantitatively different and comes at least several steps closer to that original idea—the six sets of Regulations are themselves of course fundamental *but* it may be that in practice it is the extensive guidance that assumes the dominant role, certainly in the minds of those charged with the day-to-day application of the rules. This is arguably shown simply by the form of presentation of the six ACOPs or Guidance Notes which include within them the (short) text of each regulation, which is then expanded upon to a very considerable extent.[1]

1 Prime examples of this are the Workplace (Health, Safety and Welfare) ACOP with its extensive explanations of standards etc actually required in the light of the Regulations, and the Manual Handling Guidance Notes and Display Screen Guidance Notes with their use of diagrams.

Comprehensive coverage

2.13 The comprehensive, cross-industries nature of the obligations is obvious on first reading. The Management of Health and Safety at Work Regulations could hardly have been cast in wider terms, and the same approach can be seen in for example, the definitions of 'workplace',[1] 'work equipment,[2] and VDU 'user'.[3]

1 '[A]ny premises or part of premises which are not domestic premises and are made available to any person as a place of work': Workplace (Health, Safety and Welfare) Regulations 1992, reg 2(1).
2 '[A]ny machinery, appliance, apparatus, tool or installation for use at work (whether exclusively or not)': Provision and Use of Work Equipment Regulations 1998, reg 2(1).
3 '[A]n employee who habitually uses display screen equipment as ... part of his normal work': Health and Safety (Display Screen Equipment) Regulations 1992, reg 1(2). 'Use' means 'use for or in connection with work': reg 1(2).

Maximisation of safety

2.14 The emphasis on accident *prevention* can be seen not just in the modernised forms of requirements in relation to particular machinery or processes, but more generally in the importance given to positive steps to *assess risk*, act upon such assessments and act so as to eliminate, avoid or (at the least) lessen perceived risks.[1] This can be seen particularly in the Management of Health and Safety at Work Regulations, which also reflect the modern concerns not just with accidents but also with industrial *health* risks.

1 This appears especially in the Manual Handling Operations Regulations 1992, reg 4(1): see paras **9.8–9.11**. Of major importance legally is the emphasis placed throughout the Regulations on providing, to use the old common law terminology, safe *systems* of work.

Information and written policies

2.15 The Health and Safety at Work etc Act 1974 introduced a requirement of more openness in health and safety measures, and placed a statutory obligation on employers to produce written health and safety policies.[1] A similar approach can be seen in the requirement of written risk assessments and health and safety arrangements in the Framework Regulations.[2]

1 HSWA 1974, s 2(3). Employers of less than five employees are exempt: Employers' Health and Safety Policy Statements (Exemptions) Regulations 1975, SI 1975/1584.
2 Management of Health and Safety at Work Regulations 1999, regs 3(6) and 5(2); again, employers of less than five employees are exempt. See also reg 10 on the 'comprehensible and relevant information' that must be given to employees.

Statutory duties on employees

2.16 In addition to the general duties on employees in the HSWA 1974, there are further obligations imposed by the Regulations:

(a) to use equipment in accordance with training or instructions and to inform the employer of serious and imminent dangers or shortcomings in protective arrangements;[1]

(b) to use personal protective equipment in accordance with training or instructions and to report any loss of or obvious defect in that equipment;[2] and

(c) to make full and proper use of any system provided by the employer for manual handling of loads.[3]

1 Management of Health and Safety at Work Regulations 1999, reg 14.
2 Personal Protective Equipment at Work Regulations 1992, regs 10, 11.
3 Manual Handling Operations Regulations 1992, reg 5.

HOW THE REGULATIONS WENT BEYOND THE ACT

2.17 The Regulations constituted a significant advance on the HSWA 1974 in the following ways.

Their origin in EC harmonisation

2.18 The initial reaction of the HSC to the proposed EC Directives was that the UK already complied substantially in one way or another, in other words that it was all there somewhere if you looked hard enough; in particular, much of it was already at least implicit in the wide general duties in the HSWA 1974. In the event, of course, the Directives led to a major reformulation and modernisation, in which the EC origins of the Regulations can be seen in at least two ways. First, although some of the Regulations retain traditional formulations such as 'reasonably practicable' and most are compatible with the long-standing UK approach of adopting the 'best practicable means', many of the Regulations adopt novel formulations taken from the Directives such as 'effective and suitable' and 'suitable and sufficient'. Second, harmonisation in this context includes making the Regulations harmonious with EC requirements and Directives in other areas, particularly those of product liability and industrial standards. Thus, one of the obligations of an employer under the Provision and Use of Work Equipment Regulations[1] is to ensure that work equipment complies with any applicable product Directives,[2] and personal protective equipment is to comply with relevant EC design or manufacture standards.[3]

1 Regulation 10(1). 'Existing legislation on the manufacture and supply of new work equipment is increasingly being supplemented by new and more detailed Regulations implementing EC Directives made under Article 100a' (the 'Single Market' powers): Provision and Use of Work Equipment Regulations 1998 Guidance Notes, para 201 (see p 466). On the importance of art 100a Directives, see paras 202–206 (pp 466–467).
2 The aim eventually is to have harmonised standards for the production of work equipment (bearing the 'CE Mark'); reg 10(2) provides that where equipment complies with harmonised standards the detailed requirements of regs 11–19 and 22–29 only apply to the extent that the relevant Community Directives are not applicable to that item of work equipment. Thus, the health and safety provisions and the standards provisions are meant to be integrated.
3 Personal Protective Equipment at Work Regulations 1992, reg 4(3)(e), Sch 1. Harmonised European standards (ENs) are increasingly to replace existing British standards: Guidance Notes, paras 33–35 (see Appendix 4, pp 538–539).

Greater emphasis on the concept of risk

2.19 Although the HSWA 1974 was more positive and proactive than the previous legislation, much of it was still concerned with what might be termed the 'static condition' approach to industrial safety. More advanced ideas of hazard analysis and risk assessment have already appeared in domestic legislation since the Act,[1] but they now occupy a much more central place in the scheme of the Regulations.

1 See eg the Control of Substances Hazardous to Health Regulations 1988.

Inclusion of temporary workers

2.20 In addition to the six specific Directives, the Regulations are also intended to comply with Council Directive 91/383/EEC which requires the extension of the usual level of health and safety protection to (a) employees on fixed-term contracts, and (b) temporary employees provided by an 'employment business'. There is a specific provision in the Management of Health and Safety at

Work Regulations[1] requiring the giving of health and safety information to such employees; other requirements of the Directive are presumably to be read into other parts of the Regulations by implication, at least where this is legally possible.

1 Management of Health and Safety at Work Regulations 1999, reg 15.

Nature of the obligations

2.21 In line with the EC approach generally, the modern requirements are in some ways more prescriptive than the older domestic legislation that they replace. As mentioned above, the guidance given is certainly more voluminous and detailed. With regard to the level of duty in the Regulations themselves, some vestiges remain of existing 'as far as reasonably practicable' duties, and strict liability continues to apply in certain areas where it applied before.[1] However, there is one shift of emphasis which may impart more elements of strict (though not absolute) liability. Where a new regulation is cast in terms of the provision of something to a 'suitable and sufficient level' (rather than providing it 'as far as reasonably practicable') the drafting tends to state that it *shall* be provided to that level.[2] Where this is the case, the obligation to *provide* it is strict, with any argument being as to whether that which was in fact provided was 'suitable and sufficient' (whereas before there could have been arguments as to whether it was 'reasonably practicable' to provide it *at all*); it is unlikely that the 'suitable' level of provision would be held to be nil. Moreover, in at least one place the Regulations themselves seem to accept that the new 'suitable and sufficient' type of duty is higher than the old reasonably practical duty—reg 17(2) of the Workplace (Health, Safety and Welfare) Regulations 1992 states that there must be suitable and sufficient traffic routes, but then reg 17(5) qualifies this by saying that that requirement only applies to certain premises 'as far as reasonably practicable'. Thus, one of the key areas in the legal application of the Regulations is still likely to be the interpretation to be placed by the courts on the terminology of suitability, sufficiency, effectiveness, adequacy and appropriateness. Dictionary definitions might be a starting point, but at the end of the day these terms will have to be interpreted in the light of the Regulations overall, being heavily questions of fact, on which the guidelines given in the appropriate ACOP or Guidance Notes could well be determinative in practice.

1 Thus the Provision and Use of Work Equipment Regulations, reg 11 ('Dangerous parts of machinery') is a modernised form of the strict liability regime of

the old Factories Act 1961, s 14, though of course now applicable to any workplace (unless specifically excluded); even here, however, it is arguable that the nature of the duty may have changed: see para **7.11**.

2 For example, the Provision and Use of Work Equipment Regulations, reg 21 states that 'Every employer shall ensure that suitable and sufficient lighting . . . is provided at any place where a person uses work equipment'. Another form of wording used in conjunction with a strict requirement is 'where appropriate': 'Every employer shall ensure that, where appropriate, work equipment is provided with suitable means to isolate it from its sources of energy' (reg 19(1)).

Chapter 3

The European dimension

INTRODUCTION

3.1 The Regulations were originally introduced by the UK government in order to implement the first group of Directives made under art 118a of the Treaty of Rome and adopted by the Council of Ministers of the EC for implementation by the member states by 31 December 1992. Article 118a was introduced by the Single European Act 1986 (SEA). The aim of the SEA was to transform the existing relationships between the member states of the EC into a European Union in accordance with the principles of the Stuttgart Solemn Declaration of June 1983. In support of this aim the SEA contained new policy objectives in the fields of social policy and economic and social cohesion.

Article 118a of the Treaty obliged the member states to 'pay particular attention to encouraging improvements, especially in the working environment, as regards the health and safety of workers' and provided for the adoption of Directives to achieve this purpose on the basis of majority voting in the EC Council of Ministers.

The Directives which form the subject matter of this work can therefore be seen as the initial fruits of art 118a. The wide embracing nature of art 118a was well illustrated by the decision of the ECJ in Case C-84/94 *United Kingdom v EU Council* [1996] All ER (EC) 877, [1997] IRLR 30 in which 'health' was interpreted as 'a state of complete physical, mental and social well-being that does not consist only in the absence of illness or infirmity'.

Article 118a has subsequently been replaced by art 137, which provides for 'improvements in particular of the working environment to protect workers' health and safety'.

IMPLEMENTATION IN THE UK

3.2 The UK implemented the first series of Directives by means of statutory Regulations made under s 15 of the Health and Safety at Work etc Act 1974. The Regulations were laid before Parliament by the Secretary of State for Employment on the basis of proposals submitted to her by the Health and Safety Commission, which had carried out consultations in accordance with its powers and duties under the 1974 Act.

The stated aim of the Health and Safety Commission in implementing the Directives in the UK was to propose Regulations which did not go beyond the strict requirements of the Directives in order to minimise the impact of any changes in the law. This approach left the Commission open to criticism and it was clear that, in a number of areas, the Regulations as proposed by the Commission failed to meet the standards required by the Directives, in particular with regard to the consultation of workers.

3.3 In view of the increasing influence of the EC in the area of health and safety it is important to consider the relevant Community standards and how these interact with domestic law and some consideration of the law of the EC is therefore necessary. Indeed, it is now clear that any practitioner working in this field will require a working knowledge of EC law and procedure. Although a detailed consideration of the law of the EC is beyond the scope of this work, a brief outline of the law of the Community and, in particular, the role and legal effect of Directives is of vital importance when considering the legal framework of health and safety in the UK.

THE EUROPEAN COMMUNITIES ACT 1972

3.4 EC legislation has force in the UK by virtue of the provisions of the European Communities Act 1972. The Act is of major constitutional importance and has had a number of important repercussions. It is now clearly established that the supreme judicial authority in respect of any legal issues which involve EC law and its interaction with the law of the UK is the Court of Justice of the European Community (ECJ). It is also now generally accepted that in the event of any conflict between EC law and domestic law, then EC law will normally prevail and should be enforced as such by the domestic courts.

DIRECTIVES

3.5 Article 118a, and its successor art 137, provided that its objectives should be achieved by the use of Directives. A Directive is primarily intended to create legal relationships between the EC and the member state to which it is addressed. Article 189 of the Treaty of Rome states that a Directive 'is binding as to the result to be achieved upon each member state to which it is addressed but shall leave to the national authorities choice of form and methods'.

It was originally thought that Directives did not provide any directly enforceable rights for individuals, but this view has now been revised as a result of a series of landmark decisions from the ECJ which have established that private individuals can obtain directly enforceable rights from Directives in certain circumstances. It is now possible for a private individual to invoke the provisions of a Directive directly against the state or an employer in the public sector and, to an extent which is not yet clearly defined, an individual may further rely upon the provisions of a Directive in an action against the state in circumstances where they can show that they have suffered loss or have had their rights infringed by reason of the failure of the member state to implement the Directive effectively. It needs to be remembered that this is an expanding area of jurisprudence, the limits of which are currently undefined.

3.6 When considering the practical application of the new Regulations the practitioner will be primarily concerned with the following issues:

(a) To what extent should the court look to the relevant provisions of the Directive when interpreting the Regulations?

(b) What is the position if the standards required by the Regulations appear to fall below those set in the Directive?

(c) What is the position if the standards required by the Regulations go beyond those set in the Directive?

(d) What if the Regulations fail to implement the relevant article of the Directive at all?

INTERPRETATION

3.7 It is only in a very few instances that the wording of a particular regulation has been taken directly from the wording of the Directive and, as a result, in a large number of cases there is likely to be some difference between the wording of the regulation and the equivalent article of the Directive. In the vast majority of such cases the differences can effectively be eliminated in the process of interpretation. It is now clearly established that domestic law which has been introduced to meet the UK's Community obligations is to be construed in accordance with the applicable Community law. This principle will obviously apply to the health and safety Regulations. The position has been clearly stated by the House of Lords in *Litster v Forth Dry Dock and Engineering Co Ltd* [1989] IRLR 161:

> 'The approach to the construction of primary and subordinate legislation enacted to give effect to the United Kingdom's obligations under the EEC Treaty has been the subject matter of recent authority in this House (see *Pickstone v Freemans plc* [1988] IRLR 357) and is not in doubt. If the legislation can reasonably be construed so as to conform with those obligations – obligations which are to be ascertained not only from the wording of the Directive but from the interpretation placed upon it by the European Court of Justice at Luxembourg – such a purposive construction will be applied even though, perhaps, it may involve some departure from the strict and literal application of the words which the legislature has elected to use' (per Lord Oliver at 165).

The 'purposive' approach may well create some initial difficulties for those domestic lawyers who are used to a more 'literal' approach to construction. The 'purposive' approach to construction requires the court to look at the object and purpose of the particular provision as well as the instrument as a whole. This necessarily involves some consideration of the preamble to the Directive and a willingness to depart from the literal application of the words used by the legislature even to the extent of inserting new words into the domestic provision in order to achieve the desired effect. These points are all well illustrated by *Litster*, in which the House of Lords upheld an appeal on a point of construction of a regulation giving effect to a Community Directive on the ground that the lower courts had erroneously interpreted the instrument in a literal fashion instead of adopting the proper 'purposive' approach. This approach is likely to resolve issues of interpretation of the Regulations in favour of claimants in view of the overall scheme and aim of the Directives and art 118a to promote and ensure the health and safety of workers at work.

3.8 However, there are still limits to the ambit of the 'purposive' approach to construction. In *Litster* Lord Oliver suggests that the relevant legislation must be capable of being reasonably construed in such a manner so as to conform with the necessary Community obligations, and this has been further emphasised by Lord Keith in *Webb v Emo Air Cargo (UK) Ltd* [1993] 1 WLR 49 and by Lord Templeman in *Duke v GEC Reliance Systems Ltd* [1988] AC 168. However, it is also clear from these authorities that the courts will not be willing to adopt interpretations which distort the domestic provision. It would appear to follow from this that the courts, although willing to add words to a particular domestic provision, will not be willing to strike words out even if they do not comply with Community obligations as a result. This may well lead to some difficulties in a number of areas in the Regulations.

WHAT IS THE POSITION IF THE REGULATIONS FAIL TO MEET THE STANDARDS OF THE DIRECTIVE AND ANY DIFFERENCES CANNOT BE SOLVED BY INTERPRETATION?

3.9 If the Regulations fall below the standard required by the Directive and the differences in wording cannot reasonably be resolved through the approach of interpretation, then it is necessary to consider to what extent the individual can rely directly on the Community provision. This requires some consideration of the principle of 'direct effect'. As previously noted, it is now established that Directives are capable of giving rise to enforceable individual rights in certain circumstances.

3.10 The guiding principle as to whether a provision is capable of direct effect is the extent to which the member state has been left with any discretion as regards implementation. In principle, if no real discretion has been left to the member state, then the provision is capable of direct effect. This is well illustrated by the case of *Van Duyn v Home Office*: 41/74 [1975] Ch 358, ECJ in which it was held that a Directive could be enforced directly by a private individual against a member state in certain circumstances. For such a position to arise the article of the Directive must be 'clear, precise, and permit of no exceptions' so as not to require the intervention of the national authorities in order to be enforceable. Regard must also be had to the 'nature, general scheme, and wording of the provision'.

It is, therefore, important to consider each article of the Directive in isolation. Some articles in the same Directive may be enforceable and others not. This point is well illustrated by the cases of *Marshall v Southampton and South West Hampshire Area Health Authority (Teaching)*: 152/84 [1986] IRLR 140 and *Marshall v Southampton and South West Hampshire Area Health Authority (No 2)* [1990] IRLR 481, both of which involved consideration of the direct effect of different articles of the Equal Treatment Directive (76/207). The first *Marshall* decision decided that art 5 of the Directive was capable of direct effect, whereas the *Marshall (No 2)* decision held that art 6 of the same Directive was not.

3.11 The first *Marshall* decision also established that a Directive can have direct effect as against a member state, but cannot have direct effect as between private individuals. This approach has been confirmed by the decision of the ECJ in *Officer Van Justitie v Kolpinghuis Nilmegen BV* [1987] ECR 3969, [1989] 2 CMLR 18 in which it was observed that art 189 of the Treaty provided that the obligations contained in a Directive exist only in relation to the member state to which it is addressed. It is therefore clear that a member state cannot defend itself in any proceedings on the grounds of its own failure to implement a Directive. In other words, a Directive is capable of having what is termed 'vertical effect', ie as against the state, but is not capable of 'horizontal effect', ie as between individuals.

What constitutes the state?

3.12 The *Marshall* case, therefore, raised the issue of what organisations are capable of being the state or its emanations. The leading case in this area is the decision of the ECJ in *Foster v British Gas plc* [1990] IRLR 353 in which it was held that a Directive may be directly relied upon against an entity which provides a public service under the control of the state and which has powers in excess of those which result from the normal rules applicable in relations between individuals. The legal form of the entity is irrelevant for these purposes. Whether any particular organisation falls within the definition or not is a matter for the domestic courts. As a result of the decision of the ECJ the House of Lords had little difficulty in deciding that British Gas, prior to privatisation, was an emanation of the state for these purposes.

However, it is important to remember that the control of the state is not the sole determining factor, and this is well illustrated by the case of *Doughty v Rolls-Royce plc* [1992] IRLR 126 in which it was held that Rolls-Royce was not an emanation of the state even though the state had a controlling shareholding, because it failed to meet two of the three criteria laid down by the ECJ. However, it is strongly arguable that an organisation may still fall within the definition if some but not all of the criteria are established.

3.13 The position is now relatively clear if an employee of the state or an emanation of the state brings a claim under the health and safety Regulations. If there is a difference between the wording of the Directive and the Regulations and the relevant article of the Directive is sufficiently clear and precise, then the employee is entitled to rely upon the Directive directly.

Employees in the private sector

3.14 As a result of these authorities the position emerged in which employees in the public sector appeared to be in a more advantageous position than their counterparts in the private sector because they could enforce rights contained in Directives against their employer in certain circumstances. Indeed, this apparently arbitrary and unfair distinction between private and state employees was noted in the *Marshall* decision. However, as a result of further developments, it is now strongly arguable that this distinction has been removed by the decision of the ECJ in *Francovich v Italy* [1992] IRLR 84.

3.15 The *Francovich* case establishes the principle that there is a Community law right to compensation against a member state where it has failed to implement a Directive and an individual has suffered loss or damage as a result of the state's failure to act. A claim on this basis should be brought in the domestic courts. The right to recover from the state depends upon the nature of the breach of Community law. Liability for damages will result if three conditions are satisfied:

> 'The first of these conditions is that the result required by the Directive includes the conferring of rights for the benefit of individuals. The second condition is that the content of these rights may be determined by reference to the provisions of the Directive. Finally, the third condition is the existence of a causal link between the breach of the obligation of the state and the damage suffered by the persons affected.'

Although *Francovich* involved a complete failure by the Italian government to implement the Directive, the principle would also appear to extend to cases in which ineffective attempts to implement have been made, so long as the breach of European law has been 'sufficiently serious' to warrant compensation. For an example of such a claim see *R v Secretary of State for Transport, ex p Factortame Ltd (No 5)* [2000] 1 AC 524, HL.

3.16 Although the position needs to be clarified, it would appear that:

(a) health and safety Directives confer rights on individuals;

(b) the nature of these rights is capable of identification from the Directives; and

(c) circumstances can easily be envisaged in which a causal link could be established between the breach of the member state and the damage suffered by the individual.

It would, therefore, appear that a *Francovich*-type action would be available to a claimant in an action under the Regulations if they could show that they have suffered some detriment by reason of the failure of the member state to implement the Directive adequately and that the failure to implement was sufficiently serious to justify an award of compensation.

It would also appear that the arbitrary distinction between employees in the state sector and the private sector may have been eradicated by *Francovich*, although it will still be incumbent upon the private employee to commence fresh proceedings against the member state in the domestic courts. It also needs to be borne in mind that state employees should also be able to rely upon *Francovich* in cases in which the relevant provision of Community law is not capable of direct effect.

3.17 In summary, it would appear that, if the provisions of the Regulation fall below the standards required by the Directive and the difference cannot be overcome by adopting a 'purposive' approach to construction, then a party which has a cause of action against the state or an emanation of the state can rely directly on that provision so long as it is capable of direct effect. If the provisions of the Directive are not capable of direct effect but the conditions set out in *Francovich* are satisfied, then a remedy can be obtained from the member state in a separate action in the domestic court. In all other cases a remedy can be obtained from the state if the conditions of *Francovich* have been met. There would also appear to be no reason why the principle in *Francovich* should not apply to provisions capable of direct effect.

WHAT IS THE POSITION IF THE STANDARDS REQUIRED BY THE REGULATIONS GO BEYOND THOSE REQUIRED BY THE DIRECTIVE?

3.18 If the regulation goes beyond the standards required by the Directive, then the regulation should be construed in the normal way. In other words the purposive approach to interpretation should be used so that the regulation does not fall below the standard required by the Directive.

WHAT IS THE POSITION IF THE REGULATIONS FAIL TO IMPLEMENT THE DIRECTIVE AT ALL?

3.19 In general terms, if there has been no attempt to implement the terms of the Directive at all in the Regulations, then the position is the same as if the Regulations fall below the requirements of the Directive. In other words, if the provisions of the Directive are capable of direct effect then they can be relied upon by any party which has commenced proceedings against the state or an emanation of the state. In all other circumstances an action against the member state will lie if the requirements of *Francovich* are met.

3.20 It is important that the practitioner keeps abreast with the current developments in EC law. There is some suggestion that in all cases in which there is a conflict between domestic law and Community law the domestic law provisions must be interpreted in a manner consistent with Community law (see *Marleasing SA v La Commercial Internacional de Alimentacion SA* [1990] ECR I-4135, [1992] 1 CMLR 305, although this is not the interpretation that has been put on the case by the House of Lords in *Webb v Emo Air Cargo (UK) Ltd* [1993] 1 WLR 49) even if this means striking out the relevant provisions of domestic law. If this view is confirmed by the ECJ then it would effectively mean that Directives would be capable of 'horizontal' effect by indirect means, ie through the machinery of interpretation. If this approach was confirmed by the ECJ, then in all cases it would be open to individuals to rely directly on the Directives. However, this approach would appear to conflict with the reasoning of the previous authorities and, in particular, would impose liabilities on private individuals who have never been put under any obligation to comply with Community provisions by their member state. Such a step would therefore appear to conflict with the reasoning so far adopted by the ECJ.

THE ROLE OF THE ECJ

3.21 It is also important to bear in mind that any question of the interpretation of Community law is a matter for the ECJ and not the domestic court. It is therefore only in the most simple of cases that the domestic court can interpret the meaning of a provision of Community law. If such a question is in doubt, it should be referred to the ECJ.

Chapter 4

Civil liability

4.1 Although the primary legislative aim of industrial safety legislation has always been to prevent accidents, it has to be conceded that its main significance to the personal injury lawyer has always been as a basis for civil liability once an accident has happened. The new Regulations are obviously aimed at improving industrial safety and are, as is made clear elsewhere in this book, deliberately made more pro-active and positive in their approach than much of the legislation that they replaced. At the same time, however, it is exactly that widened scope that may (1) extend the scope of the action for breach of statutory duty, (2) increase the importance of certain facets of common law negligence, and (3) have an effect on the defence of contributory negligence. These matters are now considered in turn.

BREACH OF STATUTORY DUTY

4.2 The action for breach of statutory duty is usually a more precise and focused action in an industrial accident than common law negligence, and has often also had the advantage of imposing a stricter duty on the employer.[1] The first question, therefore, is whether the Regulations themselves support civil liability. The Health and Safety at Work etc Act 1974 (HSWA 1974), s 47(2) provides:

> 'Breach of duty imposed by health and safety Regulations shall, so far as it causes damage, be actionable except in so far as the Regulations provide otherwise.'

1 For the elements of breach of statutory duty, see Munkman *Employer's Liability* (13th edn, 2001) ch 8.

4.3 The five subsidiary sets of Regulations do not have any such exclusion of civil liability and so can be relied on as the basis for actions for breach of statutory duty. However, the head Regulations of the Management of Health and Safety at Work Regulations, do provide in reg 22(1) as follows:

> 'Breach of a duty imposed by these Regulations shall not confer a right of action in any civil proceedings.'[1]

Thus, a breach of one of the very general duties in these Regulations (in particular, the obligation under reg 3 to undertake risk assessments, which is arguably central to the whole scheme of the present regulatory system, with its emphasis on *risk*) will not give rise directly to an action for breach of duty. This restriction was not contained in the original draft of the Regulations, but appeared in the final version. From the point of view of a claimant's lawyers this is unfortunate, but on the other hand it may be that this late amendment will not have the desired effect of rendering the head Regulations of criminal and administrative effect only,[2] for two reasons:

(1) It may be arguable that the insertion of this restriction is in breach of EC law, as being a failure to implement fully the Framework Directive; remedies are usually left largely to the member state, *but* there is a general requirement that whatever remedies are enacted must be realistic[3] and so, although the Management of Health and Safety at Work Regulations are amenable to criminal and administrative enforcement, it could be argued that without the possibility of a civil liability in favour of those injured by a breach of the Regulations, they do not in fact provide full and adequate remedies.

(2) In any event, it is argued below that breach of the Management Regulations may be actionable indirectly through the medium of a common law negligence action.

1 One of the changes when the Management Regulations were reissued in 1999 was to permit (in reg 22(2)) civil liability to arise under reg 16(1) (risk assessment in respect of new or expectant mothers) and reg 19 (protection of young persons).
2 Presumably the desire was to impose the same limitation on liability as has always applied to the general duties in ss 1–8 of the HSWA 1974, by virtue of s 47(1): see Chapter 2.
3 *Von Colson v Land Nordrhein-Westfalen Case* 14/83 [1984] ECR 1891, [1982] 2 CMLR 430, ECJ.

4.4 Where one of the new Regulations does support civil liability directly, it will be necessary for a claimant seeking to rely on it to prove the usual requirements of the tort of breach of statutory duty.

The duty was owed to that claimant either individually or as a member of a class meant by Parliament to be protected

4.5 This should not normally be difficult because of the breadth of coverage of the Regulations.

The duty was owed by that defendant

4.6 Again, this should not normally be difficult in the light of the extended coverage of the Regulations, which are *not* now tied to specific industries, processes or types of workplace.

The defendant was in breach of the statutory duty

4.7 The Approved Codes of Practice (ACOPs) and Guidance Notes may well give very useful material on this point, given their extensive nature and, in some cases, very detailed suggestions. They are admissible generally in civil actions as evidence. One potentially significant point relates to the burden of proof. By virtue of the HSWA 1974, s 17(2), where in any *criminal* proceedings it is proved that there was a failure to observe a provision of an ACOP[1] covering the allegation of breach in question, then that allegation is to be taken as proved 'unless the court is satisfied that the requirement or prohibition was in respect of that matter complied with otherwise than by way of observance of that provision of the code', ie there is a statutory reversal of the burden of proof. This does not apply in civil proceedings. However, it has long been held under the old factory legislation[2] that, where a duty is imposed to do or provide something 'as far as reasonably practicable' the onus is generally on the defendant employer to satisfy the court as to any question of reasonable practicability. From the claimant's point of view, it is to be hoped that this approach will now also be applied to the many new Regulations that adopt standards, not of reasonable practicability, but of the new terminology of suitability, sufficiency, adequacy, etc.

1 Note that s 17(2) does not specifically include Guidance Notes.
2 *Nimmo v Alexander Cowan & Sons Ltd* [1968] AC 107, [1967] 3 All ER 187, HL;

Jenkins v Allied Ironfounders Ltd [1969] 3 All ER 1609, [1970] 1 WLR 304, HL; *Bowes v Sedgefield District Council* [1981] ICR 234, CA. This was reaffirmed in *Lamer v British Steel plc* [1993] ICR 551, [1993] IRLR 278, CA.

Causation

4.8 There is no presumption of causation in a breach of statutory duty action.[1]

1 *Caswell v Powell Duffryn Associated Collieries Ltd* [1940] AC 152, [1939] 3 All ER 722, HL; see Munkman *Employer's Liability* (13th edn, 2001) ch 8.

The damage caused was the type to be prevented by the statute or Regulations

4.9 This requirement (usually known as the rule in *Gorris v Scott*[1]) had certain unfortunate results under the old factory legislation, especially the fencing requirements, since it could lead to a narrow and arguably over-technical interpretation of the scope of the legislative protection. In particular, it was held that the central provisions of the Factories Act 1961, s 14 on the fencing of dangerous machinery, only applied to keep the worker out, not to keep the machine in.[2] The over-subtleties of this approach were strongly criticised in later cases,[3] but by that time there was little that the courts could do about it. It is hoped that there should be less scope for such an approach under the Regulations, with their wider coverage and frequent overlap. Thus, for example, reg 11 of the Provision and Use of Work Equipment Regulations is still concerned with keeping the worker out of machinery, *but* reg 12 now goes on to cover the ejection of material from it. On the other hand, it must be assumed that the rule in *Gorris v Scott* will still apply and so it may in some cases still be important to sue under the right Regulation (or combination of Regulations).

1 (1874) LR 9 Exch 125, ExCh.
2 *Nicholls v Austin (Leyton) Ltd* [1946] AC 493, [1946] 2 All ER 92, HL.
3 See particularly the speech of Lord Hailsham LC in *FE Callow (Engineers) Ltd v Johnson* [1971] AC 335 at 341, [1970] 3 All ER 639 at 641.

NEGLIGENCE

4.10 Common law negligence has always been important in personal injury litigation, though in the context of industrial accidents it has, arguably, been secondary to the action for breach of statutory duty. Ordinary principles apply (duty—breach—damage; the standard of reasonable care, not a guarantee of safety), but in this particular area the applications of these principles have evolved in certain specific directions. Since the seminal case of *Wilsons & Clyde Coal Co Ltd v English*[1] the usual approach is that the employer is under a personal, non-delegable duty to take a reasonable care to provide:

(a) a safe workplace;

(b) safe equipment;[2]

(c) competent and safe fellow employees;

(d) a safe system of work in all the circumstances.

1 [1938] AC 57, [1937] 3 All ER 628, HL; see Munkman *Employer's Liability* (13th edn, 2001) chs 4 and 5.

2 Latent defects in machinery or tools provided by the employer caused problems where that defect could not have been discovered by reasonable inspection and the machinery or tool had been bought from a reputable supplier and was of reputable make: *Davie v New Merton Board Mills Ltd* [1959] AC 604, [1959] 1 All ER 346, HL. However, there is now a stricter duty on the employer to provide safe work equipment by virtue of the Employer's Liability (Defective Equipment) Act 1969 under which the employee injured through a defect in equipment provided by the employer may sue that employer (if he can show that the defect is attributable to the fault of a third party), leaving it to the employer then to seek to sue the manufacturer. This Act has, in the few reported cases, been applied widely: *Coltman v Bibby Tankers Ltd, The Derbyshire* [1988] AC 276, [1987] 3 All ER 1068, HL (includes a whole ship as 'equipment'); *Knowles v Liverpool City Council* [1993] IRLR 6, CA (includes *material* provided by the employer for the purpose of the work, in that case a flagstone that the employee was laying).

4.11 The first two heads concern what may be considered to be the 'static condition' of the premises and the equipment. These matters are now covered in detail by the Regulations, particularly the Workplace Regulations and the Provision and Use of Work Equipment Regulations, with their extensive ACOP and Guidance Notes; these should provide much evidential material not just for an action for breach of statutory duty but also for a common law negligence claim. Thus, as can be seen from several of the precedents in Chapter 16, it may well be possible to re-plead the particulars of breach of statutory duty also as particulars of negligence under these first two heads.

4.12 The third head, the provision of competent and safe fellow employees, is different. It is not per se covered by the new

Regulations and so the existing (sometimes quite bizarre) case law may well still be of considerable importance in any case where the claimant is injured by the negligence of a fellow employee/practical joker/homicidal maniac.[1] That fellow employee's propensity must, of course, have been known to the employer or, at the least, the employer must reasonably have been expected to have known of it in order to be liable.[2]

1 It is possible that the employer may also be vicariously liable for the acts of the fellow employee, but the problem is, of course, that the more idiotic the practical joke the more likely it is that the employee will be held to have been 'on a frolic of his own', rather than being in the course of his employment.

2 In the wider context of employment law this raises the interesting question not just whether it would be fair to dismiss the practical joker in these circumstances, but whether there would be a positive *obligation* at common law to dismiss him if that was the only effective way of protecting the workforce.

4.13 It is, however, when one turns to the fourth head, the provision of a safe *system* of work, that the possible effects of the Regulations can be seen most clearly. In the past, this head has been something of a 'makeweight', a final, residual allegation in a statement of claim. Now, however, it can be seen as having renewed emphasis for the following reasons.

(1) It is in line with the Regulations in looking not at the static condition of the workplace but instead at the way work is *organised*; it is thus capable of reflecting the more dynamic, pro-active approach to industrial safety that is to be found in the Regulations, particularly in the key concepts of *risk* and risk assessment.

(2) Moreover, it is argued in this book that, although the Management Regulations do not themselves support civil liability, there is no reason why they should not be relied upon as *evidence* of required practice, failure to follow which can constitute negligence. For one thing, the actual wording of the exclusion of civil liability in reg 22 does not preclude this indirect use of the Management Regulations in a negligence action;[1] for another, certain other parts of the Regulations as a whole assume that key concepts in the Management Regulations (especially on risk assessment) are an integral part of the overall scheme of protection.[2]

(3) Thus, although the absence of a necessary risk assessment is not itself actionable as a breach of statutory duty, it is so

central to the whole scheme that it should be considered important evidence of a failure to provide a safe system of work in all the circumstances, and therefore common law negligence. This argument is also applicable to other parts of the Management Regulations, such as health and safety arrangements, health surveillance, the need for procedures to deal with serious and imminent dangers and the principles of prevention.

(4) Further, it may even be possible to argue that if in a particular respect one of the sets of Regulations does not fully implement the appropriate EC Directive, failure by an employer to comply with the *Directive* may also be evidence of negligence.

1 The Management of Health and Safety at Work Regulations 1999, reg 22, only states that breach of a duty imposed by the Regulations is not to *confer* a right of action in any civil proceedings. It is submitted that this only means that the Regulations cannot be relied upon *directly*, in order to maintain an action for breach of statutory duty. In the case of an action for negligence, that cause of action already exists by virtue of the employer's want of care, and the Regulations are only being used to *substantiate* that want of care, not to 'confer' the right of action.

2 See eg the Guidance Notes to the Provision and Use of Work Equipment Regulations 1998, regs 11 and 12 at paras 214 and 222 (see pp 469, 472).

CONTRIBUTORY NEGLIGENCE

4.14 The partial defence of contributory negligence is, of course, of great importance in industrial accident cases, particularly as it has long been established that it applies to breach of statutory duty actions as well as negligence actions.[1] What constitutes contributory negligence and what deduction should be made from damages in respect of it are heavily questions of fact for the trial judge.

Of particular relevance here is the rule that breach by the *employee* of a statutory duty placed upon him may well constitute contributory negligence.[2] Such duties have long been found in industrial safety legislation, including the HSWA 1974, ss 7 and 8. Now, however, there are more extensive duties on employees under the new Regulations, both in reg 14 of the general Management of Health and Safety at Work Regulations 1999 and also in the other, more specific regulations.

1 See Munkman *Employer's Liability* (13th edn, 2001) ch 29.

2 *Norris v Syndic Manufacturing Co* [1952] 2 QB 135, [1952] 1 All ER 935.

4.15 These duties are mentioned to show the increased scope for reliance on them in order for a defendant to establish contributory negligence. With regard to the duties on employees in reg 14 of the Management of Health and Safety at Work Regulations 1999, that is covered by the general exclusion of civil liability by reg 22 of those Regulations. However, repeating the argument made above in relation to negligence, that only means that the employee cannot be sued *directly* for any breach; it is again submitted that it does *not* mean that breach of reg 14 cannot be used by a defendant as evidence to establish a defence of contributory negligence.

Chapter 5

The Management of Health and Safety at Work Regulations 1999

SOURCE

5.1 The Management of Health and Safety at Work Regulations 1999 replace the 1992 Regulations of the same name. These Regulations implement EC Directive 89/391/EEC in the UK and are made under the provisions of the Health and Safety at Work etc Act 1974 (HSWA 1974).

DATE OF IMPLEMENTATION

5.2 The Management of Health and Safety at Work Regulations 1999 came into effect on 29 December 1999 and were of immediate application. Regulation 30 provides that anything done under a provision of the Management of Health and Safety at Work Regulations 1992 shall have effect as if done under the corresponding provision of the 1999 Regulations.

CIVIL LIABILITY

5.3 Regulation 22 provides that a breach of duty imposed by the Regulations shall not confer a right of action in any civil proceedings except in respect of risks for new or expectant mothers and young persons under regs 16 and 19.

APPROVED CODE OF PRACTICE AND GUIDANCE NOTES

5.4 The Management of Health and Safety at Work Regulations 1999 are accompanied by a new Approved Code of Practice (ACOP) which has been supplied by the Health and Safety Executive. The paragraphs of the ACOP are interleaved with paragraphs of Guidance Notes.

OVERVIEW

5.5 The Regulations, in common with the Framework Directive to which they are directly related, have to be seen in the context of the other Directives and Regulations which have been passed in accordance with the stated aims of arts 118a and 137 of the Treaty of Rome. They essentially create a framework of provisions within which the remaining Directives and Regulations are set by creating broad and general duties on employers, employees and the self-employed which provide the backdrop against which the more specific Directives and Regulations operate.

The approach to implementation in the UK has been largely on a piecemeal basis. This is because it was felt by the Health and Safety Commission that a number of the requirements of the Framework Directive were already in force in the UK through the HSWA 1974 and it was felt that the best method of implementation was to work within the existing framework of the 1974 Act. This has meant that the obligations contained in the Framework Directive are to be found in a number of separate domestic provisions including the 1974 Act, the Safety Representatives and Safety Committees Regulations 1977 as well as the 1992 Regulations and the Provision and Use of Work Equipment Regulations 1998.

COMMENTARY

5.6 The Regulations are of very broad scope covering all employers and all forms of work apart from sea transport (see reg 2). The centrepiece of the Regulations is the obligation on all employers and the self-employed to carry out a risk assessment under reg 3. The risk assessment must be suitable and sufficient and must consider the risk to the health and safety of all employees and other persons arising out of the conduct of the business or undertaking by the employer or self-employed person. The purpose of the assessment is to identify the measures which need to be taken in order to comply with the relevant statutory provisions. The 'statutory provisions' are defined in the ACOP as including the general duties under the HSWA 1974 as well as any more specific provisions (para 12). The general principles to be applied when making a risk assessment are set out in para 18 of the ACOP.

The risk assessment has to be reviewed if there is any reason to believe that it is no longer valid or if there has been any significant change in the matters to which it relates (see reg 3(3)). There is no longer a requirement in the Regulations to record the risk assessment, as there was in the 1992 Regulations. It is only in the ACOP to the Management of Health and Safety at Work Regulations 1999 that it states that those who employ five or more employees should record the significant findings of the risk assessment. This is a step backwards.

In making or reviewing an assessment in respect of young persons, particular account must be taken of a number of factors including their inexperience and lack of awareness, the layout of the workplace and workstation, the work equipment and the health and safety training given (see reg 3(5)).

5.7 There is a new reg 4. It requires an employer who implements any preventive or protective measures to do so on the basis of principles set out in Sch 1 to the Regulations. The Schedule sets out the principles of prevention as set out in the Directive. Regard must be had to avoiding the risk, evaluating risks which cannot be avoided, combating the risks at source, adapting the work to the individual, adapting to technical progress, replacing the dangerous by the non-dangerous or the less dangerous, developing a prevention policy, giving collective protection measures priority and giving appropriate instructions to employees.

5.8 The renumbered reg 5 obliges employers to give effect to arrangements for the effective planning, organisation, control, monitoring and review of the measures which they need to take as a result of the findings of the risk assessment. These arrangements need to be recorded by any employer who employs five or more employees.

5.9 The employer is also under a duty to ensure that his employees are provided with appropriate health surveillance in accordance with reg 6. This is particularly important in cases in which the risk assessment has identified any identifiable disease or adverse health condition related to the work. The ACOP states that the minimum requirement for health surveillance is the keeping of an individual health record (see para 43).

5.10 Regulation 7 requires the employer to appoint one or more 'competent persons' to assist him in taking the relevant measures that he needs to take in order to comply with the relevant statutory obligations. The persons appointed by him must be sufficient in number and be given sufficient time and the necessary means to fulfil their obligations. If the competent person appointed is not an employee in the undertaking, then they must be given information on any special factors which may affect the risks to health and safety in the undertaking as well as information on any temporary workers who may be working in the undertaking. Where there is a competent person in the employer's employment he shall be appointed in preference to a competent person not in his employment.

A 'competent person' is defined in reg 7(5) as being someone who has sufficient training and experience to enable him to assist properly in the undertaking. The ACOP and Guidance Notes are of assistance. The Guidance Notes state that simple situations will only demand a person who has a basic understanding of current best practice, who is aware of their own limitations and who has the willingness and ability to supplement existing experience and knowledge. This definition will obviously vary depending on the nature of the undertaking and in more complex situations experts may well be required.

Regulation 7(6) and (7) provides for self-employed employers and employers to carry out the function of the competent person in very limited circumstances.

5.11 Regulations 8 and 9 need to be considered together. Regulation 8 lays down a specific requirement for employers to establish appropriate procedures which are to be followed in the 'event of serious and imminent danger to persons at work in his undertaking'. The Guidance Notes refer to the need to set out clear guidance on when employees and others at work should stop work and how they should move to a place of safety (para 56). The type of risks that need to be covered should have been revealed by the risk assessment, but would obviously include fire and bomb risks. The employer is required to nominate a sufficient number of competent persons to implement the necessary procedures and to ensure that his employees are unable to enter any dangerous areas in his undertaking without first having been given adequate health and safety training. The Guidance Notes define a danger area as a working environment 'where the level of risk is unacceptable without special precautions being taken' (para 59). This is a potentially broad definition and would clearly cover areas where there are dangerous substances present as well as areas with dangerous structures or other hazards.

Regulation 8(2) lays down some guidance on what the appropriate procedures should involve. This includes the provision of information and the opportunity for persons immediately to leave their workstation in cases of danger.

The competent person or persons appointed under reg 8(1)(b) must have sufficient training and experience and possess the necessary qualities to be able to carry out their functions properly (reg 8(3)).

Regulation 9 is new and requires an employer to ensure that any necessary contacts with external services are arranged, particularly as regards first-aid, emergency medical care and rescue work.

5.12 Regulation 10 is of importance and requires every employer to provide his employees with comprehensible and relevant information on the risks to which they are exposed and the measures which the employer has taken in accordance with the risk assessment which he has carried out. Information must also be provided on the procedures to be followed in cases of serious and imminent danger as well as any risks to which the employee may be exposed as a result of the conduct of any other employer who may be sharing the workplace.

5.13 Specific requirements are laid down in reg 11 for employers to co-operate on health and safety matters with other employers who may be sharing the same workplace. The duty also extends to the self-employed and applies whether the sharing is on a temporary or part-time basis. The respective parties must provide each other with sufficient information and should co-ordinate the steps taken when appropriate. Some assistance is given in the ACOP and Guidance Notes.

5.14 Regulation 12 extends the employer's duty to provide information, including the provision of adequate health and safety information to employers of any employees from another undertaking who are working in his undertaking. This will obviously include groups such as contractors and employees of employment businesses who are hired to work under the user's control.

Regulation 12(3) requires every employer and every self-employed person to provide any person working in his undertaking who is not his employee with adequate health and safety information and instruction on any risks which that person faces as a result of the conduct of the undertaking. This will clearly include information on the steps and procedures which are to be taken in cases of serious and imminent danger and the persons who are to operate any evacuation procedures.

5.15 Regulation 13(1) requires employers to consider the capabilities of their employees as regards health and safety before entrusting any tasks to them. This is clearly a duty which has to be met on an individual-by-individual basis. The employer must ensure that the demands of any job do not put the employee at risk. This will require consideration of the individual's capabilities, level of training, knowledge and experience (see ACOP, para 80). The employer is also under a duty to provide his employees with adequate health and safety training in a number of specific instances. These are set out in reg 13(2) and include the time of recruitment, on exposure to new risks, on introduction of new technology and on changes to any system of work. This training must be carried out in working hours.

5.16 Obligations are placed on employees by reg 14. Every employee is under a duty to use any equipment provided to him by his employer in accordance with the instruction and training that has been given to him. The employee is also under a duty to report, to his employer or to any fellow employee with specific responsibilities for health and safety, any shortcomings or defects which the employee considers exist in the health and safety measures taken by his employer.

5.17 Regulation 15 provides specific obligations on employers and the self-employed towards temporary workers. Three tiers of obligation are created which all relate to the provision of comprehensible information on any special occupational qualifications or skills which are required for the work to be carried out safely and any health surveillance which may also be necessary. The relevant information must be supplied by any employer to any of his employees who are employed on fixed-term contracts (reg 15(1)). The same obligation also applies to any employer and any self-employed person who is to use the services of any employee of an employment business. In such a case the employer or self-employed person must provide the necessary information not only to the relevant employee but also to the employment business itself (reg 15(2) and (3)). The person carrying on the employment business is also under a duty to make sure that the relevant information is supplied to their employee.

5.18 Regulation 16 provides additional protection for expectant or new mothers and their babies. Where the work could involve a risk to these categories of persons, additional to the risks outside the workplace, the assessment under reg 3 must specifically include an assessment of such a risk. Where the risk cannot be avoided, the employer shall, if it is reasonable to do so, alter the working conditions or hours of work. Alternatively the employer can suspend the employee from work for so long as is necessary to avoid the risk. The obligation on an employer to take action only arises after notification in writing from the employee that she is pregnant, has given birth in the previous six months or is breastfeeding (reg 18). Further guidance is given in the ACOP.

Regulation 17 requires an employer to suspend from work a new or expectant mother who works at night on receipt of a certificate from a registered medical practitioner or midwife showing that it is necessary for the employee's health or safety that she should not be at work for a certified period.

5.19 Regulation 19 provides for the protection of young persons. Young persons are defined by reg 2 as those not over 18. The employer must protect young persons from risks to their health or safety which are a consequence of their lack of experience, lack of awareness of risks or lack of maturity. A young person is not to be employed in work which is beyond his physical or psychological capacity. Nor should young persons have harmful exposure to toxic or carcinogenic agents, radiation, extreme heat or cold, noise or vibration. However, nothing shall prevent the employment of a young person who is over compulsory school age where it is necessary for his training, where he is supervised by a competent person and any risk is reduced to the lowest level that is reasonably practicable.

5.20 Regulation 20 permits the Secretary of State for Defence to produce exemption certificates in certain circumstances.

Regulation 22 excludes civil liability for any breach of the Regulations with exceptions for expectant mothers, new mothers and their babies and young persons.

Regulations 24–27 make amendments to other Regulations.

Regulation 28 provides that the Regulations shall have effect as if they were health and safety Regulations within the meaning of the HSWA 1974.

IMPLEMENTATION OF THE DIRECTIVE

5.21 As has previously been mentioned, it is important to note that the Management of Health and Safety at Work Regulations

1999 only implement certain sections of the Framework Directive. The view was taken by the Health and Safety Commission that some of the general duties which are set by the Directive, and in particular the general duties which are placed upon employers in art 6, are already in place in the UK under s 2 of the HSWA 1974. Similarly, the requirement set out in art 6.5 of the Directive that measures taken in relation to health and safety at work may in no circumstances involve workers in financial cost has a domestic equivalent already in place in s 9 of the HSWA 1974. Other areas of the Directive are also met by the Fire Precautions (Workplace) Regulations 1997 and the Health and Safety (First Aid) Regulations 1981.

5.22 One of the other major differences between the Management of Health and Safety at Work Regulations 1992 and 1999 and the other substantive Regulations which were introduced in 1992 is the fact that a breach of the Management Regulations will not result in civil liability save for two exceptions (although this will not prevent the Regulations from having an impact in the area of civil liability (see further Chapter 4)). The Regulations are relevant to any criminal proceedings, but it is certainly arguable that by failing to provide for civil liability the Regulations have failed to meet the requirements of the Directive. This is because it has been established that member states must provide realistic remedies in domestic law in order to make Community rights a reality. There has been some consideration of this in relation to art 6 of the Equal Treatment Directive in the case of *Von Colson and Kamann v Land Nordrhein-Westfalen*: 14/83 [1984] ECR 1891, ECJ. However, it would appear that this is a matter for the European Commission to take up through infraction proceedings against the UK government, if appropriate, under art 169 of the Treaty of Rome rather than for individuals through actions in the domestic courts. The remedies provided by the UK government are criminal and administrative in nature, but it is certainly arguable that the aim and purpose of the Directive is to provide rights for workers and others which they should be entitled to enforce directly against their employers. In principle, however, it would appear that there is no reason why an employee of the state or an emanation of the state should not be able to rely upon a breach of the Directive directly in the relevant circumstances (further reference should be made to Chapter 3). Again, this is an area in which the practitioner must keep abreast with developments as they occur.

The Framework Directive also includes a number of provisions on consultation and employment protection. These are areas which are dealt with in detail in Chapters 12 and 15 and referred to in Chapter 11 in relation to temporary workers. In all of these areas it is arguable that the Regulations fail to meet the requirements of the Directive.

5.23 Regulation 3 no longer requires a record to be kept of the risk assessment. This is a change from the 1992 Regulations where there was such a requirement for employers employing more than five persons. Regulation 5(2) only requires employers with five or more employees to record the findings of the risk assessment and the arrangements made as a result. This is contrary to the Directive, which contains no such exception. It is also important to note that art 9.2 of the Directive refers to the requirement for documents to be provided. This also appears to be correct in principle because the seriousness of the hazards faced by people at work is not necessarily linked, if at all, with the size of the undertaking.

Apart from the requirements as to the keeping of risk assessments, the current Regulations go further towards the implementation of the terms of the Directive than the original 1992 Regulations. Article 6 of the Directive laid down a hierarchy of measures and general principles of prevention which should be observed. This hierarchy and the principles are now included in reg 4 and Sch 1.

Chapter 6

The Workplace (Health, Safety and Welfare) Regulations 1992

SOURCE

6.1 The Workplace (Health, Safety and Welfare) Regulations 1992 implement Directive 89/654/EEC and are made under the Health and Safety at Work etc Act 1974.

DATE OF IMPLEMENTATION

6.2 The Regulations came into effect on 1 January 1993. The transitional provisions are now largely academic but, in the case of an accident occurring in a workplace prior to 1 January 1996 see the commencement provisions in reg 1.

CIVIL LIABILITY

6.3 Breach of the Regulations creates civil liability.

APPROVED CODE OF PRACTICE

6.4 The Approved Code of Practice (ACOP) is detailed and should be considered by the practitioner as it both clarifies and particularises the approach to be taken when considering whether a particular factual situation gives rise to breaches of the Regulations.

OVERVIEW

6.5 The Regulations do not apply to ships, building operations, works of engineering construction, mines or quarries. Aircraft, locomotives, rolling stock, or trailers and semi-trailers used as a means of transport are not within the Regulations save that reg 13 (falls or falling objects) applies when the workplace is stationary but not on a public road.

Agricultural or forestry workplaces outside a building are excluded from most of the Regulations save for provisions relating to sanitation and washing facilities.

Temporary work sites (defined as being those used only infrequently or for short periods or fairs and other structures occupying a site for a short period) are excluded save for certain sanitation and welfare provisions.

6.6 Subject to these exceptions the Regulations apply widely and encompass vastly more premises than the old Factories Act. Regulation 4 provides that the Regulations apply to all employers and occupiers of premises and to all persons having any extent or control of workplaces in connection with the carrying on or a trade or business or other undertaking (whether for profit or not).

6.7 By these Regulations administrative buildings, hotels, restaurants, schools and universities are now all encompassed within the same statutory framework as factories. Means of access are included within the definition of 'workplace' (see reg 2).

COMMENTARY

6.8 The broadest, and most often pleaded regulation, is reg 5 which provides that the workplace, and the equipment, devices and systems (eg lighting, fencing, anchorage points, escalators, powered doors) shall be maintained in an efficient state, in efficient working order and in good repair, and 'where appropriate' subject to a suitable system of maintenance. The absolute nature of the first limb will be noted and 'efficient' is defined as efficient from the point of view of health and safety, not productivity or economy (ACOP, para 20).

The Approved Code of Practice gives specific guidance on systems of maintenance and refers to further advice in Health and Safety Executive publications and British Standards.

6.9 Regulation 6 requires effective and suitable provision for ventilation of enclosed workplaces. Specific guidance is given for 'confined' spaces.

6.10 Regulation 7 provides for a reasonable temperature in the workplace. The ACOP provides useful guidance.

6.11 Regulation 8 requires workplaces to be suitably and sufficiently lit, with emergency lighting if persons at work would be 'specifically exposed to danger' in the event of the failure of the primary lighting system.

6.12 Regulation 9 relates to cleanliness of the workplace. Regulation 9(3) provides that waste materials should not be allowed to accumulate save in suitable receptacles and may be relevant to tripping cases.

6.13 Regulation 10 provides for room dimensions and working space. In practice, although guidance is provided in the ACOP, a breach of this regulation may be difficult to establish as being of causative relevance to any accident. Any criticism of the layout or space of a workplace would probably be more effectively argued as a breach of reg 5(1).

6.14 Regulation 11 provides that workstations shall be suitable for the person at work and the work likely to be done there. The regulation imports a requirement to consider the subjective needs of individual workers. Suitable seating is to be provided if work can or must be done whilst sitting.

6.15 Regulation 12 features frequently in employment accidents as it provides that floors and traffic routes have to be of suitable construction, and without holes or slopes, unevenness or slipperiness exposing persons to risk to their health or safety (reg 12(2)(a)). Regulation 2 defines 'traffic route' which includes staircases, fixed ladders, doorways, loading bays and ramps. Moveable ladders will come within the Provision and Use of Work Equipment Regulations 1998 and, in the case of falls, reg 13 of the Workplace (Health, Safety and Welfare) Regulations 1992. Regulation 12(3) provides that floors and traffic routes are kept free from obstructions and any article or substance which may cause a trip, slip or fall. Regulation 12(5) makes provision for handrails. Paragraph 103 of the ACOP should be noted as recommending that consideration should be given to providing slip resistant footwear in certain workplaces.

6.16 Regulation 13 provides that suitable and sufficient measures shall so far as reasonably practicable be taken to prevent any person falling a distance likely to cause personal injury and to prevent any person being struck by a falling object likely to cause personal injury. The measures required to be taken are, so far as reasonably practicable, measures other than personal protective equipment, information, instruction and supervision. Regulation 13(5) and (6) provides for higher degrees of protection where there is a risk of persons falling into dangerous substances. The Approved Code of Practice gives detailed guidance to the fencing required. Paragraphs 119–126 of the ACOP deal with fixed ladders and paras 127–131 with roof work. Stacking and racking recommendations appear in paras 135–137. Frequently pleaded breaches of the ACOP arise from paras 138–140 which deal with loading and unloading vehicles.

6.17 Regulation 14 relates to windows, doors, gates and walls.

Regulations 15 and 16 relate to the safe opening and cleaning of windows, skylights and ventilators.

6.18 Regulation 17(1) provides that every workplace shall be organised in such a way that pedestrians and vehicles can circulate in a safe manner. Traffic routes shall be suitable, in terms of size, position and number, for the persons or vehicles using them (reg 17(2)). More detailed requirements are contained in reg 17(3) requiring 'sufficient separation' of vehicles and pedestrians and from doors and gates. Regulation 17(4) provides that traffic routes are suitably indicated. The Approved Code of Practice makes provision for signs etc and at para 171 states that in buildings, lines should be drawn on the floor to indicate routes to be followed by vehicles such as fork lift trucks.

6.19 Regulation 19 relates to escalators and moving walkways.

Regulations 20–25 contain provisions relating to sanitation and welfare.

IMPLEMENTATION OF THE DIRECTIVE

The Directive

6.20 Article 1.2 does not exempt transport used in the undertaking, but reg 3(3) does appear to exempt such transport except when stationary and in relation to falls covered by reg 13.

In respect of the provision of information to workers (art 7) and the consultation envisaged by art 8, reference should be made to the detailed chapter covering these points.

Annex one

6.21 Point 6.2 requires air-conditioning or mechanical installation systems to operate in such a way that workers are not exposed to draughts which cause discomfort. This provision is not contained in the Regulations, although it is mentioned in the ACOP.

Point 12.5 requires that, where workplaces contain danger areas in which, owing to the nature of the work, there is the risk of the worker or objects falling, the places must be equipped, as far as possible, with devices preventing unauthorised workers from entering those areas. Regulation 13 only provides for clear indication of such an area. The Guidance Notes state that additional safeguards may be necessary in places where unauthorised entry is foreseeable.

Point 14.2 requires a loading bay to have at least one exit point. This is only referred to in the ACOP.

Point 17 requires that pregnant women and nursing mothers must be able to lie down to rest in appropriate conditions. Regulation 25(5) only requires suitable facilities to rest. The Approved Code of Practice says that where necessary this should include the facility to lie down. It is to be wondered whether the necessity is to be measured by the needs of the woman or someone else.

Point 20 states that workplaces must be organised to take account of handicapped workers, particularly in relation to doors, passage-ways, staircases, showers, washbasins, lavatories and workstations. This is not specifically mentioned in the Regulations, and it is left to interpretation of words such as 'sufficient' and 'suitable', and reference in the ACOP.

Annex two

6.22 This sets minimum standards for workplaces already in use. The following matters do not appear to be covered:

(a) Point 6 requiring that a breakdown of a forced ventilation system must be indicated by a control system where it is necessary for the worker's health.

(b) Point 10 which requires that, where workplaces contain danger areas in which, owing to the nature of the work, there is the risk of the worker or objects falling, the places must be equipped, as far as possible, with devices preventing unauthorised workers from entering those areas.

(c) Point 11 in relation to rest areas.

(d) Point 12 which requires that pregnant women and nursing mothers must be able to lie down to rest in appropriate conditions.

(e) Point 13 relating to the provision of changing-rooms and lockers.

Chapter 7

The Provision and Use of Work Equipment Regulations 1998

SOURCE

7.1 The Provision and Use of Work Equipment Regulations 1992 have now been replaced by the Provision and Use of Work Equipment Regulations 1998. There may be some cases to be litigated which will still require consideration of the 1992 Regulations, but the emphasis here is on the 1998 Regulations. The Provision and Use of Work Equipment Regulations implement EC Directive 89/655/EEC and 95/63/EC. The Regulations are made under the Health and Safety at Work etc Act 1974 (HSWA 1974).

DATE OF IMPLEMENTATION

7.2 This is defined in reg 1. The Provision and Use of Work Equipment Regulations 1998 came into force on 5 December 1998. Regulations 4–24 were of immediate effect. There is a transition period for mobile work equipment first provided for use in the premises or undertaking before 5 December 1998 in that regs 25–30 will not come into force until 5 December 2002.

CIVIL LIABILITY

7.3 Breach of the Regulations creates civil liability.

GUIDANCE NOTES

7.4 There is an Approved Code of Practice (ACOP) and the Health and Safety Executive have provided Guidance Notes. Both are set out following the text of the Regulations in Appendix 3.

OVERVIEW

7.5 The Provision and Use of Work Equipment Regulations 1998 impose new obligations. There is a requirement to inspect work equipment after installation. New sections regulate the use of mobile work equipment and power presses. The Regulations apply to employers who provide work equipment for use by an employee and to persons who have control to any extent over work equipment, control of the persons who use, supervise or manage work equipment, or control over the way it is used. The Regulations do not apply to sea-going ships, but do apply to offshore oil and gas installations and any other offshore installations. The requirements of the Regulations are also imposed on self-employed persons.

COMMENTARY

7.6 In reg 2 the definition of the words 'use' and 'work equipment' is deliberately wide. 'Work equipment' can include items as diverse as a pair of scissors, a steel rolling mill, and a lathe. Work equipment is defined as any machinery, appliance, apparatus, tool or installation for use at work. The word 'installation' replaces the phrase 'any assembly of components, which in order to achieve a common end are arranged and controlled so that they function as a whole' which was used in the 1992 Regulations. The House of Lords looked at the meaning of 'equipment' within the meaning of the Employer's Liability (Defective Equipment) Act 1969 in *Knowles v Liverpool City Council* [1994] PIQR P8. It was held that a flagstone was part of a labourer flagger's equipment within the meaning of that Act. This also demonstrates a wide interpretation.

7.7 Regulation 4(1) provides that an employer shall ensure that work equipment is so constructed or adapted as to be suitable for the purpose for which it is used or provided. By reg 10 the employer is required to ensure that any item of work equipment provided for use complies with enacted safety requirements. Regulation 4(2) requires that an employer when selecting equipment has regard to the working conditions and to the risks to health and safety posed by the use of that equipment.

Regulation 4(3) imposes an obligation upon an employer to ensure that work equipment is used only for the operations for which, and under conditions for which, it is suitable, meaning suitable in any respect which it is reasonably foreseeable will affect the health or safety of any person, not just the user. Regulations 8 and 9 require an employer to provide adequate health and safety information and training to those who use the work equipment, and to those who supervise or manage the use of work equipment. The regulation goes further than Directive 89/655/EEC in its requirements relating to supervisors and managers. It is not enough to have a foreman transferred or promoted from another department: he must have been given information and training relevant to the tasks being performed by those that he supervises.

7.8 Regulation 5 imposes a duty to ensure that work equipment is maintained in an efficient state, in efficient working order and in good repair. The Guidance Notes indicate that 'efficient' relates to conditions affecting health and safety, and is not concerned with matters of productivity. It is a pity that this was not made clear in the regulation itself. In *Stark v The Post Office* (2000) Times, 29 March, where the stirrup broke on the Post Office bicycle used by the claimant, the Court of Appeal held that the equivalently worded regulation of the Provision and Use of Work Equipment Regulations 1992 imposed an absolute duty.

7.9 Regulation 6 imposes a new obligation of inspection. The regulation requires every employer to ensure that, where the safety of work equipment depends on the installation conditions, it is inspected after installation and before being put into service for the first time. There is a similar obligation where there has been assembly at a new site or in a new location. Where work equipment is exposed to conditions causing deterioration which is liable to result in dangerous situations, there is an obligation to inspect at suitable intervals. Inspections should also be undertaken each time there are exceptional circumstances which are liable to jeopardise the safety of the work equipment. The exceptional circumstances are not defined in the Regulations, but the Directive refers to examples such as modification, accidents, natural phenomena and prolonged periods of inactivity. The Guidance Notes contain similar examples. It should be borne in mind that inspections may be required under other specific Regulations, such as COSHH or Regulations dealing with electricity, for example. Regulation 2(1)(a) requires the inspections to be carried out by a competent person. Records of the results of inspections must be kept. Where work equipment leaves an undertaking or comes from the undertaking, it must be accompanied by physical evidence that the last inspection required by the regulation has been carried out.

7.10 Where work equipment is likely to involve a specific risk to health or safety, reg 7 requires the employer to restrict its use and maintenance to specific persons.

7.11 Regulation 11 provides for protection from dangerous parts of machinery and rotating stock bars. The duty is not absolute, but provides for a hierarchy of measures governed by the standard of practicability. Thus, there has been a weakening of the protection offered by provisions such as s 14 of the Factories Act 1961 where there was an absolute duty to fence. The hierarchy of protection given by the Regulations starts with fixed guards enclosing the danger. If that is not practicable then other guards or protection devices, for example micro-switch cut-outs, may be used. If that is not practicable jigs, holders, push-sticks or similar appliances may be used. If that is not practicable then the final level of protection is the provision of information, instruction, training and supervision. Regulation 11(3) provides the standards to be met by the guards and protection devices. It appears open to argument that push-sticks, for example, are not devices which prevent access to danger zones such as those created by circular saws, because if they slip then access can be gained to the saw blade. The saw blade must be exposed to cut the wood, but is it not practicable to provide a tunnel guard through which the wood can be fed to the saw blade?

The difficulty in applying the hierarchy of protection measures arises from having provisions which cover such a wide range of work equipment outside factory premises. An oft-quoted example is the mowing machine. It cannot cut the grass if completely guarded, and there remains the risk of pulling the mower over a foot. It seems open to argument that a long handle at arm's length is what is required by the provisions of reg 11(2)(c). It is perhaps only items such as the mechanical hand-held tools that will attract the last rank of protection provided in the Regulations, namely the provision of information, instruction, training and supervision.

7.12 Regulation 12 provides measures to prevent or control hazards created by articles or substances falling or being ejected from work equipment, or by discharge from the work equipment, fire or explosion. However, the regulation does not apply where particular substances are governed by other specified Regulations, namely asbestos, ionising radiation, lead, and substances hazardous to health, noise and head protection in the construction industry. The measures required are other than the provision of personal protective equipment, or information, instruction, training and supervision. The purpose is to have the risk 'engineered out'.

7.13 Regulation 13 provides protection to prevent injury from burns, scalds, and searing from high or low temperature work equipment, parts of work equipment, and articles and substances produced, used or stored. Its universal application is a substantial increase in the protection afforded.

7.14 Regulations 14–18 relate to controls and control systems for work equipment. Previously, regulation of these matters was limited to particular areas such as woodworking machines. Regulation 14 requires the provision of controls for starting or making a significant alteration to the operating conditions of work equipment, and save in the case of the normal operation of automatic devices, deliberate action on the control is required. Regulation 15 relates to stop controls, and reg 16 relates to emergency stop controls. These provisions need to be considered in relation to each type of work equipment. Regulation 17 provides that controls shall be clearly visible and identifiable, and that its positioning shall not expose the operator to a risk to his health or safety. Regulation 18 requires that control systems shall be safe as far as is reasonably practicable, and that where reasonably practicable any failure should lead to a fail safe condition. The Guidance Notes are of particular interest when considering regs 14–18.

7.15 Regulation 19 governs isolation from sources of energy. The means of isolation must be clearly identifiable and readily accessible. Appropriate measures must be taken to ensure that reconnection does not expose any person using the work equipment to any risk to his health or safety. 'Use' is defined by reg 2 as any activity involving work equipment and includes starting, stopping, programming, setting, transporting, repairing, modifying, maintaining, servicing and cleaning.

7.16 Regulation 20 requires that work equipment or any part of it shall be stabilised by clamping or otherwise to prevent fall or collapse.

7.17 Regulation 21 provides that every employer shall ensure that suitable and sufficient lighting, which takes account of the operations to be carried out, is provided at any place where a person uses work equipment. This is more specific than the requirement for suitable and sufficient lighting for every workplace in reg 8 of the Workplace (Health, Safety and Welfare) Regulations 1992.

7.18 Regulation 22 requires appropriate measures to ensure that work equipment is so constructed or adapted that, so far as is reasonably practicable, maintenance operations can be carried out while the work equipment is shut down, or without exposing the person to risk, or taking appropriate measures for protection of the person.

7.19 Regulation 23 requires work equipment to be marked in a clearly visible manner with any marking appropriate for reasons of health and safety. Regulation 24 requires work equipment to incorporate any warnings or warning devices which are appropriate for reasons of health and safety.

7.20 Mobile work equipment is now governed by new provisions in the Provision and Use of Work Equipment Regulations 1998. There are transitional provisions hidden away at reg 37 and the regulations do not apply until 5 December 2002 for work equipment provided for use in the undertaking or establishment before 5 December 1998. There is no definition of mobile work equipment in the Regulations.

7.21 Regulation 25 requires an employer to ensure that no employee is carried by mobile work equipment unless it is suitable for carrying persons and it incorporates features for reducing risks to their safety. This includes risks arising from the wheels or tracks. The obligation to reduce the risks is subject to reasonable practicability.

7.22 Regulation 26 is concerned with the rolling over of mobile work equipment. Where there is a risk of rolling over, the employer shall ensure that the risk is minimised by stabilising the work equipment and that there is provision of a structure which ensures that the work equipment does no more than fall on its side, gives sufficient clearance to anyone being carried if it overturns further than on its side, or that it is fitted with a device giving comparable protection. Where there is a risk of anyone being carried by mobile work equipment being crushed by its rolling over, the employer shall ensure that it has a suitable restraining system for him. The restraining system is not required for a fork-lift truck which has a structure which ensures that the work equipment does no more than fall on its side and gives sufficient clearance to anyone being carried if it overturns further than on its side. Compliance with the regulation is not required where it would increase the overall risk to safety or it would not be reasonably practicable to operate the mobile work equipment. Furthermore, the regulation does not apply in relation to an item of work equipment provided for use in the undertaking or establishment before 5 December 1998 where it would not be reasonably practicable. This is in addition to the transitional provisions.

7.23 Regulation 27 requires that a fork-lift truck fitted with a structure to protect from roll-over must be adapted or equipped to reduce to as low as is reasonably practicable the risk to safety from its overturning.

7.24 Regulation 28 governs self-propelled work equipment which while in motion may involve risk to the safety of persons. The employer must ensure the equipment has facilities to prevent it being started by an unauthorised person. It must have a device for braking and stopping and there are also provisions requiring readily accessible controls or automatic systems for braking and stopping in the event of failure of the main facility. Devices for improving the driver's field of vision are also required where reasonably practicable. Lights are required at night or in dark places. Where there is more than one item of rail-mounted work equipment in motion at the same time the equipment must have appropriate facilities for minimising the consequences of a collision. Regulation 29 requires that remote-controlled, self-propelled work equipment which involves a risk to safety while in motion stops automatically once it leaves its control range and incorporates features to guard against the risk of crushing or impact.

7.25 Regulation 30 provides measures to prevent the seizure of drive shafts between mobile work equipment and its accessories or anything towed. This includes safeguarding the shaft where it could become soiled or damaged by contact with the ground while uncoupled.

7.26 Power presses are now included in the Provision and Use of Work Equipment Regulations 1998. There are some exceptions which are set out in Sch 2. Regulation 32 requires thorough examination of power presses, guards and protection devices. The obligation arises when the power press is put into service for the first time after installation, or after assembly at a new site or in a new location. Likewise, an employer must ensure that a guard or protection device is not put into service, or that a closed tool which acts as a fixed guard is not used, unless it has been thoroughly examined to ensure that it is effective for its purpose and any defect has been remedied. In order that any deterioration can be detected and remedied in good time there must be a thorough examination of the guards and protection devices at least every 12 months where it has fixed guards, or in other cases at least every six months. There is also a requirement for examination each time that exceptional circumstances have occurred which are liable to jeopardise the safety of the power press or its guards or protection devices. Any defect must be remedied before the power press is used again.

7.27 Regulation 33 requires inspection and testing of a power press after setting, re-setting or adjustment of its tools. This is not necessary where the guards and protection devices have not been altered or disturbed in the course of the adjustment of its tools. After the expiration of the fourth hour of a working period a power press shall not be used unless its guards and protection devices have been inspected and tested. The inspections must be carried out by a competent person or a person undergoing training who is acting under the immediate supervision of a competent person. The person making the inspection and test for an employer under reg 33 must forthwith notify the employer of any defect which in his opinion is or could become a danger to persons and the reason for his opinion. There is a requirement for making reports and for the keeping of the reports.

IMPLEMENTATION OF THE DIRECTIVE

7.28 The terms of the Directive are implemented only in part by these Regulations. Some of the terms of the Directive were already covered by existing legislation, in particular the HSWA 1974.

7.29 Article 4.2 requires that the employer take measures necessary to ensure that, throughout its working life, work equipment is kept, by means of adequate maintenance, at a level such that it complies with the provisions of this and other Directives. It is not clear to what extent the terms of reg 5 comply with the requirements of the Directive. The regulation imposes a duty to ensure that work equipment is maintained in an efficient state, in efficient working order and in good repair.

7.30 Article 5a (ergonomics and occupational health) which requires the working posture and position of workers and ergonomic principles to be taken fully into account, is not reproduced in these Regulations in any form. The requirement for health surveillance under reg 5 of the Management of Health and Safety at Work Regulations 1999 is not as explicit and has no civil remedy for breach. There is a requirement for a workstation which is suitable for any person likely to work there and for the work likely to be done: Workplace (Health, Safety and Welfare) Regulations 1992, reg 11. However, this is not as explicit as the new terms of the Directive.

7.31 Point 1.2 of Annex II requires that work equipment must be erected or dismantled under safe conditions, in particular observing any instructions which may have been furnished by the manufacturer. Nothing in these Regulations or s 2 of the HSWA 1974 meets these requirements. Furthermore, there is no civil remedy for any breach of the HSWA 1974. The Approved Code of Practice to the Provision and Use of Work Equipment Regulations 1998 is the only place to find the terms of the Directive.

Another use of the ACOP is to (purport to) implement Point 1.3 of Annex II concerning the protection from lightning strikes.

7.32 Point 2.3 of Annex II requires organisational measures to be taken to prevent workers on foot coming within the area of operation of self-propelled work equipment. Nothing in these Regulations or reg 17 of the Workplace (Health, Safety and Welfare) Regulations 1992 requires pedestrians to be prevented from coming within the area of operation of self-propelled work equipment. The Approved Code of Practice to the Provision and Use of Work Equipment Regulations 1998 states that measures should be taken where appropriate, but where it is not reasonably practicable the requirement is only to reduce the risks.

7.33 There is one part of the Directive that is not easy to interpret. Point 2.8 of the Annex to the Directive provides that, where there is a risk of mechanical contact with moving parts of work equipment which could lead to accidents, those parts must be provided with guards or devices to prevent access to danger zones or to halt movements of dangerous parts before the danger zones are reached. The difficulty is in the meaning to be given to the word 'mechanical'.[1] It does not seem to have a specialised meaning. If 'mechanical' is meant to include bodily contact as well as contact with clothing, as where a loose sleeve is drawn into a machine, then it has an effect similar to that given by s 14 of the Factories Act 1961. However, it could have a more restrictive meaning. Depending upon the view taken as to the meaning of the Directive, it can be argued that reg 11 either exceeds the requirements of the Directive, or falls short of full implementation.

1 The same word is used in the French version of the Directive.

7.34 Regulation 26 which deals with the risks to persons riding on mobile work equipment is qualified by reasonable practicability and the consideration as to whether it increases the overall risk to safety. The Directive states that the work equipment must be fitted out in such a way as to reduce the risks for workers during the journey. That seems to be more consistent with a qualification of practicability rather than reasonable practicability.

7.35 Regulation 27 which deals with the overturning of forklift trucks is qualified by reasonable practicability. The Directive gives a clear set of alternative means of protection but contains no such qualification.

7.36 In respect of the consultation envisaged by art 8, reference should be made to Chapter 11.

Chapter 8

The Personal Protective Equipment at Work Regulations 1992

SOURCE

8.1 The Personal Protective Equipment at Work Regulations 1992 implement EC Directive 89/656/EEC and are made under the Health and Safety at Work etc Act 1974 (HSWA 1974).

DATE OF IMPLEMENTATION

8.2 The Regulations came into force on 1 January 1993.

CIVIL LIABILITY

8.3 A breach of the Regulations creates civil liability.

GUIDANCE NOTES

8.4 Detailed Guidance Notes.

OVERVIEW

8.5 The Regulations must be considered in the light of pre-existing Regulations covering specific industries and wider Regulations, eg the Protection of Eyes Regulations 1974. Schedule 3 of the Personal Protective Equipment at Work Regulations 1992 lists those Regulations which have been revoked, in whole or in part. Schedule 2 details those Regulations (including the Control of Lead at Work Regulations 1980, the Control of Substances

Hazardous to Health Regulations 1998 and the Construction (Head Protection) Regulations 1989) which have been amended.

COMMENTARY

8.6 The Regulations apply to all workers in Great Britain except the crew of sea-going ships.[1] Personal protective equipment (PPE) is defined widely in reg 2(1) as all equipment (including clothing affording protection against the weather) which is intended to be worn or held by a person at work and which protects him against one or more risks to his health and safety. Where there are existing comprehensive Regulations which require PPE (eg ear protectors and most respiratory equipment) the Personal Protective Equipment at Work Regulations 1992 will not apply. Where there are no existing Regulations dealing with PPE the 1992 Regulations will apply. In the case of existing, but not comprehensive, Regulations requiring PPE, the 1992 Regulations will apply and will complement the requirements of the existing Regulations (see Guidance Notes, para 7—Table 1).

The broad definition of PPE is limited to a certain extent by reg 3(2) which excludes ordinary clothes and uniforms which do not specifically protect the health and safety of the wearer and equipment used during the playing of competitive sports.

Regulation 3(3) lists those pre-existing Regulations which continue to be in force and provides that the Personal Protective Equipment at Work Regulations 1992 shall not therefore apply.

1 Note the partial exemption for ships' crews set out in para 14 of the Guidance Notes.

8.7 The key to understanding the Regulations is to appreciate that they are a 'last resort' in a hierarchy of control measures which are intended to combat risks to health and safety at source. If the risks can be adequately controlled by other means, then this should be done first. The rationale is that PPE protects only the person wearing it, whereas measures controlling the risk at source can protect everyone. Further, the Guidance Notes recognise that theoretical maximum levels of PPE are seldom achieved in practice.

8.8 The core of the Regulations is reg 4(1) which provides that every employer shall ensure that suitable PPE is provided to his employees who may be exposed to a risk to their health and safety while at work 'except where and to the extent that such risk has been adequately controlled by other means which are equally or more effective'.

'Suitable' for the purposes of PPE is defined in reg 4(3) as meaning that it must be appropriate for the risks involved and the conditions at the place where the exposure to the risk may occur, it must take account of ergonomic requirements and the state of health of the person wearing it, and must be capable of fitting the wearer correctly. The PPE should, so far as is practicable, be effective to prevent or adequately control the risk without increasing the overall risk.[1]

Regulation 4(2) requires every self-employed person to ensure that he is provided with suitable PPE.

In order to provide PPE for employees, employers must do more than simply have the equipment on their premises. The employees must have the equipment readily available 'or at the very least clear instructions on where they can obtain it'.[2]

1 A note on existing British and European standards appears in Appendix 3 of the Guidance Notes.
2 Guidance Notes, para 23. See also regs 9 and 10.

8.9 Regulation 5 requires both employers and the self-employed to ensure, where more than one item of PPE is worn or used simultaneously, that such equipment is compatible and continues to be effective.

8.10 Regulation 6 requires employers and the self-employed to undertake an assessment of PPE to ensure it is suitable. This follows on from, but does not duplicate, the risk assessment provisions in the Management of Health and Safety at Work Regulations 1999. Appendix 1 of the Guidance Notes contains a specimen risk survey table (the author has yet to see one of these tables produced on disclosure) and Part 2 of the Guidance Notes provides comprehensive and detailed information on various types of PPE. Regulation 6(3) provides for the review of assessments if there is reason to suspect they are no longer valid or if there have been significant relevant changes.

8.11 Regulation 7 provides that the PPE shall be maintained, including repair or cleaning if appropriate, in an efficient state, in efficient working order and in good repair.[1]

Regulation 8 provides for the safe storage of PPE when not in use.

1 Guidance on the meaning of 'efficient' (if needed) can be gleaned from para 20 of the Workplace (Health, Safety and Welfare) Regulations 1992 Approved Code of Practice.

8.12 Regulation 9 requires employees provided with PPE to be given adequate information, instruction and training on the purpose for which and manner in which the PPE is to be used and any action to be taken by employees to ensure the PPE remains in efficient working order and in good repair. The Guidance Notes emphasise the need for management, as well as users, to be aware of and understand the proper use of PPE. Guidance is given on theoretical and practical training and the duration and frequency of training.

8.13 Regulation 10(1) provides that the employer shall take reasonable steps to ensure that PPE is properly used and reg 10(2) imposes an obligation on the employee to use PPE provided and comply with any training and instruction given in its use. The self-employed are also obliged to make full use of PPE provided.

8.14 Regulation 11 requires an employee to report the loss of or obvious defects in PPE to his employer.

IMPLEMENTATION OF THE DIRECTIVE

8.15 The Regulations appear to fall below the standards required by the Temporary Workers Directive. In a number of areas the Directive provides more detailed provisions than the Regulations. By way of example art 4.4 of the Directive provides that PPE is, in principle, intended for personal use, and, if the circumstances require that PPE be worn by more than one person, appropriate health and hygiene measures shall be taken. The Regulations do not seek to impose any principle of personal use of PPE.

8.16 Article 8 of the Directive requires consultation and participation of workers and/or their representatives in the provision and use of PPE.

8.17 Article 4.6 of the Directive states that PPE shall be provided free of charge and reiterates the need for the PPE to be maintained in a satisfactory hygienic condition. Although the Regulations do not refer to charging for PPE, the HSWA 1974, s 9 prohibits charging employees in respect of anything done or provided in pursuant of any specific statutory requirement.

Chapter 9

The Manual Handling Operations Regulations 1992

SOURCE

9.1 The Manual Handling Operations Regulations 1992 implement Directive 90/269/EEC of the Council of the European Communities. The Regulations are made under the Health and Safety at Work etc Act 1974.

DATE OF IMPLEMENTATION

9.2 This is defined in reg 1. The Regulations came into force on 1 January 1993 and were of immediate effect. There was no transition period.

CIVIL LIABILITY

9.3 Breach of the Regulations creates civil liability.

GUIDANCE NOTES

9.4 There is no Approved Code of Practice, but the Health and Safety Executive have provided very full Guidance Notes.

OVERVIEW

9.5 The Regulations apply to every employment except those on sea-going ships. Previously, legislation was limited to work in defined locations such as factories, shops, and offices, or to particular occupations such as agricultural, construction or shipyard workers. The new Regulations cover occupations as diverse as nurses, delivery drivers, and university technicians. Duties are imposed upon the self-employed. Manual handling operations are defined as any transporting or supporting of a load, including the lifting, putting down, pushing, pulling, carrying or moving of the load by hand or bodily force. Loads include animals and people.

The Regulations do not apply to a master or crew of sea-going ships. However, they will apply to other persons loading ships or working on ships in territorial waters. The Regulations apply to offshore oil and gas installations and associated vessels.

COMMENTARY

9.6 Manual handling operations are defined by reg 2 as any transporting or supporting of a load, including the lifting, putting down, pushing, pulling, carrying or moving of the load by hand or bodily force. Thus, using a foot or pushing with the back are activities covered by the Regulations. In *Purvis v Buckinghamshire County Council* [1999] Ed CR 543 it was considered that an unruly child could be a load and could attract the provisions of the Regulations. The Regulations apply to risks of injury caused by cumulative lifting events: *Stone v Metropolitan Police Comr* (1999) unreported.

It is made clear by reg 2 that these Regulations do not cover injury caused by a toxic or corrosive substance that has leaked or spilled from a load, or is present on a load, or is a constituent part of a load. The intention is to cover strain injuries, common examples of which are strains to the back, shoulder and groin.

9.7 The assessment required by the Management of Health and Safety at Work Regulations 1999 (and previously the 1992 Regulations) should have indicated work activities which entail a risk of injury from manual handling. In *Koonjul v Thameslink Healthcare Services* [2000] PIQR P123, the Court of Appeal said that an element of realism was needed in assessing whether there was a risk of injury within the meaning of the Manual Handling Operations Regulations 1992. In *Hawkes v London Borough of Southwark* (20 February 1998, unreported) CA, the test applied was whether there was a 'real' risk. However, in *Anderson v Lothian Health Board* 1996 SCLR 1086 and *Cullen v North Lanarkshire Council* 1998

SC 451 it was said that the risk need be no more than a foreseeable possibility. In *Cullen* it was held that the Regulations were sufficiently broad in their application to apply where a man caught his foot whilst unloading materials from a truck. In *Whitcombe v Baker* (February 2000, unreported) it was held that the Regulations did not apply to guarding against a fall from the place of work.

9.8 Where there is a risk of injury from a manual handling operation, reg 4(1)(a) requires the employer to avoid such an operation so far as reasonably practicable. If he fails to show that it was not reasonably practicable to avoid the need for a manual handling operation which involved a risk of injury, the claim will be successful: *Hall v City of Edinburgh Council* 1999 SLT 744. Where it is not reasonably practicable to avoid such manual handling operations, the employer must make an assessment having regard to the factors set out in the Schedule to the Regulations. There are four main factors: the task, the load, the working environment, and individual capability. In *Cullen v North Lanarkshire Council* the working environment was the back of the truck. In *Hawkes v London Borough of Southwark* the load was a wooden door and the environment was a flight of stairs. Other factors may include hindrance by clothing or protective equipment. In many ways the approach needs to be inverted. Consideration of the factors and questions set out in the Schedule will help in assessing whether there is a risk of injury, as well as identifying individual matters which can be altered or avoided so as to reduce the risk of injury. The Guidance Notes offer an assessment checklist which can be copied and completed during the assessment. There are diagrams and accompanying text in the Guidance Notes which are helpful (not reproduced in this book).

9.9 The failure to carry out a risk assessment does not absolve the employer from taking the other steps set out in reg 4(1)(b). In *Swain v Denso Martin Ltd* [2000] PIQR P51 the Court of Appeal held that an employer was required to take appropriate steps to provide an employee with general indications and, where reasonably practicable, precise information on the weight of the load, whether or not he had carried out a risk assessment. Regulation 4(1)(b)(i), (ii) and (iii) were to be read disjunctively.

9.10 The Regulations do not specify maximum weights. Nor is there any emphasis on the relative strengths of men and women, or the young and the old. The Schedule to the Regulations requires consideration of whether the job requires unusual strength, height, etc, or whether it might create a hazard for persons who are pregnant or who have a health problem. Being of small stature is not a health problem. Regulation 13 of the Management of Health and Safety at Work Regulations 1999 does require an employer to take into account their employees' capabilities when entrusting them with tasks, but there is no civil liability for breach of this regulation. Emphasis on unusual strength, height etc gives the impression that the consideration is only as to whether the task needs to be performed by a giant or by an average employee. An ordinary employee may be small or relatively weak by reason of any number of factors which do not amount to a health problem, but which may affect his lifting capability.

9.11 The employer's duty under reg 4(1)(b)(ii) is to take steps to reduce the risk of injury to the lowest level reasonably practicable, and where it is reasonably practicable to give precise information on the weight of the load and the position of the centre of gravity if it is not located centrally in the load. There is no specific obligation in these Regulations to train employees on the proper handling of loads and although reg 13(2) of the Management of Health and Safety at Work Regulations 1999 refers to training, there is no civil remedy for a failure to train. The better view seems to be that part of the obligation to reduce the risk is to provide proper training in manual handling operations.

9.12 Regulation 4(2) requires a review of the assessment if there is any reason to suspect that it is no longer valid or there has been a significant change in the manual handling operations. Thus, any injury sustained while carrying out a manual handling operation would necessitate a review. So would a relevant complaint. Any change in a manual handling operation might be significant, and thus the obligation to review the initial assessment is needed in order to assess the significance of the change.

9.13 Regulation 5 imposes a duty on employees to make a full and proper use of any system of work provided for their use by the employer following an assessment.

9.14 Regulation 6 makes provision for the Secretary of State for Defence to make any necessary exemptions in the interests of national security.

IMPLEMENTATION OF THE DIRECTIVE

9.15 There are a number of respects in which it is arguable that the Regulations fail to implement the terms of the Directive.

9.16 Article 3 provides for a hierarchy of measures. The emphasis is on avoiding manual handling, in particular by the use of mechanical equipment. It is only where the need for the manual handling of loads by workers cannot be avoided that the employer must consider organisational measures.[1] The Regulations impose the requirement of reasonable practicability.

1 See Chapter 3 for consideration of the interpretation of the terms of the Directives.

9.17 Article 6 requires that workers and/or their representatives shall be informed of all measures to be implemented, pursuant to this Directive, with regard to the protection of safety and of health. An explanatory note to the Regulations concedes that this requirement has not been included in the Regulations. It is arguable that the requirements of this article are covered by the terms of reg 8 of the Management of Health and Safety at Work Regulations 1992 and now reg 10 of the 1999 Regulations, but there is no civil liability for breach of these regulations.

9.18 Article 6.2 requires employers to ensure that workers receive proper training and information on how to handle loads correctly and the risks that arise with incorrect handling. This is not included in the Regulations as a positive requirement. The only direct reference to training comes in the Assessment Checklist in the Guidance Notes, which raises the question whether special information/training is required. There is a duty on employers under reg 4(1)(b)(ii) to take steps to reduce the risk of injury to the lowest level reasonably practicable. It is debatable whether this can be interpreted to include an obligation to train employees on the proper handling of loads. It is arguable that the requirements of this article are covered by the terms of reg 13 of the Management of Health and Safety at Work Regulations 1999, but there is no civil liability for breach of this regulation. It is regrettable that the opportunity has been missed to emphasise the obligation to train by having it spelt out in the Regulations dealing with manual handling.

9.19 In respect of the consultation envisaged by art 7, reference should be made to Chapter 11.

9.20 Annex 2 identifies workers who may be at risk if they are physically unsuited to carry out the task in question. This risk is not identified in a positive way in the Regulations, the Schedule or the Assessment Checklist contained in the Guidance Notes. The Schedule to the Regulations requires consideration of whether the job requires unusual strength, height, etc. Thus, the emphasis is on whether someone larger or stronger than the ordinary is required, rather than on the physical capabilities of the existing employee, who may be smaller or weaker than the average.

9.21 The Regulations appear to fall below the standards required by the Temporary Workers Directive.

Chapter 10

The Health and Safety (Display Screen Equipment) Regulations 1992

SOURCE

10.1 The Health and Safety (Display Screen Equipment) Regulations 1992 implement EC Directive 90/270/EEC in the UK and are made under the provisions of the Health and Safety at Work etc Act 1974 (HSWA 1974).

DATE OF IMPLEMENTATION

10.2 The Regulations came into effect on 1 January 1993 and were of immediate effect (reg 1(1)). There was a lead-in period to 31 December 1996 for meeting the minimum requirements for workstations already in use prior to 1 January 1993. The requirements applied in full to all new workstations introduced on or after 1 January 1993. Thus all workstations are now covered by the provisions of these Regulations.

CIVIL LIABILITY

10.3 Breach of the Regulations creates civil liability.

GUIDANCE NOTES

10.4 There is no Approved Code of Practice but the Health and Safety Executive have produced full Guidance Notes.

OVERVIEW

10.5 The Regulations create new duties on employers in an area of health and safety which has not been covered by any previous statutory provisions. There are, therefore, no repeals or revocations of any existing provisions and the Regulations represent a new start in this area. The Regulations are having a significant impact in the areas of repetitive strain injuries and risks to eyesight as a result of working for long periods with computers and other display screen equipment.

COMMENTARY

10.6 The Regulations cover all persons who habitually use display screen equipment as a significant part of their normal work. 'Display screen equipment' is defined in reg 1(2)(a) as any alphanumeric or graphic display screen, regardless of the display process involved. This definition therefore covers computers and microfiche but would not cover television or film pictures. Whether an individual is an 'habitual' user or not is a question of fact and degree in each individual case. The main onus should be on the relative intensity of the work that the individual is required to perform. An obvious example of an habitual user would be a word processing pool worker but a wide variety of individuals may be covered from a data input operator to a financial dealer. Useful indicators are included in the accompanying Guidance Notes. If there is any doubt as to the precise status of the individual an assessment should be carried out. The Guidance Notes give a general guide figure of three hours' use a day as being a sufficient period to qualify. However, there would appear to be no reason in principle why a shorter but more intense period may not also be sufficient.

10.7 The Regulations create different standards of protection for 'users' and 'operators'. These categories are defined in reg 1(1)(b) and (d). In essence, the 'operator' is self-employed and the 'user' is an employee. It is important to note that the duties owed by an employer to 'users' are not only duties owed to his employees but also cover employees of other employers if those employees are working in his undertaking. This, therefore, covers employees of employment businesses in accordance with the requirements of the

Temporary Workers Directive (see further para 9 of the Guidance Notes).

10.8 Regulation 2 requires the employer to carry out an analysis of the workstations in his undertaking which are used by operators or users. A workstation is broadly defined in reg 1(2)(e) as an assembly which comprises display screen equipment, any optional accessories, any disk drive, telephone, modem, printer, document holder, work chair, work desk, work surface, or other peripheral items and the immediate work environment around the equipment. The assessment must be suitable and sufficient and for the purpose of assessing the relevant risks to health and safety from the operation. The Guidance Notes recommend that the individual users and operators should take part in the assessment after they have received appropriate training. The employer is under a duty to reduce the risks revealed by the assessment to the 'lowest extent reasonably practicable' (reg 2(3)). A list of possible risks is included in Appendix B to the Guidance Notes. In a case relating to injury sustained before these Regulations came into effect, the House of Lords accepted that it was foreseeable to employers that if employees typed for excessively long hours this might produce not only backache and eyestrain but also repetitive strain injuries: *Pickford v ICI plc* [1997] ICR 566. That provides a helpful marker for defining the risks to be assessed. In the unreported case of *McPherson v London Borough of Camden* (1999) it was held that if there had been a risk assessment there would have been changes to the layout of the workstation and rest breaks and periods of alternative work would have been provided; those changes would have prevented the injury sustained.

10.9 Regulation 3 sets out minimum requirements which every workstation must meet. The requirements are set out in the Schedule to the Regulations and include provisions on the supply of equipment (including the display screen, the keyboard, the work desk and work chair), on the environment (including requirements as to space, lighting, reflections and glare, noise and heat) and the interface between computer and operator/user. The requirements in the Schedule are to be met to the extent that the 'inherent requirements or characteristics of the task make compliance appropriate'. The Guidance Notes indicate that this proviso will not affect compliance in normal office conditions.

10.10 Regulation 4 is of great importance. It requires employers to plan the activities of their 'users' at work in their undertaking so that there are periodic interruptions in their work on the display screen equipment. These interruptions can be breaks or changes of activity. This duty is not owed to the self-employed. The Guidance Notes state that shorter, more frequent breaks are to be favoured.

10.11 Regulation 5 requires employers to provide appropriate eye and eyesight tests to their employees. It is important to note that the obligation does not extend to the employees of other employers who may be working in the undertaking. The onus would appear to be on the employee to request the test. New employees should be tested before they commence work. The Guidance Notes set out the necessary requirements for the test and the competent person who is qualified to carry out the test. These should be a registered ophthalmic optician or a registered medical practitioner with suitable qualifications (see para 50 of the Guidance Notes).

Regulation 5(5) requires the employer to provide every user employed by him with appropriate special corrective appliances when any test shows that such provision is necessary. The employer should not make any charge for any such appliance unless the employee requests appliances which are more luxurious than those that are necessary on purely health and safety grounds (see further the HSWA 1974, s 9 and art 6.5 of the Framework Directive).

10.12 Regulation 6 requires the employer to provide adequate training on the health and safety aspects of the workstation and to ensure that adequate training is also given whenever the organisation of any workstation is substantially modified. The general obligation to train under reg 6(1) only applies as between an employer and his employee, whereas the obligation to train in the event of any substantial modification in circumstances under reg 6(2) applies as between the employer and any 'user' at work in his undertaking. The substantial modification of circumstances would include any alterations to software, hardware, location, furniture or tasks (see further para 64 of the Guidance Notes).

10.13 Regulation 7 imposes a requirement on employers to provide adequate information on health and safety matters and on the assessment and steps that they have taken under the Regulations to all operators and users in their undertaking.

10.14 Regulation 8 makes provision for the Secretary of State for Defence to make any necessary exemptions in the interests of national security.

IMPLEMENTATION OF THE DIRECTIVE

10.15 It is arguable that in a number of respects the Regulations fail to meet the strict requirements of the Directive. When considering these aspects, the practitioner needs to be aware of the comments made on Community law and its interrelationship with domestic law in Chapter 3.

10.16 Regulation 2(3) requires the employer to reduce the risks which are apparent from the assessment to the lowest extent reasonably practicable. Article 3.2 of the Directive, however, appears to set the higher standard of 'practicable' because it requires employers to take appropriate measures to 'remedy the risks found'. This remains a debatable point and much will depend upon the interpretation that is given to the word 'appropriate'. It is also important to remember that the preamble to the Framework Directive states that health and safety at work 'is an objective which should not be subordinated to purely economic considerations'.

It is also arguable that reg 3 fails to meet the standards of the Directive because it only requires the minimum standards for workstations to be met for those workstations which may be used by 'operators' or 'users'. In other words, the requirements only have to be met for workstations which may be used by individuals who work with display screen equipment on an habitual basis. It is arguable that art 4 of the Directive requires the standards to be met by all workstations regardless of who is using them. This is of obvious importance for workstations which may be used by individuals who do not cross the 'habitual' threshold.

10.17 It is also important to note that the extent to which workstations must comply with the minimum requirements set out in the Schedule to the Regulations arguably falls short of the requirements of the Directive. This is because para 1(c) of the Schedule refers to the extent that the 'inherent characteristics of a given task make compliance with those requirements appropriate as respects the workstation concerned' and this appears to permit some consideration of the workstation as well as the task. This is not permitted by the Directive which only refers to the task itself. It is interesting to note that para 38(c) of the Guidance Notes applies the apparently more limited test as laid down in the Directive and this will, therefore, probably be an area in which the differences between the Regulations and the Directive can be ironed out through the process of interpretation (see further Chapter 3).

10.18 Readers are referred to Chapter 11 on the question of consultation.

In general, the Regulations follow the general scheme of the Directive and have increased the levels of protection in this area.

Example pleadings are included in Chapter 16.

Chapter 11

Consultation

INTRODUCTION

11.1 The Directives contain a number of provisions regarding consultation between employers and workers on health and safety matters. The primary obligations are contained in the Framework Directive and corresponding obligations are then reflected in each of the individual Directives. The general aim of the provisions can be gleaned from the preamble to the Framework Directive which includes a reference to the need for workers and/or their representatives to take part in balanced participation on matters concerned with health and safety at work.

11.2 The UK government originally sought to implement the consultation obligations solely through amendments to the Safety Representatives and Safety Committees Regulations 1977. The original approach was therefore to work through the existing structure of health and safety representation. This approach was fundamentally flawed since it failed to recognise that the 1977 Regulations could only operate in circumstances in which employers had voluntarily accepted that workplace trade unions should take part in such consultations. The use of such voluntary recognition arrangements in order to comply with universal European obligations on worker consultation was also adopted in other areas of UK law such as consultation with regard to collective redundancies under s 188 of the Trade Union and Labour Relations (Consolidation) Act 1992 and under the Transfer of Undertakings (Protection of Employment) Regulations 1981. The government's failure to provide for some form of obligation on an unwilling employer to consult led to severe criticism and ultimately the European Commission was forced to take infraction proceedings against the UK with respect to the failure to implement the relevant obligations imposed in the Directives relating to collective redundancies and the transfer of undertakings. The government was ultimately defeated in the ECJ (*EC v UK* [1994] ICR 664) and was forced to implement some form of framework which created an obligation on employers to consult in circumstances in which voluntary recognition did not apply.

THE PRE-EXISTING POSITION

11.3 The pre-existing law on consultation on health and safety matters at work was contained in the Safety Representatives and Safety Committee Regulations 1977 (SI 1977/500) which were made under s 2(6) of the Health and Safety at Work etc Act 1974. The 1977 Regulations placed a duty on an employer to consult with appointed safety representatives from recognised unions as regards arrangements for the promotion, maintenance and monitoring of improvements in health and safety at work.

The 1977 Regulations grant to safety representatives the right to represent their fellow workers, to carry out inspections and investigations and to receive information on health and safety matters. The role of safety representatives is supposed to be on a continuing day-to-day basis and is not intended to apply only after a major accident or incident has actually taken place.

11.4 The right of appointment of safety representatives is restricted to independent unions who are already recognised by their employers for the purposes of collective bargaining. It is for the union to decide on the method of selection and appointment of safety representatives, although the employer must be informed in writing of the names of the representatives appointed and the groups they represent.

11.5 The Regulations are industrial relations orientated and there are few existing methods of legal enforcement. It would only be in the most unusual of circumstances that the Health and Safety Executive would consider prosecuting an employer for failing to consult. This has to be seen against the background of the absence of any legal method of forcing an employer to recognise the union in the first instance. This approach of keeping the law at arm's length is further illustrated by the blanket legal immunity given to safety representatives by the 1977 Regulations (reg 4(1)).

11.6 Safety representatives are also entitled to reasonable time off with pay in order to undergo training and in order to carry out their functions. If these facilities are not provided, then a complaint can be made to an industrial tribunal (reg 11). The employer is also under a general duty to provide 'such facilities and such assistance as the safety representative shall require' (reg 5(3)) and to disclose information to safety representatives which is necessary for them to carry out their functions (reg 7(2)). The employer is also under a duty to establish a Safety Committee if requested to do so, in writing, by at least two of his safety representatives (reg 9(1)).

11.7 The pre-existing law therefore provided reasonably extensive obligations on an employer to consult with his workers' representatives on health and safety matters, providing that the initial hurdle of recognition had been passed.

THE DIRECTIVES

11.8 The primary obligations are contained in art 11 of the Framework Directive and have to be read in context with the definitions of 'worker' and 'workers' representative with specific responsibility for the safety and health of workers' as set out in art 3 of the Directive. The Directive appears to provide for one level of rights for 'workers' and one level of rights for 'workers' representatives'. Article 11.1 contains the general principle that employers shall consult workers and/or their representatives and allow them to take part in discussions on all questions relating to health and safety at work. It presupposes the consultation of workers and balanced participation 'in accordance with national laws and/or practices'.

Article 11.2 provides for more specific duties on consultation and states that the employer must consult on a broad range of issues including any measure which may substantially affect health and safety, the appointment of designated workers and those with specific responsibilities in cases of serious and imminent danger as provided for in the Framework Directive, information on the risk assessment (see further art 10.3) and the planning and organisation of health and safety training. Article 11.3 provides additional rights for workers' representatives to ask the employer to take appropriate measures to mitigate hazards and remove sources of danger. Article 11.5 provides for workers' representatives to have time off without loss of pay in order to carry out their duties and art 11.6 provides for the right of appeal to the 'authority responsible for safety and health protection at work' in circumstances where workers or their representatives feel that their employer has taken inadequate steps for the purposes of ensuring health and safety at work.

IMPLEMENTATION

11.9 As has been stated above, the government's original approach to the implementation of the consultation obligations was implicitly found to have been flawed by the ECJ. However, rather than making it mandatory for an employer to recognise an independent trade union for the purposes of the Safety Representatives and Safety Committees Regulations 1977, the government introduced the Heath and Safety (Consultation with Employees) Regulations 1996. These Regulations were designed to provide for consultation to take place in circumstances in which no workplace trade unions were recognised under the 1977 Regulations. It is important to recognise that the 1996 Regulations were not designed to replace the 1977 Regulations per se and that if a trade union has been voluntarily recognised by the employer the 1977 Regulations will continue to apply. This is made clear by reg 3 of the 1996 Regulations which states that:

> 'Where there are employees who are not represented by safety representatives under the 1977 Regulations, the employer shall consult those employees in good time on matters relating to their health and safety at work . . .'

In the circumstances it can be seen that two distinct consultation codes have been established: one code for recognised unions and one code for the non-unionised workplace.

RECOGNISED UNIONS

11.10 If the relevant trade union is both recognised and independent it has the right to appoint a safety representative for any workplace where it has membership (Health and Safety at Work etc Act 1974, s 2(4) and the Safety Representatives and Safety Committees Regulations 1977, reg 2). The safety representative may conduct periodic and other inspections of the workplace (regs 5 and 6) and he is entitled to information from the employer in order to perform his functions (reg 7). If there are two or more representatives they may request that the employer establishes a safety committee and if such a request is received the committee must be established (reg 9).

11.11 The employer is placed under a general duty to consult his safety representatives 'with a view to the making and maintenance of arrangements which will enable him and his employees to co-operate effectively in promoting and developing measures to ensure the health and safety at work of the employees' (Health and Safety at Work etc Act 1974, s 2(6)). The specific issues for consultation identified by the Framework Directive were inserted in the Regulations by means of reg 4A.

In addition the employer should permit the safety representative such time off with pay as is necessary to enable him to perform his functions or to undergo reasonable training.

NON-UNIONISED WORKPLACES

11.12 As has been stated above, the government decided to implement a separate code for consultation in non-unionised workplaces. This code was put in place in the Health and Safety (Consultation with Employees) Regulations 1996. In keeping with the Safety Representatives and Safety Committees Regulations 1977, the 1996 Regulations provide for a general duty to consult on health and safety matters and then reg 3 goes on to set out a list of matters which, in particular, require consultation. The matters specifically listed include the introduction of new technology, the appointment of relevant individuals, the introduction of new measures and health and safety training and information and these are taken from the Framework Directive.

11.13 The important distinction between the Health and Safety (Consultation with Employees) Regulations 1996 and their 1977 counterparts is reflected in reg 4 which provides that the employer must either consult directly with the employees or through elected 'representatives of employee safety'. The requirement for the recognition of a trade union is therefore negated. The functions of the elected representatives are set out in reg 6. Regulation 5 sets out the duty on the employer to provide relevant health and safety information in order that consultation can take place and reg 7 sets out the entitlement of the relevant individuals to health and safety training.

VICTIMISATION

11.14 Individuals who engage in health and safety activities under either code are provided with substantial employment protection rights and these are covered in more detail in Chapter 13.

Chapter 12

Duties of the self-employed

12.1 Although s 3 of the Health and Safety at Work etc Act 1974 imposes duties on the self-employed, a breach of the section did not give rise to a right of action in civil proceedings.[1]

The 1992 Regulations (as amended) have imposed specific duties on the self-employed.

1 HSWA 1974, s 47.

12.2 Regulation 3(2) of the Management of Health and Safety at Work Regulations 1999 requires the self-employed person to make a risk assessment of his own risks to which he is exposed whilst at work and the risks to others not in his employment arising out of the conduct of his undertaking (see Chapter 5).

12.3 Regulation 11 of the Management Regulations, requiring co-operation and co-ordination where two or more employers share a workplace, also applies to the self-employed.

12.4 Regulation 12 of the Management of Health and Safety at Work Regulations 1999 requires the self-employed to ensure that employees from an outside undertaking working in the self-employed's undertaking are provided with comprehensive information on health and safety risks and the measures taken to comply with the statutory requirements.[1] Although reg 22 of the Management Regulations continues to provide that breach of the Management Regulations does not confer a right of action in civil proceedings, it is common for any such breaches to be pleaded in support of an alleged breach of the common law duty of care.

1 It should be noted that the self-employed are given the same protection as employees from an outside undertaking: reg 12(5). Also note the additional information requirements set out in reg 15.

12.5 Regulation 4(2) of the Workplace (Health, Safety and Welfare) Regulations 1992 requires the self-employed, where they have to any extent control of the workplace, to comply with the Regulations in respect of matters within their control. Regulation 4(4) of the Workplace Regulations states that reg 4(2) does not impose any requirement upon the self-employed in respect of their own work or the work of any partner in the undertaking.

12.6 Regulation 3(3) of the Provision and Use of Work Equipment Regulations 1998 provides that the requirements imposed by these Regulations on an employer shall also apply to the self-employed in respect of equipment they use at work and to persons having control of work equipment to the extent that they have control over the equipment or the way in which the equipment is used.

12.7 The Manual Handling Operations Regulations 1992 imposes, by reg 2(2), the same duties on the self-employed in respect of their own safety during manual handling as are imposed on employers, and the self-employed should take the same steps to safeguard themselves as would be expected of employers in protecting their employees in similar circumstances.

Wide obligations are imposed on the self-employed by the Personal Protective Equipment at Work Regulations 1992. Regulation 4(2) requires the self-employed person to ensure that he is provided with suitable personal protective equipment (PPE). Regulation 5(2) imposes an obligation on the self-employed to ensure that, where more than one item of PPE is worn, the PPE is compatible and continues to be effective. Regulations 6 and 7 relating to assessment and maintenance of PPE apply to the self-employed as does reg 8 providing for safe storage of PPE.

12.8 Regulation 9 of the Personal Protective Equipment at Work Regulations 1992 requires an employer (not the self-employed) to provide instruction, information and training to employees in the use of PPE but reg 10(3) requires the self-employed to make 'full and proper use of any personal protective equipment provided to him . . .' Guidance Note, para 60 states that the self-employed user 'should ensure he has been adequately trained to use PPE competently . . .' It does seem illogical that an employer may provide PPE to the self-employed yet be under no duty to train him in the proper use of it, and query the practicality of requiring the self-employed to ensure he has been adequately trained in PPE which may be entirely unfamiliar to him or may be technically complex.

12.9 A self-employed person is an operator within the meaning of the Health and Safety (Display Screen Equipment) Regulations 1992.[1]

1 See Chapter 10 for a detailed consideration of the Health and Safety (Display Screen Equipment) Regulations 1992.

12.10 In practical terms breaches by the self-employed of these Regulations in circumstances in which the breach is of causative relevance to an accident to another person will be relied upon not only as being evidence of negligence but as breaches of statutory duty by the self-employed. Although it can be argued that the majority of the Regulations applying to the self-employed are concerned with the welfare of the self-employed himself, the intention that the Regulations shall also apply to limit injury to others is illustrated by Guidance Note, para 60 to the Personal Protective Equipment at Work Regulations 1992 which states that the self-employed should ensure he is trained in the use of PPE 'to avoid creating risks to himself and others'.

Chapter 13

Employment protection

INTRODUCTION

13.1 The Framework Directive contained new protections for employees who perform health and safety duties and who leave their workstations in the event of serious and imminent danger. The relevant provisions are contained in arts 7.2, 8.4, 8.5 and 11.4 of the Directive. These measures were not implemented through the various health and safety Regulations but through the medium of the employment protection legislation. Although there is insufficient space in this book for a detailed exposition of employment law the employment protections introduced by the Directive, along with the rights to workers' consultation, are nevertheless covered because of their pivotal role in the European approach to health and safety.

13.2 Prior to the changes introduced by the Directive the vast majority of workers did not enjoy any specific employment protection with regard to health and safety matters, leaving workers to fall back on the general law of unfair dismissal. This created many difficulties and often left an employee unprotected because of the lack of the necessary qualifying service to claim unfair dismissal. There was also no specific protection available for workers who were subjected to action short of dismissal on health and safety grounds. These problems were well illustrated by the case of *Chant v Aquaboats Ltd* [1978] ICR 643. Mr Chant was a skilled shipwright who had complained to his employers that the woodworking machinery did not comply with the required health and safety standards. He subsequently organised a petition which was signed by a number of his fellow workmen and he was later dismissed for slow work. The tribunal had little difficulty in concluding that the real reason for his dismissal was his health and safety activities, but he was left with no remedy because he had insufficient service to qualify for unfair dismissal protection and his actions did not fall within the ambit of trade union activities since he was acting in an individual capacity.

13.3 Some limited specific protection was provided before the implementation of the Directive. One exception was in relation to rights for workers engaged in health and safety activities on offshore installations. The Offshore Safety (Protection against Victimisation) Act 1992 provided additional protections for safety representatives and members of safety committees. In addition, provided that the individual was acting in a union-related capacity (unlike Mr Chant), they could qualify for protection under the trade union activities legislation (see the comments of Slynn J in *Drew v St Edmundsbury Borough Council* [1980] ICR 513 at 517).

THE DIRECTIVE

13.4 The protections contained in the Directives are potentially far-reaching. However, the various provisions need to be seen in context and, in particular, against the background of other provisions in the Directives on consultation and participation of workers and/or their representatives on health and safety matters. The perspective of the European initiatives on these matters can be gleaned from the preamble to the Framework Directive which refers to the need for workers and/or their representatives to be able to contribute, through participation, to ensure that any necessary protective measures are taken and to take part in a balanced dialogue with their employers. These aims also dovetail with the provisions in the Framework Directive on the appointment of designated workers to assist in health and safety matters, the provision of information to workers and the requirement for employers to consult with workers and/or their representatives on all questions relating to health and safety at work. The method of implementation of these proposals chosen by the UK government is certainly open to question (see further Chapter 12) and this has to be borne in mind when considering the proposed employment rights protections. This is because it is certainly arguable that the level of protection offered is severely reduced because, in most cases, in order to be protected, the worker concerned has to be a safety representative from a recognised trade union and there is no existing legal obligation on an employer to recognise a trade union.

13.5 The provisions of the Directive state that workers and workers' representatives who take part in health and safety consultations with their employers 'may not be placed at a disadvantage' because of their activities (art 11.4). A similar protection is also provided for 'designated' workers (art 7.2). Further protections are also provided for workers who leave their workstation or a dangerous area in the event of serious, imminent and unavoidable danger or take steps to avoid that danger (arts 8.4 and 8.5). The wording of the Directive is somewhat inelegant and permits a lower level of protection where a worker takes steps to avoid the consequences of such danger when the immediate supervisor cannot be contacted. This is because the worker is protected from any disadvantage unless 'they acted carelessly or there was negligence on their part' (see art 8.5). The test to be used when deciding what is negligent or careless would appear to be subjective, because account should be taken of the knowledge and technical means at the disposal of the worker.

THE MODERN PROTECTIONS

13.6 The modern framework of employment protection is contained in the Employment Rights Act 1996. The protections operate at two distinct levels, namely protections against action short of dismissal and protections against dismissal. The provisions relating to protection for action short of dismissal are contained in s 44 of the 1996 Act. The general right is for an employee 'not to be subjected to any detriment by any act, or any deliberate failure to act, by his employer' on a number of different grounds. The relevant grounds are comprehensive and operate at two different levels. The first level of protections applies to those individuals who are part of the formalised structure of health and safety activities. These include those individuals who were carrying out their activities as designated persons (s 44(1)(a)), those who were representatives of workers on matters of health and safety or a member of a safety committee (s 44(1)(b)), those taking part in consultations with the employer pursuant to the Health and Safety (Consultation with Employees) Regulations 1996 and those seeking election in accordance with those Regulations (s 44(1)(ba)).

The second level of protections applies to individuals who fall outside of the scope of the formalised structure of health and safety. Accordingly, s 44(1)(c) offers protection for an employee who brings to the attention of his employer circumstances connected with his work which he reasonably believed were harmful to health or safety. The protection depends upon the individual acting by reasonable means and, if there is a safety committee in place, it has to be shown that it was not reasonably practicable for the employee to raise the matter through that mechanism. In addition it must be shown that the danger was both serious and imminent. It can therefore be seen that the protections are specifically designed to support and complement the formalised structures of health and safety activity and representation. There is no 'general charter' for all employees to freely engage in health and safety activity other than in circumstances of serious and imminent danger.

13.7 The courts have tended to adopt a broad interpretation of these provisions, a striking example being *Shillito v Van Leer (UK) Ltd* [1997] IRLR 495 in which the Employment Appeal Tribunal (EAT) confirmed that the protection afforded to safety representatives under s 44 is not dependent upon the relevant individual acting reasonably. Accordingly, protection was given to an individual who sought to conduct his activities through embarrassing his employers in the presence of outside bodies. However, this is not without limit and in *Goodwin v Cabletel UK Ltd* [1997] IRLR 665 the EAT stressed, in a case regarding dismissal for health and safety reasons, that not every act, no matter how malicious, can fall within the protection simply because it is purportedly done on health and safety grounds.

13.8 The remedy for a breach of s 44 is by way of a complaint to an employment tribunal. If the complaint is well founded the individual is entitled to a declaration from the tribunal and compensation. The award of compensation should be just and equitable in all the circumstances having regard to the right infringed and any loss which is attributable to the act or failure which infringed the relevant right (s 49).

PROTECTION AGAINST DISMISSAL

13.9 Protections against dismissals for health and safety reasons operate in two circumstances. The first is where the dismissal itself is for a prohibited reason and the second applies where the reason for dismissal is redundancy but where the selection of the particular individual for redundancy was based upon his health and safety activities. The grounds for protection are the same as for protection against action short of dismissal under s 44. Unfair selection for redundancy on health and safety grounds is covered by s 105(3) of the Employment Rights Act 1996 and dismissal per se is covered by s 100.

If the dismissal or selection for redundancy is for health and safety reasons the dismissal will be automatically unfair. The normal array of remedies for unfair dismissal is available but, in addition, the seriousness with which a dismissal for health and safety reasons is viewed is reflected by the fact that an individual who has been dismissed or selected for health and safety reasons can claim an additional award (Employment Rights Act 1996, s 118(2),(3)).

13.10 The seriousness with which health and safety based dismissals are viewed is further emphasised by the fact that the relevant protections apply regardless of length of service. In addition the Employment Relations Act 1999, s 37 provided that there should be no cap on the compensatory award for such dismissals.

13.11 Although the courts have tended to adopt a purposive approach to the health and safety protections, certain limits have been imposed. In particular it is important to note that the statutory provisions do not provide a basis for more favourable treatment for health and safety representatives (see eg the decision of the EAT in *Smiths Industries Aerospace and Defence Systems v Rawlings* [1996] IRLR 656).

Chapter 14

Practical impact upon personal injury litigation

INTRODUCTION

14.1 The 1992 Regulations came into effect under the old procedural rules. The Regulations did bring about some practical changes to litigation, but all of that has been overtaken by the advent of the Civil Procedure Rules. This book is not devoted to civil procedure and for the detailed provisions we advise reference to one of the many specialist publications. Here, we take the opportunity to highlight particular features in the hope that this will be of some assistance. The precedents provided in Chapter 16 have been drafted in accordance with the requirements of the Civil Procedure Rules.

TRANSITIONAL PROVISIONS

14.2 There were transitional provisions for the Provision and Use of Work Equipment Regulations 1992, regs 11–24; the Health and Safety (Display Screen Equipment) Regulations 1992, reg 3; and the Workplace (Health, Safety and Welfare) Regulations 1992, regs 5–27. There cannot be many cases still to be litigated where those transitional provisions are relevant. However, there are transitional provisions for the Provision and Use of Work Equipment Regulations 1998, regs 25–30 which relate to mobile work equipment.

Where there is a claim based on an accident or events occurring in a transitional period, it is important to ascertain whether the particular regulations apply. The issue should be addressed from the start in taking a statement from the claimant and any witnesses. It should be raised in the early correspondence. It may be important when considering disclosure of documents.

PARTIES

14.3 When the claimant's lawyers are satisfied that it is appropriate to rely upon the Regulations, they need to consider whether the Regulations do implement the terms of the Directives. Attention is drawn to such failures elsewhere in this book. The claimant will only be able to sue his employer, or the person in control of, or the occupier of, premises or equipment by reliance upon the Regulations, as interpreted by reference to the Directives. Where the regulation fails to implement the terms of the Directive an action may well lie against the member state for this failure. An employee of the state or an emanation of the state will be able to rely upon the provisions of a Directive directly.[1] Employees who are in the private sector will have to show that the failure to implement the terms of the Directive has resulted in loss or damage. In many cases it may be possible for the claimant to succeed against the employer, controller or occupier without reliance upon the faulty regulation. It may be necessary to take such proceedings and obtain a judgment before any right against the member state crystallises. The above matters are dealt with in greater detail in Chapter 3 on the European dimension.

1 There is an example pleading at precedent no 8 in Chapter 16.

PROTOCOLS AND PRACTICE DIRECTIONS

14.4 The pre-action protocols dictate the steps that are to be taken in the stages before proceedings are commenced. There is a specific protocol for personal injury claims. It sets out the steps that the parties should take to seek information from other parties. The protocol requires the provision of sufficient information as to the claim to be provided to the defendant so that the defendant may make a full response within three months of the letter of claim. It would be appropriate to raise the issues that are likely to arise under the Regulations. Whether litigation is fast-track or multi-track the protocol provides useful guidance.

DISCLOSURE OF DOCUMENTS

14.5 It is open to the claimant to indicate in the letter of claim which classes of documents are considered relevant for early disclosure. Where a defendant denies liability his letter of reply should contain copies of the material documents. There are specimen, but non-exhaustive, lists of documents likely to be material in different types of claim. The purpose is to promote an early exchange of relevant information to help to resolve the issues in dispute. The lists reflect the obligations imposed on employers and employees under these and other Regulations.

General workplace claims

(1) Accident book entry.

(2) First aider report.

(3) Surgery record.

(4) Foreman/supervisor accident report.

(5) Safety representatives' accident report.

(6) RIDDOR report to the Health and Safety Executive (HSE).

(7) Other communications between defendants and HSE.

(8) Minutes of Health and Safety Committee meeting(s) where accident/matter considered.

(9) Report to DSS.

(10) Documents listed above relative to any previous accident/ matter identified by the claimant and relied upon as proof of negligence.

(11) Earnings information where defendant is employer.

Management of Health and Safety Regulations 1999 (or 1992)

(1) Pre-accident Risk Assessment required by reg 3.

(2) Post-accident Re-assessment required by reg 3.

(3) Accident Investigation Report prepared in implementing the requirements of the Regulations.

(4) Health Surveillance Records in appropriate cases required by reg 6.

(5) Information provided to employees under reg 10.

(6) Documents relating to the employees' health and safety training required by reg 13.

Workplace (Health Safety and Welfare) Regulations 1992

(1) Repair and maintenance records required by reg 5.

(2) Housekeeping records to comply with the requirements of reg 9.

(3) Hazard warning signs or notices to comply with reg 17 (traffic routes).

Provision and Use of Work Equipment Regulations 1998 (or 1992)

(1) Manufacturers' specifications and instructions in respect of relevant work equipment establishing its suitability to comply with reg 4.

(2) Maintenance log/maintenance records required to comply with reg 5.

(3) Records of inspections required by reg 6.

(4) Documents providing information and instructions to employees to comply with reg 8.

(5) Documents provided to the employee in respect of training for use to comply with reg 9.

(6) Any notice, sign or document relied upon as a defence to alleged breaches of regs 14–18 dealing with controls and control systems.

(7) Instruction/training documents issued to comply with the requirements of reg 22 in so far as it deals with maintenance operations where the machinery is not shut down.

(8) Copies of markings required to comply with reg 23.

(9) Copies of warnings required to comply with reg 24.

Personal Protective Equipment at Work Regulations 1992

(1) Documents relating to the assessment of the Personal Protective Equipment to comply with reg 6.

(2) Documents relating to the maintenance and replacement of Personal Protective Equipment to comply with reg 7.

(3) Record of maintenance procedures for Personal Protective Equipment to comply with reg 7.

(4) Records of tests and examinations of Personal Protective Equipment to comply with reg 7.

(5) Documents providing information, instruction and training in relation to the Personal Protective Equipment to comply with reg 9.

(6) Instructions for use of Personal Protective Equipment to include the manufacturers' instructions to comply with reg 10.

Manual Handling Operations Regulations 1992

(1) Manual Handling Risk Assessment carried out to comply with the requirements of reg 4(1)(b)(i).

(2) Re-assessment carried out post-accident to comply with requirements of reg 4(1)(b)(i).

(3) Documents showing the information provided to the employee to give general indications related to the load and precise indications on the weight of the load and the heaviest side of the load if the centre of gravity was not positioned centrally to comply with reg 4(1)(b)(iii).

(4) Documents relating to training in respect of manual handling operations and training records.

Health and Safety (Display Screen Equipment) Regulations 1992

(1) Analysis of work stations to assess and reduce risks carried out to comply with the requirements of reg 2.

(2) Re-assessment of analysis of work stations to assess and reduce risks following development of symptoms by the claimant.

(3) Documents detailing the provision of training including training records to comply with the requirements of reg 6.

(4) Documents providing information to employees to comply with the requirements of reg 7.

STATEMENTS OF CASE

14.6 The statements of case (or pleadings) are governed by the terms of the Civil Procedure Rules. Every statement of case and defence must contain a statement in a summary form of the material facts on which the party pleading relies for his claim or defence. A party may raise any point of law. Where there are competing interpretations of the Regulations they should be clearly set out, in the alternative if necessary.

14.7 The Management Regulations require a suitable and sufficient assessment of the risks to health and safety. The Display Equipment Regulations require a suitable and sufficient analysis of workstations for the purpose of assessing the health and safety risks. The Personal Protective Equipment Regulations require an assessment to determine whether the personal protective equipment that it is proposed to provide is suitable. The Work Equipment Regulations require an employer to have regard to the risks to the health and safety of persons which exist in the premises and any additional risk posed by the use of the work equipment. The Manual Handling Regulations require assessment of the risks of injury and a suitable and sufficient assessment of manual handling operations which involve a risk of injury. An employer, or other person owing a duty under the Regulations, may purport to have carried out an assessment, but it must be sufficient and suitable or assess equipment as suitable. If the assessment does not measure up to the standard set in the Regulations it is evidence of negligence, and, save in respect of the Management Regulations, is a breach of statutory duty which is actionable if it caused or contributed to the claimant's injury. These all raise issues which are likely to require determination.

14.8 It is suggested that breaches of the Management Regulations and any Approved Code of Practice (ACOP) are pleaded as allegations of negligence.[1] On occasions the terms of the Guidance Notes may need to be pleaded in order to make clear the standard being contended for in the allegation.

1 See Chapter 16, and the example pleading at precedent no 1.

14.9 The defence must plead to any issue of practicability, reasonable practicability. The burden of proof is on the defendants, and they must plead the material facts upon which they wish to rely in support of a contention that something was not practicable, or that it was not reasonably practicable. It is likely that the courts will take the same view in relation to terms such as 'suitable and sufficient'.[1]

1 There are example pleadings at precedents no 5 and 9 in Chapter 16.

FURTHER INFORMATION

14.10 Under the Civil Procedure Rules the old Requests for Further Particulars and Interrogatories are transformed into Requests for Further Information. This is still an important tool for the claimant.

If a document is no longer in the possession, custody or control of the other party, then a Request for Further Information will be appropriate. Requests can be made as to the content of the document. Where no records were kept of an assessment, Requests can be directed to discover what matters were considered, what facts were assumed, and what factors were weighed in the balance.

Where there are pleaded issues as to whether it was practicable, or reasonably practicable to take certain measures, Further Information may be required in order to provide a fuller picture of the defendants' case.

EXPERTS

14.11 Whether jointly instructed or instructed by a particular party, the expert can only provide a full report if he is able to consider the assessments made by the defendants, and to make comparison with the Regulations, any ACOP, and the Guidance Notes. Comparison with risk assessments by others operating a similar business or undertaking may be helpful in determining whether the risk analysis or assessment was suitable and sufficient.

14.12 A party may put written questions to an expert about his report, whether that expert was instructed by another party or was a single jointly instructed expert. Written questions may be put once only and must be for the purpose only of clarification of the report; there is no reason why they should not incorporate reference to any particular regulation which it is thought might be relevant.

14.13 Where a party has access to information which is not reasonably available to the other party, the court may direct the party who has access to the information to prepare and file a document recording the information and serve a copy of that document on the other party.

REFERENCE TO THE EUROPEAN COURT

14.14 Reference is governed by RSC Ord 114, and CCR Ord 19, r 11. The Rules confer a discretion upon the judges of the county court and the High Court concerning referral. A Reference to the European Court should only be made where it is necessary to enable the national court to determine the litigation. The Reference will need to have a determination of the essential facts, and, where necessary, of the national law. However, an application for a Reference to the European Court cannot be made after the court has given judgment.[1]

1 See Chapter 3 on the European dimension.

Chapter 15

The criminal perspective

GENERAL DUTIES

15.1 The inter-relationship of the 1992 Regulations (as amended) with the relevant parts of the existing legislation, and a commentary on how the Regulations go beyond the Health and Safety at Work etc Act 1974 (HSWA 1974), are dealt with in Chapter 2. The criminal perspective is an area of the law which has been the subject of astonishing growth over the past few years, since the first edition of this book, and the Health and Safety Executive's (HSE's) policy to prosecute more companies for increasingly varied offences has also led to a recognition, not least by the HSE itself, of the importance of specialist health and safety practitioners.

The whole of this book would have to be devoted to the criminal perspective in order to provide a detailed commentary on all the relevant authorities and a consideration of criminal procedure, but some of the major authorities and principles are highlighted. In order to understand how a prosecution for an individual Regulation will be prosecuted, it is essential to consider the law relating to the 'general duties' imposed, in particular, by ss 2 and 3 of the HSWA 1974. This is necessary as, in practice, an alleged breach of one or more of the Regulations tends to lead either to a prosecution under ss 2 or 3 alone (in which case the breach of the Regulations is relied upon in support of the allegation that there has been a breach of the general duty) or informations are laid not only for breaches of the individual Regulations but accompanied by a separate charge alleging a breach of either s 2 or 3 of the HSWA 1974.

15.2 Section 2(1) of the HSWA 1974 provides:

> 'It shall be the duty of every employer to ensure, so far as is reasonably practicable, the health, safety and welfare at work of all his employees.'

Section 3(1) of the HSWA 1974 provides:

> 'It shall be the duty of every employer to conduct his undertaking in such a way as to ensure, so far as is reasonably practicable, that persons not in his employment who may be affected thereby are not thereby exposed to risks to their health or safety.'

In *R v Board of Trustees of the Science Museum* [1993] 1 WLR 1171 the Court of Appeal held, in a case involving legionella bacterium in an air conditioning cooling tower, that, on a true construction of s 3(1), it was sufficient for the prosecution to prove that members of the public were exposed to a possibility of danger. The defence submission, that the offence was not made out as there was no evidence that there had been an actual risk to members of the public, was emphatically rejected. The Court of Appeal, emphasising the preventative aims of the legislation stated:

> 'Subject only to the defence of reasonable practicability, s 3(1) is intended to be an absolute prohibition.'

15.3 Section 40 of the HSWA 1974 provides that the onus of proving practicability or reasonable practicability is on the defendant and the appropriate standard is on a balance of probabilities. The phrase 'reasonable practicability' entails a weighing up of the risk on the one hand, as against the measures required to eliminate that risk. If the defendant can prove, on a balance of probabilities, that the measures are unreasonably disproportionate to the significance of the risk then the statutory defence under ss 2(1) and 3(1) is made out. For further consideration of this issue the practitioner is referred to *Redgrave's Health and Safety* (3rd edn paras 2.54–2.56). The authors of *Redgrave* query whether the traditional interpretation should be followed in the case of the 'European' Regulations. Neither the Directives, nor the vast majority of the Regulations use the words 'reasonably practicable' or 'practicable'.[1]

1 Notable exceptions are the numerous references to reasonable practicability in the Manual Handling Operations Regulations 1992; Workplace (Health, Safety and Welfare) Regulations 1992, regs 12(3) and 13; Management of Health and Safety at Work Regulations 1999, reg 8(2)(a); and the Personal Protective Equipment at Work Regulations 1992, reg 4(3)(d). Note should also be made of the use of the phrase 'reasonable steps' in the Personal Protective Equipment at Work Regulations 1992, reg 10 and 'reasonably foreseeable' in the Provision and Use of Work Equipment Regulations 1998, reg 4(4). For a recent analysis of 'reasonable practicability' see the Court of Appeal judgment in *Hawkes v London Borough of Southwark* (20 February 1998, transcript).

15.4 Counsel for the appellants in *Nelson*,[1] commenting on the applicability to the corporate officers of a gas safety Regulation which was framed in absolute terms, submitted to the Court of Appeal that it was a fundamental principle of criminal law that statutory Regulations cannot extend the criminal liability of persons beyond that contained in the primary legislation. If this argument is sought to be transposed into a submission that all of the 'European' Regulations must also be construed as being subject to a defence of reasonable practicability (and this is an argument that has been run on occasions at first instance), then it is submitted that this submission would be wrong as a matter of law, not least as being contrary to the Directives and leading to a result which patently fails to give purposive effect to the Directives.

1 *R v Nelson Group Services (Maintenance) Ltd* [1998] 4 All ER 331 at 336B–C.

15.5 Although the importance of ss 2 and 3 has already been commented upon, there is no doubt that the HSE has been seeking to widen the responsibility of the corporate defendant by prosecuting an increasingly widening and varied number of offences under various Regulations. In the world of construction, for example, the increasing importance and emphasis placed upon the Construction (Design and Management) Regulations 1994 cannot be over-emphasised. The deliberate attempt by the HSE to tackle specific management issues has been attributed as part of the reason for its prosecution success rate having fallen to just 38% in defended cases. The HSE policy official responsible for enforcement has recently stated:

> 'There is greater incentive to fight today. Much more publicity is given to health and safety breaches. Contractors are often asked at the tender stage whether or not they have any health and safety convictions and now we are frequently up against big city legal firms who employ health and safety specialists ... We are also fighting more difficult cases, because we are trying to get convictions on underlying management issues.'[1]

1 *The Health and Safety Practitioner* August 1999.

15.6 Section 16 of the HSWA 1974 provides for the approval by the Health and Safety Commission (HSC) of Codes of Practice and s 17 of the HSWA 1974 reverses the burden of proof onto the defendant, in cases in which the Approved Code of Practice (ACOP) is not complied with, to show that the statutory requirement in question has been complied with in some other way. Practitioners familiar with the detail of the ACOPs will appreciate the potentially enormous scope that this can give to the prosecution to allege that parts of the Code have not been complied with in any particular case. It remains to be seen to what extent the prosecution will seek also to rely upon the Guidance Notes to those Regulations which do not have ACOPs.[1]

1 See paras **15.15–15.16**.

15.7 When considering whether there has been a breach of the general duties imposed by s 3 of the HSWA 1974 the question arises of defining the employer's 'undertaking'. The leading authority is the House of Lords' decision in *R v Associated Octel Co Ltd* [1996] 1 WLR 1543. Lord Hoffmann, in the leading speech, held that anything that constituted the running of the plant was part of Octel's undertaking. The issue was a factual one to be left to the jury. Although rejecting the 'extreme position' of the Crown that 'undertaking' also encompassed all works of cleaning and maintenance, wherever and whomsoever they were done by, it has to be recognised that the House of Lords, in *Octel*, have given an extremely wide definition of 'undertaking'.

One leading commentator[1] regards Lord Hoffmann's judgment as ultimately inconclusive as to the scope of the duty under s 3 to supervise the work of contractors, stating that 'the decision raises more questions than it answers'.

1 M Tyler 'Responsibility for contractors—where will it end?' (1997) 1 Journal of Institution of Occupational Safety and Health 2.

15.8 Section 4 of the HSWA 1974 imposes duties upon persons in relation to those who are not their employees but use non-domestic premises[1] made available to them as a place of work or as a place where they may use plant or substances provided for their use on the premises. The duties imposed include provision of safe access and egress to and from the premises. The section recognises that there may be more than one person exercising control over premises and the duty is imposed on those 'to any extent' controlling the premises. In essence the duty is to ensure that the place is made safe for the purposes for which visitors may be expected to use them.

1 Defined in the HSWA 1974, s 53(1).

15.9 Section 6 imposes a duty on any person who designs, manufacturers, imports or supplies any article for use at work[1] to ensure, so far as is reasonably practicable that the article is safe and imposes obligations of testing, examination and provision of adequate information about use.

Section 6(10) provides a defence in circumstances in which it can be shown that the risk to health or absence of safety could not be reasonably foreseen.

Section 6(1A) provides for detailed duties in relation to fairground equipment.

1 For the definitions of 'work' and 'at work' see HSWA 1974, s 52. For general interpretation see s 53.

PROSECUTION OF SPECIFIC REGULATIONS

Management of Health and Safety at Work Regulations 1999

15.10 The Management of Health and Safety at Work Regulations 1999 came into force on 29 December 1999 and re-enact, with certain amendments, the provisions of the 1992 Regulations. The risk assessment provisions have led to a fundamental reappraisal of the long-standing common law duties of safe system of working, safe place of work and proper plant and equipment, and have provided civil practitioners with a starting point when considering whether a potential defendant is responsible for an accident to a potential claimant.[1] In practice it is often extremely difficult for the employers to show that they have made a suitable and sufficient risk assessment pursuant to reg 3(1) as the prosecution can almost invariably suggest that some factor has not been taken adequately into account or missed out altogether. The uncertainty is compounded by the lack of any model risk assessment in the Regulations or ACOP (due, no doubt, to the perception that it would be impracticable to provide model assessments for the myriad work activities encountered in industry) and the refusal of the HSE to 'approve' risk assessments.

1 Despite reg 22 (formerly reg 15) stating that a breach of duty imposed by the Regulations does not confer a right of action in civil proceedings, many civil practitioners do plead failure to comply with the Regulations in support of allegations of common law breaches. Note, however, the exemptions for breaches of regs 16 and 19.

15.11 To defend an alleged breach of the Management of Health and Safety at Work Regulations 1999[1] the defendant is often obliged to adduce expensive expert evidence commenting on the suitability and sufficiency of the risk assessments in order to convey to a lay bench or judge the realities of industrial practice and the reasonably limited nature of most risk assessments.[2]

It should also be borne in mind that it is often possible to mitigate on the basis that the breach was not causative of the injury/incident which led to the prosecution.

1 See para **15.37** for an important removal of what had become a potentially effective line of defence in criminal prosecutions.

2 Reference can be made to the 'model' risk assessment in the Manual Handling Operations Regulations 1992, App 2, to illustrate the point.

15.12 The self-employed have duties imposed upon them by reg 3(2) and the need for a review of existing risk assessments. Regulation 3(3) has increasingly become a focus for HSE prosecutions. An inevitable difficulty, in practice, is the review of a risk assessment in the wake of an accident which is then relied upon by the prosecution as proving the inadequacy of the original risk assessment.

15.13 The Management of Health and Safety at Work Regulations 1999 provide additional protection for young persons (defined as being under the age of 18) and it is anticipated that the HSE may well be keen to prosecute incidents involving young persons under these new Regulations.

15.14 It has also been suggested by one commentator that a consequence of the *Octel* decision is that employers are obliged to perform risk assessments, regardless of whether they control the operations, if the activity is part of their undertaking.[1] Although this may well be a correct interpretation of the law, it does lead to potentially impossible, if not absurd, consequences for the employer in circumstances in which, for example, specialist contractors are brought in to carry out works at the employer's premises with no 'control' being exercised over their activities by the employer.

1 G Holgate 'The role of control in establishing criminal liability under section 3(1) of the Health and Safety at Work etc Act 1974' in *Tolley's Health and Safety*.

Manual Handling Operations Regulations 1992

15.15 Although frequently cited in civil actions, the Manual Handling Operations Regulations 1992 have yet to make any significant impact in criminal prosecutions. This is undoubtedly due to the type of accident and the likely injuries. Put bluntly, the HSE or local enforcing authority are bound to be far more interested in a serious injury involving a fall from height than a back injury caused by lifting a heavy crate.

Appendix 2 of the Regulations provides an example of an assessment checklist.

An understanding of the hierarchy of measures under the Regulations is best achieved by considering the flow chart on p 5 of the HSE Guidance booklet L23 which is reproduced on p 589 of this book.

15.16 The basic steps are to consider whether it is reasonably practicable to avoid moving loads involving a risk of injury. If not, then the Manual Handling Operations Regulations 1992 require an assessment to be carried out (reg 4(1)(b)(i)) to determine the measures to be taken to reduce the risk of injury to the lowest level reasonably practicable (reg 4(1)(b)(ii),(iii)). The measures must be implemented, and reviewed if there are any significant changes in conditions. The Regulations do not have an ACOP but are accompanied by detailed Guidance Notes. The Guidance Notes do not have statutory force but will, inevitably, be taken into account by the court. The HSE states that the Guidance Notes are not compulsory:

> 'But if you do follow the guidance you will normally be doing enough to comply with the law.'[1]

1 See foreword to the Regulation in Booklet L23. For detailed commentary on the Regulations see Chapter 9.

Workplace (Health, Safety and Welfare) Regulations 1992

15.17 The fundamental change for practitioners is the use of the word 'efficient' in these Regulations. By way of example, the Workplace (Health, Safety and Welfare) Regulations 1992, reg 5(1) provides:

> 'The workplace and the equipment, devices and systems to which this regulation applies shall be maintained (including cleaned as appropriate) in an efficient state, in efficient working order and in good repair.'

Paragraph 20 of the ACOP defines 'efficient' as meaning efficient from the point of view of health, safety and welfare and not productivity or economy. It will be noted that the wording of the Regulation imposes an absolute duty.

Personal Protective Equipment at Work Regulations 1992

15.18 For this, as with other Regulations, the reader is referred to the major chapters of the book dealing with the individual Regulations.

Recent prosecutions under the Personal Protective Equipment at Work Regulations 1992 include a window manufacturing company fined £2,000 for breach of reg 10(1) in failing to ensure that eye protection which was provided was actually worn[1] and, an example of where the Regulations are relied upon in support of a s 2 'general duties' prosecution is the case in which a metal galvanising company was fined £18,000 after a worker was splashed with molten zinc and suffered burns as he was not wearing the personal protective equipment provided for his use.[2]

1 *HSE v Starplas Ltd* Safety Management Magazine, October 1998.
2 *HSE v BE Wedge Ltd* Safety Management Magazine, December 1999.

Provision and Use of Work Equipment Regulations 1992/1998

15.19 Successful prosecutions have included fines of £10,000 for failing to provide adequate training contrary to reg 10(1) and £10,000 for failing to prevent access to dangerous parts of machinery (reg 11(1)) in a case in which a sawmill employee had an arm amputated after entering an extraction silo,[1] and fines totalling £30,000 in a fatal accident in which an inadequately trained employee was killed whilst operating a fork lift truck.[2]

See Chapter 7 for a detailed consideration of the Provision and Use of Work Equipment Regulations 1998.

1 *HSE v Wenban-Smith Ltd* Safety Management Magazine, September 1999.
2 *HSE v Frederick Potts* Safety Management Magazine, October 1999.

Construction (Design and Management) Regulations 1994

15.20 These Regulations, which came into force on 31 March 1995, have provided an increasing number of prosecutions over the past few years fuelled, to no little extent, by the HSE's determination to obtain convictions on management issues. Although decisions such as *R v Fartygsentrepenader AB* 1997[1] give examples of the wider scope for the prosecution to succeed in obtaining convictions under ss 2 or 3 of the HSWA 1974, there is an apparently increasing desire by the HSE to obtain convictions under the Construction (Design and Management) Regulations 1994, perhaps fuelled by the fact that the Regulations were the first manifestation of a 'cradle to grave' approach to health and safety issues in a particular sector.

1 *Port Ramsgate walkway collapse* Central Criminal Court, January 1997 (transcript).

15.21 The Construction (Design and Management) Regulations 1994 are complex and impose a number of detailed duties upon those involved in construction projects, namely the client, designer,[1] planning supervisor, principal contractor, and also contractors and the self-employed.

In simple terms (and the Regulations are anything but!) the client has responsibilities to be reasonably satisfied that they only use competent people as planning supervisors, designers and principal contractors and also to satisfy themselves that sufficient resources, including time, have been or will be allocated to enable the project to be carried out safely.

The planning supervisor has official responsibility for co-ordinating the health and safety aspects of the design and planning phase and ensure a health and safety plan is prepared.

Designers must design in a way which avoids, reduces or controls risks to health and safety. The principal contractor has specific responsibilities for tendering, takes over the health and safety plan and co-ordinates the activities of contractors and sub-contractors to ensure compliance with the health and safety plan and the relevant health and safety legislation.

Contractors and the self-employed should provide information on risks to health and safety created by their work and are required to co-operate with the principal contractor.

1 See *R v Paul Wurth SA* [2000] ICR 860, CA for an analysis of the obligations under the Regulations and the narrow interpretation by the Court of Appeal of 'designer'.

15.22 It should be noted that in practice the overlap of duties between the various parties involved in major construction projects is often complicated by the fact that the body corporate may be fulfilling more than one of the roles. Further, there is a perception in the construction industry that the HSE often fail to have regard to the realities of a major construction project and seek to impute duties, or methods of working, which would render the project unworkable.

15.23 Whilst it is impossible, without devoting a whole book to the Regulations, to consider the majority of the Regulations, an example of the difficulties and issues which arise in practice may be of illustrative assistance.

Regulation 16(1)(a) of the Construction (Design and Management) Regulations 1994 requires the principal contractor to take reasonable steps to ensure co-operation between contractors to enable the contractors to comply with the relevant statutory provisions. Paragraph 91 of the ACOP refers in terms to the risk assessment obligations under the Management of Health and Safety at Work Regulations, and para 92 of the ACOP states that the principal contractor needs to examine the assessments to ensure that the risks have been properly evaluated. In the case of a major construction project the principal contractors often engage a reputable and specialised firm of contractors to carry out specific complex tasks. To what extent, in practice, can the principal contractor 'examine' the assessments to confirm that the seriousness of risks has been 'properly evaluated' when the expertise and specialisation of the contractor is so much greater than that of the principal contractor? In at least one recent prosecution the HSE have sought to interpret these provisions as putting a heavy onus on the principal contractor to vet each and every individual risk assessment provided by specialised sub-contractors. It can be said, with some force, that not only would the sheer number of assessments produced by numerous sub-contractors (on major construction projects there may be thousands of workers and scores of sub-contractors) render a detailed consideration wholly impracticable, but one can also query what (if this is the obligation on the principal contractor) is the purpose of the competence and resource assessments of the contractors under regs 8 and 9? It may be that the argument will boil down to what are 'reasonable steps', but this example does highlight a significant difference in perception of the HSE and the construction industry of the proper interpretation of the Construction (Design and Management) Regulations 1994.

15.24 Other Regulations which are of particular relevance to construction work include the Construction (Health, Safety and Welfare) Regulations 1996, the Provision and Use of Work Equipment Regulations 1992, and the risk assessment provisions of the Management of Health and Safety at Work Regulations 1992. It should also be noted that, in construction prosecutions, the prosecution will often rely upon HSE Guidance documents (eg HS(G) 150 'Health and Safety in Construction' 1996).

Asbestos

15.25 The HSE have pursued a determined policy to prosecute companies, and on occasions individuals, for offences arising from the improper handling and disposal of asbestos. The Control of Asbestos at Work Regulations 1987 (as amended by the Control of Asbestos at Work (Amendment) Regulations 1992) set out detailed provisions for the assessment, notification, prevention of exposure and use of control measures in work involving asbestos. On occasions, where the acts complained of arise in the course of construction or demolition, prosecutions may be founded upon breaches of the Construction (Design and Management) Regulations 1994.

Fines have been increasing and the Court of Appeal recently dismissed an appeal against a fine of £100,000 imposed upon a carpet manufacturing firm which failed to adequately check for the presence of asbestos materials after buying a factory.[1] Despite the verbal assurances of the previous owner that the building did not contain asbestos, the company were criticised for not obtaining written assurances or having a survey performed.

1 *R v Brintons Ltd* (22 June 1999, unreported) CA.

15.26 It is strongly recommended that no defendant seeks to deny or minimise the dangers of asbestos. Doing so (apart from the difficulties posed by *R v Board of Science Museum Trustees*)[1] will lead to being bombarded with notices of additional evidence from the prosecution exhibiting detailed reports on the dangers of asbestos[2] and is likely to be regarded as a plea in aggravation. It should also be recognised that many prosecutors, as a matter of course, tend to adduce such evidence.

1 See para **15.2**.
2 PETO, for example.

PROSECUTION OF INDIVIDUALS

15.27 The self-employed may be prosecuted under s 3 of the HSWA 1974 for exposing non-employees to risks to their health and safety.

15.28 Individual employees may be prosecuted under s 7 of the HSWA 1974, which provides that it is the duty of every employee while at work to take reasonable care for the health and safety of himself and of other persons who may be affected by his acts or omissions at work. In practice such prosecutions tend to be accompanied by a prosecution of the employer as well, often on the basis that there is a flaw in the system of working or on the basis that the individual employee has been inadequately trained or instructed.

From a prosecution perspective the weakness in prosecuting an individual under s 7, without also prosecuting the employer, is the scope for the defence to run a 'scapegoat' argument and focus the defence upon attacking weaknesses in the employer's system of working. On the other hand prosecution of both employer and employee almost invariably leads to employers seeking to mitigate on the basis that the systems of work were dependent upon the co-operation of employees.

15.29 Section 8 of the HSWA 1974 provides that no person shall intentionally or recklessly interfere with or misuse anything provided in the interests of health, safety and welfare pursuant to any of the relevant statutory provisions. The author is unaware of any such prosecution having occurred and this is, no doubt, due to the relative rarity of such circumstances arising combined with the prosecution difficulty in seeking to prove intention or recklessness.

15.30 It should also be noted that certain regulations also impose duties upon employees. For example, reg 14 of the Management of Health and Safety at Work Regulations 1999 imposes obligations upon employees to use, inter alia, machinery and safety devices in accordance with instructions and training received.

15.31 Section 37 of the HSWA 1974 provides that where an offence under any of the relevant statutory provisions committed by a corporate body is proved to have been committed:

> 'With the consent or connivance of, or to have been attributable to any neglect on the part of, any director, manager, secretary or other similar officer of the body corporate or a person who was purporting to act in any such capacity, he as well as the body corporate shall be guilty of that offence and shall be liable to be proceeded against and punished accordingly.'

Strictly speaking it is submitted that it is not necessary for the body corporate to have been prosecuted, let alone convicted, for an offence to be made out against an individual officer of the company under s 37. In practice the overwhelming majority of s 37 prosecutions are accompanied by a prosecution of the corporate body, but there may be situations, for example, where the body corporate no longer exists, in which case a prosecution against the individual alone may be considered appropriate, although a more sensible alternative may be to prosecute the individual under s 7. After all, a senior manager or company officer is still an employee for the purposes of s 7.

15.32 The leading authority on s 37 is *Wotherspoon v HM Advocate* (1978) JC 74 in which Lord Justice-General Emslie held that 'neglect' pre-supposed the existence of some obligation or duty on the part of the person charged with neglect and:

> 'The search must be to discover whether the accused has failed to take some steps to prevent the commission of an offence by the corporation to which he belongs if the taking of those steps either expressly falls or should be held to fall within the scope of the functions of the office which he holds. In all cases accordingly the functions of the office of a person charged with a contravention of s 37(1) will be highly relevant consideration for any judge or jury and the question whether there was on his part, as the holder of his particular office, a failure to take a step which he could and should have taken will fall to be answered in the light of the whole circumstances of the case including his state of knowledge of the need for action, or the existence of a state of fact requiring action to be taken of which he ought to have been aware.'

The court went on to hold that, so far as 'attributable to' was concerned:

> 'In our opinion any degree of attributability will suffice . . .'

Having regard to the extremely wide scope of the words 'connivance' and 'attributable to any neglect' the defence, in s 37 cases, will undoubtedly wish to give considerable thought to the 'functions of the office' held by the defendant and also to his state of knowledge. Although it will be more difficult to disassociate the most senior employees from responsibility, the court in *Wotherspoon* highlighted the need to emphasise, in the case of more junior employees, the importance of the scope of functions of office.

RESPONSIBILITY OF A CORPORATE DEFENDANT FOR THE ACTS OR OMISSIONS OF EMPLOYEES

15.33 The Court of Appeal considered the applicability of s 3(1) of the HSWA 1974 in *R v British Steel plc* [1995] 1 WLR 1356. Counsel for British Steel submitted that the company should escape criminal liability if at 'directing mind' level the company had taken reasonable care, relying strongly upon the decision of the House of Lords in *Tesco Supermarkets Ltd v Nattrass* [1972] AC 153. The Court of Appeal rejected this submission, holding that, subject to the words 'so far as is reasonably practicable' (which were simply referable to the measures necessary to avert the risk) s 3(1) created an absolute prohibition. The defence of reasonable practicability was said to be a narrow one, analogous to the defence under s 29(1) of the Factories Act 1961.

Lord Justice Steyn stated that to accede to the company submission would 'drive a juggernaut through the legislative scheme' but did recognise that the court's strict interpretation was of concern in circumstances in which an individual junior employee was guilty of an isolated act of negligence, and this rendered the corporate employer guilty of a criminal offence.

> 'We do recognise that there may be circumstances in which it might be regarded as absurd that an employer should even be technically guilty of a criminal offence. An example might perhaps be the driver who is guilty of an error of judgment when driving his employer's lorry on his employer's business. But, in any event, so called absurdities are not peculiar to this corner of the law: at the extremities of the field of application of many rules surprising results are often to be found. That circumstance is inherent in the adoption of general rules to govern an infinity of particular circumstances. Fortunately, the cases to which counsel referred will in practice cause no real difficulty in relation to s 3(1) of the Act of 1974. Nobody has suggested that there has ever been a prosecution in such a case, and it is most unlikely that there would in future be a prosecution in such cases. Moreover, if such prosecutions are brought, they are not likely to be viewed sympathetically by a judge and jury or by magistrates. In the most unlikely event of a conviction, the judge would be entitled to impose an absolute discharge and refuse to order costs in favour of the prosecution.'

15.34 After the House of Lords' decision in *Octel* the Court of Appeal had the opportunity to reconsider this area of the law in *R v Gateway Foodmarkets Ltd* [1997] 3 All ER 78. The appellants employed lift contractors to maintain their lifts at all their stores. At one store a practice had developed in which store management would manually rectify a faulty electrical contact in the control room having been told how to do so by the lift contractors. On the day before the accident the lift contractors had carried out routine maintenance of the lift and, for no good reason, left open a trap door in the floor of the control room. The next day, when the lift jammed, the 22-year-old section manager at the store went into the control room to attempt to free the contact and fell through the trap door and was killed. The practice of management freeing the contact was unauthorised by head office and no one there was aware of it. The trial judge made a preliminary ruling that the offence under s 2(1) of the HSWA 1974 was of strict liability, whereupon the company pleaded guilty. The Court of Appeal dismissed the appeal against conviction. The 'absurdity' argument referred to by Lord Justice Steyn in *British Steel* was considered and Lord Justice Evans, whilst agreeing that it would be a 'somewhat extreme contention' that an employer should be criminally liable even for an isolated act of negligence by a junior employee under both ss 2(1) and 3(1), disagreed that the result would be 'absurd' but indicated that, whilst it was not necessary to decide the point for the purposes of the appeal, the court would have held that the extreme consequences did not result from a proper construction of the two sub-sections.

15.35 The potential absurdities and manifest potential injustice created by *R v British Steel plc* have, at least in the short term, been tempered by the Court of Appeal's judgment in *R v Nelson Group Services (Maintenance) Ltd* [1998] 4 All ER 331. In this case gas fitters employed by the defendant company exposed occupiers of private dwellings to the risks of leaking gas by failing to exercise proper skill in carrying out their gas fitting work. The company was convicted of an offence under s 3(1) of the HSWA 1974. The facts were not in dispute.

> 'In the first case it is accepted that a proper test for gas leaks in the system following installation was not carried out, and in the third case that the end of the pipe supplying gas to the fire where that pipe was broken to allow removal of the gas fire was not properly capped. It is accepted by the Respondent that in those cases the fitters concerned were properly trained, properly certificated, had as necessary undergone refresher courses, received proper instructions, had been supplied with proper equipment which would have allowed them to discharge their work as fitters satisfactorily and safely. On the other hand, it is now common ground that the gas fittings, which include gas appliances . . . were left in such a condition that the occupiers of the private houses were exposed to a risk to their health and safety.
>
> The question which arises is whether that fact alone renders the Appellants in breach of their duty under s 3(1) and guilty of a criminal offence under s 33(1)(a) . . .' (334B–E).

Roach LJ, stating that the court had derived considerable assistance from the *Gateway* case, held:

> 'We would summarise the law in this way: first, if persons not in the employment of the employer are exposed to risks to their health or safety by the conduct of the employers undertaken, the employer will be in breach of s 3(1) and guilty of an offence under s 33(1)(a) of the Act unless the employer can prove on a balance of probability that all that was reasonably practicable had been done by the employer or on the employer's behalf to ensure that such persons were not exposed to such risks. It will be question of fact for the jury in each case whether it was the conduct of the employer's undertaking which exposed the third persons to risks to their health or safety. The question what was

reasonably practicable is also a question of fact for the jury depending on the circumstances of each case. The fact that the employee who was carrying out the work, in this case the fitter installing the appliance, has done the work carelessly or omitted to take a precaution he should have taken does not of itself preclude the employer from establishing that everything that was reasonably practicable in the conduct of the employer's undertaking to ensure that third persons affected by the employer's undertaking were not exposed to risks to their health and safety had been done.

In our view this analysis is consistent with the distinction which appears in the Regulations between the duties of employers of persons and the duty of persons performing the work. It is not necessary for the adequate protection of the public that the employer should be held criminally liable even for an isolated act of negligence by the employee performing the work. Such persons are themselves liable to criminal sanctions under the Act and under the Regulations. Moreover it is a sufficient obligation to place on the employer in order to protect the public to require the employer to show that everything reasonably practicable has been done to see that a person doing the work has the appropriate skill and instruction, has had laid down for him safe systems of doing the work, has been subject to adequate supervision, and has been provided with safe plant and equipment for the proper performance of the work.'

It is submitted that this judgment accords with common sense and justice as it avoids the 'absurd' consequences recognised by the Court of Appeal as being the consequences of the strict interpretation of the sub-sections.

15.36 The alternative view is that the *Nelson* judgment undermines the protection of persons coming within the ambit of the HSWA 1974, s 3(1) and endorses or encourages low standards of health and safety.

It is submitted that this alternative view unnecessarily fears for a lowering of standards in circumstances in which, having regard to the decision in *Nelson,* it is still incumbent upon the defence to prove that everything reasonably practicable had been done to ensure that the individual employee had the appropriate skill and instruction to do the work, that a safe system of working had been laid down, and also that the individual not only had proper plant and equipment to enable him to perform the work safely but was also adequately supervised. In practice the requirement for 'adequate supervision' may cause problems for individual corporate defendants. Is it, by way of example, sufficient to say that there was 'adequate supervision' of an experienced employee who was working on his own to merely state that he was a highly experienced employee who had been adequately trained in the work that he was to do? It seems inevitable that the prosecution would suggest in such cases that, at the very least, there would need to be continual assessment and training of even the most experienced employee to ensure that he was fully aware of the systems of work and that, in practice, he was following those laid down systems of work.

15.37 Recent legislative changes provide a strong indication that the government is determined to close down possible lines of defence. Regulation 21 of the Management of Health and Safety at Work Regulations 1999[1] states:

> 'Nothing in the relevant statutory provisions shall operate so as to afford any employer a defence in any criminal proceedings for a construction of those provisions by reason of any act or default of—
>
> (a) an employee of his, or
>
> (b) a person appointed by him under Regulation 7.'[2]

1 In force 29 December 1999.
2 Regulation 7 provides for the appointment of competent persons to assist the employer in undertaking the measures required to comply with the requirements and prohibitions imposed by the Regulations.

ENFORCEMENT

15.38 In simple terms the responsibility for enforcement lies with the HSE in industrial premises, construction sites, railways, nuclear installations etc, whereas the local authorities are responsible for enforcing health and safety legislation in non-industrial (eg retail) premises.

Inspectors, from either the HSE or local authorities, are charged with monitoring premises to ensure compliance with the relevant statutory provisions and to investigate complaints and accidents.

15.39 The HSE does have a programme of preventative inspections designed to ensure, by visiting without prior warning, that industrial premises or construction sites are complying with their statutory duties. How well these unplanned visits work depends upon whom one talks to. Many visits are unannounced, but it is by no means uncommon for visits to be notified/arranged in advance. The most repeated complaint from HSE inspectors is a shortage of manpower and resources.

Nonetheless, there is no doubt that the weapons of prohibition and improvement notices are being used increasingly and effectively in order to improve standards in the workplace.

On a more positive note, there are numerous examples of excellent ongoing co-operation between the enforcing authorities and companies which, it is to be hoped, will minimise risks to health and safety and, ultimately, the prospects of being prosecuted if something does go wrong.

15.40 Section 19 of the HSWA 1974 empowers enforcing authorities to appoint inspectors and wide powers are granted to inspectors by the HSWA 1974, s 20. The inspector is empowered, inter alia, to enter premises at any reasonable time; direct that the premises or anything in them be left undisturbed pending examination or investigation, and to take samples or photographs.

Section 20(2)(j) provides that the inspector has power:

> 'To require any person whom he has reasonable cause to believe to be able to give any information relevant to any examination or investigation [pursuant to his statutory duties] . . . to answer such questions as the inspector thinks fit to ask and to sign a declaration of the truth of his answers.'

15.41 In practice the HSE inspectors either exercise their s 20 interview powers or, if they believe there may be grounds for prosecution, will interview a possible defendant (or senior manager of a possible corporate defendant) under the usual criminal caution.[1]

Section 20(7) provides that no answer given pursuant to the inspector's powers to question persons pursuant to s 20(2)(j) is admissible in evidence against that person or his/her spouse.

Section 20(2)(k) empowers the inspector to require the production of or to inspect or copy any relevant documents. This power is frequently used, with the prosecution gathering up reams of documentation.

It should also be borne in mind that there are notice requirements to be complied with by the inspector in exercising his powers (s 20(6)) which, if not complied with, would give rise to powerful arguments that evidence may be inadmissible.

1 HSWA 1974, s 33(1)(f) provides that it is an offence to prevent or attempt to prevent any person from appearing before an inspection or from answering any question to which an inspector, exercising his s 20(2) powers requires an answer.

15.42 Section 25 provides for situations of imminent danger and empowers the inspector to seize, or even in certain circumstances, destroy articles or substances.

15.43 Section 27 empowers the HSC to serve a notice requiring a person to furnish the HSC, or enforcing authority in question, with information needed for the investigation. Section 28 imposes restrictions on the disclosure of information so obtained.[1]

1 To be abolished or amended in future legislation. See paras **15.69–15.70**.

15.44 Improvement notices, made pursuant to s 21, allow an inspector, if of the opinion that a statutory provision is being contravened, or in circumstances in which provisions have been contravened and it is likely that this will continue or be repeated, to serve a notice giving particulars of the breach and requiring remedy of the breach within a specified period.

Section 22 empowers an inspector to serve prohibition notices in circumstances in which he perceives that there is a risk of serious personal injury.

Appeals against improvement and prohibition notices must be made to an employment tribunal within 21 days (s 24).

15.45 Appeals seem to be increasing, which is due in large part to the HSE's perception of the width of responsibility imposed upon companies by the various statutory provisions.

PENALTIES

15.46 Dealt with summarily the maximum fine for breach of the general duties under ss 2–6 of the HSWA 1974 is £20,000. In the Crown Court the fine is unlimited.[1]

In 1997/98 the average fine in the Crown Court per offence was £17,768. In the magistrates' court the average stood at £6,223. Increasing disquiet about the low level of fines for health and safety offences (primarily from the HSE and from the media in the wake of certain incidents) appears to have been the catalyst for the long awaited intervention of the Court of Appeal in the question of sentencing guidelines for health and safety offences.

1 See the HSWA 1974, s 33 for the detailed penalty provisions.

15.47 *R v F Howe & Son (Engineers) Ltd* [1999] 2 All ER 249 is essential reading for anyone involved in a health and safety prosecution. The Court of Appeal, recognising that it was difficult for judges and magistrates to have an instinctive feel for these relatively rarely encountered cases, stated that they would:

> 'Endeavour to outline some of the relevant factors that should be taken into account.'

Having done so the Court of Appeal were very keen to emphasise that:

> 'It is impossible to lay down any tariff or to say that the fine should bear any specific relationship to the turnover or net profit of the defendant. Each case must be dealt with according to its own particular circumstances.'

It is not proposed, in this relatively brief chapter, to outline the facts of the *Howe* decision and to consider it in more detail. It should suffice if it be said that it is somewhat unfortunate that the Court of Appeal, in setting out general guideline sentencing principles in an attempt to increase the overall level of fines, chose a case in which they allowed the appeal to the extent of actually reducing the fine imposed in the Crown Court.

15.48 The importance of *Howe* is its general observations on sentencing and guidance on the approach that the court should take. The Court of Appeal stated that:

> 'In assessing the gravity of the breach it is often helpful to look at how far short of the appropriate standard the defendant fell in failing to meet the reasonably practicable test.'

It is also stated that it is often a matter of chance whether death or serious injury results from even a serious breach but that, generally, where death was a consequence of a criminal act it should be regarded as an aggravating feature of the offence and the penalties should reflect public disquiet at the unnecessary loss of life.

The most commonly quoted and relied upon part of the *Howe* judgment is:

> 'Other matters that may be relevant to sentence are the degree of risk and extent of the danger created by the offence; the extent of the breach or breaches, for example whether it was an isolated incident or continued over a period and, importantly, the defendant's resources and the effect of the fine on its business.
>
> Particular aggravating features will include (1) a failure to heed warnings and (2) where the defendant has deliberately profited financially from a failure to take necessary health and safety steps or specifically run a risk to save money.

Particular mitigating features include (1) prompt admission of responsibility and a timely plea of guilty, (2) steps to remedy deficiencies after they are drawn to the defendant's attention and (3) a good safety record.'

15.49 There has been an increasing but by no means universal tendency for prosecutors, both in the magistrates' court and in the Crown Court, to refer to the *Howe* decision. The defence consideration of whether they wish to emphasise the absence of particular aggravating features and the presence of any particular mitigating features when presenting their plea in mitigation,[1] has been overtaken by a recent decision of the Court of Appeal.

A fundamental alteration to the quality and focus of prosecutions will occur if the recommendations of the Court of Appeal in the case of *R v Friskies Petcare UK Ltd*[2] are followed. The Court of Appeal (Beldam LJ, Silber J and the Recorder of Leeds) gave guidance on the approach to the mitigating and aggravating factors set out in *Howe*, recommending that the prosecution, when commencing proceedings, should not only set out the facts of the case but should also list the aggravating features. The defendants should be sent a copy and, in the case of a guilty plea, the defendants should submit a similar document outlining the mitigating features. The court urged, if possible, an agreement in writing between the prosecution and defence on the aggravating and mitigating features to be taken into account by the court on sentencing.

If any disagreement of substance remains, the court will have to consider whether a *Newton* hearing is required.

The Court of Appeal 'strongly' recommended that this procedure should be routinely adopted. It has to be said that, if adopted, the consequence is likely to be a more effective presentation of prosecution cases. In the past there are myriad examples of prosecutions in which the prosecution make no reference whatsoever to the aggravating features in *Howe* where the defence refer to and rely upon the mitigating features.

1 In *R v Mersey Docks and Harbour Co* (1995) 16 Cr App Rep (S) 806 the Court of Appeal held that it was no mitigation for defendants to say that they left their non-delegable duty under the HSWA 1974, s 3, to other people to fulfil.

2 10 March 2000, unreported. Transcript of Court of Appeal (Criminal Division).

15.50 It is premature for any evaluation to be made of the overall impact of the *Howe* case on the levels of sentencing save to say that experience does tend to suggest that it has tended to increase significantly the level of fines imposed upon corporate defendants in health and safety cases, particularly in the Crown Court. Magistrates' courts, for the most part (but by no means invariably), have tended to heed the Court of Appeal's exhortation in *Howe* always to think carefully before accepting jurisdiction in health and safety at work cases:

> 'Where it is arguable that the fine may exceed the limit of their jurisdiction or where death or serious injury has resulted from the offence.'

It is clear from the *Howe* judgment that the particular aggravating feature of profiting financially from the failure to take necessary health and safety steps is one that the Court of Appeal regarded as being one which seriously aggravated the offence and, in practice, is the aggravating feature most likely to result in a severe financial penalty.

15.51 It should also be noted that the Court of Appeal emphasised that the standard of care imposed by the legislation is the same regardless of the size of the defendant company. Contrasting a large corporation with an in-house Health and Safety Department with a small company with no safety officers at all, the Court of Appeal suggested that those organisations who do not have their own expertise in-house could obtain it, if necessary, by seeking assistance from the HSE. By way of comment it should be said that in practice this is unlikely to yield satisfactory results. The Health and Safety Executive have limited resources (extremely limited resources if one talks to HSE inspectors) and it is doubtful whether the HSE would become involved to the extent of being able to assist with, for instance, management procedures or systems. In practice, one knows that the HSE refuse to get involved with many aspects of systems of working. By way of example, the HSE, as a matter of policy, never 'approve' risk assessments or method statements. This is commonly encountered in the realms of construction when the Health and Safety construction inspectors attend on a building site and review the risk assessment/method statements. Although the inspector does have the power to issue a prohibition notice if he feels that the risk assessment/method statement is inadequate, he will not go to the stage further and actually 'approve' the documentation as being satisfactory. One senior HSE inspector recently said that this is the first thing he teaches new recruits to the HSE.

This should not be regarded, in itself, as a criticism of the HSE but in practical terms, is a criticism of the suggestion by the Court of Appeal that small companies should seek the assistance of the HSE in the absence of an in-house Health and Safety Department.

15.52 In *R v Rollco Screw and Rivet Co Ltd* [1999] IRLR 439 the Court of Appeal made further observations on the imposition of financial penalties on corporations in health and safety cases and stated that a longer period of payment, by instalments, was acceptable in the case of a company than would often be acceptable in the case of an individual. In *Rollco* the Court of Appeal allowed a period of five years and seven months to pay a total financial penalty of £70,000.

15.53 The Court of Appeal in *Howe* stated that the fine should reflect not only the gravity of the offence but also the means of the offender and that this applied just as much to corporate defendants as to any other.[1]

It should also be borne in mind, as this has caused certain difficulties in practice, that in *Howe* the Court of Appeal made the following observations when it came to representations being made about the means of a corporate defendant:

> 'If a defendant company wishes to make any submissions to the court about its ability to pay a fine it should supply copies of its accounts and any other financial information on which it intends to rely in good time before the hearing both to the court and to the prosecution. This will give the prosecution the opportunity to assist the court should the court wish it. Usually accounts need to be considered with some care to avoid reaching a superficial and perhaps erroneous conclusion. Where accounts or other financial information are deliberately not supplied the court will be entitled to conclude that the company is in a position to pay any financial penalty it is minded to impose.'

This would appear, at first reading, and indeed upon reading after some reflection, to be quite clear and straightforward. Unfortunately, there has been an increasing tendency for some judges to regard this part of the judgment as stating that the failure to provide company accounts is an *aggravating* feature. It is submitted that this interpretation is illogical and unsustainable and contrary to the judgment of the Court of Appeal. Nonetheless, it is a point that is worth bearing in mind from the practitioner's perspective when considering whether any accounts should be submitted prior to a sentencing hearing. It is becoming increasingly the norm to adduce some evidence of the means of a corporate defendant even if impecuniosity is not going to be argued, but this must be a matter for the individual advisers to consider in the circumstances of any particular case.

1 Reiterated by the Court of Appeal in *R v Cappagh Public Works Ltd* [1999] 2 Cr App Rep (S) 301. Further note the Criminal Justice Act 1991, s 18.

15.54 A word of caution. Although the Court of Appeal do state that the fine should reflect not only the gravity of the offence but also the means of the offender, it would appear that the Court of Appeal did recognise that there may be cases where the offences were so serious that the defendant ought not to be in business. That observation followed on from a comment that the fine needs to be large enough to bring the health and safety message home where the defendant is a company, not only to those who manage but also to its shareholders. It will be appreciated that in a case of a public limited company the level of fine required to bring this health and safety message home to its shareholders would have to be enormous.

15.55 There have been numerous significant fines imposed over the past few years in health and safety cases. Most pre-date *Howe*. It is anticipated that it is only a matter of time before a court imposes a multi-million pound fine in a health and safety prosecution, particularly in cases in which the corporate defendant may have a turnover running into hundreds of millions of pounds.[1]

As Clark J stated in *R v Fartygsentrepenader AB*:[2]

> 'The purpose of these fines is, in part, to bring it home to the boardrooms of companies and controlling minds of other entities who may be employers that the safety of the public is paramount.'

1 See also *R v Friskies Petcare UK Ltd* (10 March 2000, unreported) CA, which can be used in support of a submission that fines of £500,000 or more are only suitable for cases involving a major public disaster in which large numbers of the public are put at risk of serious injury or death.

2 Transcript. Central Criminal Court, January 1997: fines totalling £1,700,000 against four defendants.

15.56 The Health and Safety Executive, in their continuing quest to increase the levels of financial penalties imposed upon corporate defendants in health and safety cases, have produced (seemingly by their press office) a publication entitled 'Large Fines Following HSW Act Prosecutions'. This surfaces periodically during prosecutions, primarily in the Crown Court, and is sought to be used by prosecuting counsel to 'assist' the court on the appropriate level of fines in individual cases. The fines start at £100,000 and list, by way of example, over four pages of fines in which the court imposed penalties in excess of £250,000. It is submitted that the use of this document is wholly inappropriate. First, it does not assist the court in coming to a just and fair conclusion if the only examples of fines that are put forward are large fines (it is noteworthy that the heading on each page after the title page is 'Big Fines') and it is not the author's experience that the HSE have ever 'assisted' the court by putting forward lists of 'medium' or 'small' fines. Perhaps more importantly, it must be open to question as to whether it is proper for prosecution counsel, as an officer of the court, to put forward a list of severe penalties with the obvious intention of increasing the penalty upon the defendant. It is submitted that this is not, and has never been, the role of the prosecution. Quite obviously, there can be no objection whatsoever to the court being referred to the Court of Appeal decisions in *Howe* and *Rollco*.

15.57 The fines to be imposed in respect of breaches of the individual Regulations are, for the most part, £5,000 summarily and unlimited in the Crown Court. As the decision in *Howe* states in terms that the observations relate to the general duties under ss 2–6 of the HSWA 1974, it remains open to doubt the extent to which the observations in that decision can or should be applied to breaches of individual Regulations. There has been a tendency, prior to *Howe* for relatively minor fines to be imposed in respect of breach of individual Regulations. In particular, breaches of the Construction (Design and Management) Regulations 1994 have tended to attract fines of only a few thousand pounds per offence. Much depends upon whether the prosecution have chosen to prosecute or proceed with offences under the general duties or the individual Regulations. Some prosecutors wisely limit their charges to, by way of example, a charge under s 2 or 3 of the HSWA 1974, even in circumstances in which there are easily provable breaches of half a dozen Regulations. The breaches of the individual Regulations are then relied upon in support of the breach of the general duty.

15.58 A difficulty encountered by practitioners is whether to put forward a detailed plea in mitigation at the plea before venue stage in the magistrates' court. A disadvantage, if unsuccessful in keeping the case before the magistrates, is that the prosecution are fully aware and prepared for the mitigation by the time the case arrives at the Crown Court for sentence. This often tends to lead to a 'tailoring' of the prosecution opening at the higher court to anticipate and minimise the mitigation. If the prosecution state that they are content for the matter to be dealt with in the magistrates' court, as they often state they are, then it is clearly worth putting forward a powerful plea at this stage in order to try and persuade the court to sentence. If the prosecution express a view that the Crown Court is the appropriate forum (often by referring to *Howe*) experience suggests that it is difficult to persuade the magistrates to accept jurisdiction and it may often be better to 'keep your powder dry' and put forward a brief plea.

15.59 The HSE have stated that it is their routine case management procedure to 'plea bargain' (my words, not the HSE's). This would appear in part to be a reaction to the relatively low success rate in contested cases. From a corporate defendant's perspective (particularly in construction/demolition) the incentive to fight is increased by the frequent need to declare, on tender documents, convictions or even cautions.

15.60 A further power of the court, although it is not a power the author has ever seen, or even heard of being used, is contained in s 42 of the HSWA 1974 which provides that, upon conviction, the court may, in addition to, or instead of imposing any punishment, order the defendants to remedy matters arising from the conviction within a specified time limit. The reason why this power is rarely used is presumably because the defendants, albeit often prompted by prohibition and improvement notices, will already have remedied the shortcomings long before the case comes to court.

EXAMPLES OF PENALTIES IMPOSED

15.61 There is no guideline or tariff for health and safety prosecutions, and practitioners would do well to bear in mind the exhortation of the Court of Appeal in *Howe* that each case must be dealt with according to its own particular circumstances. Thomas *Current Sentencing Practice* is of no assistance. There are, however, a number of examples of penalties imposed in Croner's *Health and Safety Law*.

Probably the best source for keeping up to date with the penalties being imposed is by subscribing to *Safety Management* published monthly by the British Safety Council.

COSTS

15.62 Following the Court of Appeal decision in *Octel*[1] in which it was held that the costs of the investigation as well as the costs of the prosecution can be recovered, it is usual for a short breakdown of costs to be provided by the prosecution prior to the hearing. It is advisable to request the bill of costs as soon as it is known that suitable pleas are being tendered in order to clarify any items prior to the substantive hearing. Costs can vary from one or two thousand pounds in the most straightforward cases to several hundred thousand pounds in a complex contested action.

The investigation costs tend to be fairly modest unless the HSE has involved one or more of its laboratories to produce reports on the technical aspects of the case. In these circumstances the costs tend to rocket.

The Court of Appeal in *Octel* held that the prosecution should submit a detailed bill of costs in time for the defence to make representations and, in the case of a dispute (which can sometimes be sorted out in correspondence) the defence should give notice of the objections in writing. In the case of numerous objections it is advisable to serve a Counter Schedule.

1 [1994] 4 All ER 1051.

15.63 On appeal successful appellants to the Crown Court (having regard to the relatively restricted powers of the magistrates' court it is not the author's experience that such appeals are common) can be awarded costs from central funds (Prosecution of Offences Act 1985, s 16(3)) or, in exceptional cases where the prosecution should never have been brought, against the enforcing authority.[1]

1 *R v Lewes Crown Court, ex p Castle* (1979) 70 Cr App Rep 278.

CORPORATE KILLING

15.64 Major disasters over the course of the past 15 years or so have highlighted the fundamental inadequacies of the present law as it relates to manslaughter and corporate defendants. Under the present law to convict a corporation one has to satisfy the 'controlling mind' test. The state of mind of the directors, or exceptionally senior management, is treated in law as the state of mind of the company. This is an issue which has been referred to by commentators as the 'P&O problem' arising out of the Herald of Free Enterprise/Zeebrugge disaster. In that prosecution the judge directed acquittals before the end of the prosecution case. It was held that one cannot have an aggregation of fault and it was not enough for some directors to be aware of parts of the picture and for this knowledge of the individual directors or senior management to be aggregated together. In practice what this has meant is that in large corporations the diffuse management structure has rendered it difficult, if not impossible, to satisfy the requisite mens rea. There have been successful prosecutions and perhaps the best known was the *Lyme Bay Canoeing* tragedy in which three teenagers drowned. The managing director of the company was sentenced to an immediate period of imprisonment in a case in which there had been no proper consideration of safety and a warning letter had been received from a resigning instructor prior to the accident in which he had stated, inter alia, 'children will not be coming home'.

15.65 Most recently, in Attorney General's reference No 2/1999[1] the Court of Appeal held that the 'Identification principle remains the only basis in common law for corporate liability for gross negligence manslaughter' referring to the need for Law Commission 237's Bill conferring liability based on management failure not involving the principle of identification.

Corporate killing is defined in the draft bill as:

> '... if a management failure by the corporation is the cause or one of the causes of the person's death ... and the failure constitutes conduct falling far below what can reasonably be expected of the corporation in the circumstances.'

The issue of guilt is a question of fact and not law and is, therefore, supremely a jury question.

Management failure is defined thus:

> 'There is a management failure by a corporation if the way in which its activities are managed or organised fails to ensure the health and safety of persons employed in or affected by those activities.'

1 *Great Western Trains*: the Southall Rail disaster.

15.66 It is noteworthy that there is no requirement in the draft bill of the risk of death or serious injury being obvious. The draft bill raises a number of difficult questions. First, it is clear that the death need not be that of an employee ('persons affected by those activities') and this may give rise to difficult issues of causation and remoteness. Perhaps the most difficult question of all is what is meant by 'management'. At what level? Many commentators have sought to equate the phrase 'management failure' with high level failures. The examples given are superficial responses to known hazards, inadequate training and instruction and a blasé approach to the implementation of safety procedures of *senior* management. It has been said that prosecutors (and it should be noted that private prosecutions will not be possible under the bill) will be looking for high level management failures. This may very well be correct but it is not clear that lower level management failures would not fall within the ambit of the proposed bill. By way of example, what is the situation if a depot or workplace is badly run by an experienced manager who pays lip service to company procedures? The defence will say that the company was well run and that there was no reason to anticipate that this individual manager was doing what he was doing. The prosecution would proceed on the basis of a lack of knowledge of unsafe practices and the failure to know what was actually going on in practice. One could anticipate circumstances in which one would be drawn back, as a defendant, to considering whether it is necessary to prove the elements set out by Roach LJ in the *Nelson* case.

15.67 It is anticipated that, with workplace fatalities running at the rate of about one a day, the HSE, once the new law is in force, will actively seek to avail themselves of the new offence and prosecute corporations—particularly in circumstances in which there have been multiple fatalities and a considerable amount of media interest. It is anticipated by many practitioners that the vast majority of these prosecutions will be strongly fought.

EUROPEAN CONVENTION ON HUMAN RIGHTS

15.68 The European Convention on Human Rights came into force in October 2000. It has, at least at first instance, already given rise to submissions being made about the law relating to criminal prosecutions of health and safety offences. In September 1999, at Chelmsford Crown Court,[1] an application was made by the defendants to stay an indictment on the grounds of abuse of process. The European Convention on Human Rights guarantees a fair trial and, in particular, the presumption of innocence is paramount. The argument that was run was that s 40 of the HSWA 1974 putting the burden of proof on a balance of probabilities on the defence is repugnant to the Convention and that any conviction will eventually be set aside in due course and that, therefore, the trial ought to be stopped. Reference was made to *Kebelene*[2] to which the learned Recorder in this particular case felt himself bound, which stated that it would be contrary to the legislative intention if the provisions were treated as if they had immediate effect. The learned Recorder, in ruling against the application for an abuse of process, referred to a number of authorities in which reverse burdens had been upheld in cases in the Privy Council, in the European Commission, the European Court of Human Rights, in the English courts, and also in some Canadian authorities. The learned Recorder, setting out what were the elements of the offence, stated that the prosecution must show to the correct normal standard the employment relationship, that something was being done in the course of employment, and that a dangerous situation existed. Thereafter it was held that one moved on from what the elements of the offence were to the question of what defence was afforded to the defendants. The learned Recorder held that s 40 of the HSWA 1974 did not force the defendants to prove or disprove an essential element of the offence and also took into account the social need for the legislation and held that it was wholly reasonable to impose a positive burden on the defence and commented that, if his finding was wrong, then the entire Act would fall by the wayside. The case neatly summarises the arguments on both sides, having regard to the European Convention and the burden imposed on the defence by s 40 of the HSWA 1974. It is understood that there is a case currently before the High Court on the same point.

General considerations of the potential impact of the Convention on Criminal Procedure are outside the ambit of this work and the practitioner should refer to the specialist texts which have already been published.

1 *R v TE Scudder Ltd and S Carey* (9 September 1999, transcript).
2 *R v DPP, ex p Kebelene* [1999] 3 WLR 972, HL.

THE FUTURE

15.69 Forty-four action points are listed in the 57-page HSC/ Department of the Environment, Transport and the Regions (DETR) Strategy Statement of June 2000.[1]

The government intends to increase the maximum fine in the magistrates' courts for Regulation breaches from the present £5,000 to £20,000. Powers of imprisonment are to be extended.

Proposals are being considered to link fines to company turnover, prohibit directors' bonuses, suspending directors without pay and imposing community service orders on companies.

A 'penalty points' system may be introduced. Health and safety performance may become a significant factor in public procurement policy.

1 The statement can be read and downloaded from the DETR website: www.detr.gov.uk/hsw/pdf/strategy.pdf

15.70 The raft of proposals, some quite radical in approach, are clear signs of a significant recognition of the importance of health and safety in future legislation. If the requirement for the consent of the DPP being obtained before a private prosecution is brought is removed, which is a possibility, one can envisage a significant increase in prosecutions, which may be backed by concerned groups, victims' families or trade unions.

Chapter 16

Precedents

PRECEDENT NO 1

16.1

IN THE EVERYWHERE COUNTY COURT Case No EV2000

BETWEEN

REX HALLUX Claimant

and

NEWTON MECHANICS LIMITED Defendants

CLAIM FORM WITH PARTICULARS OF CLAIM

Concise statement

On 1 April 2000 whilst in the course of his employment with the Defendants the Claimant sustained personal injury as a result of the negligence and/or breach of statutory duty of the Defendants their servants or agents.

Remedy sought

Damages for personal injury, loss of earnings and consequential losses. Interest on general and special damages. Costs.

Statement of value

The Claimant reasonably expects to recover more than £5,000 but not more than £15,000. The Claimant reasonably expects to recover more than £1,000 as general damages for pain, suffering and loss of amenity.

PARTICULARS OF CLAIM

Concise statement of facts

1 The Claimant was employed as a finisher by the Defendants at their premises at Appletree Industrial Park, Everywhere.

2 On or about 1 April 2000 whilst in the course of that employment the Claimant had placed a casting on his workbench. The casting weighed 5 kilograms. The casting was cylindrical. The casting rolled along the workbench and fell off the edge onto the Claimant's foot, causing injury to the Claimant.

Legal principles relied upon

3 The provisions of the Workplace (Health, Safety and Welfare) Regulations 1992, the Provision and Use of Work Equipment Regulations 1998 and the Personal Protective Equipment at Work Regulations 1992 applied.

4 The Defendants were in breach of the following statutory duties:

(a) contrary to regulation 11(1) of the Workplace (Health, Safety and Welfare) Regulations 1992 the Claimant's workstation was not so arranged that it was suitable for the work that was done there; the workbench was not provided with a rail or raised edge to prevent components rolling off.

(b) contrary to regulations 13(1) and 13(3)(b) of the Workplace Regulations 1992 failed to take suitable and effective measures to prevent any person being struck by a falling object likely to cause personal injury.

(c) contrary to regulation 4(1) of the Provision and Use of Work Equipment Regulations 1998 failed to ensure that the work equipment, namely the workbench, was so constructed or adapted as to be suitable for the purpose for which it was used or provided.

(d) contrary to regulation 4(2) of the Provision and Use of Work Equipment Regulations 1998 in selecting work equipment, namely the workbench, failed to have regard to the working conditions and to the risks to health and safety of persons which existed in the premises in which the work equipment was used and any additional risk posed by the use of that work equipment.

(e) contrary to regulation 4(3) of the Provision and Use of Work Equipment Regulations 1998 failed to ensure that the work equipment, namely the workbench, was used only for operations for which and under conditions for which it was suitable.

(f) contrary to regulation 4(1) of the Personal Protective Equipment at Work Regulations 1992 failed to provide suitable personal protective equipment, namely boots with steel toe-caps, to the Claimant.

(g) contrary to regulation 6(2) of the Personal Protective Equipment at Work Regulations 1992 failed to make a reassessment of the suitability of the boots provided when cylindrical castings were first worked upon in about early March 2000.

5 The Defendants were negligent in the following respects:

(a) the Claimant repeats as allegations of negligence the allegations of breach of statutory duty.

(b) failed to carry out a suitable and sufficient assessment of the risks to the health and safety of their employees to which they were exposed whilst at work; the Defendants had a duty to carry out an assessment under regulation 3(1)(a) of the Management of Health and Safety at Work Regulations 1999.

(c) failed to provide the Claimant with information upon the weight of the casting and the risk to his health of dropping the casting or letting the casting fall on his foot; the Defendants had a duty to provide information under regulation 10(1) of the Management of Health and Safety at Work Regulations 1999.

(d) caused, permitted or suffered the Claimant to work whilst exposed to hazards in the workplace and in the system of work.

(e) failed until after the Claimant's accident, to heed the hazardous working conditions to which workers were exposed.

(f) failed, until after the Claimant's accident to modify or alter the system of working.

(g) failed to devise and operate a safe system of work.

(h) failed to make and keep safe the Claimant's place of work.

Particulars of damages claim

6 The Claimant was born on 1 April 1962.

7 The Claimant suffered a fracture of the right great toe. The Claimant has experienced pain, suffering and disability.

8 The Claimant is handicapped on the labour market.

9 A report about these personal injuries from a medical practitioner is served with these Particulars of Claim.

10 A schedule giving details of the past and future expenses and losses claimed is served with these Particulars of Claim.

Claim for interest

11 The Claimant claims interest on damages. Interest is claimed pursuant to section 69 of the County Courts Act 1984.

12 Interest on general damages is claimed from the date of notification of the claim until the date of settlement or Judgment at the annual rate of 3%.

13 Interest on special damages is claimed at the full Special Account rate from the date when each loss was sustained until the date of settlement or Judgment on the grounds that the Claimant has lost the benefit of the money from that date when the loss was first sustained.

P I HACKE

Statement of truth

The Claimant believes that the facts stated in the Claim Form and these Particulars of Claim are true.

Signed

Dated

The Solicitors to the Claimant are KLAIM, KLAIM & PARTNERS, Station Road, Everywhere, EV1 5CH.

To the District Judge and to the Defendants.

PRECEDENT NO 2

16.2

IN THE EVERYWHERE COUNTY COURT Case No EV2002

BETWEEN

ROBERT CLUMSEY Claimant

and

MILLENIUM WIDGETS plc Defendants

PARTICULARS OF CLAIM

Concise statement of facts

1 The Claimant was employed by the Defendants as a machine operator at their premises at the Riverside Industrial Estate, Everywhere.

2 On 2 April 2000 whilst in the course of that employment the Claimant was working a drilling machine, drilling holes in widgets. As the Claimant was placing a widget in the machine the splonger descended and crushed the Claimant's hand. The guard descended and trapped the Claimant's arm. The drill bit descended and drilled a 1mm hole in the Claimant's thumb.

3 As a consequence the claimant suffered multiple personal injuries.

Legal principles relied upon

4 The Provision and Use of Work Equipment Regulations 1998 applied.

5 The Defendants were in breach of the following statutory duties:

(a) contrary to regulation 11(1) of the Provision and Use of Work Equipment Regulations 1998 effective measures were not taken to prevent access to any dangerous part of the machinery, namely the splonger.

(b) contrary to regulation 11(1) of the Provision and Use of Work Equipment Regulations 1998 effective measures were not taken to prevent access to any dangerous part of machinery, namely the drill.

(c) contrary to regulation 11(3)(c) of the Provision and Use of Work Equipment Regulations 1998 the guard provided for the drilling machine was not maintained in an efficient state in that it tended to stick and did not descend before the operational cycle of the machine commenced.

(d) contrary to regulation 11(3)(d) of the Provision and Use of Work Equipment Regulations 1998 the guard provided gave rise to an increased risk to health or safety in that it trapped and held the Claimant's arm.

(e) contrary to regulation 15(1) of the Provision and Use of Work Equipment Regulations 1998 the machine was not provided with one or more readily accessible controls the operation of which would have brought the drilling machine to a safe condition in a safe manner. The control to stop the operating cycle of the drilling machine was situated at the side of the machine, and could not be reached by the Claimant when his arm was trapped by the guard.

(f) contrary to regulation 16(1) of the Provision and Use of Work Equipment Regulations 1998 the machine was not provided with a readily accessible emergency stop control.

(g) contrary to regulation 8(1) of the Provision and Use of Work Equipment Regulations 1998 failed to ensure that the Claimant had available to him adequate health and safety information or written instructions pertaining to the use of the work equipment.

6 The Defendants were negligent in the following respects:

(a) the Claimant repeats as allegations of negligence the allegations of breach of statutory duty.

(b) failed to carry out a suitable and sufficient assessment of the risks to the health and safety of their employees to which they were exposed whilst at work; the Defendants had a duty to carry out an assessment under regulation 3(1)(a) of the Management of Health and Safety at Work Regulations 1999.

(c) caused, permitted or suffered the Claimant to work whilst exposed to hazards in the workplace.

(d) failed until after the Claimant's accident, to heed the hazardous working conditions to which workers were exposed.

(e) failed to make and keep safe the Claimant's place of work.

(TO BE COMPLETED AS IN PRECEDENT NO 1)

PRECEDENT NO 3

16.3

IN THE EVERYWHERE COUNTY COURT Case No EV2002

BETWEEN

ROBERT CLUMSEY Claimant

and

MILLENIUM WIDGETS plc Defendants

DEFENCE

1 It is admitted that from 1992 to 4 July 2000 the Claimant was employed by the Defendants as a machine operator at their premises at the Riverside Industrial Estate, Everywhere.

2 It is admitted that on 2 April 2000 the Claimant was working at the Defendant's premises.

3 It is admitted that the Claimant's work included working a drilling machine, drilling holes in widgets.

4 It is denied that the Defendants, their servants or agents were in breach of statutory duty or negligent whether as alleged in the Particulars of Claim or at all. Causation is denied.

5 The application of the Provision and Use of Work Equipment Regulations 1998 is admitted.

6 The Defendants will contend as follows:

(a) a guard was provided to prevent access to the splonger and drill.

(b) the guard was maintained in an efficient state; any sticking of the guard was caused by the application of a wedge by the Claimant.

(c) the guard only held the Claimant's arm because the Claimant had interfered with the mechanism. The splonger and drill should not have descended unless the guard was fully closed and activating a micro-switch. The Claimant had inserted a match in the micro-switch.

(d) a readily accessible stop control would not have prevented the accident.

(e) the Claimant was an experienced machine operator who had used machines with an almost identical function and operation for many years.

7 Any injury suffered by the Claimant was caused wholly or in part by his own negligence.

PARTICULARS OF NEGLIGENCE

(a) attempted to wedge open the machine guard.

(b) inserted a matchstick in the micro-switch.

(c) failed to use the machinery and equipment and any safety device provided to him by the Defendants in accordance both with any training in the use of the equipment concerned which has been received by him and the instructions respecting that use which have been provided to him by the Defendants in compliance with the requirements and prohibitions imposed upon the Defendants by or under the relevant statutory provisions; the Claimant had such a duty under regulation 14 of the Management of Health and Safety at Work Regulations 1999. Failure to comply with that duty is evidence of negligence.

(d) failed to inform the Defendants or any other employee of the Defendants with specific responsibility for the health and safety of his fellow employees –

(i) of any work situation which a person with the Claimant's training and instruction would reasonably consider represented a serious and immediate danger to health and safety;

(ii) of any matter which a person with the Claimant's training and instruction would reasonably consider represented a shortcoming in the Defendant's protection arrangements for health and safety.

8 The Claimant had a duty to inform the Defendants under regulation 14 of the Management of Health and Safety at Work Regulations 1999. Failure to comply with that duty is evidence of negligence.

9 It is admitted that the Claimant suffered some injury to his hand and arm. No admissions are made as to the nature and extent of the alleged injuries; the Defendants are awaiting a medical report from their nominated expert Mr Bert Grass.

10 It is admitted that the Claimant ceased work on 14 April 2000. It is not admitted that any injury sustained by the Claimant at work was the cause of the cessation of work. It is not admitted that the Claimant is handicapped on the labour market. The Defendants are awaiting a medical report from their nominated expert Mr Bert Grass. The Defendants contend that the Claimant was fit for any work within three months of the alleged accident.

11 An entitlement to interest is admitted.

ROD IRONGATE

Dated 21 April 2000

The solicitors to the Defendants are DOUGHTER DENAI & DAUGHTERS, Thomasville, SN1 1PE

PRECEDENT NO 4

16.4

IN THE EVERYWHERE COUNTY COURT Case No EV2003

BETWEEN

ADIE TRIPPER Claimant

and

THE EVERYWHERE HOSPITAL TRUST Defendants

PARTICULARS OF CLAIM

Concise statement of facts

1 At all material times the Claimant was employed as a nurse by the Defendants at their premises at the Everywhere Hospital, Caring Road, Everywhere.

2 On or about 1 February 2000 whilst in the course of that employment the Claimant was walking from the car park along the path towards the Maternity Wing. There was a pothole in the path. The Claimant tripped in the pothole and fell, and as a consequence she sustained injury.

Legal principles relied upon

3 The Workplace (Health, Safety and Welfare) Regulations 1992 applied to the Defendant's premises.

4 The Defendants were in breach of the following statutory duties:

(a) contrary to regulation 5(1) of the Workplace (Health, Safety and Welfare) Regulations 1992 the workplace was not maintained in an efficient state, in efficient working order or in good repair.

(b) contrary to regulation 12(2)(a) of the Workplace (Health, Safety and Welfare) Regulations 1992 the path was a traffic route which contained a hole so as to expose a person to a risk to his health or safety.

5 The Defendants were negligent in the following respects:

(a) the Claimant repeats as allegations of negligence the allegations of breach of statutory duty.

(b) failed to warn the Claimant by means of a sign, notice or otherwise of the presence of the pothole in the path.

(c) failed to prevent the Claimant walking along the path by the use of barriers, fencing, or cones, or by marking off the area.

(d) failed to inspect, repair or maintain the path.

(e) failed to provide and maintain safe means of access to the Claimant's place of work.

(f) failed to carry out a suitable and sufficient assessment of the risks to the health and safety of their employees to which they were exposed whilst at work; the Defendants had a duty to carry out an assessment under regulation 3(1)(a) of the Management of Health and Safety at Work Regulations 1999.

(TO BE COMPLETED AS IN PRECEDENT NO 1)

PRECEDENT NO 5

16.5

IN THE EVERYWHERE COUNTY COURT Case No EV2003

BETWEEN

ADIE TRIPPER Claimant

and

THE EVERYWHERE HOSPITAL TRUST Defendants

DEFENCE

1 It is admitted that from 1992 the Claimant has been employed by the Defendants as a nurse at their premises at the Everywhere Hospital.

2 It is admitted that on 1 February 2000 the Claimant reported that she fell whilst walking from the car park towards the Maternity Wing. The Defendants cannot admit or deny the cause of the fall as there were no witnesses other than the Claimant.

3 It is admitted that there was a pothole in the path from the car park to the Maternity Wing. The Defendants will contend as follows:

(a) any hole on the path was not of a size or depth to expose any person to a risk to his health or safety.

(b) the hole had been marked with yellow paint and that was an adequate measure to prevent persons falling within the meaning of regulation 12(4)(a) of the Workplace (Health, Safety and Welfare) Regulations 1992.

4 Without prejudice to the general denial of negligence, the Defendants will contend that it was not reasonably practicable to fill every hole within a period of a week and that any hole had only become apparent some three days before the Claimant's alleged accident.

5 It is denied that the Defendants, their servants or agents were in breach of statutory duty or negligent whether as alleged in the Particulars of Claim or at all. Causation is denied.

6 Any injury suffered by the Claimant was caused wholly or in part by her own negligence.

PARTICULARS OF NEGLIGENCE

(a) failed to keep any or any proper lookout.

(b) failed to observe the hole or its markings.

(c) failed to watch where or how she placed her feet.

(d) failed to keep her balance.

(e) failed to inform the Defendants or any other employee of the Defendants with specific responsibility for the health and safety of his fellow employees –

(i) of any work situation which a person with the Claimant's training and instruction would reasonably consider represented a serious and immediate danger to health and safety;

(ii) of any matter which a person with the Claimant's training and instruction would reasonably consider represented a shortcoming in the Defendant's protection arrangements for health and safety.

7 The Claimant had a duty to inform the Defendants under regulation 14 of the Management of Health and Safety at Work Regulations 1999. Failure to comply with that duty is evidence of negligence.

(*TO BE COMPLETED AS IN PRECEDENT NO 3*)

PRECEDENT NO 6

16.6

IN THE EVERYWHERE COUNTY COURT Case No EV2005

BETWEEN

BASIL DONNE Claimant

and

SCROOGE LIMITED Defendants

PARTICULARS OF CLAIM

Concise statement of facts

1 At all material times the Claimant was employed by the Defendants as a typist using a word-processor at their premises at Canary Wharf.

2 The Claimant worked in an office at one of two glass-topped tables which faced a window. The Claimant sat on a four-legged wooden chair. The window was without curtains or blinds.

3 The Claimant worked from 9am to 5pm with a break of one hour for lunch. His work was brought to his table, and completed work was collected from his table.

4 The Claimant commenced work on 4 January 2000. The Claimant experienced eye strain and tiredness after the first week. In about April 2000 the Claimant began to experience aching in his upper back, neck and shoulders, and suffered headaches and irritability.

5 The Claimant's symptoms were caused by the breach of statutory duty and/or negligence of the Defendants, their servants or agents.

Legal principles relied upon

6 The provisions of the Health and Safety (Display Screen Equipment) Regulations 1992 and the Workplace (Health, Safety and Welfare) Regulations 1992 applied.

7 The Defendants were in breach of the following statutory duties:

(a) contrary to regulation 2(1) of the Health and Safety (Display Screen Equipment) Regulations 1992 failed to perform any or any suitable and sufficient analysis of the Claimant's workstation for the purpose of assessing the health and safety risks to which the Claimant was exposed in consequence of his use of the workstation. The risks which should have been identified were that his seat was not adjustable in height, that the table caused distracting glare and reflections, and that the windows were not fitted with a system to attenuate the daylight that fell on the workstation.

(b) contrary to regulation 2(3) of the Health and Safety (Display Screen Equipment) Regulations 1992 failed to reduce the risks identified in consequence of any assessment made under regulation 2(1).

(c) contrary to regulation 3(1) of the Health and Safety (Display Screen Equipment) Regulations 1992 failed to ensure that the Claimant's workstation met the requirements laid down in the Schedule to the Regulations. The requirements that were not met were the provision of a seat which was adjustable in height, the provision of a table which did not cause distracting glare and reflections, and the windows were not fitted with a system to attenuate the daylight that fell on the workstation.

(d) contrary to regulation 4 of the Health and Safety (Display Screen Equipment) Regulations 1992 failed to plan the activities of the Claimant at his work so that his daily work on display screen equipment was periodically interrupted by such breaks or changes of activity as to reduce his workload at that equipment.

(e) contrary to regulation 5(1) of the Health and Safety (Display Screen Equipment) Regulations 1992 failed to provide the Claimant with an appropriate eyesight test by a competent person following the request made by the Claimant on 13 January 2000.

(f) contrary to regulation 6 of the Health and Safety (Display Screen Equipment) Regulations 1992 failed to provide the Claimant with adequate health and safety training in the use of the workstation upon which he was required to work. The Claimant should have been trained to adopt and maintain a comfortable position.

(g) contrary to regulation 7 of the Health and Safety (Display Screen Equipment) Regulations 1992 failed to provide the Claimant with adequate information about all aspects of health and safety relating to the workstation upon which he was required to work. The Claimant should have been informed that the work could cause eye symptoms and musculo-skeletal symptoms and that he should report the first onset of any such symptoms to the Defendants.

(h) contrary to regulation 5(1) of the Workplace (Health, Safety and Welfare) Regulations 1992 the workplace was not maintained in an efficient state, in efficient working order or in good repair.

8 The Defendants were negligent in the following respects:

(a) the Claimant repeats as allegations of negligence the allegations of breach of statutory duty.

(b) failed to provide safe plant and equipment; the Claimant will rely upon the matters pleaded above.

(c) in the above respects failed to make and keep safe the Claimant's place of work and workstation.

(d) failed to carry out a suitable and sufficient assessment of the risks to the health and safety of their employees to which they were exposed whilst at work; the Defendants had a duty to carry out an assessment under regulation 3(1)(a) of the Management of Health and Safety at Work Regulations 1999.

(e) caused, permitted or suffered the Claimant to work whilst exposed to hazards in the workplace and in the system of work.

(f) failed until after the Claimant's accident, to heed the hazardous working conditions to which workers were exposed.

(g) failed, until after the Claimant's accident to modify or alter the system of working and the layout of the workstation.

(h) failed to devise and operate a safe system of work.

(TO BE COMPLETED AS IN PRECEDENT NO 1)

PRECEDENT NO 7

16.7

IN THE EVERYWHERE COUNTY COURT Case No EV2007

BETWEEN

SHARON SHOEBURY Claimant

and

TEMPS UNLIMITED First Defendants

and

SCROOGE LIMITED Second Defendants

PARTICULARS OF CLAIM

Concise statement of facts

1 At all material times the Claimant was employed by the First Defendants as a 'Temp' typist using a word-processor and was contracted to work at the premises of the Second Defendants at Canary Wharf.

2 The Claimant worked in a small office at one of two glass-topped tables which faced a window. The Claimant sat on a four-legged wooden chair. The window was without curtains or blinds. The screen display was of poor contrast.

3 The Claimant worked from 9am to 5pm with a break of one hour for lunch. Her work was brought to her table, and completed work was collected from her table.

4 The Claimant commenced work at the premises of the Second Defendants on 4 January 2000. The Claimant experienced eye

strain and tiredness after the first week. In about April 2000 the Claimant began to experience aching in her upper back, neck and shoulders, and suffered headaches and irritability.

5 The Claimant's symptoms were caused by the breach of statutory duty and/or negligence of the First and/or Second Defendants, their servants or agents.

Legal principles relied upon

6 The provisions of the Health and Safety (Display Screen Equipment) Regulations 1992 and the Workplace (Health, Safety and Welfare) Regulations 1992 applied.

7 The First Defendants were in breach of the following statutory duties:

(a) contrary to regulation 5(1) of the Health and Safety (Display Screen Equipment) Regulations 1992 failed to provide the Claimant with an appropriate eyesight test by a competent person following the request made by the Claimant on 13 January 2000.

(b) contrary to regulation 6(1) of the Health and Safety (Display Screen Equipment) Regulations 1992 failed to provide the Claimant with adequate health and safety training in the use of the workstation upon which she was required to work. The Claimant should have been trained how to adjust the screen display.

8 The First Defendants were negligent in the following respects:

(a) the Claimant repeats as allegations of negligence the allegations of breach of statutory duty.

(b) failed to ensure that the Second Defendants had performed a suitable and sufficient analysis of the Claimant's workstation for the purpose of assessing the health and safety risks to which the Claimant was exposed in consequence of her use of the workstation. The risks which should have been identified were that her seat was not adjustable in height, that the table caused distracting glare and reflections, and that the windows were not fitted with a system to attenuate the daylight that fell on the workstation.

(c) failed to ensure that the Second Defendants had reduced the risks identified in consequence of the analysis made under regulation 2(1).

(d) failed to provide the Claimant with adequate health and safety training in the use of the workstation upon which she was required to work. The Claimant should have been trained how to adjust the screen display.

(e) failed to provide the Claimant with adequate information about all aspects of health and safety relating to her workstation upon which she was required to work. The Claimant should have been informed that the work could cause eye symptoms and musculo-skeletal symptoms and that she should report the first onset of any such symptoms to the First and/or Second Defendants.

(f) failed to warn or instruct the Claimant to plan her activities at her work so that her daily work on display screen equipment was periodically interrupted by such breaks or changes of activity as to reduce her workload at that equipment.

(g) failed to devise and operate a safe system of work.

9 The Second Defendants were in breach of the following statutory duties:

(a) contrary to regulation 2(1) of the Health and Safety (Display Screen Equipment) Regulations 1992 failed to perform any or any suitable and sufficient analysis of the Claimant's workstation for the purpose of assessing the health and safety risks to which the Claimant was exposed in consequence of her use of the workstation. The risks which should have been identified were that her seat was not adjustable in height, that the table caused distracting glare and reflections, and that the windows were not fitted with a system to attenuate the daylight that fell on the workstation.

(b) contrary to regulation 2(3) of the Health and Safety (Display Screen Equipment) Regulations 1992 failed to reduce the risks identified in consequence of any assessment made under regulation 2(1).

(c) contrary to regulation 3(1) of the Health and Safety (Display Screen Equipment) Regulations 1992 failed to ensure that the Claimant's workstation met the requirements laid down in the Schedule to the Regulations. The requirements that were not met were the provision of a seat which was adjustable in height, the provision of a table which did not cause distracting glare and reflections, and the windows were not fitted with a system to attenuate the daylight that fell on the workstation.

(d) contrary to regulation 4 of the Health and Safety (Display Screen Equipment) Regulations 1992 failed to plan the activities of the Claimant at her work so that her daily work on display screen equipment was periodically interrupted by such breaks or changes of activity as to reduce her workload at that equipment.

(e) contrary to regulation 6 of the Health and Safety (Display Screen Equipment) Regulations 1992 failed to provide the Claimant with adequate health and safety training in the use of the workstation upon which she was required to work. The Claimant should have been trained how to adjust the screen display.

(f) contrary to regulation 7 of the Health and Safety (Display Screen Equipment) Regulations 1992 failed to provide the Claimant with adequate information about all aspects of health and safety relating to the workstation upon which she was required to work. The Claimant should have been informed that the work could cause eye symptoms and musculo-skeletal symptoms and that she should report the first onset of any such symptoms to the Defendants.

(g) contrary to regulation 5(1) of the Workplace (Health, Safety and Welfare) Regulations 1992 the workplace was not maintained in an efficient state, in efficient working order or in good repair.

10 The Second Defendants were negligent in the following respects:

(a) the Claimant repeats as allegations of negligence the allegations of breach of statutory duty made against the Second Defendants.

(b) failed to ensure that the Claimant's workstation met the requirements laid down in the Schedule to the Health and Safety (Display Screen Equipment) Regulations 1992. The

Claimant contends that it would have been reasonable for the Second Defendants to meet the requirements of the Schedule because the Second Defendants only operated three workstations with display screen equipment and the costs and time expended to meet the requirements would have been modest.

(c) failed to plan the activities of the Claimant at her work so that her daily work on display screen equipment was periodically interrupted by such breaks or changes of activity as to reduce her workload at that equipment.

(d) failed to provide safe plant and equipment; the Claimant will rely upon the matters pleaded above.

(e) failed to carry out a suitable and sufficient assessment of the risks to the health and safety of persons not in their employment arising out of or in connection with the conduct by them of their undertaking; the Defendants had a duty to carry out an assessment under regulation 3(1)(b) of the Management of Health and Safety at Work Regulations 1999.

(f) in the above respects failed to make and keep safe the Claimant's workstation.

(g) failed to devise and operate a safe system of work.

(*TO BE COMPLETED AS IN PRECEDENT NO 1*)

PRECEDENT NO 8

16.8
IN THE EVERYWHERE COUNTY COURT Case No EV2008

BETWEEN

ABLE LYFTER Claimant

and

THE STATE RAMBLER MOTOR COMPANY Defendants

PARTICULARS OF CLAIM

Concise statement of facts

1 At all material times the Claimant was employed as a production worker by the Defendants at their workshop premises at Shortbridge Close, Everywhere.

2 On or about 1 April 2000 whilst in the course of that employment the Claimant was attempting to move a K9 casting from a pallet beside his workbench onto his workbench. The casting weighed 30 kilograms. The pallet had been placed on the floor by a forklift truck. The surface of the workbench was 80 centimetres above the level of the floor.

3 As the Claimant was lifting the casting from the pallet by holding it at each end he suffered a strain to his back.

4 The injury to the Claimant was caused by the breach of statutory duty and/or negligence of the Defendants, their servants or agents.

Legal principles relied upon

5 The provisions of the Manual Handling Operations Regulations 1992 and the Workplace (Health, Safety and Welfare) Regulations 1992 applied.

6 The Defendants were in breach of the following statutory duties:

(a) contrary to regulation 4(1)(a) of the Manual Handling Operations Regulations 1992 failed to avoid the need for the Claimant to undertake a manual handling operation which involved a risk of the Claimant being injured.

(b) contrary to regulation 4(1)(b)(i) of the Manual Handling Operations Regulations 1992 failed to make a suitable and sufficient assessment of the manual handling operations to be undertaken by their employees.

(c) contrary to regulation 4(1)(b)(ii) of the Manual Handling Operations Regulations 1992 failed to take appropriate steps to reduce to the lowest level reasonably practicable the risk of injury to employees arising out of manual handling operations; without prejudice to regulation 4(1)(a), the pallet should have been placed at a height level with the surface of the Claimant's workbench.

(d) contrary to regulation 4(2) of the Manual Handling Operations Regulations 1992 failed to make a reassessment of the manual handling operations in relation to castings when the Claimant's workbench was moved away from the conveyor; it had been possible to slide castings from the conveyor onto the workbench.

(e) contrary to regulation 4(2) of the Manual Handling Operations Regulations 1992, if there was a reassessment of the manual handling operations in relation to castings when the Claimants workbench was moved away from the conveyor, failed to make any changes to the manual handling operations.

(f) contrary to regulation 11(1) of the Workplace (Health, Safety and Welfare) Regulations 1992 the Claimant's workstation was not so arranged that it was suitable for the work that was done there; it should have been arranged so as to avoid the need for the Claimant to undertake a lift from close to the floor; the pallet should have been placed at a height level with the surface of the Claimant's workbench.

7 The Defendants are an emanation of the state and the provisions of the European Council Directive 90/269/EEC of 29 May 1990 on the minimum health and safety requirements for the manual handling of loads applied. The Defendants failed to heed the terms of the Directive in the following respects:

(a) Article 3(1): appropriate organisational measures should be taken, or appropriate means in particular mechanical equipment used in order to avoid the need for the manual handling of loads.

(b) Article 3(2): appropriate organisational measures should be taken, or appropriate means used to reduce the risk from the manual handling of loads.

(c) Article 4: assess the risks and take into account the characteristics of the load and of the working environment and take appropriate measures.

8 The Defendants were negligent in the following respects:

(a) the Claimant repeats as allegations of negligence the allegations of breach of statutory duty.

(b) failed to provide the Claimant with information upon the weight of the casting and the risk to his health in attempting a manual lift of the casting; the Defendants had a duty to provide information under regulation 10 of the Management of Health and Safety at Work Regulations 1999.

(c) failed to provide adequate health and safety training for the Claimant in the aspects of safe lifting and handling of heavy loads; the Defendants had a duty to provide information under regulation 13(2) of the Management of Health and Safety at Work Regulations 1999.

(d) failed to carry out a suitable and sufficient assessment of the risks to the health and safety of their employees to which they were exposed whilst at work; the Defendants had a duty to carry out an assessment under regulation 3(1)(a) of the Management of Health and Safety at Work Regulations 1999.

(e) caused, permitted or suffered the Claimant to work whilst exposed to hazards in the workplace and in the system of work.

(f) failed until after the Claimant's accident, to heed the hazardous working conditions to which workers were exposed.

(g) failed, until after the Claimant's accident to modify or alter the system of working.

(h) failed to devise and operate a safe system of work.

(i) failed to make and keep safe the Claimant's place of work.

(TO BE COMPLETED AS IN PRECEDENT NO 1)

PRECEDENT NO 9

16.9

IN THE EVERYWHERE COUNTY COURT Case No EV2008

BETWEEN

ABLE LYFTER Claimant

and

THE STATE RAMBLER MOTOR COMPANY Defendants

DEFENCE

1 It is admitted that at the material time the Claimant was employed as a production worker by the Defendants at their workshop premises at Shortbridge Close, Everywhere.

2 It is admitted that on or about 1 April 2000 whilst in the course of that employment the Claimant was attempting to move a K9 casting from a pallet beside his workbench onto his workbench.

3 It is denied that the casting weighed 30 kilograms. It will be contended that the casting weighed 14 kilograms.

4 It is admitted that the surface of the workbench was 80 centimetres above the level of the floor.

5 The top of the pallet was 20 centimetres above the level of the floor.

6 It is admitted that the Claimant subsequently complained that as he was lifting the casting from the pallet he suffered a strain to his back. It is denied that the Claimant suffered injury when lifting the casting. The Claimant injured his back the previous weekend when laying a patio in his back garden.

7 It is denied that the Defendants, their servants or agents were in breach of statutory duty or negligent whether as alleged in the Particulars of Claim or at all. Causation is denied.

8 The application of the Manual Handling Operations Regulations 1992 is admitted. The application of Council Directive 90/269/EEC is admitted.

9 It is denied that the Defendants, their servants or agents were in breach of statutory duty, breach of Directive or negligent whether as alleged in the Particulars of Claim or at all. Causation is denied.

10 The lifting of the 20 kilogram weight of the K9 casting from the pallet to the surface of the workbench did not create a risk of injury to the Claimant.

11 Alternatively, if the lifting of the K9 casting did create a risk of injury it was not reasonably practicable to avoid manual lifting of the casting for the following reasons:

(a) the Defendants worked upon a K9 casting only once every six months.

(b) the overhead hoist did not travel to the Claimant's workbench.

(c) the Claimant's workbench needed to be sited near the exhaust appliances for the purpose of his other work.

(d) there were other employees in the area who could and would assist the Claimant in lifting the casting onto the workbench.

12 The Defendants took steps to reduce to the lowest level reasonably practicable the risk of injury to the Claimant from handling a K9 casting by:

(a) delivering the casting on a pallet to a position beside the Claimant's workbench.

(b) instructing the Claimant only to lift with the assistance of one or more fellow employees.

13 Any injury suffered by the Claimant in lifting the K9 casting was caused wholly or in part by his own breach of statutory duty and/or negligence.

(a) contrary to regulation 5 of the Manual Handling Operations Regulations 1992 failed to make full and proper use of the system of work provided by the Defendants, namely the assistance of one or more fellow employees in lifting a K9 casting.

(b) failed to heed the training given by the Defendants upon the aspects of safe lifting and handling of heavy loads.

(c) failed to inform the Defendants or any other employee of the Defendants with specific responsibility for the health and safety of his fellow employees –

(i) of any work situation which a person with the Claimant's training and instruction would reasonably consider represented a serious and immediate danger to health and safety;

(ii) of any matter which a person with the Claimant's training and instruction would reasonably consider represented a shortcoming in the Defendant's protection arrangements for health and safety.

14 The Claimant had a duty to inform the Defendants under regulation 14 of the Management of Health and Safety at Work Regulations 1999. Failure to comply with that duty is evidence of negligence.

(TO BE COMPLETED AS IN PRECEDENT NO 3)

Chapter 17

Directives

17.1 COUNCIL DIRECTIVE
of 12 June 1989

on the introduction of measures to encourage improvements in the safety and health of workers at work
(89/391/EEC)

THE COUNCIL OF THE EUROPEAN COMMUNITIES,

Having regard to the Treaty establishing the European Economic Community, and in particular Article 118a thereof,

Having regard to the proposal from the Commission, drawn up after consultation with the Advisory Committee on Safety, Hygiene and Health Protection at Work,

In cooperation with the European Parliament, Having regard to the opinion of the Economic and Social Committee,

Whereas Article 118a of the Treaty provides that the Council shall adopt, by means of Directives, minimum requirements for encouraging improvements, especially in the working environment, to guarantee a better level of protection of the safety and health of workers;

Whereas this Directive does not justify any reduction in levels of protection already achieved in individual Member States, the Member State being committed, under the Treaty, to encouraging improvements in conditions in this area and to harmonising conditions while maintaining the improvements made;

Whereas it is known that workers can be exposed to the effects of dangerous environmental factors at the work place during the course of their working life;

Whereas, pursuant to Article 118a of the Treaty, such Directives must avoid imposing administrative, financial and legal constraints which would hold back the creation and development of small and medium-sized undertakings;

Whereas the communication from the Commission on its programme concerning safety, hygiene and health at work provides for the adoption of Directives designed to guarantee the safety and health of workers;

Whereas the Council, in its resolution of 21 December 1987 on safety, hygiene and health at work, took note of the Commission's intention to submit to the Council in the near future a Directive on the organisation of the safety and health of workers at the workplace;

Whereas in February 1988 the European Parliament adopted four resolutions following the debate on the internal market and worker protection; whereas these resolutions specifically invited the Commission to draw up a framework Directive to serve as a basis for more specific Directives covering all the risks connected with safety and health at the workplace;

Whereas Member States have a responsibility to encourage improvements in the safety and health of workers on their territory; whereas taking measures to protect the health and safety of workers at work also helps, in certain cases, to preserve the health and possibly the safety of persons residing with them;

Whereas Member States' legislative systems covering safety and health at the workplace differ widely and need to be improved; whereas national provisions on the subject, which often include technical specifications and/or self-regulatory standards, may result in different levels of safety and health protection and allow competition at the expense of safety and health;

Whereas the incident of accidents at work and occupational diseases is still too high; whereas preventive measures must be introduced or improved without delay in order to safeguard the safety and health of workers and ensure a higher degree of protection;

Whereas, in order to ensure an improved degree of protection, workers and/or their representatives must be informed of the risks to their safety and health and of the measures required to reduce or eliminate these risks; whereas they must also be in a position to contribute, by means of balanced participation in accordance with national laws and/or practices, to seeing that the necessary protective measures are taken;

Whereas information, dialogue and balanced participation on safety and health at work must be developed between employers and workers and/or their representatives by means of appropriate procedures and instruments, in accordance with national laws and/or practices;

Whereas the improvement of workers' safety, hygiene and health at work is an objective which should not be subordinated to purely economic considerations;

Whereas employers shall be obliged to keep themselves informed of the latest advances in technology and scientific findings concerning workplace design, account being taken of the inherent dangers in their undertaking, and to inform accordingly the workers' representatives exercising participation rights under this Directive, so as to be able to guarantee a better level of protection of workers' health and safety;

Whereas the provisions of this Directive apply, without prejudice to more stringent present or future Community provisions, to all risks, and in particular to those arising from the use at work of chemical, physical and biological agents covered by Directive 80/1107/EEC, as last amended by Directive 88/642/EEC;

Whereas, pursuant to Decision 74/325/EEC, the Advisory Committee on Safety, Hygiene and Health Protection at Work is consulted by the Commission on the drafting of proposals in this field;

Whereas a Committee composed of members nominated by the Member States needs to be set up to assist the Commission in making the technical adaptations to the individual Directives provided for in this Directive,

HAS ADOPTED THIS DIRECTIVE:

SECTION I
GENERAL PROVISIONS

Article 1

Object

1. The object of this Directive is to introduce measures to encourage improvements in the safety and health of workers at work.

2. To that end it contains general principles concerning the prevention of occupational risks, the protection of safety and health, the elimination of risk and accident factors, the informing, consultation, balanced participation in accordance with national laws and/or practices and training of workers and their representatives, as well as general guidelines for the implementation of the said principles.

3. This Directive shall be without prejudice to existing or future national and Community provisions which are more favourable to protection of the safety and health of workers at work.

Article 2

Scope

1. This Directive shall apply to all sectors of activity, both public and private (industrial, agricultural, commercial, administrative, service, educational, cultural, leisure, etc).
2. This Directive shall not be applicable where characteristics peculiar to certain specific public service activities, such as the armed forces or the police, or to certain specific activities in the civil protection services inevitably conflict with it.

In that event, the safety and health of workers must be ensured as far as possible in the light of the objectives of this Directive.

Article 3

Definitions

For the purposes of this Directive, the following terms shall have the following meanings:

(a) 'worker': any person employed by an employer, including trainee and apprentices but excluding domestic servants;

(b) 'employer': any natural or legal person who has an employment relationship with the worker and has responsibility for the undertaking and/or establishment;

(c) 'workers' representative with specific responsibility for the safety and health of workers': any person elected, chosen or designated in accordance with national laws and/or practices to represent workers where problems arise relating to the safety and health protection of workers at work;

(d) 'prevention': all the steps or measures taken or planned at all stages of work in the undertaking to prevent or reduce occupational risks.

Article 4

1. Member States shall take the necessary steps to ensure that employers, workers and workers' representatives are subject to

the legal provisions necessary for the implementation of this Directive.

2. In particular, Member States shall ensure adequate control and supervision.

SECTION II
EMPLOYERS' OBLIGATIONS

Article 5

General provision

1. The employer shall have a duty to ensure the safety and health of workers in every aspect related to the work.

2. Where, pursuant to Article 7(3), an employer enlists competent external services or persons, this shall not discharge him from his responsibilities in this area.

3. The workers' obligations in the field of safety and health at work shall not affect the principle of the responsibility of the employer.

4. This Directive shall not restrict the option of Member States to provide for the exclusion or the limitation of employers' responsibility where occurrences are due to unusual and unforeseeable circumstances, beyond the employers' control, or to exceptional events, the consequences of which could not have been avoided despite the exercise of all due care.

Member States need not exercise the option referred to in the first subparagraph.

Article 6

General obligations on employers

1. Within the context of his responsibilities, the employer shall take the measures necessary for the safety and health protection of workers, including prevention of occupational risks and provision of information and training, as well as provision of the necessary organisation and means.

The employer shall be alert to the need to adjust these measures to take account of changing circumstances and aim to improve existing situations.

2. The employer shall implement the measures referred to in the first subparagraph of paragraph 1 on the basis of the following general principles of prevention:

(a) avoiding risks;

(b) evaluating the risks which cannot be avoided;

(c) combating the risks at source;

(d) adapting the work to the individual, especially as regards the design of workplaces, the choice of work equipment and the choice of working and production methods, with a view, in particular, to alleviating monotonous work and work at a predetermined work-rate and to reducing their effect on health;

(e) adapting to technical progress;

(f) replacing the dangerous by the non-dangerous or the less dangerous;

(g) developing a coherent overall prevention policy which covers technology, organisation of work, working conditions, social relationships and the influence of factors related to the working environment;

(h) giving collective protective measures priority over individual protective measures;

(i) giving appropriate instructions to the workers.

3. Without prejudice to the other provisions of this Directive, the employer shall, taking into account the nature of the activities of the enterprise and/or establishment:

(a) evaluate the risks to the safety and health of workers, inter alia in the choice of work equipment, the chemical substances or preparations used, and the fitting-out of workplaces.

Subsequent to this evaluation and as necessary, the preventive measures and the working and production methods implemented by the employer must:

— assure an improvement in the level of protection afforded to workers with regard to safety and health,

— be integrated into all the activities of the undertaking and/or establishment and at all hierarchical levels;

(b) where he entrusts tasks to a worker, take into consideration the worker's capabilities as regards health and safety;

(c) ensure that the planning and introduction of new technologies are the subject of consultation with the workers and/or their representatives, as regards the consequences of the choice of equipment, the working conditions and the working environment for the safety and health of workers;

(d) take appropriate steps to ensure that only workers who have received adequate instructions may have access to areas where there is serious and specific danger.

4. Without prejudice to the other provisions of this Directive, where several undertakings share a workplace, the employers shall cooperate in implementing the safety, health and occupational hygiene provisions and, taking into account the nature of the activities, shall coordinate their actions in matters of the protection and prevention of occupational risks, and shall inform one another and their respective workers and/or workers' representatives of these risks.

5. Measures related to safety, hygiene and health at work may in no circumstances involve the workers in financial cost.

Article 7

Protective and preventive services

1. Without prejudice to the obligations referred to in Articles 5 and 6, the employer shall designate one or more workers to carry out activities related to the protection and prevention of occupational risks for the undertaking and/or establishment.

2. Designated workers may not be placed at any disadvantage because of their activities related to the protection and prevention of occupational risks.

Designated workers shall be allowed adequate time to enable them to fulfil their obligations arising from this Directive.

3. If such protective and preventive measures cannot be organised for lack of competent personnel in the undertaking and/or establishment, the employer shall enlist competent external services or persons.

4. Where the employer enlists such services or persons, he shall inform them of the factors known to affect, or suspected of affecting, the safety and health of the workers and they must have access to the information referred to in Article 10(2).

5. In all cases:

— the workers designated must have the necessary capabilities and the necessary means,

— the external services or persons consulted must have the necessary aptitudes and the necessary personal and professional means, and

— the workers designated and the external services or persons consulted must be sufficient in number,

to deal with the organisation of protective and preventive measures, taking into account the size of the undertaking and/or establishment and/or the hazards to which the workers are exposed and their distribution throughout the entire undertaking and/or establishment.

6. The protection from, and prevention of, the health and safety risks which form the subject of this Article shall be the responsibility of one or more workers, of one service or of separate services whether from inside or outside the undertaking and/or establishment.

The worker(s) and/or agency(ies) must work together whenever necessary.

7. Member States may define, in the light of the nature of the activities and size of the undertakings, the categories of undertakings in which the employer, provided he is competent, may himself take responsibility for the measures referred to in paragraph 5.

8. Member States shall define the necessary capabilities and aptitudes referred to in paragraph 5.

They may determine the sufficient number referred to in paragraph 5.

Article 8

First aid, fire-fighting and evacuation of workers, serious and imminent danger

1. The employer shall:

— take the necessary measures for first aid, fire-fighting and evacuation of workers, adapted to the nature of the activities and the size of the undertaking and/or establishment and taking into account other persons present,

— arrange any necessary contacts with external services, particularly as regards first aid, emergency medical care, rescue work and fire-fighting.

2. Pursuant to paragraph 1, the employer shall, inter alia, for first aid, fire-fighting and the evacuation of workers, designate the workers required to implement such measures.

The number of such workers, their training and the equipment available to them shall be adequate, taking account of the size and/or specific hazards of the undertaking and/or establishment.

3. The employer shall:

(a) as soon as possible, inform all workers who are, or may be, exposed to serious and imminent danger of the risk involved and of the steps taken or to be taken as regards protection;

(b) take action and give instructions to enable workers in the event of serious, imminent and unavoidable danger to stop work and/or immediately to leave the workplace and proceed to a place of safety;

(c) save in exceptional cases for reasons duly substantiated, refrain from asking workers to resume work in a working situation where there is still a serious and imminent danger.

4. Workers who, in the event of serious, imminent and unavoidable danger, leave their workstation and/or a dangerous area may not be placed at any disadvantage because of their action and must be protected against any harmful and unjustified consequences, in accordance with national laws and/or practices.

5. The employer shall ensure that all workers are able, in the event of serious and imminent danger to their own safety and/or that of other persons, and where the immediate superior responsible cannot be contacted, to take the appropriate steps in the light of their knowledge and the technical means at their disposal, to avoid the consequences of such danger.

Their actions shall not place them at any disadvantage, unless they acted carelessly or there was negligence on their part.

Article 9

Various obligations on employers

1. The employer shall:

(a) be in possession of an assessment of the risks to safety and health at work, including those facing groups of workers exposed to particular risks;

(b) decide on the protective measures to be taken and, if necessary, the protective equipment to be used;

(c) keep a list of occupational accidents resulting in a worker being unfit for work for more than three working days;

(d) draw up, for the responsible authorities and in accordance with national laws and/or practices, reports on occupational accidents suffered by his workers.

2. Member States shall define, in the light of the nature of the activities and size of the undertakings, the obligations to be met by the different categories of undertakings in respect of the drawing-up of the documents provided for in paragraph 1(a) and (b) and when preparing the documents provided for in paragraph 1(c) and (d).

Article 10

Worker information

1. The employer shall take appropriate measures so that workers and/or their representatives in the undertaking and/or establishment receive, in accordance with national laws and/or practices which may take account, inter alia, of the size of the undertaking and/or establishment, all the necessary information concerning:

(a) the safety and health risks and protective and preventive measures and activities in respect of both the undertaking and/or establishment in general and each type of workstation and/or job;

(b) the measures taken pursuant to Article 8(2).

2. The employer shall take appropriate measures so that employers of workers from any outside undertakings and/or establishments engaged in work in his undertaking and/or establishment receive, in accordance with national laws and/or practices, adequate information concerning the points referred to in paragraph 1(a) and (b) which is to be provided to the workers in question.

3. The employer shall take appropriate measures so that workers with specific functions in protecting the safety and health of

workers, or workers' representatives with specific responsibility for the safety and health of workers shall have access, to carry out their functions and in accordance with national laws and/or practices, to:

(a) the risk assessment and protective measures referred to in Article 9(1)(a) and (b);

(b) the list and reports referred to in Article 9(1)(c) and (d);

(c) the information yielded by protective and preventive measures, inspection agencies and bodies responsible for safety and health.

Article 11

Consultation and participation of workers

1. Employers shall consult workers and/or their representatives and allow them to take part in discussions on all questions relating to safety and health at work.

This presupposes:

— the consultation of workers,

— the right of workers and/or their representatives to make proposals,

— balanced participation in accordance with national laws and/or practices.

2. Workers or workers' representatives with specific responsibility for the safety and health of workers shall take part in a balanced way, in accordance with national laws and/or practices, or shall be consulted in advance and in good time by the employer with regard to:

(a) any measure which may substantially affect safety and health;

(b) the designation of workers referred to in Articles 7(1) and 8(2) and the activities referred to in Article 7(1);

(c) the information referred to in Articles 9(1) and 10;

(d) the enlistment, where appropriate, of the competent services or persons outside the undertaking and/or establishment, as referred to in Article 7(3);

(e) the planning and organisation of the training referred to in Article 12.

3. Workers' representatives with specific responsibility for the safety and health of workers shall have the right to ask the employer to take appropriate measures and to submit proposals to him to that end to mitigate hazards for workers and/or to remove sources of danger.

4. The workers referred to in paragraph 2 and the workers' representatives referred to in paragraphs 2 and 3 may not be placed at a disadvantage because of their respective activities referred to in paragraphs 2 and 3.

5. Employers must allow workers' representatives with specific responsibility for the safety and health of workers adequate time off work, without loss of pay, and provide them with the necessary means to enable such representatives to exercise their rights and functions deriving from this Directive.

6. Workers and/or their representatives are entitled to appeal, in accordance with national law and/or practice, to the authority responsible for safety and health protection at work if they consider that the measures taken and the means employed by the employer are inadequate for the purposes of ensuring safety and health at work.

Workers' representatives must be given the opportunity to submit their observations during inspection visits by the competent authority.

Article 12

Training of workers

1. The employer shall ensure that each worker receives adequate safety and health training, in particular in the form of information and instructions specific to his workstation or job:

— on recruitment,

— in the event of a transfer or a change of job,

— in the event of the introduction of new work equipment or a change in equipment,

— in the event of the introduction of any new technology.

The training shall be:

— adapted to take account of new or changed risks, and

— repeated periodically if necessary.

2. The employer shall ensure that workers from outside undertakings and/or establishments engaged in work in his undertaking and/or establishment have in fact received appropriate instructions regarding health and safety risks during their activities in his undertaking and/or establishment.

3. Workers' representatives with a specific role in protecting the safety and health of workers shall be entitled to appropriate training.

4. The training referred to in paragraphs 1 and 3 may not be at the workers' expense or at that of the workers' representatives.

The training referred to in paragraph 1 must take place during working hours.

The training referred to in paragraph 3 must take place during working hours or in accordance with national practice either within or outside the undertaking and/or the establishment.

SECTION III
WORKERS' OBLIGATIONS

Article 13

1. It shall be the responsibility of each worker to take care as far as possible of his own safety and health and that of other persons affected by his acts or commissions at work in accordance with his training and the instructions given by his employer.

2. To this end, workers must in particular, in accordance with their training and the instructions given by their employer:

(a) make correct use of machinery, apparatus, tools, dangerous substances, transport equipment and other means of production;

(b) make correct use of the personal protective equipment supplied to them and, after use, return it to its proper place;

(c) refrain from disconnecting, changing or removing arbitrarily safety devices fitted, eg to machinery, apparatus, tools, plant and buildings, and use such safety devices correctly;

(d) immediately inform the employer and/or the workers with specific responsibility for the safety and health of workers of any work situation they have reasonable grounds for considering represents a serious and immediate danger to safety and health and of any shortcomings in the protection arrangements;

(e) cooperate, in accordance with national practice, with the employer and/or workers with specific responsibility for the safety and health of workers, for as long as may be necessary to enable any tasks or requirements imposed by the competent authority to protect the safety and health of workers at work to be carried out;

(f) cooperate, in accordance with national practice, with the employer and/or workers with specific responsibility for the safety and health of workers, for as long as may be necessary to enable the employer to ensure that the working environment and working conditions are safe and pose no risk to safety and health within their field of activity.

SECTION IV
MISCELLANEOUS PROVISIONS

Article 14

Health surveillance

1. To ensure that workers receive health surveillance appropriate to the health and safety risks they incur at work, measures shall be introduced in accordance with national law and/or practices.

2. The measures referred to in paragraph 1 shall be such that each worker, if he so wishes, may receive health surveillance at regular intervals.

3. Health surveillance may be provided as part of a national health system.

Article 15

Risk groups

Particularly sensitive risk groups must be protected against the dangers which specifically affect them.

Article 16

Individual Directives—Amendments—General scope of this Directive

1. The Council, acting on a proposal from the Commission based on Article 118a of the Treaty, shall adopt individual Directives, inter alia, in the areas listed in the Annex.

2. This Directive and, without prejudice to the procedure referred to in Article 17 concerning technical adjustments, the individual Directives may be amended in accordance with the procedure provided for in Article 118a of the Treaty.

3. The provisions of this Directive shall apply in full to all the areas covered by the individual Directives, without prejudice to more stringent and/or specific provisions contained in these individual Directives.

Article 17

Committee

1. For the purely technical adjustments to the individual Directives provided for in Article 16(1) to take account of:

— the adoption of Directives in the field of technical harmonisation and standardisation, and/or

— technical progress, changes in international regulations or specifications, and new findings,

the Commission shall be assisted by a committee composed of the representatives of the Member States and chaired by the representative of the Commission.

2. The representative of the Commission shall submit to the committee a draft of the measures to be taken.

The committee shall deliver its opinion on the draft within a time limit which the chairman may lay down according to the urgency of the matter.

The opinion shall be delivered by the majority laid down in Article 148(2) of the Treaty in the case of decisions which the Council is required to adopt on a proposal from the Commission.

The votes of the representatives of the Member States within the committee shall be weighted in the manner set out in that Article. The chairman shall not vote.

3. The Commission shall adopt the measures envisaged if they are in accordance with the opinion of the committee.

If the measures envisaged are not in accordance with the opinion of the committee, or if no opinion is delivered, the Commission shall, without delay, submit to the Council a proposal relating to the measures to be taken. The Council shall act by a qualified majority.

If, on the expiry of three months from the date of the referral to the Council, the Council has not acted, the proposed measures shall be adopted by the Commission.

Article 18

Final provisions

1. Member States shall bring into force the laws, regulations and administrative provisions necessary to comply with this Directive by 31 December 1992.

They shall forthwith inform the Commission thereof.

2. Member States shall communicate to the Commission the texts of the provisions of national law which they have already adopted or adopt in the field covered by this Directive.

3. Member States shall report to the Commission every five years on the practical implementation of the provisions of this Directive, indicating the points of view of employers and workers.

The Commission shall inform the European Parliament, the Council, the Economic and Social Committee and the Advisory Committee on Safety, Hygiene and Health Protection at Work.

4. The Commission shall submit periodically to the European Parliament, the Council and the Economic and Social Committee a report on the implementation of this Directive, taking into account paragraphs 1 to 3.

Article 19

This Directive is addressed to the Member States.

Done at Luxembourg, 12 June 1989.

For the Council

The President

M CHAVES GONZALES

ANNEX
LIST OF AREAS REFERRED TO IN ARTICLE 16(1)

— Workplaces

— Work equipment

— Personal protective equipment

— Work with visual display units

— Handling of heavy loads involving risk of back injury

— Temporary or mobile work sites

— Fisheries and agriculture

17.2 COUNCIL DIRECTIVE
of 30 November 1989

concerning the minimum safety and health requirements for the workplace (first individual Directive within the meaning of Article 16(1) of Directive 89/391/EEC)
(89/654/EEC)

THE COUNCIL OF THE EUROPEAN COMMUNITIES,

Having regard to the Treaty establishing the European Economic Community, and in particular Article 118a thereof,

Having regard to the proposal from the Commission, submitted after consulting the Advisory Committee on Safety, Hygiene and Health Protection at Work,

In cooperation with the European Parliament,

Having regard to the opinion of the Economic and Social Committee,

Whereas Article 118a of the Treaty provides that the Council shall adopt, by means of Directives, minimum requirements for encouraging improvements, especially in the working environment, to ensure a better level of protection of the safety and health of workers;

Whereas, under the terms of that Article, those Directives are to avoid imposing administrative, financial and legal constraints in a way which would hold back the creation and development of small and medium-sized undertakings;

Whereas the communication from the Commission on its programme concerning safety, hygiene and health at work provides for the adoption of a Directive designed to guarantee the safety and health of workers at the workplace;

Whereas, in its resolution of 21 December 1987 on safety, hygiene and health at work, the Council took note of the Commission's intention of submitting to the Council in the near future minimum requirements concerning the arrangement of the place of work;

Whereas compliance with the minimum requirements designed to guarantee a better standard of safety and health at work is essential to ensure the safety and health of workers;

Whereas this Directive is an individual Directive within the meaning of Article 16(1) of Council Directive 89/391/EEC of 12 June 1989 on the introduction of measures to encourage improvements in the safety and health of workers at work; whereas the provisions of the latter are therefore fully applicable to the workplace without prejudice to more stringent and/or specific provisions contained in the present Directive;

Whereas this Directive is a practical contribution towards creating the social dimension of the internal market;

Whereas, pursuant to Decision 74/325/EEC, as last amended by the 1985 Act of Accession, the Advisory Committee on Safety, Hygiene and Health Protection at Work is consulted by the Commission on the drafting of proposals in this field,

HAS ADOPTED THIS DIRECTIVE:

SECTION I
GENERAL PROVISIONS

Article 1

Subject

1. This Directive, which is the first individual Directive within the meaning of Article 16(1) of Directive 89/391/EEC, lays down minimum requirements for safety and health at the workplace, as defined in Article 2.

2. This Directive shall not apply to:

(a) means of transport used outside the undertaking and/or the establishment, or workplaces inside means of transport;

(b) temporary or mobile work sites;

(c) extractive industries;

(d) fishing boats;

(e) fields, woods and other land forming part of an agricultural or forestry undertaking but situated away from the undertaking's buildings.

3. The provisions of Directive 89/391/EEC are fully applicable to the whole scope referred to in paragraph 1, without prejudice to more stringent and/or specific provisions contained in this Directive.

Article 2

Definition

For the purposes of this Directive, 'workplace' means the place intended to house workstations on the premises of the undertaking and/or establishment and any other place within the area of the undertaking and/or establishment to which the worker has access in the course of his employment.

SECTION II
EMPLOYERS' OBLIGATIONS

Article 3

Workplaces used for the first time

Workplaces used for the first time after 31 December 1992 must satisfy the minimum safety and health requirements laid down in Annex I.

Article 4

Workplaces already in use

Workplaces already in use before 1 January 1993 must satisfy the minimum safety and health requirements laid down in Annex II at the latest three years after that date.

However, as regards the Portuguese Republic, workplaces used before 1 January 1993 must satisfy, at the latest four years after that date, the minimum safety and health requirements appearing in Annex II.

Article 5

Modifications to workplaces

When workplaces undergo modifications, extensions and/or conversions after 31 December 1992, the employer shall take the measures necessary to ensure that those modifications, extensions and/or conversions are in compliance with the corresponding minimum requirements laid down in Annex I.

Article 6

General requirements

To safeguard the safety and health of workers, the employer shall see to it that:

— traffic routes to emergency exits and the exits themselves are kept clear at all times,

— technical maintenance of the workplace and of the equipment and devices, and in particular those referred to in Annexes I and II, is carried out and any faults found which are liable to affect the safety and health of workers are rectified as quickly as possible,

— the workplace and the equipment and devices, and in particular those referred to in Annex I, point 6, and Annex II, point 6, are regularly cleaned to an adequate level of hygiene,

— safety equipment and devices intended to prevent or eliminate hazards, and in particular those referred to in Annexes I and II, are regularly maintained and checked.

Article 7

Information of workers

Without prejudice to Article 10 of Directive 89/391/EEC, workers and/or their representatives shall be informed of all measures to be taken concerning safety and health at the workplace.

Article 8

Consultation of workers and workers' participation

Consultation and participation of workers and/or of their representatives shall take place in accordance with Article 11 of

Directive 89/391/EEC on the matters covered by this Directive, including the Annexes thereto.

SECTION III
MISCELLANEOUS PROVISIONS

Article 9

Amendments to the Annexes

Strictly technical amendments to the Annexes as a result of:

— the adoption of Directives on technical harmonisation and standardisation of the design, manufacture or construction of parts of workplaces, and/or

— technical progress, changes in international regulations or specifications and knowledge with regard to workplaces,

shall be adopted in accordance with the procedure laid down in Article 17 of Directive 89/391/EEC.

Article 10

Final provisions

1. Member States shall bring into force the laws, regulations and administrative provisions necessary to comply with this Directive by 31 December 1992. They shall forthwith inform the Commission thereof.

However, the date applicable for the Hellenic Republic shall be 31 December 1994.

2. Member States shall communicate to the Commission the texts of the provisions of national law which they have already adopted or adopt in the field governed by this Directive.

3. Member States shall report to the Commission every five years on the practical implementation of the provisions of this Directive, indicating the points of view of employers and workers.

The Commission shall inform the European Parliament, the Council, the Economic and Social Committee and the Advisory Council on Safety, Hygiene and Health Protection at Work.

4. The Commission shall submit periodically to the European Parliament, the Council and the Economic and Social Committee a report on the implementation of this Directive, taking into account paragraphs 1 to 3.

Article 11

This Directive is addressed to the Member States.

Done at Brussels, 30 November 1989.

For the Council

The President

J P SOISSON

ANNEX I
MINIMUM SAFETY AND HEALTH REQUIREMENTS FOR WORKPLACES USED FOR THE FIRST TIME, AS REFERRED TO IN ARTICLE 3 OF THE DIRECTIVE

1. Preliminary note

The obligations laid down in this Annex apply whenever required by the features of the workplace, the activity, the circumstances or a hazard.

2. Stability and solidity

Buildings which house workplaces must have a structure and solidity appropriate to the nature of their use.

3. Electrical installations

Electrical installations must be designed and constructed so as not to present a fire or explosion hazard; persons must be adequately protected against the risk of accidents caused by direct or indirect contact.

The design, construction and choice of material and protection devices must be appropriate to the voltage, external conditions and the competence of persons with access to parts of the installation.

4. Emergency routes and exits

4.1. Emergency routes and exits must remain clear and lead as directly as possible to the open air or to a safe area.

4.2. In the event of danger, it must be possible for workers to evacuate all workstations quickly and as safely as possible.

4.3. The number, distribution and dimensions of the emergency routes and exits depend on the use, equipment and dimensions of the workplaces and the maximum number of persons that may be present.

4.4. Emergency doors must open outwards.

Sliding or revolving doors are not permitted if they are specifically intended as emergency exits.

Emergency doors should not be so locked or fastened that they cannot be easily and immediately opened by any person who may require to use them in an emergency.

4.5. Specific emergency routes and exits must be indicated by signs in accordance with the national regulations transposing Directive 77/576/EEC into law.

Such signs must be placed at appropriate points and be made to last.

4.6. Emergency doors must not be locked.

The emergency routes and exits, and the traffic routes and doors giving access to them, must be free from obstruction so that they can be used at any time without hindrance.

4.7. Emergency routes and exits requiring illumination must be provided with emergency lighting of adequate intensity in case the lighting fails.

5. Fire detection and fire fighting

5.1. Depending on the dimensions and use of the buildings, the equipment they contain, the physical and chemical properties of the substances present and the maximum potential number of people present, workplaces must be equipped with appropriate fire-fighting equipment and, as necessary, with fire detectors and alarm systems.

5.2. Non-automatic fire-fighting equipment must be easily accessible and simple to use.

The equipment must be indicated by signs in accordance with the national regulations transposing Directive 77/576/EEC into law.

Such signs must be placed at appropriate points and be made to last.

6. Ventilation of enclosed workplaces

6.1. Steps shall be taken to see to it that there is sufficient fresh air in enclosed workplaces, having regard to the working methods used and the physical demands placed on the workers.

If a forced ventilation system is used, it shall be maintained in working order.

Any breakdown must be indicated by a control system where this is necessary for workers' health.

6.2. If air-conditioning or mechanical ventilation installations are used, they must operate in such a way that workers are not exposed to draughts which cause discomfort.

Any deposit or dirt likely to create an immediate danger to the health of workers by polluting the atmosphere must be removed without delay.

7. Room temperature

7.1. During working hours, the temperature in rooms containing workstations must be adequate for human beings, having regard to the working methods being used and the physical demands placed on the workers.

7.2. The temperature in rest areas, rooms for duty staff, sanitary facilities, canteens and first aid rooms must be appropriate to the particular purpose of such areas.

7.3. Windows, skylights and glass partitions should allow excessive effects of sunlight in workplaces to be avoided, having regard to the nature of the work and of the workplace.

8. Natural and artificial room lighting

8.1. Workplaces must as far as possible receive sufficient natural light and be equipped with artificial lighting adequate for the protection of workers' safety and health.

8.2. Lighting installations in rooms containing workstations and in passageways must be placed in such a way that there is no risk of accident to workers as a result of the type of lighting fitted.

8.3. Workplaces in which workers are especially exposed to risks in the event of failure of artificial lighting must be provided with emergency lighting of adequate intensity.

9. Floors, walls, ceilings and roofs of rooms

9.1. The floors of rooms must have no dangerous bumps, holes or slopes and must be fixed, stable and not slippery.

Workplaces containing workstations must be adequately thermally insulated, bearing in mind the type of undertaking involved and the physical activity of the workers.

9.2. The surfaces of floors, walls and ceilings in rooms must be such that they can be cleaned or refurbished to an appropriate standard of hygiene.

9.3. Transparent or translucent walls, in particular all glass partitions, in rooms or in the vicinity of workstations and traffic routes must be clearly indicated and made of safety material or be shielded from such places or traffic routes to prevent workers from coming into contact with walls or being injured should the walls shatter.

9.4. Access to roofs made of materials of insufficient strength must not be permitted unless equipment is provided to ensure that the work can be carried out in a safe manner.

10. Windows and skylights

10.1. It must be possible for workers to open, close, adjust or secure windows, skylights and ventilators in a safe manner. When open, they must not be positioned so as to constitute a hazard to workers.

10.2. Windows and skylights must be designed in conjunction with equipment or otherwise fitted with devices allowing them to be cleaned without risk to the workers carrying out this work or to workers present in and around the building.

11. Doors and gates

11.1. The position, number and dimensions of doors and gates, and the materials used in their construction, are determined by the nature and use of the rooms or areas.

11.2. Transparent doors must be appropriately marked at a conspicuous level.

11.3. Swing doors and gates must be transparent or have see-through panels.

11.4. If transparent or translucent surfaces in doors and gates are not made of safety material and if there is a danger that workers may be injured if a door or gate should shatter, the surfaces must be protected against breakage.

11.5. Sliding doors must be fitted with a safety device to prevent them from being derailed and falling over.

11.6. Doors and gates opening upwards must be fitted with a mechanism to secure them against falling back.

11.7. Doors along escape routes must be appropriately marked.

It must be possible to open them from the inside at any time without special assistance.

It must be possible to open the doors when the workplaces are occupied.

11.8. Doors for pedestrians must be provided in the immediate vicinity of any gates intended essentially for vehicle traffic, unless it is safe for pedestrians to pass through; such doors must be clearly marked and left permanently unobstructed.

11.9. Mechanical doors and gates must function in such a way that there is no risk of accident to workers.

They must be fitted with easily identifiable and accessible emergency shutdown devices and, unless they open automatically in the event of a power failure, it must also be possible to open them manually.

12. Traffic routes—danger areas

12.1. Traffic routes, including stairs, fixed ladders and loading bays and ramps, must be located and dimensioned to ensure easy, safe and appropriate access for pedestrians or vehicles in such a way as not to endanger workers employed in the vicinity of these traffic routes.

12.2. Routes used for pedestrian traffic and/or goods traffic must be dimensioned in accordance with the number of potential users and the type of undertaking.

If means of transport are used on traffic routes, a sufficient safety clearance must be provided for pedestrians.

12.3. Sufficient clearance must be allowed between vehicle traffic routes and doors, gates, passages for pedestrians, corridors and staircases.

12.4. Where the use and equipment of rooms so requires for the protection of workers, traffic routes must be clearly identified.

12.5. If the workplaces contain danger areas in which, owing to the nature of the work, there is a risk of the worker or objects falling, the places must be equipped, as far as possible, with devices preventing unauthorised workers from entering those areas.

Appropriate measures must be taken to protect workers authorised to enter danger areas.

Danger areas must be clearly indicated.

13. Specific measures for escalators and travelators

Escalators and travelators must function safely.

They must be equipped with any necessary safety devices.

They must be fitted with easily identifiable and accessible emergency shut-down devices.

14. Loading bays and ramps

14.1. Loading bays and ramps must be suitable for the dimensions of the loads to be transported.

14.2. Loading bays must have at least one exit point.

Where technically feasible, bays over a certain length must have an exit point at each end.

14.3. Loading ramps must as far as possible be safe enough to prevent workers from falling off.

15. Room dimensions and air space in rooms—freedom of movement at the workstation

15.1. Workrooms must have sufficient surface area, height and airspace to allow workers to perform their work without risk to their safety, health or well-being.

15.2. The dimensions of the free unoccupied area at the workstation must be calculated to allow workers sufficient freedom of movement to perform their work.

If this is not possible for reasons specific to the workstation, the worker must be provided with sufficient freedom of movement near his workstation.

16. Rest rooms

16.1. Where the safety or health of workers, in particular because of the type of activity carried out or the presence of more than a certain number of employees, so require, workers must be provided with an easily accessible rest room.

This provision does not apply if the workers are employed in offices or similar workrooms providing equivalent relaxation during breaks.

16.2. Rest rooms must be large enough and equipped with an adequate number of tables and seats with backs for the number of workers.

16.3. In rest rooms appropriate measures must be introduced for the protection of non-smokers against discomfort caused by tobacco smoke.

16.4. If working hours are regularly and frequently interrupted and there is no rest room, other rooms must be provided in which workers can stay during such interruptions, wherever this is required for the safety or health of workers.

Appropriate measures should be taken for the protection of non-smokers against discomfort caused by tobacco smoke.

17. Pregnant women and nursing mothers

Pregnant women and nursing mothers must be able to lie down to rest in appropriate conditions.

18. Sanitary equipment

18.1. *Changing rooms and lockers*

18.1.1. Appropriate changing rooms must be provided for workers if they have to wear special work clothes and where, for reasons of health or propriety, they cannot be expected to change in another room.

Changing rooms must be easily accessible, be of sufficient capacity and be provided with seating.

18.1.2. Changing rooms must be sufficiently large and have facilities to enable each worker to lock away his clothes during working hours.

If circumstances so require (eg dangerous substances, humidity, dirt), lockers for work clothes must be separate from those for ordinary clothes.

18.1.3. Provision must be made for separate changing rooms or separate use of changing rooms for men and women.

18.1.4. If changing rooms are not required under 18.1.1, each worker must be provided with a place to store his clothes.

18.2. *Showers and washbasins*

18.2.1. Adequate and suitable showers must be provided for workers if required by the nature of the work or for health reasons.

Provision must be made for separate shower rooms or separate use of shower rooms for men and women.

18.2.2. The shower rooms must be sufficiently large to permit each worker to wash without hindrance in conditions of an appropriate standard of hygiene.

The showers must be equipped with hot and cold running water.

18.2.3. Where showers are not required under the first subparagraph of 18.2.1, adequate and suitable washbasins with running water (hot water if necessary) must be provided in the vicinity of the workstations and the changing rooms.

Such washbasins must be separate for, or used separately by, men and women when so required for reasons of propriety.

18.2.4. Where the rooms housing the showers or washbasins are separate from the changing rooms, there must be easy communication between the two.

18.3. *Lavatories and washbasins*

Separate facilities must be provided in the vicinity of workstations, rest rooms, changing rooms and rooms housing showers or washbasins, with an adequate number of lavatories and washbasins.

Provision must be made for separate lavatories or separate use of lavatories for men and women.

19. First aid rooms

19.1. One or more first aid rooms must be provided where the size of the premises, type of activity being carried out and frequency of accidents so dictate.

19.2. First aid rooms must be fitted with essential first aid installations and equipment and be easily accessible to stretchers.

They must be signposted in accordance with the national regulations transposing Directive 77/576/EEC into law.

19.3. In addition, first aid equipment must be available in all places where working conditions require it.

This equipment must be suitably marked and easily accessible.

20. Handicapped workers

Workplaces must be organised to take account of handicapped workers, if necessary.

This provision applies in particular to the doors, passageways, staircases, showers, washbasins, lavatories and workstations used or occupied directly by handicapped persons.

21. Outdoor workplaces (special provisions)

21.1. Workstations, traffic routes and other areas or installations outdoors which are used or occupied by the workers in the course of their activity must be organised in such a way that pedestrians and vehicles can circulate safely.

Sections 12, 13 and 14 also apply to main traffic routes on the site of the undertaking (traffic routes leading to fixed workstations), to traffic routes used for the regular maintenance and supervision of the undertaking's installations and to loading bays.

Section 12 is also applicable to outdoor workplaces.

21.2. Workplaces outdoors must be adequately lit by artificial lighting if daylight is not adequate.

21.3. When workers are employed at workstations outdoors, such workstations must as far as possible be arranged so that workers:

(a) are protected against inclement weather conditions and if necessary against falling objects;

(b) are not exposed to harmful noise levels nor to harmful outdoor influences such as gases, vapours or dust;

(c) are able to leave their workstations swiftly in the event of danger or are able to be rapidly assisted;

(d) cannot slip or fall.

ANNEX II
MINIMUM HEALTH AND SAFETY REQUIREMENTS FOR WORKPLACES ALREADY IN USE, AS REFERRED TO IN ARTICLE 4 OF THE DIRECTIVE

1. Preliminary note

The obligations laid down in this Annex apply wherever required by the features of the workplace, the activity, the circumstances or a hazard.

2. Stability and solidity

Buildings which have workplaces must have a structure and solidity appropriate to the nature of their use.

3. Electrical installations

Electrical installations must be designed and constructed so as not to present a fire or explosion hazard; persons must be adequately protected against the risk of accidents caused by direct or indirect contact.

Electrical installations and protection devices must be appropriate to the voltage, external conditions and the competence of persons with access to parts of the installation.

4. Emergency routes and exits

4.1. Emergency routes and exits must remain clear and lead as directly as possible to the open air or to a safe area.

4.2. In the event of danger, it must be possible for workers to evacuate all workstations quickly and as safely as possible.

4.3. There must be an adequate number of escape routes and emergency exits.

4.4. Emergency exit doors must open outwards.

Sliding or revolving doors are not permitted if they are specifically intended as emergency exits.

Emergency doors should not be so locked or fastened that they cannot be easily and immediately opened by any person who may require to use them in an emergency.

4.5. Specific emergency routes and exits must be indicated by signs in accordance with the national regulations transposing Directive 77/576/EEC into law.

Such signs must be placed at appropriate points and be made to last.

4.6. Emergency doors must not be locked.

The emergency routes and exits, and the traffic routes and doors giving access to them, must be free from obstruction so that they can be used at any time without hindrance.

4.7. Emergency routes and exits requiring illumination must be provided with emergency lighting of adequate intensity in case the lighting fails.

5. Fire detection and fire fighting

5.1. Depending on the dimensions and use of the buildings, the equipment they contain, the physical and chemical characteristics of the substances present and the maximum potential number of people present, workplaces must be equipped with appropriate fire-fighting equipment, and, as necessary, fire detectors and an alarm system.

5.2. Non-automatic fire-fighting equipment must be easily accessible and simple to use.

It must be indicated by signs in accordance with the national regulations transposing Directive 77/576/EEC into law.

Such signs must be placed at appropriate points and be made to last.

6. Ventilation of enclosed workplaces

Steps shall be taken to see to it that there is sufficient fresh air in enclosed workplaces, having regard to the working methods used and the physical demands placed on the workers.

If a forced ventilation system is used, it shall be maintained in working order.

Any breakdown must be indicated by a control system where this is necessary for the workers' health.

7. Room temperature

7.1. During working hours, the temperature in rooms containing workplaces must be adequate for human beings, having regard to the working methods being used and the physical demands placed on the workers.

7.2. The temperature in rest areas, rooms for duty staff, sanitary facilities, canteens and first aid rooms must be appropriate to the particular purpose of such areas.

8. Natural and artificial room lighting

8.1. Workplaces must as far as possible receive sufficient natural light and be equipped with artificial lighting adequate for workers' safety and health.

8.2. Workplaces in which workers are especially exposed to risks in the event of failure of artificial lighting must be provided with emergency lighting of adequate intensity.

9. Doors and gates

9.1. Transparent doors must be appropriately marked at a conspicuous level.

9.2. Swing doors and gates must be transparent or have see-through panels.

10. Danger areas

If the workplaces contain danger areas in which, owing to the nature of the work, there is a risk of the worker or objects falling, the places must be equipped, as far as possible, with devices preventing unauthorised workers from entering those areas.

Appropriate measures must be taken to protect workers authorised to enter danger areas.

Danger areas must be clearly indicated.

11. Rest rooms and rest areas

11.1. Where the safety or health of workers, in particular because of the type of activity carried out or the presence of more than a certain number of employees, so require, workers must be provided with an easily accessible rest room or appropriate rest area.

This provision does not apply if the workers are employed in offices or similar workrooms providing equivalent relaxation during breaks.

11.2. Rest rooms and rest areas must be equipped with tables and seats with backs.

11.3. In rest rooms and rest areas appropriate measures must be introduced for the protection of non-smokers against discomfort caused by tobacco smoke.

12. Pregnant women and nursing mothers

Pregnant women and nursing mothers must be able to lie down to rest in appropriate conditions.

13. Sanitary equipment

13.1. *Changing rooms and lockers*

13.1.1. Appropriate changing rooms must be provided for workers if they have to wear special work clothes and where, for reasons of health or propriety, they cannot be expected to change in another room.

Changing rooms must be easily accessible and of sufficient capacity.

13.1.2. Changing rooms must have facilities to enable each worker to lock away his clothes during working hours.

If circumstances so require (eg dangerous substances, humidity, dirt), lockers for work clothes must be separate from those for ordinary clothes.

13.1.3. Provision must be made for separate changing rooms or separate use of changing rooms for men and women.

13.2. *Showers, lavatories and washbasins*

13.2.1. Workplaces must be fitted out in such a way that workers have in the vicinity:

— showers, if required by the nature of their work,

— special facilities equipped with an adequate number of lavatories and washbasins.

13.2.2. The showers and washbasins must be equipped with running water (hot water if necessary).

13.2.3. Provision must be made for separate showers or separate use of showers for men and women.

Provision must be made for separate lavatories or separate use of lavatories for men and women.

14. First aid equipment

Workplaces must be fitted with first aid equipment.

The equipment must be suitably marked and easily accessible.

15. Handicapped workers

Workplaces must be organised to take account of handicapped workers, if necessary.

This provision applies in particular to the doors, passageways, staircases, showers, washbasins, lavatories and workstations used or occupied directly by handicapped persons.

16. Movement of pedestrians and vehicles

Outdoor and indoor workplaces must be organised in such a way that pedestrians and vehicles can circulate in a safe manner.

17. Outdoor workplaces (special provisions)

When workers are employed at workstations outdoors, such workstations must as far as possible be organised so that workers:

(a) are protected against inclement weather conditions and if necessary against falling objects;

(b) are not exposed to harmful noise levels nor to harmful external influences such as gases, vapours or dust;

(c) are able to leave their workstations swiftly in the event of danger or are able to be rapidly assisted;

(d) cannot slip or fall.

17.3 COUNCIL DIRECTIVE
of 30 November 1989

concerning the minimum safety and health requirements for the use of work equipment by workers at work (second individual Directive within the meaning of Article 16(1) of Directive 89/391/EEC)

(89/655/EEC, as amended by 95/63/EC)

THE COUNCIL OF THE EUROPEAN COMMUNITIES,

Having regard to the Treaty establishing the European Economic Community, and in particular Article 118a thereof,

Having regard to the proposal from the Commission, submitted after consulting the Advisory Committee on Safety, Hygiene and Health Protection at Work,

In cooperation with the European Parliament, Having regard to the opinion of the Economic and Social Committee,

Whereas Article 118a of the Treaty provides that the Council shall adopt, by means of Directives, minimum requirements for encouraging improvements, especially in the working environment, to guarantee a better level of protection of the safety and health of workers;

Whereas, pursuant to the said Article, such Directives must avoid imposing administrative, financial and legal constraints in a way which would hold back the creation and development of small and medium-sized undertakings;

Whereas the communication from the Commission on its programme concerning safety, hygiene and health at work provides for the adoption of a Directive on the use of work equipment at work;

Whereas, in its resolution of 21 December 1987 on safety, hygiene and health at work, the Council took note of the Commission's intention of submitting to the Council in the near future minimum requirements concerning the organisation of safety and health at work;

Whereas compliance with the minimum requirements designed to guarantee a better standard of safety and health in the use of work equipment is essential to ensure the safety and health of workers;

Whereas this Directive is an individual Directive within the meaning of Article 16(1) of Council Directive 89/391/EEC of 12 June 1989 on the introduction of measures to encourage improvements in the safety and health of workers at work;

whereas, therefore, the provisions of the said Directive are fully applicable to the scope of the use of work equipment by workers at work without prejudice to more restrictive and/or specific provisions contained in this Directive;

Whereas this Directive constitutes a practical aspect of the realisation of the social dimension of the internal market;

Whereas, pursuant to Decision 83/189/EEC, Member States are required to notify the Commission of any draft technical regulations relating to machines, equipment and installations;

Whereas, pursuant to Decision 74/325/EEC, as last amended by the 1985 Act of Accession, the Advisory Committee on Safety, Hygiene and Health Protection at Work is consulted by the Commission on the drafting of proposals in this field,

HAS ADOPTED THIS DIRECTIVE:

SECTION I
GENERAL PROVISIONS

Article 1
Subject

1. This Directive, which is the second individual Directive within the meaning of Article 16(1) of Directive 89/391/EEC, lays down minimum safety and health requirements for the use of work equipment by workers at work, as defined in Article 2.

2. The provisions of Directive 89/391/EEC are fully applicable to the whole scope referred to in paragraph 1, without prejudice to more restrictive and/or specific provisions contained in this Directive.

Article 2

Definitions

For the purposes of this Directive, the following terms shall have the following meanings:

(a) 'work equipment': any machine, apparatus, tool or installation used at work;

(b) 'use of work equipment': any activity involving work equipment such as starting or stopping the equipment, its use, transport, repair, modification, maintenance and servicing, including, in particular, cleaning;

(c) 'danger zone': any zone within and/or around work equipment in which an exposed worker is subject to a risk to his health or safety;

(d) 'exposed worker': any worker wholly or partially in a danger zone;

(e) 'operator': the worker or workers given the task of using work equipment.

SECTION II
EMPLOYERS' OBLIGATIONS

Article 3

General obligations

1. The employer shall take the measures necessary to ensure that the work equipment made available to workers in the undertaking and/or establishment is suitable for the work to be carried out or properly adapted for that purpose and may be used by workers without impairment to their safety or health.

In selecting the work equipment which he proposes to use, the employer shall pay attention to the specific working conditions and characteristics and to the hazards which exist in the undertaking and/or establishment, in particular at the workplace, for the safety and health of the workers, and/or any *additional hazards posed by the use of work equipment in question*.

2. Where it is not possible fully so to ensure that work equipment can be used by workers without risk to their safety or health, the employer shall take appropriate measures to minimise the risks.

Article 4

Rules concerning work equipment

1. Without prejudice to Article 3, the employer must obtain and/or use:

(a) work equipment which, if provided to workers in the undertaking and/or establishment for the first time after

31 December 1992, complies with:

(i) the provisions of any relevant Community Directive which is applicable;

(ii) the minimum requirements laid down [in Annex I], to the extent that no other Community Directive is applicable or is so only partially;

(b) work equipment which, if already provided to workers in the undertaking and/or establishment by 31 December 1992, complies with the minimum requirements laid down in [Annex I] no later than four years after that date.

[(c) without prejudice to point (a)(i), and notwithstanding point (a)(ii) and point (b), specific work equipment subject to the requirements of point 3 of Annex I, which, if already provided to workers in the undertaking and/or establishment by 5 December 1998, complies with the minimum requirements laid down in Annex I, no later than four years after that date.]

2. The employer shall take the measures necessary to ensure that, throughout its working life, work equipment is kept, by means of adequate maintenance, at a level such that it complies with the provisions of paragraph (1)(a) or (b) as applicable.

[3. Member States shall, after consultation with both sides of industry, and with due allowance for national legislation and/or practice, establish procedures whereby a level of safety may be attained corresponding to the objectives indicated by the provisions of Annex II.]

[Article 4a

Inspection of work equipment

1. The employer shall ensure that where the safety of work equipment depends on the installation conditions, it shall be subject to an initial inspection (after installation and before first being put into service) and an inspection after assembly at a new site or in a new location by competent persons within the meaning of national laws and/or practices, to ensure that the work equipment has been installed correctly and is operating properly.

2. The employer shall ensure that work equipment exposed to conditions causing deterioration which is liable to result in dangerous situations is subject to:

— periodic inspections and, where appropriate, testing by competent persons within the meaning of national laws and/or practices,

— special inspections by competent persons within the meaning of national laws and/or practices each time that exceptional circumstances which are liable to jeopardise the safety of the work equipment have occurred, such as modification work, accidents, natural phenomena or prolonged periods of inactivity,

to ensure that health and safety conditions are maintained and that the deterioration can be detected and remedied in good time.

3. The results of inspections must be recorded and kept at the disposal of the authorities concerned. They must be kept for a suitable period of time.

When work equipment is used outside the undertaking it must be accompanied by physical evidence that the last inspection has been carried out.

4. Member States shall determine the conditions under which such inspections are made.]

Article 5

Work equipment involving specific risks

When the use of work equipment is likely to involve a specific risk to the safety or health of workers, the employer shall take the measures necessary to ensure that:

— the use of work equipment is restricted to those persons given the task of using it;

— in the case of repairs, modifications, maintenance or servicing, the workers concerned are specifically designated to carry out such work.

[Article 5a

Ergonomics and occupational health

The working posture and position of workers while using work equipment and ergonomic principles must be taken fully into account by the employer when applying minimum health and safety requirements.]

Article 6

Informing workers

1. Without prejudice to Article 10 of Directive 89/391/EEC, the employer shall take the measures necessary to ensure that workers have at their disposal adequate information and, where appropriate, written instructions on the work equipment used at work.

2. The information and the written instructions must contain at least adequate safety and health information concerning:

— the conditions of use of work equipment,

— foreseeable abnormal situations,

— the conclusions to be drawn from experience, where appropriate, in using work equipment.

[Workers must be made aware of dangers relevant to them, work equipment present in the work area or site, and any changes affecting them, in as much as they affect work equipment situated in their immediate work area or site, even if they do not use such equipment directly.]

3. The information and the written instructions must be comprehensible to the workers concerned.

Article 7

Training of workers

Without prejudice to Article 12 of Directive 89/391/EEC, the employer shall take the measures necessary to ensure that:

— workers given the task of using work equipment receive adequate training, including training on any risks which such use may entail,

— workers referred to in the second indent of Article 5 receive adequate specific training.

Article 8

Consultation of workers and workers' participation

Consultation and participation of workers and/or of their representatives shall take place in accordance with Article 11 of

Directive 89/391/EEC on the matters covered by this Directive, including [the Annexes] thereto.

SECTION III
MISCELLANEOUS PROVISIONS

Article 9

Amendment [of the Annexes]

1. Addition to [Annex I] of the supplementary minimum requirements applicable to specific work equipment referred to in point 3 thereof shall be adopted by the Council in accordance with the procedure laid down in Article 118a of the Treaty.

2. Strictly technical adaptations of [the Annexes] as a result of:

— the adoption of Directives on technical harmonisation and standardisation of work equipment, and/or

— technical progress, changes in international regulations or specifications or knowledge in the field of work equipment,

shall be adopted in accordance with the procedure laid down in Article 17 of Directive 89/391/EEC.

Article 10

Final provisions

1. Member States shall bring into force the laws, regulations and administrative provisions necessary to comply with this Directive by 31 December 1992. They shall forthwith inform the Commission thereof.

2. Member States shall communicate to the Commission the texts of the provisions of national law which they have already adopted or adopt in the field governed by this Directive.

3. Member States shall report to the Commission every five years on the practical implementation of the provisions of this Directive, indicating the points of view of employers and workers.

The Commission shall accordingly inform the European Parliament, the Council, the Economic and Social Committee, and the Advisory Council on Safety, Hygiene and Health Protection at Work.

4. The Commission shall submit periodically to the European Parliament, the Council and the Economic and Social Committee a report on the implementation of this Directive, taking into account paragraphs 1 to 3.

Article 11

This Directive is addressed to the Member States.

Done at Brussels, 30 November 1989.

For the Council

The President

J P SOISSON

ANNEX [I]
MINIMUM REQUIREMENTS REFERRED TO IN ARTICLE 4(1)(A)(II) AND (B)

1. General comment

The obligations laid down in this Annex apply having regard to the provisions of the Directive and where the corresponding risk exists for the work equipment in question.

[The following minimum requirements, in as much as they apply to work equipment in use, do not necessarily call for the same measures as the essential requirements concerning new work equipment.]

2. General minimum requirements applicable to work equipment

2.1. Work equipment control devices which affect safety must be clearly visible and identifiable and appropriately marked where necessary.

Except where necessary for certain devices, control devices must be located outside danger zones and in such a way that their operation cannot cause additional hazard. They must not give rise to any hazard as a result of any unintentional operation.

If necessary, from the main control position, the operator must be able to ensure that no person is present in the danger zones. If this is impossible, a safe system such as an audible and/or visible warning signal must he given automatically whenever the machinery is about to start. An exposed worker must have the time and/or the means quickly to avoid hazards caused by the starting and/or stopping of the work equipment.

[Control systems must be safe and must be chosen making due allowance for the failures, faults and constraints to be expected in the planned circumstances of use.]

2.2. It must be possible to start work equipment only by deliberate action on a control provided for the purpose.

The same shall apply:

— to restart it after a stoppage for whatever reason,

— for the control of a significant change in the operating conditions (eg speed, pressure, etc),

unless such a restart or change does not subject exposed workers to any hazard.

This requirement does not apply to restarting or a change in operating conditions as a result of the normal operating cycle of an automatic device.

2.3. All work equipment must be fitted with a control to stop it completely and safely.

Each work station must be fitted with a control to stop some or all of the work equipment, depending on the type of hazard, so that the equipment is in a safe state. The equipment's stop control must have priority over the start controls. When the work equipment or the dangerous parts of it have stopped, the energy supply of the actuators concerned must be switched off.

2.4. Where appropriate, and depending on the hazards the equipment presents and its normal stopping time, work equipment must be fitted with an emergency stop device.

2.5. Work equipment presenting risk due to falling objects or projections must be fitted with appropriate safety devices corresponding to the risk.

Work equipment presenting hazards due to emissions of gas, vapour, liquid or dust must be fitted with appropriate containment and/or extraction devices near the sources of the hazard.

2.6. Work equipment and parts of such equipment must, where necessary for the safety and health of workers, be stabilised by clamping or some other means.

2.7. Where there is a risk of rupture or disintegration of parts of the work equipment, likely to pose significant danger to the safety and health of workers, appropriate protection measures must be taken.

2.8. Where there is a risk of mechanical contact with moving parts of work equipment which could lead to accidents, those parts must be provided with guards or devices to prevent access to danger zones or to halt movements of dangerous parts before the danger zones are reached.

The guards and protection devices must:

— be of robust construction,

— not give rise to any additional hazard,

— not be easily removed or rendered inoperative,

— be situated at a sufficient distance from the danger zone,

— not restrict more than necessary the view of the operating cycle of the equipment,

— allow operations necessary to fit or replace parts and for maintenance work, restricting access only to the area where the work is to be carried out and, if possible, without removal of the guard or protection device.

2.9. Areas and points for working on, or maintenance of, work equipment must be suitably lit in line with the operation to be carried out.

2.10. Work equipment parts at high or very low temperature must, where appropriate, be protected to avoid the risk of workers coming into contact or coming too close.

2.11. Warning devices on work equipment must be unambiguous and easily perceived and understood.

2.12. Work equipment may be used only for operations and under conditions for which it is appropriate.

2.13. It must be possible to carry out maintenance operations when the equipment is shut down. If this is not possible, it must be possible to take appropriate protection measures for the carrying out of such operations or for such operations to be carried out outside the danger zones.

If any machine has a maintenance log, it must be kept up to date.

2.14. All work equipment must be fitted with clearly identifiable means to isolate it from all energy sources.

Reconnection must be presumed to pose no risk to the workers concerned.

2.15. Work equipment must bear the warnings and markings essential to ensure the safety of workers.

2.16. Workers must have safe means of access to, and be able to remain safely in, all the areas necessary for production, adjustment and maintenance operations.

2.17. All work equipment must be appropriate for protecting workers against the risk of the work equipment catching fire or overheating, or of discharges of gas, dust, liquid, vapour or other substances produced, used or stored in the work equipment.

2.18. All work equipment must be appropriate for preventing the risk of explosion of the work equipment or substances produced, used or stored in the work equipment.

2.19. All work equipment must be appropriate for protecting exposed workers against the risk of direct or indirect contact with electricity.

[3. Additional minimum requirements applicable to specific types of work equipment

3.1. *Minimum requirements for mobile work equipment, whether or not self-propelled*

3.1.1. Work equipment with ride-on workers must be fitted out in such a way as to reduce the risks for workers during the journey.

Those risks must include the risks of contact with or trapping the wheels or tracks.

3.1.2. Where an inadvertent seizure of the drive unit between an item of mobile work equipment and its accessories and/or anything towed might create a specific risk, such work equipment must be equipped or adapted to prevent blockages of the drive units.

Where such seizure cannot be avoided, every possible measure must be taken to avoid any adverse effects on workers.

3.1.3. Where drive shafts for the transmission of energy between mobile items of work equipment can become soiled or damaged by trailing on the ground, facilities must be available for fixing them.

3.1.4. Mobile work equipment with ride-on workers must be designed to restrict, under actual conditions of use, the risks arising from work equipment rollover:

— either by a protection structure to ensure that the equipment does not tilt by more than a quarter turn, or

— a structure giving sufficient clearance around the ride-on workers if the tilting movement can continue beyond a quarter turn, or

— by some other device of equivalent effect.

These protection structures may be an integral part of the work equipment.

These protection structures are not required when the work equipment is stabilised during operation or where the design makes rollover impossible.

Where there is a risk of a ride-on worker being crushed between parts of the work equipment and the ground, should the equipment roll over, a restraining system for the ride-on workers must be installed.

3.1.5. Fork-lift trucks carrying one or more workers must be adapted or equipped to limit the risk of the fork-lift truck overturning, eg:

— by the installation of an enclosure for the driver, or

— by a structure preventing the fork-lift truck from overturning, or

— by a structure ensuring that, if the fork-lift truck overturns, sufficient clearance remains between the ground and certain parts of the fork-lift truck for the workers carried, or

— by a structure restraining the workers on the driving seat so as to prevent them from being crushed by parts of the fork-lift truck which overturns.

3.1.6. Self-propelled work equipment which may, in motion, engender risks for persons must fulfil the following conditions:

(a) the equipment must have facilities for preventing unauthorised start-up;

(b) it must have appropriate facilities for minimising the consequences of a collision where there is more than one item of track-mounted work equipment in motion at the same time;

(c) there must be a device for braking and stopping equipment. Where safety constraints so require, emergency facilities operated by readily accessible controls or automatic systems must be available for braking and stopping equipment in the event of failure of the main facility;

(d) where the driver's direct field of vision is inadequate to ensure safety, adequate auxiliary devices must be installed to improve visibility;

(e) work equipment designed for use at night or in dark places must be equipped with lighting appropriate to the work to be carried out and must ensure sufficient safety for workers;

(f) work equipment which constitutes a fire hazard, either on its own or in respect of whatever it is towing and/or carrying and which is liable to endanger workers, must be equipped with appropriate fire-fighting appliances where such appliances are not available sufficiently nearby at the place of use;

(g) remote-controlled work equipment must stop automatically once it leaves the control range;

(h) remote-controlled work equipment which may in normal conditions engender a crushing or impact hazard must have facilities to guard against this risk, unless other appropriate devices are present to control the impact risk.

3.2. *Minimum requirements for work equipment for lifting loads*

3.2.1. When work equipment for lifting loads is installed permanently, its strength and stability during use must be assured, having regard, in particular, to the loads to be lifted and the stress induced at the mounting or fixing point of the structures.

3.2.2. Machinery for lifting loads must be clearly marked to indicate its nominal load, and where appropriate a load plate giving the nominal load for each configuration of the machinery.

Accessories for lifting must be marked in such a way that it is possible to identify the characteristics essential for safe use.

Work equipment which is not designed for lifting persons but which might be so used in error must be appropriately and clearly marked to this effect.

3.2.3. Permanently installed work equipment must be installed in such a way as to reduce the risk of the load:

(a) striking workers;

(b) drifting dangerously or falling freely;

(c) being released unintentionally.

3.2.4. Work equipment for lifting or moving workers must be such as to:

(a) prevent the risk of the car falling, where one exists, by suitable devices;

(b) prevent the risk of the user himself falling from the car, where one exists;

(c) prevent the risk of the user being crushed, trapped or struck, in particular through inadvertent contact with objects;

(d) ensure that persons trapped in the car in the event of an incident are not exposed to danger and can be freed.

If, for reasons inherent in the site and height differences, the risks referred to in point (a) cannot be avoided by any safety measures, an enhanced safety coefficient suspension rope must be installed and checked every working day.]

[ANNEX II
PROVISIONS CONCERNING THE USE OF
WORK EQUIPMENT REFERRED TO IN ARTICLE 4(3)

General comment

The obligations laid down in this Annex apply having regard to the provisions of this Directive and where the corresponding risk exists for the work equipment in question.

1. General provisions for all work equipment

1.1. Work equipment must be installed, located and used in such a way as to reduce risks to users of the work equipment and for other workers, for example by ensuring that there is sufficient space between the moving parts of work equipment and fixed or moving parts in its environment and that all forms of energy and substances used or produced can be supplied and/or removed in a safe manner.

1.2. Work equipment must be erected or dismantled under safe conditions, in particular observing any instructions which may have been furnished by the manufacturer.

1.3. Work equipment which may be struck by lightning while being used must be protected by devices or appropriate means against the effects of lightning.

2. Provisions concerning the use of mobile equipment; whether or not self-propelled

2.1. Self-propelled work equipment shall be driven only by workers who have been appropriately trained in the safe driving of such equipment.

2.2. If work equipment is moving around in a work area, appropriate traffic rules must be drawn up and followed.

2.3. Organisational measures must be taken to prevent workers on foot coming within the area of operation of self-propelled work equipment.

If work can be done properly only if workers on foot are present, appropriate measures must be taken to prevent them from being injured by the equipment.

2.4. The transport of workers on mechanically driven mobile work equipment is authorised only where safe facilities are provided to this effect. If work must be carried out during the journey, speeds must be adjusted as necessary.

2.5. Mobile work equipment with a combustion engine may not be used in working areas unless sufficient quantities of air presenting no health or safety risk to workers can be guaranteed.

3. Provisions concerning the use of work equipment for lifting loads

3.1. *General considerations*

3.1.1. Work equipment which is mobile or can be dismantled and which is designed for lifting loads must be used in such a way as to ensure the stability of the work equipment during use under all foreseeable conditions, taking into account the nature of the ground.

3.1.2. Persons may be lifted only by means of work equipment and accessories provided for this purpose.

Without prejudice to Article 5 of Directive 89/391/EEC, exceptionally, work equipment which is not specifically designed for the purpose of lifting persons may be used to this effect, provided appropriate action has been taken to ensure safety in accordance with national legislation and/or practice laying down appropriate supervision.

While workers are on work equipment designed for lifting loads the control position must be manned at all times. Persons being lifted must have reliable means of communication. In the event of danger, there must be reliable means of evacuating them.

3.1.3. Unless required for the effective operation of the work, measures must be taken to ensure that workers are not present under suspended loads.

Loads may not be moved above unprotected workplaces usually occupied by workers.

Where that is the case, if work cannot be carried out properly any other way, appropriate procedures must be laid down and applied.

3.1.4. Lifting accessories must be selected as a function of the loads to be handled, gripping points, attachment tackle and the atmospheric conditions having regard to the mode and configuration of slinging. Lifting accessory tackle must be clearly marked so that users are aware of its characteristics where such tackle is not dismantled after use.

3.1.5. Lifting accessories must be stored in a way that ensures that they will not be damaged or degraded.

3.2. *Work equipment for lifting non-guided loads*

3.2.1. When two or more items of work equipment used for lifting non-guided loads are installed or erected on a site in such a way that their working radii overlap, appropriate measures must be taken to avoid collision between loads and/or the work equipment parts themselves.

3.2.2. When using mobile work equipment for lifting non-guided loads, measures must be taken to prevent the equipment from tilting, overturning or, if necessary, moving or slipping. Checks must be made to ensure that these measures are executed properly.

3.2.3. If the operator of work equipment designed for lifting non-guided loads cannot observe the full path of the load either directly or by means of auxiliary equipment providing the necessary information, a competent person must be in communication with the operator to guide him and organisational measures must be taken to prevent collisions of the load which could endanger workers.

3.2.4. Work must be organised in such a way that when a worker is attaching or detaching a load by hand, it can be done safely, in particular through the worker retaining direct or indirect control of the work equipment.

3.2.5. All lifting operations must be properly planned, appropriately supervised and carried out to protect the safety of workers.

In particular, if a load has to be lifted by two or more pieces of work equipment for lifting non-guided loads simultaneously, a procedure must be established and applied to ensure good co-ordination on the part of the operators.

3.2.6. If work equipment designed for lifting non-guided loads cannot maintain its hold on the load in the event of a complete or partial power failure, appropriate measures must be taken to avoid exposing workers to any resultant risks.

Suspended loads must not be left without surveillance unless access to the danger zone is prevented and the load has been safely suspended and is safely held.

3.2.7. Open-air use of work equipment designed for lifting non-guided loads must be halted when meteorological conditions deteriorate to the point of jeopardising the safe use of the equipment and exposing workers to risks. Adequate protection measures, in particular, to avoid work equipment turning over must be taken to avoid any risks to workers.]

17.4 COUNCIL DIRECTIVE

of 30 November 1989

on the minimum health and safety requirements for the use by workers of personal protective equipment at the workplace (third individual Directive within the meaning of Article 16(1) of Directive 89/391/EEC)

(89/656/EEC)

THE COUNCIL OF THE EUROPEAN COMMUNITIES,

Having regard to the Treaty establishing the European Economic Community, and in particular Article 118a thereof,

Having regard to the Commission proposal, submitted after consultation with the Advisory Committee on Safety, Hygiene and Health Protection at Work,

In cooperation with the European Parliament,

Having regard to the opinion of the Economic and Social Committee,

Whereas Article 118a of the Treaty provides that the Council shall adopt, by means of Directives, minimum requirements designed to encourage improvements, especially in the working environment, to guarantee greater protection of the health and safety of workers;

Whereas, under the said Article, such Directives shall avoid imposing administrative, financial and legal constraints in a way which would hold back the creation and development of small and medium-sized undertakings;

Whereas the Commission communication on its programme concerning safety, hygiene and health at work provides for the adoption of a Directive on the use of personal protective equipment at work;

Whereas the Council, in its resolution of 21 December 1987 concerning safety, hygiene and health at work, noted the Commission's intention of submitting to it in the near future minimum requirements concerning the organisation of the safety and health of workers at work;

Whereas compliance with the minimum requirements designed to guarantee greater health and safety for the user of personal protective equipment is essential to ensure the safety and health of workers;

Whereas this Directive is an individual Directive within the meaning of Article 16(1) of Council Directive 89/391/EEC of 12 June 1989 on the introduction of measures to encourage improvements in the safety and health of workers at work; whereas, consequently, the provisions of the said Directive apply fully to the use by workers of personal protective equipment at the workplace, without prejudice to more stringent and/or specific provisions contained in this Directive;

Whereas this Directive constitutes a practical step towards the achievement of the social dimension of the internal market;

Whereas collective means of protection shall be accorded priority over individual protective equipment; whereas the employer shall be required to provide safety equipment and take safety measures;

Whereas the requirements laid down in this Directive should not entail alterations to personal protective equipment whose design and manufacture complied with Community Directives relating to safety and health at work;

Whereas provision should be made for descriptions which Member States may use when laying down general rules for the use of individual protective equipment;

Whereas, pursuant to Decision 74/325/EEC, as last amended by the 1985 Act of Accession, the Advisory Committee on Safety, Hygiene and Health Protection at Work is consulted by the Commission with a view to drawing up proposals in this field,

HAS ADOPTED THIS DIRECTIVE:

SECTION I
GENERAL PROVISIONS

Article 1
Subject

1. This Directive, which is the third individual Directive within the meaning of Article 16(1) of Directive 89/391/EEC, lays down minimum requirements for personal protective equipment used by workers at work.

2. The provisions of Directive 89/391/EEC are fully applicable to the whole scope referred to in paragraph 1, without prejudice to more restrictive and/or specific provisions contained in this Directive.

Article 2
Definition

1. For the purposes of this Directive, personal protective equipment shall mean all equipment designed to be worn or held by the worker to protect him against one or more hazards likely to

endanger his safety and health at work, and any addition or accessory designed to meet this objective.

2. The definition in paragraph 1 excludes:

(a) ordinary working clothes and uniforms not specifically designed to protect the safety and health of the worker;

(b) equipment used by emergency and rescue services;

(c) personal protective equipment worn or used by the military, the police and other public order agencies;

(d) personal protective equipment for means of road transport;

(e) sports equipment;

(f) self-defence or deterrent equipment;

(g) portable devices for detecting and signalling risks and nuisances.

Article 3

General rule

Personal protective equipment shall be used when the risks cannot be avoided or sufficiently limited by technical means of collective protection or by measures, methods or procedures of work organisation.

SECTION II
EMPLOYERS' OBLIGATIONS

Article 4

General provisions

1. Personal protective equipment must comply with the relevant Community provisions on design and manufacture with respect to safety and health.

All personal protective equipment must:

(a) be appropriate for the risks involved, without itself leading to any increased risk;

(b) correspond to existing conditions at the workplace;

(c) take account of ergonomic requirements and the worker's state of health;

(d) fit the wearer correctly after any necessary adjustment.

2. Where the presence of more than one risk makes it necessary for a worker to wear simultaneously more than one item of personal protective equipment, such equipment must be compatible and continue to be effective against the risk or risks in question.

3. The conditions of use of personal protective equipment, in particular the period for which it is worn, shall be determined on the basis of the seriousness of the risk, the frequency of exposure to the risk, the characteristics of the workstation of each worker and the performance of the personal protective equipment.

4. Personal protective equipment is, in principle, intended for personal use.

If the circumstances require personal protective equipment to be worn by more than one person, appropriate measures shall be taken to ensure that such use does not create any health or hygiene problem for the different users.

5. Adequate information on each item of personal protective equipment, required under paragraphs 1 and 2, shall be provided and made available within the undertaking and/or establishment.

6. Personal protective equipment shall be provided free of charge by the employer, who shall ensure its good working order and satisfactory hygienic condition by means of the necessary maintenance, repair and replacements.

However, Member States may provide, in accordance with their national practice, that the worker be asked to contribute towards the cost of certain personal protective equipment in circumstances where use of the equipment is not exclusive to the workplace.

7. The employer shall first inform the worker of the risks against which the wearing of the personal protective equipment protects him.

8. The employer shall arrange for training and shall, if appropriate, organise demonstrations in the wearing of personal protective equipment.

9. Personal protective equipment may be used only for the purposes specified, except in specific and exceptional circumstances.

It must be used in accordance with instructions.

Such instructions must be understandable to the workers.

Article 5

Assessment of personal protective equipment

1. Before choosing personal protective equipment, the employer is required to assess whether the personal protective equipment he intends to use satisfies the requirements of Article 4(1) and (2).

This assessment shall involve:

(a) an analysis and assessment of risks which cannot be avoided by other means;

(b) the definition of the characteristics which personal protective equipment must have in order to be effective against the risks referred to in (a), taking into account any risks which this equipment itself may create;

(c) comparison of the characteristics of the personal protective equipment available with the characteristics referred to in (b).

2. The assessment provided for in paragraph 1 shall be reviewed if any changes are made to any of its elements.

Article 6

Rules for use

1. Without prejudice to Articles 3, 4 and 5, Member States shall ensure that general rules are established for the use of personal protective equipment and/or rules covering cases and situations where the employer must provide the personal protective equipment, taking account of Community legislation on the free movement of such equipment.

These rules shall indicate in particular the circumstances or the risk situations in which, without prejudice to the priority to be given to collective means of protection, the use of personal protective equipment is necessary.

Annexes I, II and III, which constitute a guide, contain useful information for establishing such rules.

2. When Member States adapt the rules referred to in paragraph 1, they shall take account of any significant changes to the risk, collective means of protection and personal protective equipment brought about by technological developments.

3. Member States shall consult the employers' and workers' organisation on the rules referred to in paragraphs 1 and 2.

Article 7

Information for workers

Without prejudice to Article 10 of Directive 89/391/EEC, workers and/or their representatives shall be informed of all measures to be taken with regard to the health and safety of workers when personal protective equipment is used by workers at work.

Article 8

Consultation of workers and workers' participation

Consultation and participation of workers and/or of their representatives shall take place in accordance with Article 11 of Directive 89/391/EEC on the matters covered by this Directive, including the Annexes thereto.

SECTION III
MISCELLANEOUS PROVISIONS

Article 9

Adjustment of the Annexes

Alterations of a strictly technical nature to Annexes I, II and III resulting from:

— the adoption of technical harmonisation and standardisation Directives relating to personal protective equipment, and/or

— technical progress and changes in international regulations and specifications or knowledge in the field of personal protective equipment,

shall be adopted in accordance with the procedure provided for in Article 17 of Directive 89/391/EEC.

Article 10

Final provisions

1. Member States shall bring into force the laws, regulations and administrative provisions necessary to comply with this Directive not later than 31 December 1992. They shall immediately inform the Commission thereof.

2. Member States shall communicate to the Commission the text of the provisions of national law which they adopt, as well as those already adopted, in the field covered by this Directive.

3. Member States shall report to the Commission every five years on the practical implementation of the provisions of this Directive, indicating the points of view of employers and workers.

The Commission shall inform the European Parliament, the Council, the Economic and Social Committee, and the Advisory Committee on Safety, Hygiene and Health Protection at Work.

4. The Commission shall report periodically to the European Parliament, the Council and the Economic and Social Committee on the implementation of the Directive in the light of paragraphs 1, 2 and 3.

Article 11

This Directive is addressed to the Member States.

Done at Brussels, 30 November 1989.

For the Council

The President

J P SOISSON

ANNEX I
SPECIMEN RISK SURVEY TABLE FOR THE USE OF PERSONAL PROTECTIVE EQUIPMENT

ANNEX II
NON-EXHAUSTIVE GUIDE LIST OF ITEMS OF PERSONAL PROTECTIVE EQUIPMENT

Head protection

— Protective helmets for use in industry (mines, building sites, other industrial uses).

— Scalp protection (caps, bonnets, hairnets—with or without eye shade).

— Protective headgear (bonnets, caps, sou'westers, etc in fabric with proofing, etc).

Hearing protection

— Earplugs and similar devices.

— Full acoustic helmets.

— Earmuffs which can be fitted to industrial helmets.

— Ear defenders with receiver for LF induction loop.

— Ear protection with intercom equipment.

Eye and face protection

— Spectacles.

— Goggles.

— X-ray goggles, laser-beam goggles, ultra-violet, infra-red, visible radiation goggles.

— Face shields.

— Arc-welding masks and helmets (hand masks, headband masks or masks which can be fitted to protective helmets).

Respiratory protection

— Dust filters, gas filters and radioactive dust filters.

— Insulating appliances with an air supply.

— Respiratory devices including a removable welding mask.

— Diving equipment.

— Diving suits.

Hand and arm protection

— Gloves to provide protection:

  — from machinery (piercing, cuts, vibrations, etc),

  — from chemicals,

  — for electricians and from heat.

— Mittens.

— Finger stalls.

— Oversleeves.

— Wrist protection for heavy work.

— Fingerless gloves.

— Protective gloves.

Foot and leg protection

— Low shoes, ankle boots, calf-length boots, safety boots.

— Shoes which can be unlaced or unhooked rapidly.

— Shoes with additional protective toe-cap.

— Shoes and overshoes with heat-resistant soles.

— Heat-resistant shoes, boots and overboots.

— Thermal shoes, boots and overboots.

— Vibration-resistant shoes, boots and overboots.

— Anti-static shoes, boots and overboots.

— Insulating shoes, boots and overboots.

— Protective boots for chain saw operators.

— Clogs.

— Kneepads.

— Removable instep protectors.

— Gaiters.

— Removable soles (heat-proof, pierce-proof or sweat-proof).

— Removable spikes for ice, snow or slippery flooring.

Skin protection

— Barrier creams/ointments.

Trunk and abdomen protection

— Protective waistcoats, jackets and aprons to provide protection from machinery (piercing, cutting, molten metal splashes, etc).

— Protective waistcoats, jackets and aprons to provide protection from chemicals.

— Heated waistcoats.

— Lifejackets.

— Protective X-ray aprons.

— Body belts.

Whole body protection

Equipment designed to prevent falls

— Fall-prevention equipment (full equipment with all necessary accessories).

— Braking equipment to absorb kinetic energy (full equipment with all necessary accessories).

— Body-holding devices (safety harness).

Protective clothing

— Safety working clothing (two-piece and overalls).

— Clothing to provide protection from machinery (piercing, cutting, etc).

— Clothing to provide protection from chemicals.

— Clothing to provide protection from molten splashes and infra-red radiation.

— Heat-resistant clothing.

— Thermal clothing.

— Clothing to provide protection from radioactive contamination.

— Dust-proof clothing.

— Gas-proof clothing.

— Fluorescent signalling, retro-reflecting clothing and accessories (arm-bands, gloves, etc).

— Protective coverings.

ANNEX III
NON-EXHAUSTIVE GUIDE LIST OF ACTIVITIES AND SECTORS OF ACTIVITY WHICH MAY REQUIRE THE PROVISION OF PERSONAL PROTECTIVE EQUIPMENT

1. Head protection (skull protection)

Protective helmets

— Building work, particularly work on, underneath or in the vicinity of scaffolding and elevated workplaces, erection and stripping of formwork, assembly and installation work, work on scaffolding and demolition work.

— Work on steel bridges, steel building construction, masts, towers, steel hydraulic structures, blast furnaces, steel works and rolling mills, large containers, large pipelines, boiler plants and power stations.

— Work in pits, trenches, shafts and tunnels.

— Earth and rock works.

— Work in underground workings, quarries, open diggings, coal stock removal.

— Work with bolt-driving tools.

— Blasting work.

— Work in the vicinity of lifts, lifting gear, cranes and conveyors.

— Work with blast furnaces, direct reduction plants, steelworks, rolling mills, metalworks, forging, drop forging and casting.

— Work with industrial furnaces, containers, machinery, silos, bunkers and pipelines.

— Shipbuilding.

— Railway shunting work.

— Slaughterhouses.

2. Foot protection

Safety shoes with puncture-proof soles

— Carcase work, foundation work and roadworks.

— Scaffolding work.

— The demolition of carcase work.

— Work with concrete and prefabricated parts involving formwork erection and stripping.

— Work in contractors' yards and warehouses.

— Roof work.

Safety shoes without pierce-proof soles

— Work on steel bridges, steel building construction, masts, towers, lifts, steel hydraulic structures, blast furnaces, steelworks and rolling mills, large containers, large pipelines, cranes, boiler plants and power stations.

— Furnace construction, heating and ventilation installation and metal assembly work.

— Conversion and maintenance work.

— Work with blast furnaces, direct reduction plants, steelworks, rolling mills, metalworks, forging, drop forging, hot pressing and drawing plants.

— Work in quarries and open diggings, coal stock removal.

— Working and processing of rock.

— Flat glass products and container glassware manufacture, working and processing.

— Work with moulds in the ceramics industry.

— Lining of kilns in the ceramics industry.

— Moulding work in the ceramic ware and building materials industry.

— Transport and storage.

— Work with frozen meat blocks and preserved foods packaging.

— Shipbuilding.

— Railway shunting work.

Safety shoes with heels or wedges and pierce-proof soles

— Roof work.

Protective shoes with insulated soles

— Work with and on very hot or very cold materials.

Safety shoes which can easily be removed

— Where there is a risk of penetration by molten substances.

3. Eye or face protection

Protective goggles, face shields or screens

— Welding, grinding and separating work.

— Caulking and chiselling.

— Rock working and processing.

— Work with bolt-driving tools.

— Work on stock removing machines for small chippings.

— Drop forging.

— The removal and breaking up of fragments.

— Spraying of abrasive substances.

— Work with acids and caustic solutions, disinfectants and corrosive cleaning products.

— Work with liquid sprays.

— Work with and in the vicinity of molten substances.

— Work with radiant heat.

— Work with lasers.

4. Respiratory protection

Respirators/breathing apparatus

— Work in containers, restricted areas and gas-fired industrial furnaces where there may be gas or insufficient oxygen.

— Work in the vicinity of the blast furnace charge.

— Work in the vicinity of gas converters and blast furnace gas pipes.

— Work in the vicinity of blast furnace taps where there may be heavy metal fumes.

— Work on the lining of furnaces and ladles where there may be dust.

— Spray painting where dedusting is inadequate.

— Work in shafts, sewers and other underground areas connected with sewage.

— Work in refrigeration plants where there is a danger that the refrigerant may escape.

5. Hearing protection

Ear protectors

— Work with metal presses.

— Work with pneumatic drills.

— The work of ground staff at airports.

— Pile-driving work.

— Wood and textile working.

6. Body arm and hand protection

Protective clothing

— Work with acids and caustic solutions, disinfectants and corrosive cleaning substances.

— Work with or in the vicinity of hot materials and where the effects of heat are felt.

— Work on flat glass products.

— Shot blasting.

— Work in deep-freeze rooms.

Fire-resistant protective clothing

— Welding in restricted areas.

Pierce-proof aprons

— Boning and cutting work.

— Work with hand knives involving drawing the knife towards the body.

Leather aprons

— Welding.

— Forging.

— Casting.

Forearm protection

— Boning and cutting.

Gloves

— Welding.

— Handling of sharp-edged objects, other than machines where there is a danger of the gloves being caught.

— Unprotected work with acids and caustic solutions.

Metal mesh gloves

— Boning and cutting.

— Regular cutting using a hand knife for production and slaughtering.

— Changing the knives of cutting machines.

7. Weatherproof clothing

— Work in the open air in rain and cold weather.

8. Reflective clothing

— Work where the workers must be clearly visible.

9. Safety harnesses

— Work on scaffolding.

— Assembly of prefabricated parts.

— Work on masts.

10. Safety ropes

— Work in high crane cabs.

— Work in high cabs of warehouse stacking and retrieval equipment.

— Work in high sections of drilling towers.

— Work in shafts and sewers.

11. Skin protection

— Processing of coating materials.

— Tanning.

17.5 COUNCIL DIRECTIVE
of 29 May 1990

on the minimum health and safety requirements for the manual handling of loads where there is a risk particularly of back injury to workers (fourth individual Directive within the meaning of Article 16(1) of Directive 89/391/EEC)

(90/269/EEC)

THE COUNCIL OF THE EUROPEAN COMMUNITIES,

Having regard to the Treaty establishing the European Economic Community, and in particular Article 118a thereof,

Having regard to the Commission proposal submitted after consultation with the Advisory Committee on Safety, Hygiene and Health Protection at Work,

In cooperation with the European Parliament,

Having regard to the opinion of the Economic and Social Committee,

Whereas Article 118a of the Treaty provides that the Council shall adopt, by means of Directives, minimum requirements for encouraging improvements, especially in the working environment, to guarantee a better level of protection of the health and safety of workers;

Whereas, pursuant to that Article, such Directives must avoid imposing administrative, financial and legal constraints in a way which would hold back the creation and development of small and medium-sized undertakings;

Whereas the Commission communication on its programme concerning safety, hygiene and health at work, provides for the adoption of Directives designed to guarantee the health and safety of workers at the workplace;

Whereas the Council, in its resolution of 21 December 1987 on safety, hygiene and health at work, took note of the Commission's intention of submitting to the Council in the near future a Directive on protection against the risks resulting from the manual handling of heavy loads;

Whereas compliance with the minimum requirements designed to guarantee a better standard of health and safety at the workplace is essential to ensure the health and safety of workers;

Whereas this Directive is an individual Directive within the meaning of Article 16(1) of Council Directive 89/391/EEC of 12 June 1989 on the introduction of measures to encourage improvements in the health and safety of workers at work; whereas therefore the provisions of the said Directive are fully applicable to the field of the manual handling of loads where there is a risk particularly of back injury to workers, without prejudice to more stringent and/or specific provisions set out in this Directive;

Whereas this Directive constitutes a practical step towards the achievement of the social dimensions of the internal market;

Whereas, pursuant to Decision 74/325/EEC, the Advisory Committee on Safety, Hygiene and Health Protection at Work shall be consulted by the Commission with a view to drawing up proposals in this field,

HAS ADOPTED THIS DIRECTIVE:

SECTION I
GENERAL PROVISIONS

Article 1
Subject

1. This Directive, which is the fourth individual Directive within the meaning of Article 16(1) of Directive 89/391/EEC, lays down minimum health and safety requirements for the manual handling of loads where there is a risk particularly of back injury to workers.

2. The provisions of Directive 89/391/EEC shall be fully applicable to the whole sphere referred to in paragraph 1, without prejudice to more restrictive and/or specific provisions contained in this Directive.

Article 2

Definition

For the purposes of this Directive, 'manual handling of loads' means any transporting or supporting of a load, by one or more workers, including lifting, putting down, pushing, pulling, carrying or moving of a load, which, by reason of its characteristics or of unfavourable ergonomic conditions, involves a risk particularly of back injury to workers.

SECTION II
EMPLOYERS' OBLIGATIONS

Article 3

General provision

1. The employer shall take appropriate organisational measures, or shall use the appropriate means, in particular mechanical equipment, in order to avoid the need for the manual handling of loads by workers.

2. Where the need for the manual handling of loads by workers cannot be avoided, the employer shall take the appropriate organisational measures, use the appropriate means or provide workers with such means in order to reduce the risk involved in the manual handling of such loads, having regard to Annex I.

Article 4

Organisation of workstations

Wherever the need for manual handling of loads by workers cannot be avoided, the employer shall organise workstations in such a way as to make such handling as safe and healthy as possible and:

(a) assess, in advance if possible, the health and safety conditions of the type of work involved, and in particular examine the characteristics of loads, taking account of Annex I;

(b) take care to avoid or reduce the risk particularly of back injury to workers, by taking appropriate measures, considering in particular the characteristics of the working environment and the requirements of the activity, taking account of Annex I.

Article 5

Reference to Annex II

For the implementation of Article 6(3)(b) and Articles 14 and 15 of Directive 89/391/EEC, account should be taken of Annex II.

Article 6

Information for, and training of, workers

1. Without prejudice to Article 10 of Directive 89/391/EEC, workers and/or their representatives shall be informed of all measures to be implemented, pursuant to this Directive, with regard to the protection of safety and of health.

Employers must ensure that workers and/or their representatives receive general indications and, where possible, precise information on:

— the weight of a load,

— the centre of gravity of the heaviest side when a package is eccentrically loaded.

2. Without prejudice to Article 12 of Directive 89/391/EEC, employers must ensure that workers receive in addition proper training and information on how to handle loads correctly and the risks they might be open to particularly if these tasks are not performed correctly, having regard to Annexes I and II.

Article 7

Consultation of workers and workers' participation

Consultation and participation of workers and/or of their representatives shall take place in accordance with Article 11 of Directive 89/391/EEC on matters covered by this Directive, including the Annexes thereto.

SECTION III
MISCELLANEOUS PROVISIONS

Article 8

Adjustment of the Annexes

Alterations of a strictly technical nature to Annexes I and II resulting from technical progress and changes in international regulations and specifications or knowledge in the field of the manual handling of loads shall be adopted in accordance with the procedure provided for in Article 17 of Directive 89/391/EEC.

Article 9

Final provisions

1. Member States shall bring into force the laws, regulations and administrative provisions needed to comply with this Directive not later than 31 December 1992.

They shall forthwith inform the Commission thereof.

2. Member States shall communicate to the Commission the text of the provisions of national law which they adopt, or have adopted, in the field covered by this Directive.

3. Member States shall report to the Commission every four years on the practical implementation of the provisions of this Directive, indicating the points of view of employers and workers.

The Commission shall inform the European Parliament, the Council, the Economic and Social Committee and the Advisory Committee on Safety, Hygiene and Health Protection at Work thereof.

4. The Commission shall report periodically to the European Parliament, the Council and the Economic and Social Committee on the implementation of the Directive in the light of paragraphs 1, 2 and 3.

Article 10

This Directive is addressed to the Member States.

Done at Brussels, 29 May 1990.

For the Council

The President

B AHERN

ANNEX I
REFERENCE FACTORS

(Article 3(2), Article 4(a) and (b) and Article 6(2))

1. Characteristics of the load

The manual handling of a load may present a risk particularly of back injury if it is:

— too heavy or too large,

— unwieldy or difficult to grasp,

— unstable or has contents likely to shift,

— positioned in a manner requiring it to be held or manipulated at a distance from the trunk, or with a bending or twisting of the trunk,

— likely, because of its contours and/or consistency, to result in injury to workers, particularly in the event of a collision.

2. Physical effort required

A physical effort may present a risk particularly of back injury if it is:

— too strenuous,

— only achieved by a twisting movement of the trunk,

— likely to result in a sudden movement of the load,

— made with the body in an unstable posture.

3. Characteristics of the working environment

The characteristics of the work environment may increase a risk particularly of back injury if:

— there is not enough room, in particular vertically, to carry out the activity,

— the floor is uneven, thus presenting tripping hazards, or is slippery in relation to the worker's footwear,

— the place of work or the working environment prevents the handling of loads at a safe height or with good posture by the worker,

— there are variations in the level of the floor or the working surface, requiring the load to be manipulated on different levels,

— the floor or foot rest is unstable,

— the temperature, humidity or ventilation is unsuitable.

4. Requirements of the activity

The activity may present a risk particularly of back injury if it entails one or more of the following requirements:

— over-frequent or over-prolonged physical effort involving in particular the spine,

— an insufficient bodily rest or recovery period,

— excessive lifting, lowering or carrying distances,

— a rate of work imposed by a process which cannot be altered by the worker.

ANNEX II
INDIVIDUAL RISK FACTORS

(Articles 5 and 6(2))

The worker may be at risk if he/she:

— is physically unsuited to carry out the task in question,

— is wearing unsuitable clothing, footwear or other personal effects,

— does not have adequate or appropriate knowledge or training.

17.6 COUNCIL DIRECTIVE
of 29 May 1990

on the minimum safety and health requirements for work with display screen equipment (fifth individual Directive within the meaning of Article 16(1) of Directive 87/391/EEC) (90/270/EEC)

THE COUNCIL OF THE EUROPEAN COMMUNITIES,

Having regard to the Treaty establishing the European Economic Community, and in particular Article 118a thereof,

Having regard to the Commission proposal drawn up after consultation with the Advisory Committee on Safety, Hygiene and Health Protection at Work,

In cooperation with the European Parliament,

Having regard to the opinion of the Economic and Social Committee,

Whereas Article 118a of the Treaty provides that the Council shall adopt by means of Directives, minimum requirements designed to encourage improvements, especially in the working environment, to ensure a better level of protection of workers' safety and health;

Whereas, under the terms of that Article, those Directives shall avoid imposing administrative, financial and legal constraints, in a way which would hold back the creation and development of small and medium-sized undertakings;

Whereas the communication from the Commission on its programme concerning safety, hygiene and health at work provides for the adoption of measures in respect of new technologies; whereas the Council has taken note thereof in its resolution of 21 December 1987 on safety, hygiene and health at work;

Whereas compliance with the minimum requirements for ensuring a better level of safety at workstations with display screens is essential for ensuring the safety and health of workers;

Whereas this Directive is an individual Directive within the meaning of Article 16(1) of Council Directive 89/391/EEC of 12 June 1989 on the introduction of measures to encourage improvements in the safety and health of workers at work; whereas the provisions of the latter are therefore fully applicable to the use by workers of display screen equipment, without prejudice to more stringent and/or specific provisions contained in the present Directive;

Whereas employers are obliged to keep themselves informed of the latest advances in technology and scientific findings concerning workstation design so that they can make any changes necessary so as to be able to guarantee a better level of protection of workers safety and health;

Whereas the ergonomic aspects are of particular importance for a workstation with display screen equipment;

Whereas this Directive is a practical contribution towards creating the social dimension of the internal market;

Whereas, pursuant to Decision 74/325/EEC, the Advisor Committee on Safety, Hygiene and Health Protection at Work shall be consulted by the Commission on the drawing-up of proposals in this field,

HAS ADOPTED THIS DIRECTIVE:

SECTION I
GENERAL PROVISIONS

Article 1

Subject

1. This Directive, which is the fifth individual Directive within the meaning of Article 16(1) of Directive 89/391/EEC, lays down minimum safety and health requirements for work with display screen equipment as defined in Article 2.

2. The provisions of Directive 89/391/EEC are fully applicable to the whole field referred to in paragraph 1, without prejudice to more stringent and/or specific provisions contained in the present Directive.

3. This Directive shall not apply to:

(a) drivers' cabs or control cabs for vehicles or machinery;

(b) computer systems on board a means of transport;

(c) computer systems mainly intended for public use;

(d) 'portable' systems not in prolonged use at a workstation;

(e) calculators, cash registers and any equipment having a small data or measurement display required for direct use of the equipment;

(f) typewriters of traditional design, of the type known as 'typewriter with window'.

Article 2

Definitions

For the purpose of this Directive, the following terms shall have the following meanings:

(a) *display screen equipment*: an alphanumeric or graphic display screen, regardless of the display process employed;

(b) *workstation*: an assembly comprising display screen equipment, which may be provided with a keyboard or input device and/or software determining the operator/machine interface, optional accessories, peripherals including the diskette drive, telephone, modem, printer, document holder, work chair and work desk or work surface, and the immediate work environment;

(c) *worker*: any worker as defined in Article 3(a) of Directive 89/391/EEC who habitually uses display screen equipment as a significant part of his normal work.

SECTION II
EMPLOYERS' OBLIGATIONS

Article 3

Analysis of workstations

1. Employers shall be obliged to perform an analysis of workstations in order to evaluate the safety and health conditions to which they give rise for their workers, particularly as regards possible risks to eyesight, physical problems and problems of mental stress.

2. Employers shall take appropriate measures to remedy the risks found, on the basis of the evaluation referred to in paragraph 1, taking account of the additional and/or combined effects of the risks so found.

Article 4

Workstations put into service for the first time

Employers must take the appropriate steps to ensure that workstations first put into service after 31 December 1992 meet the minimum requirements laid down in the Annex.

Article 5

Workstations already put into service

Employers must take the appropriate steps to ensure that workstations already put into service on or before 31 December 1992 are adapted to comply with the minimum requirements laid down in the Annex not later than four years after that date.

Article 6

Information for, and training of, workers

1. Without prejudice to Article 10 of Directive 89/391/EEC, workers shall receive information on all aspects of safety and health relating to their workstation, in particular information on such measures applicable to workstations as are implemented under Articles 3, 7 and 9.

In all cases, workers or their representatives shall be informed of any health and safety measure taken in compliance with this Directive.

2. Without prejudice to Article 12 of Directive 89/391/EEC, every worker shall also receive training in use of the workstation before commencing this type of work and whenever the organisation of the workstation is substantially modified.

Article 7

Daily work routine

The employer must plan the worker's activities in such a way that daily work on a display screen is periodically interrupted by breaks or changes of activity reducing the workload at the display screen.

Article 8

Worker consultation and participation

Consultation and participation of workers and/or their representatives shall take place in accordance with Article 11 of Directive 89/391/EEC on the matters covered by this Directive, including its Annex.

Article 9

Protection of workers' eyes and eyesight

1. Workers shall be entitled to an appropriate eye and eyesight test carried out by a person with the necessary capabilities:

— before commencing display screen work,

— at regular intervals thereafter, and

— if they experience visual difficulties which may be due to display screen work.

2. Workers shall be entitled to an ophthalmological examination if the results of the test referred to in paragraph 1 show that this is necessary.

3. If the result of the test referred to in paragraph 1 or of the examination referred to in paragraph 2 show that it is necessary and if normal corrective appliances cannot be used, workers must be provided with special corrective appliances appropriate for the work concerned.

4. Measures taken pursuant to this Article may in no circumstances involve workers in additional financial cost.

5. Protection of workers' eyes and eyesight may be provided as part of a national health system.

SECTION III
MISCELLANEOUS PROVISIONS

Article 10

Adaptations to the Annex

The strictly technical adaptations to the Annex to take account of technical progress, developments in international regulations and specifications and knowledge in the field of display screen equipment shall be adopted in accordance with the procedure laid down in Article 17 of Directive 89/391/EEC.

Article 11

Final provisions

1. Member States shall bring into force the laws, regulations and administrative provisions necessary to comply with this Directive

by 31 December 1992.

They shall forthwith inform the Commission thereof.

2. Member States shall communicate to the Commission the texts of the provisions of national law which they adopt, or have already adopted, in the field covered by this Directive.

3. Member States shall report to the Commission every four years on the practical implementation of the provisions of this Directive, indicating the points of view of employers and workers.

The Commission shall inform the European Parliament, the Council, the Economic and Social Committee and the Advisory Committee on Safety, Hygiene and Health Protection at Work.

4. The Commission shall submit a report on the implementation of this Directive at regular intervals to the European Parliament, the Council and the Economic and Social Committee, taking into account paragraphs 1, 2 and 3.

Article 12

This Directive is addressed to the Member States.

Done at Brussels, 29 May 1990.

For the Council

The President

B AHERN

ANNEX
MINIMUM REQUIREMENTS

(Articles 4 and 5)

Preliminary remark

The obligations laid down in this Annex shall apply in order to achieve the objectives of this Directive and to the extent that, firstly, the components concerned are present at the workstation, and secondly, the inherent requirements or characteristics of the task do not preclude it.

1. Equipment

(a) *General comment*

The use as such of the equipment must not be a source of risk for workers.

(b) *Display screen*

The characters on the screen shall be well-defined and clearly formed, of adequate size and with adequate spacing between the characters and lines.

The image on the screen should be stable, with no flickering or other forms of instability.

The brightness and/or the contrast between the characters and the background shall be easily adjustable by the operator, and also be easily adjustable to ambient conditions.

The screen must swivel and tilt easily and freely to suit the needs of the operator.

It shall be possible to use a separate base for the screen or an adjustable table.

The screen shall be free of reflective glare and reflections liable to cause discomfort to the user.

(c) *Keyboard*

The keyboard shall be tiltable and separate from the screen so as to allow the worker to find a comfortable working position avoiding fatigue in the arms or hands.

The space in front of the keyboard shall be sufficient to provide support for the hands and arms of the operator.

The keyboard shall have a matt surface to avoid reflective glare.

The arrangement of the keyboard and the characteristics of the keys shall be such as to facilitate the use of the keyboard.

The symbols on the keys shall be adequately contrasted and legible from the design working position.

(d) *Work desk or work surface*

The work desk or work surface shall have a sufficiently large, low-reflectance surface and allow a flexible arrangement of the screen, keyboard, documents and related equipment.

The document holder shall be stable and adjustable and shall be positioned so as to minimise the need for uncomfortable head and eye movements.

There shall be adequate space for workers to find a comfortable position.

(e) *Work chair*

The work chair shall be stable and allow the operator easy freedom of movement and a comfortable position.

The seat shall be adjustable in height.

The seat back shall be adjustable in both height and tilt.

A footrest shall be made available to anyone who wishes for one.

2. Environment

(a) *Space requirements*

The workstation shall be dimensioned and designed so as to provide sufficient space for the user to change position and vary movements.

(b) *Lighting*

Room lighting and/or spotlighting (work lamps) shall ensure satisfactory lighting conditions and an appropriate contrast between the screen and the background environment, taking into account the type of work and the user's vision requirements.

Possible disturbing glare and reflections on the screen or other equipment shall be prevented by coordinating workplace and workstation layout with the positioning and technical characteristics of the artificial light sources.

(c) *Reflections and glare*

Workstations shall be so designed that sources of light, such as windows and other openings, transparent or translucid walls, and brightly coloured fixtures or walls cause no direct glare and [no distracting][1] reflections on the screen.

Windows shall be fitted with a suitable system of adjustable covering to attenuate the daylight that falls on the workstation.

1 Correction indicated here taken from Official Journal, L 171/30.

(d) *Noise*

Noise emitted by equipment belonging to workstation(s) shall be taken into account when a workstation is being equipped, in particular so as not to distract attention or disturb speech.

(e) *Heat*

Equipment belonging to workstation(s) shall not produce excess heat which could cause discomfort to workers.

(f) *Radiation*

All radiation with the exception of the visible part of the electromagnetic spectrum shall be reduced to negligible levels from the point of view of the protection of workers' safety and health.

(g) *Humidity*

An adequate level of humidity shall be established and maintained.

3. Operator/computer interface

In designing, selecting, commissioning and modifying software, and in designing tasks using display screen equipment, the employer shall take into account the following principles:

(a) software must be suitable for the task;

(b) software must be easy to use and, where appropriate, adaptable to the operator's level of knowledge or experience; no quantitative or qualitative checking facility may be used without the knowledge of the workers;

(c) systems must provide feedback to workers on their performance;

(d) systems must display information in a format and at a pace which are adapted to operators;

(e) the principles of software ergonomics must be applied, in particular to human data processing.

17.7 COUNCIL DIRECTIVE
of 25 June 1991

supplementing the measures to encourage improvements in the safety and health at work of workers with a fixed-duration employment relationship or a temporary employment relationship
(91/383/EEC)

THE COUNCIL OF THE EUROPEAN COMMUNITIES,

Having regard to the Treaty establishing the European Economic Community, and in particular Article 118a thereof,

Having regard to the proposal from the Commission,

In cooperation with the European Parliament,

Having regard to the opinion of the Economic and Social Committee,

Whereas Article 118a of the Treaty provides that the Council shall adopt, by means of Directives, minimum requirements for encouraging improvements, especially in the working environment, to guarantee a better level of protection of the safety and health of workers;

Whereas, pursuant to the said Article, Directives must avoid imposing administrative, financial and legal constraints which would hold back the creation and development of small and medium-sized undertakings;

Whereas recourse to forms of employment such as fixed-duration employment and temporary employment has increased considerably;

Whereas research has shown that in general workers with a fixed-duration employment relationship or temporary employment relationship are, in certain sectors, more exposed to the risk of accidents at work and occupational diseases than other workers;

Whereas these additional risks in certain sectors are in part linked to certain particular modes of integrating new workers into the undertaking; whereas these risks can be reduced through adequate provision of information and training from the beginning of employment;

Whereas the Directives on health and safety at work, notably Council Directive 89/391/EEC of 12 June 1989 on the introduction of measures to encourage improvements in the safety and health of workers at work, contain provisions intended to improve the safety and health of workers in general;

Whereas the specific situation of workers with a fixed-duration employment relationship or a temporary employment relationship and the special nature of the risks they face in certain sectors calls for special additional rules, particularly as regards the provision of information, the training and the medical surveillance of the workers concerned;

Whereas this Directive constitutes a practical step within the framework of the attainment of the social dimension of the internal market,

HAS ADOPTED THIS DIRECTIVE:

SECTION I
SCOPE AND OBJECT

Article 1

Scope

This Directive shall apply to:

1. employment relationships governed by a fixed-duration contract of employment concluded directly between the employer and the worker, where the end of the contract is established by objective conditions such as: reaching a specific date, completing a specific task or the occurrence of a specific event;

2. temporary employment relationships between a temporary employment business which is the employer and the worker, where the latter is assigned to work for and under the control of an undertaking and/or establishment making use of his services.

Article 2

Object

1. The purpose of this Directive is to ensure that workers with an employment relationship as referred to in Article 1 are afforded, as regards safety and health at work, the same level of protection as that of other workers in the user undertaking and/or establishment.

2. The existence of an employment relationship as referred to in Article 1 shall not justify different treatment with respect to working conditions inasmuch as the protection of safety and health at work are involved, especially as regards access to personal protective equipment.

3. Directive 89/391/EEC and the individual Directives within the meaning of Article 16(1) thereof shall apply in full to workers with an employment relationship as referred to in Article 1, without prejudice to more binding and/or more specific provisions set out in this Directive.

SECTION II
GENERAL PROVISIONS

Article 3

Provision of information to workers

Without prejudice to Article 10 of Directive 89/391/EEC, Member States shall take the necessary steps to ensure that:

1. before a worker with an employment relationship as referred to in Article 1 takes up any activity, he is informed by the undertaking and/or establishment making use of his services of the risks which he faces;

2. such information:
   - — covers, in particular, any special occupational qualifications or skills or special medical surveillance required, as defined in national legislation, and
   - — states clearly any increased specific risks, as defined in national legislation, that the job may entail.

Article 4

Workers' training

Without prejudice to Article 12 of Directive 89/391/EEC, Member States shall take the necessary measures to ensure that, in the cases referred to in Article 3, each worker receives sufficient training appropriate to the particular characteristics of the job, account being taken of his qualifications and experience.

Article 5

Use of workers' services and medical surveillance of workers

1. Member States shall have the option of prohibiting workers with an employment relationship as referred to in Article 1 from

being used for certain work as defined in national legislation, which would be particularly dangerous to their safety or health, and in particular for certain work which requires special medical surveillance, as defined in national legislation.

2. Where Member States do not avail themselves of the option referred to in paragraph 1, they shall, without prejudice to Article 14 of Directive 89/391/ EEC, take the necessary measures to ensure that workers with an employment relationship as referred to in Article 1 who are used for work which requires special medical surveillance, as defined in national legislation, are provided with appropriate special medical surveillance.

3. It shall be open to Member States to provide that the appropriate special medical surveillance referred to in paragraph 2 shall extend beyond the end of the employment relationship of the worker concerned.

Article 6

Protection and prevention services

Member States shall take the necessary measures to ensure that workers, services or persons designated, in accordance with Article 7 of Directive 89/ 391/EEC, to carry out activities related to protection from and the prevention of occupational risks are informed of the assignment of workers with an employment relationship as referred to in Article 1, to the extent necessary for the workers, services or persons designated to be able to carry out adequately their protection and prevention activities for all the workers in the undertaking and/or establishment.

SECTION III
SPECIAL PROVISIONS

Article 7

Temporary employment relationships: information

Without prejudice to Article 3, Member States shall take the necessary steps to ensure that:

1. before workers with an employment relationship as referred to in Article 1(2) are supplied, a user undertaking and/or establishment shall specify to the temporary employment

business, inter alia, the occupational qualifications required and the specific features of the job to be filled;

2. the temporary employment business shall bring all these facts to the attention of the workers concerned.

Member States may provide that the details to be given by the user undertaking and/or establishment to the temporary employment business in accordance with point 1 of the first subparagraph shall appear in a contract of assignment.

Article 8

Temporary employment relationships: responsibility

Member States shall take the necessary steps to ensure that:

1. without prejudice to the responsibility of the temporary employment business as laid down in national legislation, the user undertaking and/or establishment is/are responsible, for the duration of the assignment, for the conditions governing performance of the work;

2. for the application of point 1, the conditions governing the performance of the work shall be limited to those connected with safety, hygiene and health at work.

SECTION IV
MISCELLANEOUS PROVISIONS

Article 9

More favourable provisions

This Directive shall be without prejudice to existing or future national or Community provisions which are more favourable to the safety and health protection of workers with an employment relationship as referred to in Article 1.

Article 10

Final provisions

1. Member States shall bring into force the laws, regulations and administrative provisions necessary to comply with this Directive by 31 December 1992 at the latest. They shall forthwith inform the Commission thereof.

When Member States adopt these measures, the latter shall contain a reference to this Directive or shall be accompanied by such reference on the occasion of their official publication. The methods of making such a reference shall be laid down by the Member States.

2. Member States shall forward to the Commission the texts of the provisions of national law which they have already adopted or adopt in the field covered by this Directive.

3. Member States shall report to the Commission every five years on the practical implementation of this Directive, setting out the points of view of workers and employers.

The Commission shall bring the report to the attention of the European Parliament, the Council, the Economic and Social Committee and the Advisory Committee on Safety, Hygiene and Health Protection at Work.

4. The Commission shall submit to the European Parliment, the Council and the Economic and Social Committee a regular report on the implementation of this Directive, due account being taken of paragraphs 1, 2 and 3.

Article 11

This Directive is addressed to the Member States.

Done at Luxembourg, 25 June 1991

For the Council

The President

J-C JUNCKER

Appendix 1

Management of Health and Safety at Work Regulations 1999

THE REGULATIONS (SI 1999/3242)

Made 3 December 1999

Laid before Parliament 8 December 1999

Initial Commencement 29 December 1999

The Secretary of State, being a Minister designated for the purposes of section 2(2) of the European Communities Act 1972 in relation to measures relating to employers' obligations in respect of the health and safety of workers and in relation to measures relating to the minimum health and safety requirements for the workplace that relate to fire safety and in exercise of the powers conferred on him by the said section 2 and by sections 15(1), (2), (3)(a), (5), and (9), 47(2), 52(2), and (3), 80(1) and 82(3)(a) of and paragraphs 6(1), 7, 8(1), 10, 14, 15, and 16 of Schedule 3 to, the Health and Safety at Work etc Act 1974 ('the 1974 Act') and of all other powers enabling him in that behalf—

(a) for the purpose of giving effect without modifications to proposals submitted to him by the Health and Safety Commission under section 11(2)(d) of the 1974 Act after the carrying out by the Commission of consultations in accordance with section 50(3) of that Act; and

(b) it appearing to him that the modifications to the Regulations marked with an asterisk in Schedule 2 are expedient and that it also appearing to him not to be appropriate to consult bodies in respect of such modifications in accordance with section 80(4) of the 1974 Act,

hereby makes the following Regulations:

Regulation 1

Citation, commencement and interpretation

(1) These Regulations may be cited as the Management of Health and Safety at Work Regulations 1999 and shall come into force on 29 December 1999.

(2) In these Regulations—

— 'the 1996 Act' means the Employment Rights Act 1996;

— 'the assessment' means, in the case of an employer or self-employed person, the assessment made or changed by him in accordance with regulation 3;

— 'child'—

(a) as respects England and Wales, means a person who is not over compulsory school age, construed in accordance with section 8 of the Education Act 1996; and

(b) as respects Scotland, means a person who is not over school age, construed in accordance with section 31 of the Education (Scotland) Act 1980;

— 'employment business' means a business (whether or not carried on with a view to profit and whether or not carried on in conjunction with any other business) which supplies persons (other than seafarers) who are employed in it to work for and under the control of other persons in any capacity;

— 'fixed-term contract of employment' means a contract of employment for a specific term which is fixed in advance or which can be ascertained in advance by reference to some relevant circumstance;

— 'given birth' means delivered a living child or, after twenty-four weeks of pregnancy, a stillborn child;

— 'new or expectant mother' means an employee who is pregnant; who has given birth within the previous six months; or who is breastfeeding;

— 'the preventive and protective measures' means the measures which have been identified by the employer or by the self-employed person in consequence of the assessment as the measures he needs to take to comply with the requirements and prohibitions imposed upon him by or under the relevant statutory provisions and by Part II of the Fire Precautions (Workplace) Regulations 1997;

— 'young person' means any person who has not attained the age of eighteen.

(3) Any reference in these Regulations to—

(a) a numbered regulation or Schedule is a reference to the regulation or Schedule in these Regulations so numbered; or

(b) a numbered paragraph is a reference to the paragraph so numbered in the regulation in which the reference appears.

Regulation 2 *(See Guidance Notes, pp 291–292)*

Disapplication of these Regulations

(1) These Regulations shall not apply to or in relation to the master or crew of a sea-going ship or to the employer of such persons in respect of the normal ship-board activities of a ship's crew under the direction of the master.

(2) Regulations 3(4), (5), 10 (2) and 19 shall not apply to occasional work or short-term work involving—

(a) domestic service in a private household; or

(b) work regulated as not being harmful, damaging or dangerous to young people in a family undertaking.

Regulation 3 *(See Approved Code of Practice, pp 270–278)*

Risk assessment

(1) Every employer shall make a suitable and sufficient assessment of—

(a) the risks to the health and safety of his employees to which they are exposed whilst they are at work; and

(b) the risks to the health and safety of persons not in his employment arising out of or in connection with the conduct by him of his undertaking,

for the purpose of identifying the measures he needs to take to comply with the requirements and prohibitions imposed upon him by or under the relevant statutory provisions and by Part II of the Fire Precautions (Workplace) Regulations 1997.

(2) Every self-employed person shall make a suitable and sufficient assessment of—

(a) the risks to his own health and safety to which he is exposed whilst he is at work; and

(b) the risks to the health and safety of persons not in his employment arising out of or in connection with the conduct by him of his undertaking,

for the purpose of identifying the measures he needs to take to comply with the requirements and prohibitions imposed upon him by or under the relevant statutory provisions.

(3) Any assessment such as is referred to in paragraph (1) or (2) shall be reviewed by the employer or self-employed person who made it if—

(a) there is reason to suspect that it is no longer valid; or

(b) there has been a significant change in the matters to which it relates; and where as a result of any such review changes to an assessment are required, the employer or self-employed person concerned shall make them.

(4) An employer shall not employ a young person unless he has, in relation to risks to the health and safety of young persons, made or reviewed an assessment in accordance with paragraphs (1) and (5).

(5) In making or reviewing the assessment, an employer who employs or is to employ a young person shall take particular account of—

(a) the inexperience, lack of awareness of risks and immaturity of young persons;

(b) the fitting-out and layout of the workplace and the workstation;

(c) the nature, degree and duration of exposure to physical, biological and chemical agents;

(d) the form, range, and use of work equipment and the way in which it is handled;

(e) the organisation of processes and activities;

(f) the extent of the health and safety training provided or to be provided to young persons; and

(g) risks from agents, processes and work listed in the Annex to Council Directive 94/33/EC on the protection of young people at work.

(6) Where the employer employs five or more employees, he shall record—

(a) the significant findings of the assessment; and

(b) any group of his employees identified by it as being especially at risk.

Regulation 4 *(See Approved Code of Practice, p 278; Guidance Notes, pp 292–293)*

Principles of prevention to be applied

Where an employer implements any preventive and protective measures he shall do so on the basis of the principles specified in Schedule 1 to these Regulations.

Regulation 5 *(See Approved Code of Practice, pp 278–281; Guidance Notes, p 294)*

Health and safety arrangements

(1) Every employer shall make and give effect to such arrangements as are appropriate, having regard to the nature of his activities and the size of his undertaking, for the effective planning, organisation, control, monitoring and review of the preventive and protective measures.

(2) Where the employer employs five or more employees, he shall record the arrangements referred to in paragraph (1).

Regulation 6 *(See Approved Code of Practice, pp 281–282; Guidance Notes, pp 294–295)*

Health surveillance

Every employer shall ensure that his employees are provided with such health surveillance as is appropriate having regard to the risks to their health and safety which are identified by the assessment.

Regulation 7 *(See Approved Code of Practice, p 282;*
Guidance Notes, pp 295–296)

Health and safety assistance

(1) Every employer shall, subject to paragraphs (6) and (7), appoint one or more competent persons to assist him in undertaking the measures he needs to take to comply with the requirements and prohibitions imposed upon him by or under the relevant statutory provisions and by Part II of the Fire Precautions (Workplace) Regulations 1997.

(2) Where an employer appoints persons in accordance with paragraph (1), he shall make arrangements for ensuring adequate co-operation between them.

(3) The employer shall ensure that the number of persons appointed under paragraph (1), the time available for them to fulfil their functions and the means at their disposal are adequate having regard to the size of his undertaking, the risks to which his employees are exposed and the distribution of those risks throughout the undertaking.

(4) The employer shall ensure that—

(a) any person appointed by him in accordance with paragraph (1) who is not in his employment—

(i) is informed of the factors known by him to affect, or suspected by him of affecting, the health and safety of any other person who may be affected by the conduct of his undertaking, and

(ii) has access to the information referred to in regulation 10; and

(b) any person appointed by him in accordance with paragraph (1) is given such information about any person working in his undertaking who is—

(i) employed by him under a fixed-term contract of employment, or

(ii) employed in an employment business,

as is necessary to enable that person properly to carry out the function specified in that paragraph.

(5) A person shall be regarded as competent for the purposes of paragraphs (1) and (8) where he has sufficient training and

experience or knowledge and other qualities to enable him properly to assist in undertaking the measures referred to in paragraph (1).

(6) Paragraph (1) shall not apply to a self-employed employer who is not in partnership with any other person where he has sufficient training and experience or knowledge and other qualities properly to undertake the measures referred to in that paragraph himself.

(7) Paragraph (1) shall not apply to individuals who are employers and who are together carrying on business in partnership where at least one of the individuals concerned has sufficient training and experience or knowledge and other qualities—

(a) properly to undertake the measures he needs to take to comply with the requirements and prohibitions imposed upon him by or under the relevant statutory provisions; and

(b) properly to assist his fellow partners in undertaking the measures they need to take to comply with the requirements and prohibitions imposed upon them by or under the relevant statutory provisions.

(8) Where there is a competent person in the employer's employment, that person shall be appointed for the purposes of paragraph (1) in preference to a competent person not in his employment.

Regulation 8

Procedures for serious and imminent danger and for danger areas

(1) Every employer shall—

(a) establish and where necessary give effect to appropriate procedures to be followed in the event of serious and imminent danger to persons at work in his undertaking;

(b) nominate a sufficient number of competent persons to implement those procedures in so far as they relate to the evacuation from premises of persons at work in his undertaking; and

(c) ensure that none of his employees has access to any area occupied by him to which it is necessary to restrict access on grounds of health and safety unless the employee concerned has received adequate health and safety instruction.

(2) Without prejudice to the generality of paragraph (1)(a), the procedures referred to in that sub-paragraph shall—

(a) so far as is practicable, require any persons at work who are exposed to serious and imminent danger to be informed of the nature of the hazard and of the steps taken or to be taken to protect them from it;

(b) enable the persons concerned (if necessary by taking appropriate steps in the absence of guidance or instruction and in the light of their knowledge and the technical means at their disposal) to stop work and immediately proceed to a place of safety in the event of their being exposed to serious, imminent and unavoidable danger; and

(c) save in exceptional cases for reasons duly substantiated (which cases and reasons shall be specified in those procedures), require the persons concerned to be prevented from resuming work in any situation where there is still a serious and imminent danger.

(3) A person shall be regarded as competent for the purposes of paragraph (1)(b) where he has sufficient training and experience or knowledge and other qualities to enable him properly to implement the evacuation procedures referred to in that sub-paragraph.

Regulation 9 *(See Approved Code of Practice, pp 283–284; Guidance Notes, pp 296–298)*

Contacts with external services

Every employer shall ensure that any necessary contacts with external services are arranged, particularly as regards first-aid, emergency medical care and rescue work.

Regulation 10 *(See Guidance Notes, pp 298–299)*

Information for employees

(1) Every employer shall provide his employees with comprehensible and relevant information on—

(a) the risks to their health and safety identified by the assessment;

(b) the preventive and protective measures;

(c) the procedures referred to in regulation 8(1)(a) and the measures referred to in regulation 4(2)(a) of the Fire Precautions (Workplace) Regulations 1997;

(d) the identity of those persons nominated by him in accordance with regulation 8(1)(b) and regulation 4(2)(b) of the Fire Precautions (Workplace) Regulations 1997; and

(e) the risks notified to him in accordance with regulation 11(1)(c).

(2) Every employer shall, before employing a child, provide a parent of the child with comprehensible and relevant information on—

(a) the risks to his health and safety identified by the assessment;

(b) the preventive and protective measures; and

(c) the risks notified to him in accordance with regulation 11(1)(c).

(3) The reference in paragraph (2) to a parent of the child includes—

(a) in England and Wales, a person who has parental responsibility, within the meaning of section 3 of the Children Act 1989, for him; and

(b) in Scotland, a person who has parental rights, within the meaning of section 8 of the Law Reform (Parent and Child) (Scotland) Act 1986 for him.

Regulation 11 *(See Approved Code of Practice, pp 284–285; Guidance Notes, pp 299–301)*

Co-operation and co-ordination

(1) Where two or more employers share a workplace (whether on a temporary or a permanent basis) each such employer shall—

(a) co-operate with the other employers concerned so far as is necessary to enable them to comply with the requirements and prohibitions imposed upon them by or under the relevant statutory provisions and by Part II of the Fire Precautions (Workplace) Regulations 1997;

(b) (taking into account the nature of his activities) take all reasonable steps to co-ordinate the measures he takes to comply with the requirements and prohibitions imposed upon him by or under the relevant statutory provisions and by Part II of the Fire Precautions (Workplace) Regulations 1997 with the measures the other employers concerned are taking to comply with the requirements and prohibitions imposed upon them by that legislation; and

(c) take all reasonable steps to inform the other employers concerned of the risks to their employees' health and safety arising out of or in connection with the conduct by him of his undertaking.

(2) Paragraph (1) (except in so far as it refers to Part II of the Fire Precautions (Workplace) Regulations 1997) shall apply to employers sharing a workplace with self-employed persons and to self-employed persons sharing a workplace with other self-employed persons as it applies to employers sharing a workplace with other employers; and the references in that paragraph to employers and the reference in the said paragraph to their employees shall be construed accordingly.

Regulation 12 *(See Approved Code of Practice, pp 285–286; Guidance Notes, p 301)*

Persons working in host employers' or self-employed persons' undertakings

(1) Every employer and every self-employed person shall ensure that the employer of any employees from an outside undertaking who are working in his undertaking is provided with comprehensible information on—

(a) the risks to those employees' health and safety arising out of or in connection with the conduct by that first-mentioned employer or by that self-employed person of his undertaking; and

(b) the measures taken by that first-mentioned employer or by that self-employed person in compliance with the requirements and prohibitions imposed upon him by or under the relevant statutory provisions and by Part II of the Fire Precautions (Workplace) Regulations 1997 in so far as the said requirements and prohibitions relate to those employees.

(2) Paragraph (1) (except in so far as it refers to Part II of the Fire Precautions (Workplace) Regulations 1997) shall apply to a

self-employed person who is working in the undertaking of an employer or a self-employed person as it applies to employees from an outside undertaking who are working therein; and the reference in that paragraph to the employer of any employees from an outside undertaking who are working in the undertaking of an employer or a self-employed person and the references in the said paragraph to employees from an outside undertaking who are working in the undertaking of an employer or a self-employed person shall be construed accordingly.

(3) Every employer shall ensure that any person working in his undertaking who is not his employee and every self-employed person (not being an employer) shall ensure that any person working in his undertaking is provided with appropriate instructions and comprehensible information regarding any risks to that person's health and safety which arise out of the conduct by that employer or self-employed person of his undertaking.

(4) Every employer shall—

(a) ensure that the employer of any employees from an outside undertaking who are working in his undertaking is provided with sufficient information to enable that second-mentioned employer to identify any person nominated by that first mentioned employer in accordance with regulation 8(1)(b) to implement evacuation procedures as far as those employees are concerned; and

(b) take all reasonable steps to ensure that any employees from an outside undertaking who are working in his undertaking receive sufficient information to enable them to identify any person nominated by him in accordance with regulation 8(1)(b) to implement evacuation procedures as far as they are concerned.

(5) Paragraph (4) shall apply to a self-employed person who is working in an employer's undertaking as it applies to employees from an outside undertaking who are working therein; and the reference in that paragraph to the employer of any employees from an outside undertaking who are working in an employer's undertaking and the references in the said paragraph to employees from an outside undertaking who are working in an employer's undertaking shall be construed accordingly.

Regulation 13 *(See Approved Code of Practice, p 286; Guidance Notes, pp 301–302)*

Capabilities and training

(1) Every employer shall, in entrusting tasks to his employees, take into account their capabilities as regards health and safety.

(2) Every employer shall ensure that his employees are provided with adequate health and safety training—

(a) on their being recruited into the employer's undertaking; and

(b) on their being exposed to new or increased risks because of—

(i) their being transferred or given a change of responsibilities within the employer's undertaking,

(ii) the introduction of new work equipment into or a change respecting work equipment already in use within the employer's undertaking,

(iii) the introduction of new technology into the employer's undertaking, or

(iv) the introduction of a new system of work into or a change respecting a system of work already in use within the employer's undertaking.

(3) The training referred to in paragraph (2) shall—

(a) be repeated periodically where appropriate;

(b) be adapted to take account of any new or changed risks to the health and safety of the employees concerned; and

(c) take place during working hours.

Regulation 14 *(See Approved Code of Practice, pp 286–287; Guidance Notes, p 302)*

Employees' duties

(1) Every employee shall use any machinery, equipment, dangerous substance, transport equipment, means of production or safety device provided to him by his employer in accordance both with any training in the use of the equipment concerned which has been received by him and the instructions respecting that use which have been provided to him by the said employer in compliance

with the requirements and prohibitions imposed upon that employer by or under the relevant statutory provisions.

(2) Every employee shall inform his employer or any other employee of that employer with specific responsibility for the health and safety of his fellow employees—

(a) of any work situation which a person with the first-mentioned employee's training and instruction would reasonably consider represented a serious and immediate danger to health and safety; and

(b) of any matter which a person with the first-mentioned employee's training and instruction would reasonably consider represented a shortcoming in the employer's protection arrangements for health and safety,

in so far as that situation or matter either affects the health and safety of that first mentioned employee or arises out of or in connection with his own activities at work, and has not previously been reported to his employer or to any other employee of that employer in accordance with this paragraph.

Regulation 15 *(See Guidance Notes, pp 303–304)*

Temporary workers

(1) Every employer shall provide any person whom he has employed under a fixed-term contract of employment with comprehensible information on—

(a) any special occupational qualifications or skills required to be held by that employee if he is to carry out his work safely; and

(b) any health surveillance required to be provided to that employee by or under any of the relevant statutory provisions,

and shall provide the said information before the employee concerned commences his duties.

(2) Every employer and every self-employed person shall provide any person employed in an employment business who is to carry out work in his undertaking with comprehensible information on—

(a) any special occupational qualifications or skills required to be held by that employee if he is to carry out his work safely; and

(b) health surveillance required to be provided to that employee by or under any of the relevant statutory provisions.

(3) Every employer and every self-employed person shall ensure that every person carrying on an employment business whose employees are to carry out work in his undertaking is provided with comprehensible information on—

(a) any special occupational qualifications or skills required to be held by those employees if they are to carry out their work safely; and

(b) the specific features of the jobs to be filled by those employees (in so far as those features are likely to affect their health and safety);

and the person carrying on the employment business concerned shall ensure that the information so provided is given to the said employees.

Regulation 16

Risk assessment in respect of new or expectant mothers

(1) Where—

(a) the persons working in an undertaking include women of child-bearing age; and

(b) the work is of a kind which could involve risk, by reason of her condition, to the health and safety of a new or expectant mother, or to that of her baby, from any processes or working conditions, or physical, biological or chemical agents, including those specified in Annexes I and II of Council Directive 92/85/EEC on the introduction of measures to encourage improvements in the safety and health at work of pregnant workers and workers who have recently given birth or are breastfeeding,

the assessment required by regulation 3(1) shall also include an assessment of such risk.

(2) Where, in the case of an individual employee, the taking of any other action the employer is required to take under the relevant statutory provisions would not avoid the risk referred to in paragraph (1) the employer shall, if it is reasonable to do so, and would avoid such risks, alter her working conditions or hours of work.

(3) If it is not reasonable to alter the working conditions or hours of work, or if it would not avoid such risk, the employer shall, subject to section 67 of the 1996 Act suspend the employee from work for so long as is necessary to avoid such risk.

(4) In paragraphs (1) to (3) references to risk, in relation to risk from any infectious or contagious disease, are references to a level of risk at work which is in addition to the level to which a new or expectant mother may be expected to be exposed outside the workplace.

Regulation 17

Certificate from registered medical practitioner in respect of new or expectant mothers

Where—

(a) a new or expectant mother works at night; and

(b) a certificate from a registered medical practitioner or a registered midwife shows that it is necessary for her health or safety that she should not be at work for any period of such work identified in the certificate,

the employer shall, subject to section 67 of the 1996 Act, suspend her from work for so long as is necessary for her health or safety.

Regulation 18 *(See Approved Code of Practice, pp 287–288)*

Notification by new or expectant mothers

(1) Nothing in paragraph (2) or (3) of regulation 16 shall require the employer to take any action in relation to an employee until she has notified the employer in writing that she is pregnant, has given birth within the previous six months, or is breastfeeding.

(2) Nothing in paragraph (2) or (3) of regulation 16 or in regulation 17 shall require the employer to maintain action taken in relation to an employee—

(a) in a case—

(i) to which regulation 16(2) or (3) relates; and

(ii) where the employee has notified her employer that she is pregnant, where she has failed, within a reasonable time of being requested to do so in writing by her employer, to produce for the employer's inspection a certificate from a registered medical practitioner or a registered midwife showing that she is pregnant;

(b) once the employer knows that she is no longer a new or expectant mother; or

(c) if the employer cannot establish whether she remains a new or expectant mother.

Regulation 19 *(See Approved Code of Practice, pp 288–289)*

Protection of young persons

(1) Every employer shall ensure that young persons employed by him are protected at work from any risks to their health or safety which are a consequence of their lack of experience, or absence of awareness of existing or potential risks or the fact that young persons have not yet fully matured.

(2) Subject to paragraph (3), no employer shall employ a young person for work—

(a) which is beyond his physical or psychological capacity;

(b) involving harmful exposure to agents which are toxic or carcinogenic, cause heritable genetic damage or harm to the unborn child or which in any other way chronically affect human health;

(c) involving harmful exposure to radiation;

(d) involving the risk of accidents which it may reasonably be assumed cannot be recognised or avoided by young persons owing to their insufficient attention to safety or lack of experience or training; or

(e) in which there is a risk to health from—

(i) extreme cold or heat;

(ii) noise; or

(iii) vibration,

and in determining whether work will involve harm or risks for the purposes of this paragraph, regard shall be had to the results of the assessment.

(3) Nothing in paragraph (2) shall prevent the employment of a young person who is no longer a child for work—

(a) where it is necessary for his training;

(b) where the young person will be supervised by a competent person; and

(c) where any risk will be reduced to the lowest level that is reasonably practicable.

(4) The provisions contained in this regulation are without prejudice to—

(a) the provisions contained elsewhere in these Regulations; and

(b) any prohibition or restriction, arising otherwise than by this regulation, on the employment of any person.

Regulation 20

Exemption certificates

(1) The Secretary of State for Defence may, in the interests of national security, by a certificate in writing exempt—

(a) any of the home forces, any visiting force or any headquarters from those requirements of these Regulations which impose obligations other than those in regulations 16–18 on employers; or

(b) any member of the home forces, any member of a visiting force or any member of a headquarters from the requirements imposed by regulation 14;

and any exemption such as is specified in sub-paragraph (a) or (b) of this paragraph may be granted subject to conditions and to a limit of time and may be revoked by the said Secretary of State by a further certificate in writing at any time.

(2) In this regulation—

(a) 'the home forces' has the same meaning as in section 12(1) of the Visiting Forces Act 1952;

(b) 'headquarters' means a headquarters for the time being specified in Schedule 2 to the Visiting Forces and International Headquarters (Application of Law) Order 1999;

(c) 'member of a headquarters' has the same meaning as in paragraph 1(1) of the Schedule to the International Headquarters and Defence Organisations Act 1964; and

(d) 'visiting force' has the same meaning as it does for the purposes of any provision of Part I of the Visiting Forces Act 1952.

Regulation 21 *(See Approved Code of Practice, p 289; Guidance Notes, p 305)*

Provisions as to liability

Nothing in the relevant statutory provisions shall operate so as to afford an employer a defence in any criminal proceedings for a contravention of those provisions by reason of any act or default of—

(a) an employee of his, or

(b) a person appointed by him under regulation 7.

Regulation 22

Exclusion of civil liability

(1) Breach of a duty imposed by these Regulations shall not confer a right of action in any civil proceedings.

(2) Paragraph (1) shall not apply to any duty imposed by these Regulations on an employer—

(a) to the extent that it relates to risk referred to in regulation 16(1) to an employee; or

(b) which is contained in regulation 19.

Regulation 23 *(See Guidance Notes, p 305)*

Extension outside Great Britain

(1) These Regulations shall, subject to regulation 2, apply to and in relation to the premises and activities outside Great Britain to which sections 1 to 59 and 80 to 82 of the Health and Safety at Work etc Act 1974 apply by virtue of the Health and Safety at Work etc Act 1974 (Application Outside Great Britain) Order 1995 as they apply within Great Britain.

(2) For the purposes of Part I of the 1974 Act, the meaning of 'at work' shall be extended so that an employee or a self-employed person shall be treated as being at work throughout the time that he is present at the premises to and in relation to which these

Regulations apply by virtue of paragraph (1); and, in that connection, these Regulations shall have effect subject to the extension effected by this paragraph.

Regulation 24 *(See Guidance Notes, p 305)*

Amendment of the Health and Safety (First-Aid) Regulations 1981

Regulation 6 of the Health and Safety (First-Aid) Regulations 1981 is hereby revoked.

Regulation 25 *(See Guidance Notes, pp 305–306)*

Amendment of the Offshore Installations and Pipeline Works (First-Aid) Regulations 1989

(1) The Offshore Installations and Pipeline Works (First-Aid) Regulations 1989 shall be amended in accordance with the following provisions of this regulation.

(2) In regulation 7(1) for the words 'from all or any of the requirements of these Regulations', there shall be substituted 'the words from regulation 5(1)(b) and (c) and (2)(a) of these Regulations'.

(3) After regulation 7(2) the following paragraph shall be added—

> '(3) An exemption granted under paragraph (1) above from the requirements in regulation 5(2)(a) of these Regulations shall be subject to the condition that a person provided under regulation 5(1)(a) of these Regulations shall have undergone adequate training.'

Regulation 26 *(See Guidance Notes, p 306)*

Amendment of the Mines Miscellaneous Health and Safety Provisions Regulations 1995

(1) The Mines Miscellaneous Health and Safety Provisions Regulations 1995 shall be amended in accordance with the following provisions of this regulation.

(2) Paragraph (2)(b) of regulation 4 shall be deleted.

(3) After paragraph (4) of regulation 4 there shall be added the following paragraph—

> '(5) In relation to fire, the health and safety document prepared pursuant to paragraph (1) shall—

(a) include a fire protection plan detailing the likely sources of fire, and the precautions to be taken to protect against, to detect and combat the outbreak and spread of fire; and

(b) in respect of every part of the mine other than any building on the surface of that mine—

(i) include the designation of persons to implement the plan, ensuring that the number of such persons, their training and the equipment available to them is adequate, taking into account the size of, and the specific hazards involved in the mine concerned; and

(ii) include the arrangements for any necessary contacts with external emergency services, particularly as regards rescue work and fire-fighting; and

(iii) be adapted to the nature of the activities carried on at that mine, the size of the mine and take account of the persons other than employees who may be present.'

Regulation 27 *(See Guidance Notes, p 306)*

Amendment of the Construction (Health, Safety and Welfare) Regulations 1996

(1) The Construction (Health, Safety and Welfare) Regulations 1996 shall be amended in accordance with the following provisions of this regulation.

(2) Paragraph (2) of regulation 20 shall be deleted and the following substituted—

'(2) Without prejudice to the generality of paragraph (1), arrangements prepared pursuant to that paragraph shall—

(a) have regard to those matters set out in paragraph (4) of regulation 19;

(b) designate an adequate number of persons who will implement the arrangements; and

(c) include any necessary contacts with external emergency services, particularly as regards rescue work and fire-fighting.'

Regulation 28

Regulations to have effect as health and safety regulations

Subject to regulation 9 of the Fire Precautions (Workplace) Regulations 1997, these Regulations shall, to the extent that they would not otherwise do so, have effect as if they were health and safety regulations within the meaning of Part I of the Health and Safety at Work etc Act 1974.

Regulation 29

Revocations and consequential amendments

(1) The Management of Health and Safety at Work Regulations 1992, the Management of Health and Safety at Work (Amendment) Regulations 1994, the Health and Safety (Young Persons) Regulations 1997 and Part III of the Fire Precautions (Workplace) Regulations 1997 are hereby revoked.

(2) The instruments specified in column 1 of Schedule 2 shall be amended in accordance with the corresponding provisions in column 3 of that Schedule.

Regulation 30

Transitional provision

The substitution of provisions in these Regulations for provisions of the Management of Health and Safety at Work Regulations 1992 shall not affect the continuity of the law; and accordingly anything done under or for the purposes of such provision of the 1992 Regulations shall have effect as if done under or for the purposes of any corresponding provision of these Regulations.

SCHEDULE
GENERAL PRINCIPLES OF PREVENTION

Regulation 4

(This Schedule specifies the general principles of prevention set out in Article 6(2) of Council Directive 89/391/EEC)

(a) avoiding risks;

(b) evaluating the risks which cannot be avoided;

(c) combating the risks at source;

(d) adapting the work to the individual, especially as regards the design of workplaces, the choice of work equipment and the choice of working and production methods, with a view, in particular, to alleviating monotonous work and work at a predetermined work-rate and to reducing their effect on health;

(e) adapting to technical progress;

(f) replacing the dangerous by the non-dangerous or the less dangerous;

(g) developing a coherent overall prevention policy which covers technology, organisation of work, working conditions, social relationships and the influence of factors relating to the working environment;

(h) giving collective protective measures priority over individual protective measures; and

(i) giving appropriate instructions to employees.

SCHEDULE 2
CONSEQUENTIAL AMENDMENTS

Regulation 29

Column 1 *Description of Instrument*	*Column 2* *References*	*Column 3* *Extent of Modification*
The Safety Representatives and Safety Committees Regulations 1977	SI 1977/500; amended by SI 1992/2051; SI 1996/1513; SI 1997/1840; SI 1999/860 and by section 1(1) and (2) of the Employment Rights (Dispute Resolution) Act 1998.	In regulation 4A(1)(b) for 'regulations 6(1) and 7(1)(b) of the Management of Health and Safety at Work Regulations 1992', there shall be substituted 'regulations 7(1) and 8(1)(b) of the Management of Health and Safety at Work Regulations 1999'.

The Offshore Installations (Safety Representatives and Safety Committees) Regulations 1989	SI 1989/971; amended by SI 1992/2885; SI 1993/1823; SI 1995/738; SI 1995/743; and SI 1995/3163.	In regulation 23(4) for 'regulation 6(1) of the Management of Health and Safety at Work Regulations 1992', there shall be substituted 'regulation 7(1) of the Management of Health and Safety at Work Regulations 1999'.
The Railways (Safety Case) Regulations 1994	SI 1994/237; amended by SI 1996/1592.	In paragraph 6 of Schedule 1 for 'regulation 3 of the Management of Health and Safety at Work Regulations 1992 and particulars of the arrangements he has made pursuant to regulation 4(1) thereof', there shall be substituted 'regulation 3 of the Management of Health and Safety at Work Regulations 1999 and particulars of the arrangements he has made in accordance with regulation 5(1) thereof'.
The Suspension from Work (on Maternity Grounds) Order 1994*	SI 1994/2930.	In article 1(2)(b) for '"the 1992 Regulations" means the Management of Health and Safety at Work Regulations 1992', there shall be substituted, '"the 1999 Regulations" means the Management of Health and Safety at Work Regulations 1999'; and In article 2(b) for 'regulation 13B of the 1992 Regulations', there shall be substituted 'regulation 17 of the 1999 Regulations'.

* Note: The Regulations marked with an asterisk are referred to in the Preamble to these Regulations.

The Construction (Design and Management) Regulations 1994	SI 1994/3140; amended by SI 1996/1592.	In regulation 16(1)(a) for 'regulation 9 of the Management of Health and Safety at Work Regulations 1992', there shall be substituted 'regulation 11 of the Management of Health and Safety at Work Regulations 1999'; In regulation 17(2)(a) for 'regulation 8 of the Management of Health and Safety at Work Regulations 1992', there shall be substituted 'regulation 10 of the Management of Health and Safety at Work Regulations 1999'; In regulation 17(2)(b) for 'regulation 11(2)(b) of the Management of Health and Safety at Work Regulations 1992', there shall be substituted 'regulation 13(2)(b) of the Management of Health and Safety at Work Regulations 1999'; and In regulation 19(1)(b) for 'the Management of Health and Safety at Work Regulations 1992', there shall be substituted 'the Management of Health and Safety at Work Regulations 1999'.
The Escape and Rescue from Mines Regulations 1995	SI 1995/2870.	In regulation 2(1) for '"the 1992 Regulations" means the Management of Health and Safety at Work Regulations 1992', there shall be substituted '"the 1999

		Regulations" means the Management of Health and Safety at Work Regulations 1999'; and In regulation 4(2) for 'regulation 3 of the 1992 Regulations', there shall be substituted 'regulation 3 of the 1999 Regulations'.
The Mines Miscellaneous Health and Safety Provisions Regulations 1995	SI 1995/2005.	In regulation 2(1) for '"the 1992 Regulations" means the Management of Health and Safety at Work Regulations 1992', there shall be substituted '"the 1999 Regulations" means the Management of Health and Safety at Work Regulations 1999'; and In regulation 4(1)(a) for 'regulation 3 of the 1992 Regulations', there shall be substituted 'regulation 3 of the 1999 Regulations'.
The Quarries Miscellaneous Health and Safety Provisions Regulations 1995	SI 1995/2036.	In regulation 2(1) for '"the 1992 Regulations" means the Management of Health and Safety at Work Regulations 1992', there shall be substituted '"the 1999 Regulations" means the Management of Health and Safety at Work Regulations 1999'; and In regulation 4(1)(a) for 'regulation 3 of the 1992 Regulations', there shall be substituted 'regulation 3 of the 1999 Regulations'.

The Borehole Sites and Operations Regulations 1995	SI 1995/2038.	In regulation 7(5) for '"the Management Regulations" means the Management of Health and Safety at Work Regulations 1992', there shall be substituted '"the Management Regulations" means the Management of Health and Safety at Work Regulations 1999'.
The Gas Safety (Management) Regulations 1996	SI 1996/551.	In paragraph 5 of Schedule 1 for 'regulation 3 of the Management of Health and Safety at Work Regulations 1992, and particulars of the arrangements he has made in accordance with regulation 4(1) thereof', there shall be substituted 'regulation 3 of the Management of Health and Safety at Work Regulations 1999, and particulars of the arrangements he has made in accordance with regulation 5(1) thereof'.
The Health and Safety (Safety Signs and Signals) Regulations 1996	SI 1996/341.	In regulation 4(1) for 'paragraph (1) of regulation 3 of the Management of Health and Safety at Work Regulations 1992', there shall be substituted 'paragraph (1) of regulation 3 of the Management of Health and Safety at Work Regulations 1999'.

The Health and Safety (Consultation with Employees) Regulations 1996	SI 1996/1513.	In regulation 3(b) for 'regulations 6(1) and 7(1)(b) of the Management of Health and Safety at Work Regulations 1992', there shall be substituted 'regulations 7(1) and 8(1)(b) of the Management of Health and Safety at Work Regulations 1999'.
The Fire Precautions (Workplace) Regulations 1997	SI 1997/1840; amended by SI 1999/1877.	In regulation 2(1) for '"the 1992 Management Regulations" means the Management of Health and Safety at Work Regulations 1992', there shall be substituted '"the 1999 Management Regulations" means the Management of Health and Safety at Work Regulations 1999'; In regulation 2(1) in the definitions of 'employee' and 'employer' for '1992' substitute '1999'; and In regulation 9(2)(b) for the words 'regulations 1 to 4, 6 to 10 and 11(2) and (3) of the 1992 Management Regulations (as amended by Part III of these Regulations)', there shall be substituted 'regulations 1 to 5, 7 to 12 and 13(2) and (3) of the 1999 Management Regulations'.
The Control of Lead at Work Regulations 1998	SI 1998/543.	In regulation 5 for 'regulation 3 of the Management of Health and Safety at Work Regulations 1992', there shall

		be substituted 'regulation 3 of the Management of Health and Safety at Work Regulations 1999'.
The Working Time Regulations 1998*	SI 1998/1833.	In regulation 6(8)(b) for 'regulation 3 of the Management of Health and Safety at Work Regulations 1992', there shall be substituted 'regulation 3 of the Management of Health and Safety at Work Regulations 1999'.
The Quarries Regulations 1999	SI 1999/2024.	In regulation 2(1) for '"the 1992 Regulations" means the Management of Health and Safety at Work Regulations 1992', there shall be substituted '"the 1999 Regulations" means the Management of Health and Safety at Work Regulations 1999'. In regulation 7(1)(a) for 'paragraphs (1) to (3c) of regulation 3 of the 1992 Regulations', there shall be substituted 'regulation 3 of the Management of Health and Safety at Work Regulations 1999'. In regulation 43 for 'regulation 5 of the 1992 Regulations' there shall be substituted 'regulation 6 of the 1999 Regulations'.

APPROVED CODE OF PRACTICE

Introduction

1–6 *(See Guidance Notes, pp 289–291)*

Regulation 2

Disapplication of these Regulations

7–8 *(See Guidance Notes, pp 291–292)*

Regulation 3

Risk assessment

General principles and purpose of risk assessment

9 This regulation requires all employers and self-employed people to assess the risks to workers and any others who may be affected by their work or business. This will enable them to identify the measures they need to take to comply with health and safety law. All employers should carry out a systematic general examination of the effect of their undertaking, their work activities and the condition of the premises. Those who employ five or more employees should record the significant findings of that risk assessment.

10 A risk assessment is carried out to identify the risks to health and safety to any person arising out of, or in connection with, work or the conduct of their undertaking. It should identify how the risks arise and how they impact on those affected. This information is needed to make decisions on how to manage those risks so that the decisions are made in an informed, rational and structured manner, and the action taken is proportionate.

11 A risk assessment should usually involve identifying the hazards present in any working environment or arising out of commercial activities and work activities, and evaluating the extent of the risks involved, taking into account existing precautions and their effectiveness. In this Approved Code of Practice:

(a) a hazard is something with the potential to cause harm (this can include articles, substances, plant or machines, methods of work, the working environment and other aspects of work organisation);

(b) a risk is the likelihood of potential harm from that hazard being realised. The extent of the risk will depend on:

(i) the likelihood of that harm occurring;

(ii) the potential severity of that harm, ie of any resultant injury or adverse health effect; and

(iii) the population which might be affected by the hazard, ie the number of people who might be exposed.

12 The purpose of the risk assessment is to help the employer or self-employed person to determine what measures should be taken to comply with the employer's or self-employed person's duties under the 'relevant statutory provisions' and Part II of the Fire Regulations. This covers the general duties in the HSW Act and the requirements of Part II of the Fire Regulations and the more specific duties in the various acts and regulations (including these Regulations) associated with the HSW Act. Once the measures have been determined in this way, the duty to put them into effect will be defined in the statutory provisions. For example a risk assessment on machinery would be undertaken under these Regulations, but the Provision and Use of Work Equipment Regulations (PUWER 1998)[1] determine what precautions must be carried out. A risk assessment carried out by a self-employed person in circumstances where he or she does not employ others does not have to take into account duties arising under Part II of the Fire Regulations.

Suitable and sufficient

13 A suitable and sufficient risk assessment should be made. 'Suitable and sufficient' is not defined in the Regulations. In practice it means the risk assessment should do the following:

(a) The risk assessment should identify the risks arising from or in connection with work. The level of detail in a risk assessment should be proportionate to the risk. Once the risks are assessed and taken into account, insignificant risks can usually be ignored, as can risks arising from routine activities associated with life in general, unless the work activity compounds or significantly alters those risks. The level of risk arising from the work activity should determine the degree of sophistication of the risk assessment.

1 SI 1998/2306.

(i) For small businesses presenting few or simple hazards a suitable and sufficient risk assessment can be a very straightforward process based on informed judgement and reference to appropriate guidance. Where the hazards and risks are obvious, they can be addressed directly. No complicated processes or skills will be required.

(ii) In many intermediate cases the risk assessment will need to be more sophisticated. There may be some areas of the assessment for which specialist advice is required; for example risks which require specialist knowledge such as a particularly complex process or technique, or risks which need specialist analytical techniques such as being able to measure air quality and to assess its impact. Whenever specialist advisers are used, employers should ensure that the advisers have sufficient understanding of the particular work activity they are advising on, this will often require effective involvement of everyone concerned—employer, employees and specialist.

(iii) Large and hazardous sites will require the most developed and sophisticated risk assessments, particularly where there are complex or novel processes. In the case of certain manufacturing sites who use or store bulk hazardous substances, large scale mineral extraction or nuclear plant, the risk assessment will be a significant part of the safety case or report which is legally required and may incorporate such techniques as quantified risk assessment. A number of other statutory requirements exist (eg the Control of Major Accident Hazards (COMAH), and Nuclear Installations licensing arrangements) which include more specific and detailed arrangements for risk assessment.

(iv) Risk assessments must also consider all those who might be affected by the undertaking, whether they are workers or others such as members of the public. For example, the risk assessment produced by a railway company will, inter alia, have to consider the hazards and risks which arise from the operation and maintenance of rail vehicles and train services and which might adversely affect workers (their own employees and others), passengers and any member of the public who could foreseeably be affected (eg level crossing users).

(b) Employers and the self-employed are expected to take reasonable steps to help themselves identify risks, eg by looking at appropriate sources of information, such as relevant legislation, appropriate guidance, supplier manuals and manufacturers' instructions and reading trade press, or seeking advice from competent sources. They should also look at and use relevant examples of good practice from within their industry. The risk assessment should include only what an employer or self-employed person could reasonably be expected to know; they would not be expected to anticipate risks that were not foreseeable;

(c) The risk assessment should be appropriate to the nature of the work and should identify the period of time for which it is likely to remain valid. This will enable management to recognise when short-term control measures need to be reviewed and modified, and to put in place medium and long-term controls where these are necessary.

14 For activities where the nature of the work may change fairly frequently or the workplace itself changes and develops (such as a construction site), or where workers move from site to site, the risk assessment might have to concentrate more on the broad range of risks that can be foreseen. When other less common risks arise, detailed planning and employee training will be needed to take account of those risks and enable them to be controlled.

Risk assessment in practice

15 There are no fixed rules about how a risk assessment should be carried out; indeed it will depend on the nature of the work or business and the types of hazards and risks. Paragraph 18 does, however, set out the general principles that should be followed. The risk assessment process needs to be practical and take account of the views of employees and their safety representatives who will have practical knowledge to contribute. It should involve management, whether or not advisers or consultants assist with the detail. Employers should ensure that those involved take all reasonable care in carrying out the risk assessment. For further guidance see HSE's publication *Five steps to risk assessment*.

16 Where employees of different employers work in the same workplace, their respective employers may have to co-operate to produce an overall risk assessment. Detailed requirements on co-operation and co-ordination are covered by Regulation 11.

17 In some cases employers may make a first rough assessment, to eliminate from consideration those risks on which no further action is needed. This should also show where a fuller assessment is needed, if appropriate, using more sophisticated techniques. Employers who control a number of similar workplaces containing similar activities may produce a 'model' risk assessment reflecting the core hazards and risks associated with these activities. 'Model' assessments may also be developed by trade associations, employers' bodies or other organisations concerned with a particular activity. Such 'model' assessments may be applied by employers or managers at each workplace, but only if they:

(a) satisfy themselves that the 'model' assessment is appropriate to their type of work; and

(b) adapt the 'model' to the detail of their own actual work situations, including any extension necessary to cover hazards and risks not referred to in the 'model'.

18 A risk assessment should:

(a) ensure the significant risks and hazards are addressed;

(b) ensure all aspects of the work activity are reviewed, including routine and non-routine activities. The assessment should cover all parts of the work activity, including those that are not under the immediate supervision of the employer, such as employees working off site as contractors, workers from one organisation temporarily working for another organisation, self-employed people, homeworkers and mobile employees. Details of where to find additional guidance on homeworkers and volunteers is given in the References and further reading section (not reproduced in this book). Where workers visit members of the public in the home, eg nurses, employers should consider any risks arising from potential dangers;

(c) take account of the non-routine operations, eg maintenance, cleaning operations, loading and unloading of vehicles, changes in production cycles, emergency response arrangements;

(d) take account of the management of incidents such as interruptions to the work activity, which frequently cause accidents, and consider what procedures should be followed to mitigate the effects of the incident;

(e) be systematic in identifying hazards and looking at risks, whether one risk assessment covers the whole activity or the assessment is divided up. For example, it may be necessary to look at activities in groups such as machinery, transport, substances, electrical etc, or to divide the work site on a geographical basis. In other cases, an operation by operation approach may be needed, dealing with materials in production, dispatch, offices etc. The employer or self-employed person should always adopt a structured approach to risk assessment to ensure all significant risks or hazards are addressed. Whichever method is chosen, it should reflect the skills and abilities of the individuals carrying out that aspect of the assessment;

(f) take account of the way in which work is organised, and the effects this can have on health;

(g) take account of risks to the public;

(h) take account of the need to cover fire risks. The guide *Fire safety: An employer's guide* tells you how to comply with the law relating to fire issues and how to carry out a fire risk assessment.

Identifying the hazards

19 First, identify what the hazards are.

20 If there are specific acts or regulations to be complied with, these may help to identify the hazards. Some regulations require the assessment of particular risks or types of risks. If these particular risks are present, they must all be addressed in a risk assessment process for the purpose of these Regulations.

Identifying who might be harmed and how

21 Identify people who might be harmed by the hazard, including employees, other workers in the workplace and members of the public. Do not forget office staff, night cleaners, maintenance staff, security guards, visitors and members of the public. You should identify groups of workers who might be particularly at risk, such as young or inexperienced workers, new and expectant mothers, night workers, homeworkers, those who work alone and disabled staff.

Evaluating the risks from the identified hazards

22 You need to evaluate the risks from the identified hazards. Of course, if there are no hazards, there are no risks. Where risks are

already controlled in some way, the effectiveness of those controls needs to be considered when assessing the extent of risk which remains. You also need to:

(a) observe the actual practice; this may differ from the works manual, and the employees concerned or their safety representatives should be consulted;

(b) address what actually happens in the workplace or during the work activity; and

(c) take account of existing preventive or precautionary measures; if existing measures are not adequate, ask yourself what more should be done to reduce risk sufficiently.

Recording

23 All employers and self-employed people are required to make a risk assessment. The regulation also provides that employers with five or more employees must record the significant findings of their risk assessment. This record should represent an effective statement of hazards and risks which then leads management to take the relevant actions to protect health and safety. The record should be retrievable for use by management in reviews and for safety representatives or other employee representatives and visiting inspectors. Where appropriate, it should be linked to other health and safety records or documents such as the record of health and safety arrangements required by regulation 5 and the written health and safety policy statement required by section 2(3) of the HSW Act. It may be possible to combine these documents into one health and safety management document.

24 This record may be in writing or recorded by other means (eg electronically) as long as it is retrievable and remains retrievable even when, for example, the technology of electronic recording changes. The record will often refer to other documents and records describing procedures and safeguards.

25 The significant findings should include:

(a) a record of the preventive and protective measures in place to control the risks;

(b) what further action, if any, needs to be taken to reduce risk sufficiently;

(c) proof that a suitable and sufficient assessment has been made.

In many cases, employers (and the self-employed) will also need to record sufficient detail of the assessment itself, so that they can demonstrate (eg to an inspector or to safety representatives or other employee representatives) that they have carried out a suitable and sufficient assessment. This record of the significant findings will also form a basis for a revision of the assessment.

Review and revision

26 The regulation requires employers and the self-employed to review and, if necessary, modify their risk assessments, since assessment should not be a once-and-for-all activity. HSE's guide *Successful health and safety management* provides sound guidance on good practice. The following sub-paragraphs identify particular examples of review and revision.

(a) As the nature of work changes, the appreciation of hazards and risks may develop. Monitoring under the arrangements required by regulation 5 may reveal near misses or defects in plant or equipment. The risk assessment may no longer be valid because of, for example, the results of health surveillance, or a confirmed case of occupationally induced disease. Adverse events such as an accident, ill health or dangerous occurrence may take place even if a suitable and sufficient risk assessment has been made and appropriate preventive and protective measures taken. Such events should be a trigger for reviewing the original assessment.

(b) The employer or self-employed person needs to review the risk assessment if developments suggest that it may no longer be valid (or can be improved). In most cases, it is prudent to plan to review risk assessments at regular intervals. The time between reviews is dependent on the nature of the risks and the degree of change likely in the work activity. Such reviews should form part of standard management practice.

Assessment under other regulations

27 Other regulations also contain requirements for risk assessment specific to the hazards and risks they cover. Where an employer is assessing a work situation or activity for the first time, an assessment is particularly useful to identify where a more

detailed risk assessment is needed to fulfil the requirements of other regulations.

28 An assessment made for the purpose of other regulations will partly cover the obligation to make assessments under these regulations. Where employers have already carried out assessments under other regulations, they need not repeat those assessments as long as they remain valid; but they do need to ensure that they cover all significant risks.

Regulation 4

Principles of prevention to be applied

29 Employers and the self-employed need to introduce preventive and protective measures to control the risks identified by the risk assessment in order to comply with the relevant legislation. A set of principles to be followed in identifying the appropriate measures are set out in Schedule 1 to the Regulations and are described below. Employers and the self-employed should use these to direct their approach to identifying and implementing the necessary measures.

30–31 *(See Guidance Notes, pp 292–293)*

Regulation 5

Health and safety arrangements

32 This regulation requires employers to have arrangements in place to cover health and safety. Effective management of health and safety will depend, amongst other things, on a suitable and sufficient risk assessment being carried out and the findings being used effectively. The health and safety arrangements can be integrated into the management system for all other aspects of the organisation's activities. The management system adopted will need to reflect the complexity of the organisation's activities and working environment. Where the work process is straightforward and the risks generated are relatively simple to control, then very straightforward management systems may be appropriate. For large complicated organisations more complex systems may be appropriate. Although the principles of the management arrangements are the same irrespective of the size of an organisation. The key elements of such effective systems can be found in *Successful*

health and safety management or the British Standard for health and safety management systems BS8800. A successful health and safety management system will include all the following elements.

Planning

33 Employers should set up an effective health and safety management system to implement their health and safety policy which is proportionate to the hazards and risks. Adequate planning includes:

(a) adopting a systematic approach to the completion of a risk assessment. Risk assessment methods should be used to decide on priorities and to set objectives for eliminating hazards and reducing risks. This should include a programme, with deadlines for the completion of the risk assessment process, together with suitable deadlines for the design and implementation of the preventive and protective measures which are necessary;

(b) selecting appropriate methods of risk control to minimise risks;

(c) establishing priorities and developing performance standards both for the completion of the risk assessment(s) and the implementation of preventive and protective measures, which at each stage minimises the risk of harm to people. Wherever possible, risks are eliminated through selection and design of facilities, equipment and processes.

Organisation

34 This includes:

(a) involving employees and their representatives in carrying out risk assessments, deciding on preventive and protective measures and implementing those requirements in the workplace. This may be achieved by the use of formal health and safety committees where they exist, and by the use of teamworking, where employees are involved in deciding on the appropriate preventive and protective measures and written procedures etc;

(b) establishing effective means of communication and consultation in which a positive approach to health and safety is visible and clear. The employer should have adequate health and safety information and make sure it is communicated to employees and their representatives, so informed decisions can be made about the choice of preventive and protective measures. Effective communication will ensure that employees are provided with sufficient information so that control measures can be implemented effectively;

(c) securing competence by the provision of adequate information, instruction and training and its evaluation, particularly for those who carry out risk assessments and make decisions about preventive and protective measures. Where necessary this will need to be supported by the provision of adequate health and safety assistance or advice.

Control

35 Establishing control includes:

(a) clarifying health and safety responsibilities and ensuring that the activities of everyone are well co-ordinated;

(b) ensuring everyone with responsibilities understands clearly what they have to do to discharge their responsibilities, and ensure they have the time and resources to discharge them effectively;

(c) setting standards to judge the performance of those with responsibilities and ensure they meet them. It is important to reward good performance as well as to take action to improve poor performance; and

(d) ensuring adequate and appropriate supervision, particularly for those who are learning and who are new to a job.

Monitoring

36 Employers should measure what they are doing to implement their health and safety policy, to assess how effectively they are controlling risks, and how well they are developing a positive health and safety culture. Monitoring includes:

(a) having a plan and making adequate routine inspections and checks to ensure that preventive and protective measures are in place and effective. Active monitoring reveals how effectively the health and safety management system is functioning;

(b) adequately investigating the immediate and underlying causes of incidents and accidents to ensure that remedial action is taken, lessons are learnt and longer term objectives are introduced.

37 In both cases it may be appropriate to record and analyse the results of monitoring activity, to identify any underlying themes or trends which may not be apparent from looking at events in isolation.

Review

38 Review involves:

(a) establishing priorities for necessary remedial action that were discovered as a result of monitoring to ensure that suitable action is taken in good time and is completed;

(b) periodically reviewing the whole of the health and safety management system including the elements of planning, organisation, control and monitoring to ensure that the whole system remains effective.

39–40 *(See Guidance Notes, p 294)*

Regulation 6

Health surveillance

41 The risk assessment will identify circumstances in which health surveillance is required by specific health and safety regulations, eg COSHH. Health surveillance should also be introduced where the assessment shows the following criteria to apply:

(a) there is an identifiable disease or adverse health condition related to the work concerned; and

(b) valid techniques are available to detect indications of the disease or condition; and

(c) there is a reasonable likelihood that the disease or condition may occur under the particular conditions of work; and

(d) surveillance is likely to further the protection of the health and safety of the employees to be covered.

42 Those employees concerned and their safety or other representatives should be given an explanation of, and opportunity to

comment on, the nature and proposed frequency of such health surveillance procedures and should have access to an appropriately qualified practitioner for advice on surveillance.

43 The appropriate level, frequency and procedure of health surveillance should be determined by a competent person acting within the limits of their training and experience. This could be determined on the basis of suitable general guidance (eg regarding skin inspection for dermal effects) but in certain circumstances this may require the assistance of a qualified medical practitioner. The minimum requirement for health surveillance is keeping a health record. Once it is decided that health surveillance is appropriate, it should be maintained throughout an employee's employment unless the risk to which the worker is exposed and associated health effects are rare and short term.

44–45 *(See Guidance Notes, pp 294–295)*

Regulation 7

Health and safety assistance

46 Employers are solely responsible for ensuring that those they appoint to assist them with health and safety measures are competent to carry out the tasks they are assigned and are given adequate information and support. In making decisions on who to appoint, employers themselves need to know and understand the work involved, the principles of risk assessment and prevention, and current legislation and health and safety standards. Employers should ensure that anyone they appoint is capable of applying the above to whatever task they are assigned.

47 Employers must have access to competent help in applying the provisions of health and safety law, including these Regulations. In particular they need competent help in devising and applying protective measures, unless they are competent to undertake the measures without assistance. Appointment of competent people for this purpose should be included among the health and safety arrangements recorded under regulation 5(2). Employers are required by the Safety Representatives and Safety Committees Regulations 1977 to consult safety representatives in good time on arrangements for the appointment of competent assistance.

48–52 *(See Guidance Notes, pp 295–296)*

Regulation 9

Contacts with external services

53 Employers should establish procedures for any worker to follow if situations presenting serious and imminent danger were to arise, eg a fire, or for the police and emergency services on outbreak of public disorder. The procedures should set out:

(a) the nature of the risk (eg a fire in certain parts of a building where substances might be involved), and how to respond to it;

(b) additional procedures needed to cover risks beyond those posed by fire and bombs. These procedures should be geared, as far as is practicable, to the nature of the serious and imminent danger that those risks might pose;

(c) the additional responsibilities of any employees, or groups of employees, who may have specific tasks to perform in the event of emergencies (eg to shut down a plant that might otherwise compound the danger); or who have had training so that they can seek to bring an emergency event under control. Police officers, fire-fighters and other emergency service workers, for example, may sometimes need to work in circumstances of serious or imminent danger in order to fulfil their commitment to the public service. The procedures should reflect these responsibilities, and the time delay before such workers can move to a place of safety. Appropriate preventive and protective measures should be in place for these employees;

(d) the role, responsibilities and authority of the competent people nominated to implement the detailed actions. The procedures should also ensure that employees know who the relevant competent people are and understand their role;

(e) any requirements laid on employers by health and safety regulations which cover some specific emergency situations;

(f) details of when and how the procedures are to be activated so that employees can proceed in good time to a place of safety. Procedures should cater for the fact that emergency events can occur and develop rapidly, thus requiring employees to act without waiting for further guidance. It may be necessary to commence evacuation while attempts to control an emergency (eg a process in danger of running out of control) are still under way, in case those attempts fail.

54 Emergency procedures should normally be written down as required by regulation 5(2), clearly setting out the limits of actions to be taken by employees. Information on the procedures should be made available to all employees (under regulation 10), to any external health and safety personnel appointed under regulation 7(1), and where necessary to other workers and/or their employers under regulation 12. Induction training, carried out under regulation 13, should cover emergency procedures and should familiarise employees with those procedures.

55 Work should not be resumed after an emergency if a serious danger remains. If there are any doubts, expert assistance should be sought, eg from the emergency services and others. There may, for certain groups of workers, be exceptional circumstances when re-entry to areas of serious danger may be deemed necessary, eg police officers, fire-fighters and other emergency service workers, where, for example, human life is at risk. When such exceptional circumstances can be anticipated, the procedures should set out the special protective measures to be taken (and the pre-training required) and the steps to be taken for authorisation of such actions.

56–62 *(See Guidance Notes, pp 296–298)*

Regulation 10

Information for employees

63–66 *(See Guidance Notes, pp 298–299)*

Regulation 11

Co-operation and co-ordination

67 To meet the requirements of these Regulations, such as carrying out a risk assessment under regulation 3 and establishing procedures to follow serious and dangerous situations under regulation 8, it is necessary to cover the whole workplace to be fully effective. When the workplace is occupied by more than one employer, this will require some degree of co-ordination and co-operation. All employers and self-employed people involved should satisfy themselves that the arrangements adopted are adequate. Employers should ensure that all their employees, but especially the competent people appointed under regulations 7 and 8, are aware of and take full part in the arrangements. Specific co-ordination arrangements may be required by other regulations.

68 Where a particular employer controls the workplace, others should assist the controlling employer in assessing the shared risks and co-ordinating any necessary measures. In many situations providing information may be sufficient. A controlling employer who has established site-wide arrangements will have to inform new employers or self-employed people of those arrangements so that they can integrate themselves into the co-operation and co-ordination procedures.

69–74 *(See Guidance Notes, pp 299–301)*

Regulation 12

Persons working in host employers' or self-employed persons' undertakings

75 The risk assessment carried out under regulation 3 will identify risks to people other than the host employer's employees. This will include other employers' employees and self-employed people working in that business. Employers and self-employed people need to ensure that comprehensive information on those risks, and the measures taken to control them is given to other employers and self-employed people. Further guidance can be found under regulation 10.

76 Host employers and self-employed people must ensure that people carrying out work on their premises receive relevant information. This may be done by either providing them with information directly or by ensuring that their employers provide them with the relevant information. If you rely on their employers to provide information to the visiting employees, then adequate checks should be carried out to ensure that the information is passed on. The information should be sufficient to allow the employer of the visiting employee to comply with their statutory duties, and should include the identity of people nominated by the host employer to help with an emergency evacuation under regulation 8.

77 Information may be provided through a written permit-to-work system. Where the visiting employees are specialists, brought in to do specialist tasks, the host employer's instructions need to be concerned with those risks which are peculiar to the activity and premises. The visiting employee may also introduce risks to the permanent workforce (eg from equipment or substances they may bring with them). Their employers have a general duty under

section 3 of the HSW Act to inform the host employer of such risks and to co-operate and coordinate with the host employer to the extent needed to control those risks.

78 The guidance on information for employees under regulation 10 applies equally to information provided under regulation 12.

79 *(See Guidance Notes, p 301)*

Regulation 13

Capabilities and training

80 When allocating work to employees, employers should ensure that the demands of the job do not exceed the employees' ability to carry out the work without risk to themselves or others. Employers should take account of the employees' capabilities and the level of their training, knowledge and experience. Managers should be aware of relevant legislation and should be competent to manage health and safety effectively. Employers should review their employees' capabilities to carry out their work, as necessary. If additional training, including refresher training, is needed, it should be provided.

81 Health and safety training should take place during working hours. If it is necessary to arrange training outside an employee's normal hours, this should be treated as an extension of time at work. Employees are not required to pay for their own training. Section 9 of the HSW Act prohibits employers from charging employees for anything they have to do or are required to do in respect of carrying out specific requirements of the relevant statutory provisions. The requirement to provide health and safety training is such a provision.

82–84 *(See Guidance Notes, pp 301–302)*

Regulation 14

Employees' duties

85 Employees' duties under section 7 of the HSW Act include co-operating with their employer to enable the employer to comply with statutory duties for health and safety. Under these Regulations, employers or those they appoint (eg under regulation 7) to assist them with health and safety matters need to be informed

without delay of any work situation which might present a serious and imminent danger. Employees should also notify any shortcomings in the health and safety arrangements, even when no immediate danger exists, so that employers can take remedial action if needed.

86 The duties placed on employees do not reduce the responsibility of the employer to comply with duties under these Regulations and the other relevant statutory provisions. In particular, employers need to ensure that employees receive adequate instruction and training to enable them to comply with their duties.

87 *(See Guidance Notes, p 302)*

Regulation 15

Temporary workers

88–93 *(See Guidance Notes, pp 303–304)*

Regulation 18

Notification by new or expectant mothers

94 Where the risk assessment identifies risks to new and expectant mothers and these risks cannot be avoided by the preventive and protective measures taken by an employer, the employer will need to:

(a) alter her working conditions or hours of work if it is reasonable to do so and would avoid the risks or, if these conditions cannot be met;

(b) identify and offer her suitable alternative work that is available, and if that is not feasible;

(c) suspend her from work. The Employment Rights Act 1996 (which is the responsibility of the Department of Trade and Industry) requires that this suspension should be on full pay. Employment rights are enforced through the employment tribunals.

95 All employers should take account of women of child-bearing age when carrying out the risk assessment and identify the preventive and protective measures that are required in

Regulation 3. The additional steps of altering working conditions or hours of work, offering suitable alternative work or suspension as outlined above may be taken once an employee has given her employer notice in writing that she is pregnant, has given birth within the last six months or is breastfeeding. If the employee continues to breastfeed for more than six months after the birth she should ensure the employer is informed of this, so that the appropriate measures can continue to be taken. Employers need to ensure that those workers who are breastfeeding are not exposed to risks that could damage their health and safety as long as they breastfeed. If the employee informs her employer that she is pregnant for the purpose of any other statutory requirements, such as statutory maternity pay, this will be sufficient for the purpose of these Regulations.

96 Once an employer has been informed in writing that an employee is a new or expectant mother, the employer needs to immediately put into place the steps described in paragraph 94 and 95. The employer may request confirmation of the pregnancy by means of a certificate from a registered medical practitioner or a registered midwife in writing. If this certificate has not been produced within a reasonable period of time, the employer is not bound to maintain changes to working hours or conditions or to maintain paid leave. A reasonable period of time will allow for all necessary medical examinations and tests to be completed.

97 Further guidance on new and expectant mothers is contained in *New and expectant mothers at work: A guide for employers*. The table of hazards identified in the EC Directive on Pregnant Workers (92/85/EEC) is given in this publication, along with the risks and ways to avoid them. The DTI booklets *PL705 Suspension from work on medical or maternity grounds* and *PL958 Maternity rights*, both of which cover the maternity suspension provisions, are available from DTI.

Regulation 19

Protection of young persons

98 The employer needs to carry out the risk assessment before young workers start work and to see where risk remains, taking account of control measures in place, as described in regulation 3. For young workers, the risk assessment needs to pay attention to areas of risk described in regulation 19(2). For several of these

areas the employer will need to assess the risks with the control measures in place under other statutory requirements.

99 When control measures have been taken against these risks and if a significant risk still remains, no child (young worker under the compulsory school age) can be employed to do this work. A young worker, above the minimum school leaving age, cannot do this work unless:

(a) it is necessary for his or her training; and

(b) she or he is supervised by a competent person; and

(c) the risk will be reduced to the lowest level reasonably practicable.

100 Further guidance on young workers is contained in *Young people at work: A guide for employers*. The table on hazards, risks and ways of avoiding them from the EC Directive on the protection of Young People at Work (94/33/EEC) is given in this publication.

Regulation 21

Provisions as to liability

Employers' liability

101 An employer is not to be afforded a defence for any contravention of his health and safety obligations by reason of any act or default caused by an employee or by a person appointed to give competent advice. It does not affect employees' duties to take reasonable care of their own health and safety and that of others affected by their work activity.

GUIDANCE NOTES

Introduction

1 The original Management of Health and Safety at Work Regulations ('the Management Regulations') came into force in 1993[2] as the principal method of implementing the EC Framework Directive (89/391/EEC), adopted in 1989. The Regulations were supported by an Approved Code of Practice. The original Regulations have had to be amended four times since 1992 by the

2 SI 1992/2051.

Management of Health and Safety at Work (Amendment) Regulations 1994,[3] which relates to new or expectant mothers, the Health and Safety (Young Persons) Regulations 1997,[4] the Fire Precautions (Workplace) Regulations 1997[5] and by the Management of Health and Safety at Work Regulations 1999.[6] Because the original Regulations have been so significantly amended, they have been revised and published with this new Approved Code of Practice.

2 The Fire Precautions (Workplace) Regulations 1997, as amended by the Fire Precautions (Workplace) (Amendment) Regulations 1999[7] ('the Fire Regulations'), introduced by the Home Office, amend the Management Regulations in several respects. The amendments made by the Fire Regulations make explicit the risk assessment requirement, in so far as it relates to fire safety. They directly require employers to take account of their general fire precautions requirements in Part II of the Fire Regulations (concerning fire-fighting, fire detection, emergency routes and exits and their maintenance) in their assessments. The amendments made by the Fire Regulations affect employers but not self-employed people who do not employ others. The Fire Regulations also introduce:

(a) a requirement for competent assistance to deal with general fire safety risks;

(b) a requirement to provide employees with information on fire provisions; and

(c) a requirement on employers and self-employed people in a shared workplace to co-operate and co-ordinate with others on fire provisions and to provide outside employers with comprehensive information on fire provisions.

Further guidance on fire precautions is available in *Fire safety: An employer's guide.*

3 The duties of the Management Regulations overlap with other regulations because of their wide-ranging general nature. Where duties overlap, compliance with the more specific regulation will

3 SI 1994/2865.
4 SI 1997/135.
5 SI 1997/1840.
6 The Management of Health and Safety at Work Regulations 1999 introduced amendments proposed by the Health and Safety (Miscellaneous Modification) Regulations 1999.
7 SI 1999/1877.

normally be sufficient to comply with the corresponding duty in the Management Regulations. For example, the Control of Substances Hazardous to Health Regulations (COSHH) require employers and the self-employed to assess the risks from exposure to substances hazardous to health. An assessment made for the purposes of COSHH will not need to be repeated for the purposes of the Management Regulations. Other instances where overlap may occur include the appointment of people to carry out specific tasks or arrangements for emergencies. However, where the duties in the Management Regulations go beyond those in the more specific regulations, additional measures will be needed to comply fully with the Management of Health and Safety at Work Regulations.

4 Although only the courts can give an authoritative interpretation of law, in considering the application of these regulations and guidance to people working under another's direction, the following should be considered.

5 If people working under the control and direction of others are treated as self-employed for tax and national insurance purposes they may nevertheless be treated as their employees for health and safety purposes. It may therefore be necessary to take appropriate action to protect them. If any doubt exists about who is responsible for the health and safety of a worker this could be clarified and included in the terms of a contract. However, remember, a legal duty under section 3 of the Health and Safety at Work etc Act 1974 (HSW Act)[8] cannot be passed on by means of a contract and there will still be duties towards others under section 3 of HSW Act. If such workers are employed on the basis that they are responsible for their own health and safety, legal advice should be sought before doing so.

6 Words or expressions which are defined in the Management Regulations or in the HSW Act have the same meaning in this Code unless the context requires otherwise.

Regulation 2

Disapplication of these Regulations

7 Regulation 2(1) excludes the master and crew of a sea-going ship, as similar duties are placed on them by the Merchant Shipping and Fishing Vessels (Health and Safety at Work)

8 SI 1974 c 37.

Regulations 1997.[9] However, when a ship is in a port in Great Britain and shoreside workers and the ship's crew work together, eg in dock operations, or in carrying out repairs to the ship, these Regulations may apply. Dock operations, ship construction, ship repair carried out in port with shoreside assistance, and work connected to construction or the offshore industry (other than navigation, pollution prevention and other aspects of the operation of the ship, which are subject to international shipping standards) are not considered as 'normal ship-board activities' and are therefore subject to these Regulations.

8 Regulation 2(2) clarifies that these Regulations do not apply to domestic services in a private household.

Regulation 3

Risk assessment

General principles and purpose of risk assessment

9–28 *(See Approved Code of Practice, pp 270–278)*

Regulation 4

Principles of prevention to be applied

29 *(See Approved Code of Practice, p 278)*

30 In deciding which preventive and protective measures to take, employers and self-employed people should apply the following principles of prevention:

(a) if possible avoid a risk altogether, eg do the work in a different way, taking care not to introduce new hazards;

(b) evaluate risks that cannot be avoided by carrying out a risk assessment;

(c) combat risks at source, rather than taking palliative measures. So, if the steps are slippery, treating or replacing them is better than displaying a warning sign;

(d) adapt work to the requirements of the individual (consulting those who will be affected when designing workplaces, selecting work and personal protective equipment and

9 SI 1997/2962.

drawing up working and safety procedures and methods of production). Aim to alleviate monotonous work and paced working at a predetermined rate, and increase the control individuals have over work they are responsible for;

(e) take advantage of technological and technical progress, which often offers opportunities for improving working methods and making them safer;

(f) implement risk prevention measures to form part of a coherent policy and approach. This will progressively reduce those risks that cannot be prevented or avoided altogether, and will take account of the way work is organised, the working conditions, the environment and any relevant social factors. Health and safety policy statements required under section 2(3) of the HSW Act should be prepared and applied by reference to these principles;

(g) give priority to those measures which protect the whole workplace and everyone who works there, and so give the greatest benefit (ie give collective protective measures priority over individual measures);

(h) ensure that workers, whether employees or self-employed, understand what they must do;

(i) the existence of a positive health and safety culture should exist within an organisation. That means the avoidance, prevention and reduction of risks at work must be accepted as part of the organisation's approach and attitude to all its activities. It should be recognised at all levels of the organisation, from junior to senior management.

31 These are general principles rather than individual prescriptive requirements. They should, however, be applied wherever it is reasonable to do so. Experience suggests that, in the majority of cases, adopting good practice will be enough to ensure risks are reduced sufficiently. Authoritative sources of good practice are prescriptive legislation, Approved Codes of Practice and guidance produced by Government and HSE inspectors. Other sources include standards produced by standard-making organisations and guidance agreed by a body representing an industrial or occupational sector, provided the guidance has gained general acceptance. Where established industry practices result in high levels of health and safety, risk assessment should not be used to justify reducing current control measures.

Regulation 5

Health and safety arrangements

32–38 *(See Approved Code of Practice, pp 278–281)*

39 Consulting employees or their representatives about matters to do with their health and safety is good management practice, as well as being a requirement under health and safety law. Employees are a valuable source of information and can provide feedback about the effectiveness of health and safety management arrangements and control measures. Where safety representatives exist, they can act as an effective channel for employees' views.

40 Safety representatives' experience of workplace conditions and their commitment to health and safety means they often identify potential problems, allowing the employer to take prompt action. They can also have an important part to play in explaining safety measures to the workforce and gaining commitment.

Regulation 6

Health surveillance

41–43 *(See Approved Code of Practice, pp 281–282)*

44 Where appropriate, health surveillance may also involve one or more health surveillance procedures depending on suitability in the circumstances (a non-exhaustive list of examples of diseases is included in the footnote for guidance).[10] Such procedures can include:

(a) inspection of readily detectable conditions by a responsible person acting within the limits of their training and experience;

(b) enquiries about symptoms, inspection and examination by a qualified person such as an Occupational Health Nurse;

(c) medical surveillance, which may include clinical examination and measurement of physiological or psychological effects by an appropriately qualified person;

10 If the worker is exposed to noise or hand-arm vibrations, health surveillance may be needed under these regulations. If the worker is exposed to hazardous substances such as chemicals, solvents, fumes, dusts, gases and vapours, aerosols, biological agents (micro-organisms), health surveillance may be needed under COSHH. If the worker is exposed to asbestos, lead, work in compressed air, medical examinations may be needed under specific regulations.

(d) biological effect monitoring, ie the measurement and assessment of early biological effects such as diminished lung function in exposed workers; and

(e) biological monitoring, ie the measurement and assessment of workplace agents or their metabolites either in tissues, secreta, excreta, expired air or any combination of these in exposed workers.

45 The primary benefit, and therefore objective of health surveillance should be to detect adverse health effects at an early stage, thereby enabling further harm to be prevented. The results of health surveillance can provide a means of:

(a) checking the effectiveness of control measures;

(b) providing feedback on the accuracy of the risk assessment; and

(c) identifying and protecting individuals at increased risk because of the nature of their work.

Regulation 7

Health and safety assistance

46–47 *(See Approved Code of Practice, p 282)*

48 When seeking competent assistance employers should look to appoint one or more of their employees, with the necessary means, or themselves, to provide the health and safety assistance required. If there is no relevant competent worker in the organisation or the level of competence is insufficient to assist the employer in complying with health and safety law, the employer should enlist an external service or person. In some circumstances a combination of internal and external competence might be appropriate, recognising the limitations of the internal competence. Some regulations contain specific requirements for obtaining advice from competent people to assist in complying with legal duties. For example, the Ionising Radiation Regulations require the appointment of a radiation protection adviser in many circumstances, where work involves ionising radiations.

49 Employers who appoint doctors, nurses or other health professionals to advise them of the effects of work on employee health, or to carry out certain procedures, for example health surveillance, should first check that such providers can offer

evidence of a sufficient level of expertise or training in occupational health. Registers of competent practitioners are maintained by several professional bodies, and are often valuable.

50 The appointment of such health and safety assistants or advisers does not absolve the employer from responsibilities for health and safety under the HSW Act and other relevant statutory provisions and under Part II of the Fire Regulations. It can only give added assurance that these responsibilities will be discharged adequately. Where external services are employed, they will usually be appointed in an advisory capacity only.

51 Competence in the sense it is used in these Regulations does not necessarily depend on the possession of particular skills or qualifications. Simple situations may require only the following:

(a) an understanding of relevant current best practice;

(b) an awareness of the limitations of one's own experience and knowledge; and

(c) the willingness and ability to supplement existing experience and knowledge, when necessary by obtaining external help and advice.

52 More complicated situations will require the competent assistant to have a higher level of knowledge and experience. More complex or highly technical situations will call for specific applied knowledge and skills which can be offered by appropriately qualified specialists. Employers are advised to check the appropriate health and safety qualifications (some of which may be competence-based and/or industry specific), or membership of a professional body or similar organisation (at an appropriate level and in an appropriate part of health and safety) to satisfy themselves that the assistant they appoint has a sufficiently high level of competence. Competence-based qualifications accredited by the Qualifications and Curriculum Authority and the Scottish Qualifications Authority may also provide a guide.

Regulation 9

Contacts with external services

53–55 *(See Approved Code of Practice, pp 283–284)*

56 The procedure for any worker to follow in serious and imminent danger has to be clearly explained by the employer.

Employees and others at work need to know when they should stop work and how they should move to a place of safety. In some cases this will require full evacuation of the workplace; in others it might mean some or all of the workforce moving to a safer part of the workplace.

57 The risk assessment should identify the foreseeable events that need to be covered by these procedures. For some employers, fire (and possibly bomb) risks will be the only ones that need to be covered. For others, additional risks will be identified.

58 Where different employers (or self-employed people) share a workplace, their separate emergency procedures will need to take account of everyone in the workplace, and as far as is appropriate the procedures should be coordinated. Detailed requirements on co-operation and co-ordination are covered by regulation 11.

Danger areas

59 A danger area is a work environment which must be entered by an employee where the level of risk is unacceptable without taking special precaution. Such areas are not necessarily static in that minor alterations or an emergency may convert a normal working environment into a danger area. The hazard involved need not occupy the whole area, as in the case of a toxic gas, but can be localised, eg where there is a risk of an employee coming into contact with bare live electrical conductors. The area must he restricted to prevent inadvertent access.

60 This regulation does not specify the precautions to take to ensure safe working in the danger area—this is covered by other legislation. However, once the employer has established suitable precautions the relevant employees must receive adequate instruction and training in those precautions before entering any such danger area.

Contacts with external services

61 The employer should ensure that appropriate external contacts are in place to make sure there are effective provisions for first aid, emergency medical care and rescue work, for incidents and accidents which may require urgent action, and/or medical attention beyond the capabilities of on-site personnel. This may only mean making sure that employees know the necessary

telephone numbers and, where there is a significant risk, that they are able to contact any help they need. This requirement does not in any way reduce employers' duty to prevent accidents as the first priority.

62 Where a number of employers share a workplace and their employees face the same risks, it would be possible for one employer to arrange contacts on behalf of themselves and the other employers. In these circumstances it would be for the other employers to ensure that the contacts had been made. In hazardous or complex workplaces, employers should designate appropriate staff to routinely contact the emergency services to give them sufficient knowledge of the risks they need to take appropriate action in emergencies, including those likely to happen outside normal working hours. This will help these services in planning for providing first aid, emergency medical care and rescue work, and to take account of risks to everyone involved, including rescuers. Contacts and arrangements with external services should be recorded, and should be reviewed and revised as necessary, in the light of changes to staff, processes and plant, and revisions to health and safety procedures.

Regulation 10

Information for employees

63 The risk assessment will help identify information which has to be provided to employees under specific regulations, as well as any further information relevant to risks to employees' health and safety. Relevant information on risks and on preventive and protective measures will be limited to what employees need to know to ensure their own health and safety and not to put others at risk. This regulation also requires information to be provided on the emergency arrangements established under regulation 8, including the identity of staff nominated to help if there is an evacuation.

64 The information provided should be pitched appropriately, given the level of training, knowledge and experience of the employee. It should be provided in a form which takes account of any language difficulties or disabilities. Information can be provided in whatever form is most suitable in the circumstances, as long as it can be understood by everyone. For employees with little or no understanding of English, or who cannot read English,

employers may need to make special arrangements. These could include providing translation, using interpreters, or replacing written notices with clearly understood symbols or diagrams.

65 This regulation applies to all employees, including trainees and those on fixed-duration contracts. Additional information for employees on fixed-duration contracts is contained in regulation 15. Specific requirements relate to the provision of information to safety representatives, and enabling full and effective consultation of employees.[11]

66 While a child (below minimum school leaving age) is at work, the requirements to provide information are the same as for other employees. There is, however, an extra requirement on the employer to provide the parents or guardians of children at work (including those on work experience) with information on the key findings of the risk assessment and the control measures taken, before the child starts work. This information can be provided in any appropriate form, including verbally or directly to the parents or guardians, or in the case of work experience, via an organisation such as the school, the work experience agency, or, if agreed with the parents, via the child him or herself, as long as this is considered a reliable method.

Regulation 11

Co-operation and co-ordination

67–68 *(See Approved Code of Practice, pp 284–285)*

69 Where the activities of different employers and self-employed people interact, for example where they share premises or workplaces, they may need to co-operate with each other to make sure their respective obligations are met. This regulation does not extend to the relationship between a host employer and a contractor, which will be covered by regulation 12.

70 The duties to co-operate and co-ordinate measures relate to all statutory duties, except for Part II of the Fire Regulations, in the case of people who are self-employed and are not employers. Therefore, they concern all people who may be at risk, both on and

11 Safety Representatives and Safety Committees Regulations 1977, SI 1977/500 and the Health and Safety (Consultation with Employees) Regulations 1996, SI 1996/1513.

off site, and not just where employers and self-employed people share workplaces all the time. They also include situations where an employer may not be physically present at the workplace.

Appointment of health and safety co-ordinator

71 Where there is no controlling employer, the employers and self-employed people present will need to agree any joint arrangements needed to meet the requirements of the Regulations, such as appointing a health and safety co-ordinator. This will be particularly useful in workplaces where management control is fragmented and employment is largely casual or short term (eg in construction). The Construction (Design and Management) Regulations 1994[12] require principal contractors to ensure co-operation between all contractors. In workplaces which are complex or contain significant hazards, the controlling employer or health and safety co-ordinator (on behalf of the employers etc present) may need to seek competent advice in making or assisting with the risk assessment and determining appropriate measures. Employers do not absolve themselves of their legal responsibilities by appointing such co-ordinators who provide competent advice.

Person in control

72 The person in control of a multi-occupancy workplace may not always be an employer of the people working in that workplace or be self-employed, but will still need to co-operate with those occupying the workplace under their control. For example, procedures for authorising or carrying out repairs and modifications will have to take account of the need for co-operation and exchanges of information. Co-operation is needed to effectively carry out the general duties placed on those people under section 4 of the HSW Act, as well as more specific duties under other Regulations (eg in offshore health and safety legislation or in relation to welfare facilities provided under the Workplace (Health, Safety and Welfare) Regulations 1992).[13]

73 People who control the premises and make arrangements to co-ordinate health and safety activities, particularly for

12 SI 1994/3140.
13 SI 1992/3004.

emergencies, may help employers and self-employed people who participate in those arrangements to comply with regulation 11(1)(b).

74 This regulation does not apply to multi-occupancy buildings or sites, where each unit under the control of an individual tenant employer or self-employed person is regarded as a separate workplace. In some cases, however, the common parts of such multi-occupancy sites may be shared workplaces (eg a common reception area in an office building) or may be under the control of a person to whom section 4 of the HSW Act applies and suitable arrangements may need to be put in place for these areas.

Regulation 12

Persons working in host employers' or self-employed persons' undertakings

75–78 *(See Approved Code of Practice, pp 285–286)*

79 This regulation applies where employees or self-employed people carry out work for an employer other than their own or of another self-employed person. This will include contractors' employees carrying out cleaning, repair, or maintenance work under a service contract; and employees in temporary employment businesses, hired to work under a host employer's control (additional requirements for information to employment businesses are under regulation 15). Safety representatives and other employee representatives are often used to ensure information is supplied to everyone who comes on site.

Regulation 13

Capabilities and training

80–81 *(See Approved Code of Practice, p 286)*

82 The risk assessment and subsequent reviews of the risk assessment will help determine the level of training and competence needed for each type of work. Competence is the ability to do the work required to the necessary standard. All employees, including senior management, should receive relevant training. This may need to include basic skills training, specific on-the-job training and training in health and safety or emergency procedures. There may be a need for further training, eg about specific risks, required by

other legislation. For those working towards National and Scottish Vocational Qualifications, the Employment National Training Organisation has designed stand-alone training units in health, safety and the environment. These vocational units are for people at work who are not health and safety professionals/specialists.

83 Training needs are likely to be greatest for new employees on recruitment. They should receive basic induction training on health and safety, including arrangements for first-aid, fire and evacuation. Particular attention should be given to the needs of young workers. The risk assessment should identify further specific training needs. In some cases, training may be required even though an employee already holds formal qualifications (eg for an update on new technology). Training and competence needs will have to be reviewed if the work activity a person is involved in or the working environment changes. This may include a change of department or the introduction of new equipment, processes or tasks.

84 An employee's competence will decline if skills are not used regularly (eg in emergency procedures, operating a particular item of equipment or carrying out a task). Training therefore needs to be repeated periodically to ensure continued competence. This will be particularly important for employees who occasionally deputise for others, home workers and mobile employees. Information from personal performance monitoring, health and safety checks, accident investigations and near-miss incidents can help to establish a suitable period for re-training. Employers are required by the Safety Representatives and Safety Committees Regulations 1997 to consult safety representatives in good time about the planning and organisation of health and safety training required for the employees they represent.

Regulation 14

Employees' duties

85–86 *(See Approved Code of Practice, pp 286–287)*

87 Employees have a duty under section 7 of the HSW Act to take reasonable care for their own health and safety and that of others who may be affected by their actions or omissions at work. Therefore, employees must use all work items provided by their employer correctly, in accordance with their training and the instructions they received to use them safely.

Regulation 15

Temporary workers

88 This regulation supplements previous regulations requiring the provision of information with additional requirements on temporary workers (ie those employed on fixed-duration contracts and those employed in employment businesses, but working for a user company). The use of temporary workers needs to be notified to health and safety staff under regulation 7(4)(b), as necessary.

Fixed-duration contracts

89 Regulation 10 deals with the provision of information by employers to their employees. This includes those on fixed-duration contracts. Under regulation 15(1), employees on fixed-duration contracts also have to be informed of any special occupational qualifications or skills required to carry out the work safely and whether the job is subject to statutory health surveillance (the latter being a protective measure covered in general by regulation 10(1)(b)).

Employment businesses

90 Regulation 12(4) deals with the provision of information by a first employer to a second employer, whose employees are working on the premises. This includes employees of people who have an employment business. Under regulation 15(3), an employment business has to be informed of any special occupational qualifications or skills required to carry out the work safely and the specific features of the job which might affect health and safety (eg work at heights).

91 The person who has an employment business and the user employer both have duties to provide information to the employee. The person with the employment business has a duty under regulation 10 (as an employer) and a duty under regulation 15(3) to ensure that the information provided by the user employer is given to the employee. The user employer has a duty under regulation 12(4) to check that information provided to an employer (including someone carrying on an employment business) is received by the employee. In addition, regulations 15(1) and (2) require that information on qualifications,

skills and health surveillance are given directly to employees in an employment business.

92 These duties overlap to make sure the information needs of those working for, but not employed by, user employers are not overlooked. User employers and people carrying on employment businesses should therefore make suitable arrangements to satisfy themselves that information is provided. In most cases, it may be enough for user employees to provide information directly to employees. Those carrying on employment businesses will need to satisfy themselves that arrangements for this are adequate. However, basic information on job demands and risks should be supplied to the employment business at an early stage to help select those most suitable to carry out the work (in accordance with regulation 15(3)).

Self-employed

93 Self-employed people have similar duties under regulations 11(2), 12, 15(2) and 15(3) to inform employment businesses and the employees of employment businesses who carry out work on their premises. They may also need to agree arrangements with the employment businesses concerned. Self-employed workers hired through employment businesses are entitled to receive health and safety information from the employers or self-employed people for whom they carry out work under regulation 12(2). There is a full definition of how these Regulations apply to self-employed workers in paragraphs 4 and 5. It explains how people working under the control and direction of others, but are treated as self-employed for other reasons such as tax and national insurance, are nevertheless treated as their employees for health and safety purposes.

Regulation 18

Notification by new or expectant mothers

94–97 *(See Approved Code of Practice, pp 287–288)*

Regulation 19

Protection of young persons

98–100 *(See Approved Code of Practice, pp 288–289)*

Regulation 21

Provisions as to liability

Employers' liability

101 *(See Approved Code of Practice, p 289)*

102 In practice enforcers will take account of the circumstances of each case before deciding on the appropriateness of any enforcement action, and who this should be taken against. Where the employer has taken reasonable steps to satisfy him or herself of the competency of the employee or person appointed to provide competent advice or services, this will be taken into account.

Regulation 23

Extension outside Great Britain

103 The 1989 Order has been replaced by the Health and Safety at Work etc Act 1974 (Application outside Great Britain) Order 1995. This Order applies the Act to offshore installations, wells, pipelines and pipeline works, and to connected activities within the territorial waters of Great Britain or in designated areas of the United Kingdom Continental Shelf, plus certain other activities within territorial waters. Regulation 23(1) applies the Management of Health and Safety at Work Regulations to these places and activities. Regulation 23(2) ensures that workers are protected even while off duty offshore.

Regulation 24

Amendment of the Health and Safety (First-Aid) Regulations 1981

104 Regulation 24 revokes the Health and Safety Executive's powers to grant exemptions from the Health and Safety (First-Aid) Regulations 1981.

Regulation 25

Amendment of the Offshore Installations and Pipeline Works (First-Aid) Regulations 1989

105 Regulation 25 limits the scope of the exemptions that can be granted by the Health and Safety Executive to those specified in

regulations 5(1)(b)(c) and (2)(a) of the Offshore Installations and Pipeline Works (First Aid) Regulations 1989. It also requires that where an exemption is granted the person provided under regulation 5(1)(a) shall have undergone adequate training.

Regulation 26

Amendment of the Mines Miscellaneous Health and Safety Provisions Regulations 1995

106 Regulation 26 introduces a new paragraph into regulation 4 of the Mines Miscellaneous Health and Safety Provisions Regulations 1995. This requires that a fire protection plan be included in all cases in the health and safety document required by these Regulations. In all parts of the mine other than buildings on the surface, the mine owner is required to designate in the document those who are to implement the plan and include the arrangements for the necessary contacts with external services, especially rescue work and fire-fighting.

Regulation 27

Amendment of the Construction (Health, Safety and Welfare) Regulations 1996

107 This regulation amends regulation 20 of the Construction (Health, Safety and Welfare) Regulations 1996, so that arrangements for dealing with foreseeable emergencies on construction sites include designating people to implement the arrangements and arranging necessary contacts with external services, especially rescue work and fire-fighting.

REFERENCES AND FURTHER READING

This section is not reproduced in this book.

Appendix 2

Workplace (Health, Safety and Welfare) Regulations 1992

THE REGULATIONS (SI 1992/3004)

Made 1 December 1992

The Secretary of State, in exercise of the powers conferred on her by sections 15(1), (2), (3)(a) and (5)(b), and 82(3)(a) of, and paragraphs 1(2), 9 and 10 of Schedule 3 to, the Health and Safety at Work etc Act 1974 ('the 1974 Act') and of all other powers enabling her in that behalf and for the purpose of giving effect without modifications to proposals submitted to her by the Health and Safety Commission under section 11(2)(d) of the 1974 Act after the carrying out by the said Commission of consultations in accordance with section 50(3) of that Act, hereby makes the following Regulations:

Regulation 1 *(See Guidance Notes, p 360)*

Citation and commencement

(1) These Regulations may be cited as the Workplace (Health, Safety and Welfare) Regulations 1992.

(2) Subject to paragraph (3), these Regulations shall come into force on 1 January 1993.

(3) Regulations 5 to 27 and the Schedules shall come into force on 1 January 1996 with respect to any workplace or part of a workplace which is not—

(a) a new workplace; or

(b) a modification, an extension or a conversion.

Regulation 2 *(See Guidance Notes, pp 360–362)*

Interpretation

(1) In these Regulations, unless the context otherwise requires—

— ['mine' means a mine within the meaning of the Mines and Quarries Act 1954;][1]

— 'new workplace' means a workplace used for the first time as a workplace after 31 December 1992;

— 'public road' means (in England and Wales) a highway maintainable at public expense within the meaning of section 329 of the Highways Act 1980 and (in Scotland) a public road within the meaning assigned to that term by section 151 of the Roads (Scotland) Act 1984;

— ['quarry' means a quarry within the meaning of the Quarries Regulations 1999;][2]

— 'traffic' route means a route for pedestrian traffic, vehicles or both and includes any stairs, staircase, fixed ladder, doorway, gateway, loading bay or ramp;

— 'workplace' means, subject to paragraph (2), any premises or part of premises which are not domestic premises and are made available to any person as a place of work, and includes—

(a) any place within the premises to which such person has access while at work; and

(b) any room, lobby, corridor, staircase, road or other place used as a means of access to or egress from that place of work or where facilities are provided for use in connection with the place of work other than a public road;

but shall not include a modification, an extension or a conversion of any of the above until such modification, extension or conversion is completed.

(2) Any reference in these Regulations, except in paragraph (1), to a modification, an extension or a conversion is a reference, as the case may be, to a modification, an extension or a conversion of a workplace started after 31 December 1992.

1 Definition 'mine' inserted by SI 1995/2036, reg 11, Sch 3. Date in force: 26 July 1998: see SI 1995/2036, reg 1(2).

2 Definition 'quarry' substituted by SI 1999/2024, reg 48(2), Sch 5, Pt II. Date in force: 1 January 2000: see SI 1999/2024, reg 1(1).

(3) Any requirement that anything done or provided in pursuance of these Regulations shall be suitable shall be construed to include a requirement that it is suitable for any person in respect of whom such thing is so done or provided.

(4) Any reference in these Regulations to—

(a) a numbered regulation or Schedule is a reference to the regulation in or Schedule to these Regulations so numbered; and

(b) a numbered paragraph is a reference to the paragraph so numbered in the regulation in which the reference appears.

Regulation 3 *(See Guidance Notes, pp 362–363)*

Application of these Regulations

(1) These Regulations apply to every workplace but shall not apply to—

(a) a workplace which is or is in or on a ship within the meaning assigned to that word by regulation 2(1) of the Docks Regulations 1988;

(b) a workplace where the only activities being undertaken are building operations or works of engineering construction within, in either case, section 176 of the Factories Act 1961 and activities for the purpose of or in connection with the first-mentioned activities; or

[(c) a workplace located below ground at a mine][3]

(d) . . .[4]

(2) In their application to temporary work sites, any requirement to ensure a workplace complies with any of regulations 20 to 25 shall have effect as a requirement to so ensure so far as is reasonably practicable.

(3) As respects any workplace which is or is in or on an aircraft, locomotive or rolling stock, trailer or semi-trailer used as a means of transport or a vehicle for which a licence is in force under the

3 Substituted by SI 1995/2036, reg 11, Sch 3. Date in force: 26 July 1998: see SI 1995/2036, reg 1(2).

4 Revoked by SI 1995/2036, reg 11, Sch 3. Date in force: 26 July 1998: see SI 1995/2036, reg 1(2).

Vehicles (Excise) Act 1971 or a vehicle exempted from duty under that Act—

(a) regulations 5 to 12 and 14 to 25 shall not apply to any such workplace; and

(b) regulation 13 shall apply to any such workplace only when the aircraft, locomotive or rolling stock, trailer or semi-trailer or vehicle is stationary inside a workplace and, in the case of a vehicle for which a licence is in force under the Vehicles (Excise) Act 1971, is not on a public road.

(4) As respects any workplace which is in fields, woods or other land forming part of an agricultural or forestry undertaking but which is not inside a building and is situated away from the undertakings main buildings—

(a) regulations 5 to 19 and 23 to 25 shall not apply to any such workplace; and

(b) any requirement to ensure that any such workplace complies with any of regulations 20 to 22 shall have effect as a requirement to so ensure so far as is reasonably practicable.

[(5) As respects any workplace which is at a quarry or above ground at a mine regulation 12 shall only apply to a floor or traffic route which is located inside a building.][5]

Regulation 4 *(See Guidance Notes, pp 363–365)*

Requirements under these Regulations

(1) Every employer shall ensure that every workplace, modification, extension or conversion which is under his control and where any of his employees works complies with any requirement of these Regulations which—

(a) applies to that workplace or, as the case may be, to the workplace which contains that modification, extension or conversion; and

(b) is in force in respect of the workplace, modification, extension or conversion.

5 Inserted by SI 1995/2036, reg 11, Sch 3. Date in force: 26 July 1998: see SI 1995/2036, reg 1(2).

(2) Subject to paragraph (4), every person who has, to any extent, control of a workplace, modification, extension or conversion shall ensure that such workplace, modification, extension or conversion complies with any requirement of these Regulations which—

(a) applies to that workplace or, as the case may be, to the workplace which contains that modification, extension or conversion;

(b) is in force in respect of the workplace, modification, extension, or conversion; and

(c) relates to matters within that person's control.

(3) Any reference in this regulation to a person having control of any workplace, modification, extension or conversion is a reference to a person having control of the workplace, modification, extension or conversion in connection with the carrying on by him of a trade, business or other undertaking (whether for profit or not).

(4) Paragraph (2) shall not impose any requirement upon a self-employed person in respect of his own work or the work of any partner of his in the undertaking.

(5) Every person who is deemed to be the occupier of a factory by virtue of section 175(5) of the Factories Act 1961 shall ensure that the premises which are so deemed to be a factory comply with these Regulations.

Regulation 5 *(See Approved Code of Practice, pp 329–330; Guidance Notes, pp 365–366)*

Maintenance of workplace, and of equipment, devices and systems

(1) The workplace and the equipment, devices and systems to which this regulation applies shall be maintained (including cleaned as appropriate) in an efficient state, in efficient working order and in good repair.

(2) Where appropriate, the equipment, devices and systems to which this regulation applies shall be subject to a suitable system of maintenance.

(3) The equipment, devices and systems to which this regulation applies are—

(a) equipment and devices a fault in which is liable to result in a failure to comply with any of these Regulations; and

(b) mechanical ventilation systems provided pursuant to regulation 6 (whether or not they include equipment or devices within sub-paragraph (a) of this paragraph).

Regulation 6 *(See Approved Code of Practice, pp 330–331; Guidance Notes, p 366)*

Ventilation

(1) Effective and suitable provision shall be made to ensure that every enclosed workplace is ventilated by a sufficient quantity of fresh or purified air.

(2) Any plant used for the purpose of complying with paragraph (1) shall include an effective device to give visible or audible warning of any failure of the plant where necessary for reasons of health or safety.

(3) This regulation shall not apply to any enclosed workplace or part of a workplace which is subject to the provisions of—

(a) section 30 of the Factories Act 1961;

(b) regulations 49 to 52 of the Shipbuilding and Ship-Repairing Regulations 1960;

(c) regulation 21 of the Construction (General Provisions) Regulations 1961;

(d) regulation 18 of the Docks Regulations 1988.

Regulation 7 *(See Approved Code of Practice, pp 331–333; Guidance Notes, p 367)*

Temperature in indoor workplaces

(1) During working hours, the temperature in all workplaces inside buildings shall be reasonable.

(2) A method of heating or cooling shall not be used which results in the escape into a workplace of fumes, gas or vapour of such character and to such extent that they are likely to be injurious or offensive to any person.

(3) A sufficient number of thermometers shall be provided to enable persons at work to determine the temperature in any workplace inside a building.

Regulation 8 *(See Approved Code of Practice, pp 333–334; Guidance Notes, pp 367–368)*

Lighting

(1) Every workplace shall have suitable and sufficient lighting.

(2) The lighting mentioned in paragraph (1) shall, so far as is reasonably practicable, be by natural light.

(3) Without prejudice to the generality of paragraph (1), suitable and sufficient emergency lighting shall be provided in any room in circumstances in which persons at work are specially exposed to danger in the event of failure of artificial lighting.

Regulation 9 *(See Approved Code of Practice, pp 334–335; Guidance Notes, p 368)*

Cleanliness and waste materials

(1) Every workplace and the furniture, furnishings and fittings therein shall be kept sufficiently clean.

(2) The surfaces of the floors, walls and ceilings of all workplaces inside buildings shall be capable of being kept sufficiently clean.

(3) So far as is reasonably practicable, waste materials shall not be allowed to accumulate in a workplace except in suitable receptacles.

Regulation 10 *(See Approved Code of Practice, pp 335–336; Guidance Notes, pp 368–369)*

Room dimensions and space

(1) Every room where persons work shall have sufficient floor area, height and unoccupied space for purposes of health, safety and welfare.

(2) It shall be sufficient compliance with this regulation in a workplace which is not a new workplace, a modification, an extension or a conversion and which, immediately before this regulation came into force in respect of it, was subject to the provisions of the

Factories Act 1961, if the workplace does not contravene the provisions of Part I of Schedule 1.

Regulation 11 *(See Approved Code of Practice, pp 336–337; Guidance Notes, p 369)*

Workstations and seating

(1) Every workstation shall be so arranged that it is suitable both for any person at work in the workplace who is likely to work at that workstation and for any work of the undertaking which is likely to be done there.

(2) Without prejudice to the generality of paragraph (1), every workstation outdoors shall be so arranged that—

(a) so far as is reasonably practicable, it provides protection from adverse weather;

(b) it enables any person at the workstation to leave it swiftly or, as appropriate, to be assisted in the event of an emergency; and

(c) it ensures that any person at the workstation is not likely to slip or fall.

(3) A suitable seat shall be provided for each person at work in the workplace whose work includes operations of a kind that the work (or a substantial part of it) can or must be done sitting.

(4) A seat shall not be suitable for the purpose of paragraph (3) unless—

(a) it is suitable for the person for whom it is provided as well as for the operations to be performed; and

(b) a suitable footrest is also provided where necessary.

Regulation 12 *(See Approved Code of Practice, pp 337–339; Guidance Notes, pp 369–370)*

Condition of floors and traffic routes

(1) Every floor in a workplace and the surface of every traffic route in a workplace shall be of a construction such that the floor or surface of the traffic route is suitable for the purpose for which it is used.

(2) Without prejudice to the generality of paragraph (1), the requirements in that paragraph shall include requirements that—

(a) the floor, or surface of the traffic route, shall have no hole or slope, or be uneven or slippery so as, in each case, to expose any person to a risk to his health or safety; and

(b) every such floor shall have effective means of drainage where necessary.

(3) So far as is reasonably practicable, every floor in a workplace and the surface of every traffic route in a workplace shall be kept free from obstructions and from any article or substance which may cause a person to slip, trip or fall.

(4) In considering whether for the purposes of paragraph (2)(a) a hole or slope exposes any person to a risk to his health or safety—

(a) no account shall be taken of a hole where adequate measures have been taken to prevent a person falling; and

(b) account shall be taken of any handrail provided in connection with any slope.

(5) Suitable and sufficient handrails and, if appropriate, guards shall be provided on all traffic routes which are staircases except in circumstances in which a handrail can not be provided without obstructing the traffic route.

Regulation 13 *(See Approved Code of Practice, pp 339–343; Guidance Notes, pp 370–373)*

Falls or falling objects

(1) So far as is reasonably practicable, suitable and effective measures shall be taken to prevent any event specified in paragraph (3).

(2) So far as is reasonably practicable, the measures required by paragraph (1) shall be measures other than the provision of personal protective equipment, information, instruction, training or supervision.

(3) The events specified in this paragraph are—

(a) any person falling a distance likely to cause personal injury;

(b) any person being struck by a falling object likely to cause personal injury.

(4) Any area where there is a risk to health or safety from any event mentioned in paragraph (3) shall be clearly indicated where appropriate.

(5) So far as is practicable, every tank, pit or structure where there is a risk of a person in the workplace falling into a dangerous substance in the tank, pit or structure, shall be securely covered or fenced.

(6) Every traffic route over, across or in an uncovered tank, pit or structure such as is mentioned in paragraph (5) shall be securely fenced.

(7) In this regulation, 'dangerous substance' means—

(a) any substance likely to scald or burn;

(b) any poisonous substance;

(c) any corrosive substance;

(d) any fume, gas or vapour likely to overcome a person; or

(e) any granular or free-flowing solid substance, or any viscous substance which, in any case, is of a nature or quantity which is likely to cause danger to any person.

Regulation 14 *(See Approved Code of Practice, pp 344–345; Guidance Notes, p 373)*

Windows, and transparent or translucent doors, gates and walls

(1) Every window or other transparent or translucent surface in a wall or partition and every transparent or translucent surface in a door or gate shall, where necessary for reasons of health or safety—

(a) be of safety material or be protected against breakage of the transparent or translucent material; and

(b) be appropriately marked or incorporate features so as, in either case, to make it apparent.

Regulation 15 *(See Approved Code of Practice, p 345; Guidance Notes, p 373)*

Windows, skylights and ventilators

(1) No window, skylight or ventilator which is capable of being opened shall be likely to be opened, closed or adjusted in a

manner which exposes any person performing such operation to a risk to his health or safety.

(2) No window, skylight or ventilator shall be in a position when open which is likely to expose any person in the workplace to a risk to his health or safety.

Regulation 16 *(See Approved Code of Practice, pp 345–346; Guidance Notes, p 373)*

Ability to clean windows etc safely

(1) All windows and skylights in a workplace shall be of a design or be so constructed that they may be cleaned safely.

(2) In considering whether a window or skylight is of a design or so constructed as to comply with paragraph (1), account may be taken of equipment used in conjunction with the window or skylight or of devices fitted to the building.

Regulation 17 *(See Approved Code of Practice, pp 346–350; Guidance Notes, p 374)*

Organisation etc of traffic routes

(1) Every workplace shall be organised in such a way that pedestrians and vehicles can circulate in a safe manner.

(2) Traffic routes in a workplace shall be suitable for the persons or vehicles using them, sufficient in number, in suitable positions and of sufficient size.

(3) Without prejudice to the generality of paragraph (2), traffic routes shall not satisfy the requirements of that paragraph unless suitable measures are taken to ensure that—

(a) pedestrians or, as the case may be, vehicles may use a traffic route without causing danger to the health or safety of persons at work near it;

(b) there is sufficient separation of any traffic route for vehicles from doors or gates or from traffic routes for pedestrians which lead onto it; and

(c) where vehicles and pedestrians use the same traffic route, there is sufficient separation between them.

(4) All traffic routes shall be suitably indicated where necessary for reasons of health or safety.

(5) Paragraph (2) shall apply so far as is reasonably practicable, to a workplace which is not a new workplace, a modification, an extension or a conversion.

Regulation 18 *(See Approved Code of Practice, pp 351–352; Guidance Notes, pp 374–375)*

Doors and gates

(1) Doors and gates shall be suitably constructed (including being fitted with any necessary safety devices).

(2) Without prejudice to the generality of paragraph (1), doors and gates shall not comply with that paragraph unless—

(a) any sliding door or gate has a device to prevent it coming off its track during use;

(b) any upward opening door or gate has a device to prevent it falling back;

(c) any powered door or gate has suitable and effective features to prevent it causing injury by trapping any person;

(d) where necessary for reasons of health or safety, any powered door or gate can be operated manually unless it opens automatically if the power fails; and

(e) any door or gate which is capable of opening by being pushed from either side is of such a construction as to provide, when closed, a clear view of the space close to both sides.

Regulation 19 *(See Guidance Notes, p 375)*

Escalators and moving walkways

Escalators and moving walkways shall:

(a) function safely;

(b) be equipped with any necessary safety devices;

(c) be fitted with one or more emergency stop controls which are easily identifiable and readily accessible.

Regulation 20

Sanitary conveniences

(1) Suitable and sufficient sanitary conveniences shall be provided at readily accessible places.

(2) Without prejudice to the generality of paragraph (1), sanitary conveniences shall not be suitable unless—

(a) the rooms containing them are adequately ventilated and lit;

(b) they and the rooms containing them are kept in a clean and orderly condition; and

(c) separate rooms containing conveniences are provided for men and women except where and so far as each convenience is in a separate room the door of which is capable of being secured from inside.

(3) It shall be sufficient compliance with the requirement in paragraph (1) to provide sufficient sanitary conveniences in a workplace which is not a new workplace, a modification, an extension or a conversion and which, immediately before this regulation came into force in respect of it, was subject to the provisions of the Factories Act 1961, if sanitary conveniences are provided in accordance with the provisions of Part II of Schedule 1.

Regulation 21 *(See Approved Code of Practice, pp 352–356; Guidance Notes, p 375)*

Washing facilities

(1) Suitable and sufficient washing facilities, including showers if required by the nature of the work or for health reasons, shall be provided at readily accessible places.

(2) Without prejudice to the generality of paragraph (1), washing facilities shall not be suitable unless—

(a) they are provided in the immediate vicinity of every sanitary convenience, whether or not provided elsewhere as well;

(b) they are provided in the vicinity of any changing rooms required by these Regulations, whether or not provided elsewhere as well;

(c) they include a supply of clean hot and cold, or warm, water (which shall be running water so far as is practicable);

(d) they include soap or other suitable means of cleaning;

(e) they include towels or other suitable means of drying;

(f) the rooms containing them are sufficiently ventilated and lit;

(g) they and the rooms containing them are kept in a clean and orderly condition; and

(h) separate facilities are provided for men and women, except where and so far as they are provided in a room the door of which is capable of being secured from inside and the facilities in each such room are intended to be used by only one person at a time.

(3) Paragraph (2)(h) shall not apply to facilities which are provided for washing hands, forearms and face only.

Regulation 22 *(See Approved Code of Practice, pp 356–357; Guidance Notes, p 375)*

Drinking water

(1) An adequate supply of wholesome drinking water shall be provided for all persons at work in the workplace.

(2) Every supply of drinking water required by paragraph (1) shall—

(a) be readily accessible at suitable places; and

(b) be conspicuously marked by an appropriate sign where necessary for reasons of health or safety.

(3) Where a supply of drinking water is required by paragraph (1), there shall also be provided a sufficient number of suitable cups or other drinking vessels unless the supply of drinking water is in a jet from which persons can drink easily.

Regulation 23 *(See Approved Code of Practice, p 357; Guidance Notes, p 376)*

Accommodation for clothing

(1) Suitable and sufficient accommodation shall be provided—

(a) for the clothing of any person at work which is not worn during working hours; and

(b) for special clothing which is worn by any person at work but which is not taken home.

(2) Without prejudice to the generality of paragraph (1), the accommodation mentioned in that paragraph shall not be suitable unless—

(a) where facilities to change clothing are required by regulation 24, it provides suitable security for the clothing mentioned in paragraph (1)(a);

(b) where necessary to avoid risks to health or damage to the clothing, it includes separate accommodation for clothing worn at work and for other clothing;

(c) so far as is reasonably practicable, it allows or includes facilities for drying clothing; and

(d) it is in a suitable location.

Regulation 24 *(See Approved Code of Practice, pp 357–358; Guidance Notes, p 376)*

Facilities for changing clothing

(1) Suitable and sufficient facilities shall be provided for any person at work in the workplace to change clothing in all cases where—

(a) the person has to wear special clothing for the purpose of work; and

(b) the person can not, for reasons of health or propriety, be expected to change in another room.

(2) Without prejudice to the generality of paragraph (1), the facilities mentioned in that paragraph shall not be suitable unless they include separate facilities for, or separate use of facilities by, men and women where necessary for reasons of propriety.

Regulation 25 *(See Approved Code of Practice, pp 358–360; Guidance Notes, pp 376–377)*

Facilities for rest and to eat meals

(1) Suitable and sufficient rest facilities shall be provided at readily accessible places.

(2) Rest facilities provided by virtue of paragraph (1) shall—

(a) where necessary for reasons of health or safety include, in the case of a new workplace an extension or a conversion, rest facilities provided in one or more rest rooms, or, in other cases, in rest rooms or rest areas;

(b) include suitable facilities to eat meals where food eaten in the workplace would otherwise be likely to become contaminated.

(3) Rest rooms and rest areas shall include suitable arrangements to protect non-smokers from discomfort caused by tobacco smoke.

(4) Suitable facilities shall be provided for any person at work who is a pregnant woman or nursing mother to rest.

(5) Suitable and sufficient facilities shall be provided for persons at work to eat meals where meals are regularly eaten in the workplace.

Regulation 26

Exemption certificates

(1) The Secretary of State for Defence may, in the interests of national security, by a certificate in writing exempt any of the home forces, any visiting force or any headquarters from the requirements of these Regulations and any exemption may be granted subject to conditions and to a limit of time and may be revoked by the said Secretary of State by a further certificate in writing at any time.

(2) In this regulation—

(a) 'the home forces' has the same meaning as in section 12(1) of the Visiting Forces Act 1952;

(b) 'headquarters' has the same meaning as in article 3(2) of the Visiting Forces and International Headquarters (Application of Law) Order 1965;

(c) 'visiting force' has the same meaning as it does for the purposes of any provision of Part I of the Visiting Forces Act 1952.

Regulation 27

Repeals, saving and revocations

(1) The enactments mentioned in column 2 of Part I of Schedule 2 are repealed to the extent specified in column 3 of that Part.

(2) Nothing in this regulation shall affect the operation of any provision of the Offices, Shops and Railway Premises Act 1963 as that provision has effect by virtue of section 90(4) of that Act.

(3) The instruments mentioned in column 1 of Part II of Schedule 2 are revoked to the extent specified in column 3 of that Part.

SCHEDULE 1
PROVISIONS APPLICABLE TO FACTORIES WHICH ARE NOT NEW WORKPLACES, MODIFICATIONS, EXTENSIONS OR CONVERSIONS

Regulations 10, 20

Part I
Space

1

No room in the workplace shall be so overcrowded as to cause risk to the health or safety of persons at work in it.

2

Without prejudice to the generality of paragraph 1, the number of persons employed at a time in any workroom shall not be such that the amount of cubic space allowed for each is less than 11 cubic metres.

3

In calculating for the purposes of this Part of this Schedule the amount of cubic space in any room no space more than 4.2 metres from the floor shall be taken into account and, where a room contains a gallery, the gallery shall be treated for the purposes of this Schedule as if it were partitioned off from the remainder of the room and formed a separate room.

Part II
Number of Sanitary Conveniences

4

In workplaces where females work, there shall be at least one suitable water closet for use by females only for every 25 females.

5

In workplaces where males work, there shall be at least one suitable water closet for use by males only for every 25 males.

6

In calculating the number of males or females who work in any workplace for the purposes of this Part of this Schedule, any number not itself divisible by 25 without fraction or remainder shall be treated as the next number higher than it which is so divisible.

SCHEDULE 2
REPEALS AND REVOCATIONS

Regulation 27

Part I
Repeals

1 *Chapter*	2 *Short title*	3 *Extent of repeal*
1961 c 34	The Factories Act 1961	Sections 1 to 7, 18, 28, 29, 57 to 60 and 69
1963 c 41	The Offices, Shops and Railway Premises Act 1963	Sections 4 to 16
1956 c 49	The Agriculture (Safety, Health and Welfare Provisions) Act 1956	Sections 3 and 5 and, in section 25, sub-sections (3) and (6)

Part II
Revocations

(1) *Title*	(2) *Reference*	(3) *Extent of revocation*
The Flax and Tow Spinning and Weaving Regulations 1906	SR & O 1906/177, amended by SI 1988/1657	Regulations 3, 8, 10, 11 and 14
The Hemp Spinning and Weaving Regulations 1907	SR & O 1907/660, amended by SI 1988/1657	Regulations 3 to 5 and 8
Order dated 5 October 1917 (the Tin or Terne Plates Manufacture Welfare Order 1917)	SR & O 1917/1035	The whole Order
Order dated 15 May 1918 (the Glass Bottle, etc Manufacture Welfare Order 1918)	SR & O 1918/558	The whole Order
Order dated 15 August 1919 (the Fruit Preserving Welfare Order 1919)	SR & O 1919/1136, amended by SI 1988/1657	The whole Order
Order dated 23 April 1920 (the Laundries Welfare Order 1920)	SR & O 1920/654	The whole Order
Order dated 28 July 1920 (the Gut Scraping, Tripe Dressing, etc Welfare Order 1920)	SR & O 1920/1437	The whole Order
Order dated 9 September 1920 (the Herring Curing (Norfolk and Suffolk) Welfare Order 1920)	SR & O 1920/1662	The whole Order
Order dated 3 March 1921 (the Glass Bevelling Welfare Order 1921)	SR & O 1921/288	The whole Order

The Herring Curing (Scotland) Welfare Order 1926	SR & O 1926/535 (S 24)	The whole Order
The Herring Curing Welfare Order 1927	SR & O 1927/813, amended by SI 1960/1690 and 917	The whole Order
The Sacks (Cleaning and Repairing) Welfare Order 1927	SR & O 1927/860	The whole Order
The Horizontal Milling Machines Regulations 1928	SR & O 1928/548	The whole Regulations
The Cotton Cloth Factories Regulations 1929	SI 1929/300	Regulations 5 to 10, 11 and 12
The Oil Cake Welfare Order 1929	SR & O 1929/534	Articles 3 to 6
The Cement Works Welfare Order 1930	SR & O 1930/94	The whole Order
The Tanning Welfare Order 1930	SR & O 1930/312	The whole Order
The Kiers Regulations 1938	SR & O 1938/106, amended by SI 1981/1152	Regulations 12 to 15
The Sanitary Accommodation Regulations 1938	SR & O 1938/611	The whole Regulations
The Clay Works (Welfare) Special Regulations 1948	SI 1948/1547	Regulations 3, 4, 6, 8 and 9
The Jute (Safety, Health and Welfare) Regulations 1948	SI 1948/1696, amended by SI 1988/1657	Regulations 11, 13, 14 to 16 and 19 to 26

The Pottery (Health and Welfare) Special Regulations 1950	SI 1950/65, amended by SI 1963/879, 1973/36, 1980/1248, 1982/877, 1988/1657, 1989/2311 and 1990/305	Regulation 15
The Iron and Steel Foundries Regulations 1953	SI 1953/1464, amended by SI 1974/1681 and 1981/1332	The whole Regulations
The Washing Facilities (Running Water) Exemption Regulations 1960	SI 1960/1029	The whole Regulations
The Washing Facilities (Miscellaneous Industries) Regulations 1960	SI 1960/1214	The whole Regulations
The Factories (Cleanliness of Walls and Ceilings) Order 1960	SI 1960/1794, amended by SI 1974/427	The whole Order
The Non-ferrous Metals (Melting and Founding) Regulations 1962	SI 1962/1667, amended by SI 1974/1681, 1981/1332 and 1988/165	Regulations 5, 6 to 10, 14 to 17 and 20
The Offices, Shops and Railway Premises Act 1963 (Exemption No 1) Order 1964	SI 1964/964	The whole Order
The Washing Facilities Regulations 1964	SI 1964/965	The whole Regulations
The Sanitary Conveniences Regulations 1964	SI 1964/966, amended by SI 1982/827	The whole Regulations

The Offices, Shops and Railway Premises Act 1963 (Exemption No 7) Order 1968	SI 1968/1947, amended by SI 1982/827	The whole Order
The Abrasive Wheels Regulations 1970	SI 1970/535	Regulation 17
The Sanitary Accommodation (Amendment) Regulations 1974	SI 1974/426	The whole Regulations
The Factories (Cleanliness of Walls and Ceilings) (Amendment) Regulations 1974	SI 1974/427	The whole Regulations
The Woodworking Machines Regulations 1974	SI 1974/903, amended by SI 1978/1126	Regulations 10 to 12
The Offices, Shops and Railway Premises Act 1963 etc (Metrication) Regulations 1982	SI 1982/827	The whole Regulations

APPROVED CODE OF PRACTICE

Regulation 1

Citation and commencement

1 *(See Guidance Notes, p 360)*

Regulation 2

Interpretation

2–8 *(See Guidance Notes, pp 360–362)*

Regulation 3

Application of these Regulations

9–13 *(See Guidance Notes, pp 362–363)*

Regulation 4

Requirements under these Regulations

14–19 *(See Guidance Notes, pp 363–365)*

Regulation 5

Maintenance of workplace, and of equipment, devices and systems

20 The workplace, and the equipment and devices mentioned in these Regulations, should be maintained in an efficient state, in efficient working order and in good repair. 'Efficient' in this context means efficient from the view of health, safety and welfare (not productivity or economy). If a potentially dangerous defect is discovered, the defect should be rectified immediately or steps should be taken to protect anyone who might be put at risk, for example by preventing access until the work can be carried out or the equipment replaced. Where the defect does not pose a danger but makes the equipment unsuitable for use, for example a sanitary convenience with a defective flushing mechanism, it may be taken out of service until it is repaired or replaced, but if this would result in the number of facilities being less than that required by the Regulations, the defect should be rectified without delay.

21 Steps should be taken to ensure that repair and maintenance work is carried out properly.

22 Regulation 5(2) requires a system of maintenance where appropriate, for certain equipment and devices and for ventilation systems. A suitable system of maintenance involves ensuring that:

(a) regular maintenance (including, as necessary, inspection, testing, adjustment, lubrication and cleaning) is carried out at suitable intervals;

(b) any potentially dangerous defects are remedied, and that access to defective equipment is prevented in the meantime;

(c) regular maintenance and remedial work is carried out properly; and

(d) a suitable record is kept to ensure that the system is properly implemented and to assist in validating maintenance programmes.

23 Examples of equipment and devices which require a system of maintenance include emergency lighting, fencing, fixed equipment

used for window cleaning, anchorage points for safety harnesses, devices to limit the opening of windows, powered doors, escalators and moving walkways.

24–27 *(See Guidance Notes, pp 365–366)*

Regulation 6

Ventilation

28 Enclosed workplaces should be sufficiently well ventilated so that stale air, and air which is hot or humid because of the processes or equipment in the workplace, is replaced at a reasonable rate.

29 The air which is introduced should, as far as possible, be free of any impurity which is likely to be offensive or cause ill health. Air which is taken from the outside can normally be considered to be 'fresh', but air inlets for ventilation systems should not be sited where they may draw in excessively contaminated air (for example close to a flue, an exhaust ventilation system outlet, or an area in which vehicles manoeuvre). Where necessary the inlet air should be filtered to remove particulates.

30 In many cases, windows or other openings will provide sufficient ventilation in some or all parts of the workplace. Where necessary, mechanical ventilation systems should be provided for parts or all of the workplace, as appropriate.

31 Workers should not be subject to uncomfortable draughts. In the case of mechanical ventilation systems it may be necessary to control the direction or velocity of air flow. Workstations should be re-sited or screened if necessary.

32 In the case of mechanical ventilation systems which recirculate air, including air-conditioning systems, recirculated air should be adequately filtered to remove impurities. To avoid air becoming unhealthy, purified air should have some fresh air added to it before being recirculated. Systems should therefore be designed with fresh air inlets which should be kept open.

33 Mechanical ventilation systems (including air-conditioning systems) should be regularly and properly cleaned, tested and maintained to ensure that they are kept clean and free from anything which may contaminate the air.

34 The requirement of regulation 6(2) for a device to give warning of breakdowns applies only 'where necessary for reasons

of health or safety'. It will not apply in most workplaces. It will, however, apply to 'dilution ventilation' systems used to reduce concentrations of dust or fumes in the atmosphere, and to any other situation where a breakdown in the ventilation system would be likely to result in harm to workers.

35–41 *(See Guidance Notes, p 366)*

Regulation 7

Temperature in indoor workplaces

42 The temperature in workrooms should provide reasonable comfort without the need for special clothing. Where such a temperature is impractical because of hot or cold processes, all reasonable steps should be taken to achieve a temperature which is as close as possible to comfortable. 'Workroom' in paragraphs 43 to 49 means a room where people normally work for more than short periods.

43 The temperature in workrooms should normally be at least 16 degrees Celsius unless much of the work involves severe physical effort in which case the temperature should be at least 13 degrees Celsius. These temperatures may not, however, ensure reasonable comfort, depending on other factors such as air movement and relative humidity. These temperatures refer to readings taken using an ordinary dry bulb thermometer, close to workstations, at working height and away from windows.

44 Paragraph 43 does not apply to rooms or parts of rooms where it would be impractical to maintain those temperatures, for example in rooms which have to be open to the outside, or where food or other products have to be kept cold. In such cases the temperature should be as close to those mentioned in paragraph 43 as is practical. In rooms where food or other products have to be kept at low temperatures this will involve such measures as:

(a) enclosing or insulating the product;

(b) pre-chilling the product;

(c) keeping chilled areas as small as possible;

(d) exposing the product to workroom temperatures as briefly as possible.

45 Paragraphs 43 and 44 do not apply to rooms where lower maximum room temperatures are required in other laws. It should be noted that general Food Hygiene Regulations do not specify maximum room temperatures.

46 Where the temperature in a workroom would otherwise be uncomfortably high, for example because of hot processes or the design of the building, all reasonable steps should be taken to achieve a reasonably comfortable temperature, for example by:

(a) insulating hot plants or pipes;

(b) providing air-cooling plant;

(c) shading windows;

(d) siting workstations away from places subject to radiant heat.

47 Where a reasonably comfortable temperature cannot be achieved throughout a workroom, local heating or cooling (as appropriate) should be provided. In extremely hot weather fans and increased ventilation may be used instead of local cooling. Insulated duckboards or other floor coverings should be provided where workers have to stand for long periods on cold floors unless special footwear is provided which prevents discomfort. Draughts should be excluded and self-closing doors installed where such measures are practical and would reduce discomfort.

48 Where, despite the provision of local heating or cooling, workers are exposed to temperatures which do not give reasonable comfort, suitable protective clothing and rest facilities should be provided. Where practical there should be systems of work (for example, task rotation) to ensure that the length of time for which individual workers are exposed to uncomfortable temperatures is limited.

49 In parts of the workplace other than workrooms, such as sanitary facilities or rest facilities, the temperature should be reasonable in all the circumstances including the length of time people are likely to be there. Changing rooms and shower rooms should not be cold.

50 Where persons are required to work in normally unoccupied rooms such as storerooms, other than for short periods, temporary heating should be provided if necessary to avoid discomfort.

51–55 *(See Guidance Notes, p 367)*

Injurious or offensive fumes

56 Fixed heating systems should be installed and maintained in such a way that the products of combustion do not enter the workplace. Any heater which produces heat by combustion should have a sufficient air supply to ensure complete combustion. Care should be taken that portable paraffin and liquefied petroleum gas heaters do not produce fumes which will be harmful or offensive (see paragraph 52).

Thermometers

57 Thermometers should be available at a convenient distance from every part of the workplace to persons at work to enable temperatures to be measured throughout the workplace, but need not be provided in each workroom.

Regulation 8
Lighting

58 Lighting should be sufficient to enable people to work, use facilities and move from place to place safely and without experiencing eye-strain. Stairs should be well lit in such a way that shadows are not cast over the main part of the treads. Where necessary, local lighting should be provided at individual workstations, and at places of particular risk such as pedestrian crossing points on vehicular traffic routes. Outdoor traffic routes used by pedestrians should be adequately lit after dark.

59 Dazzling lights and annoying glare should be avoided. Lights and light fittings should be of a type, and so positioned, that they do not cause a hazard (including electrical, fire, radiation or collision hazards). Light switches should be positioned so that they may be found and used easily and without risk.

60 Lights should not be allowed to become obscured, for example by stacked goods, in such a way that the level of light becomes insufficient. Lights should be replaced, repaired or cleaned, as necessary, before the level of lighting becomes insufficient. Fittings or lights should be replaced immediately if they become dangerous, electrically or otherwise.

61–62 *(See Guidance Notes, p 367)*

Natural lighting

63 Windows and skylights should where possible be cleaned regularly and kept free from unnecessary obstructions to admit maximum daylight. Where this would result in excessive heat or glare at a workstation, however, the workstation should be repositioned or the window or skylight should be shaded.

64 *(See Guidance Notes, pp 367–368)*

Emergency lighting

65 The normal precautions required by these and other Regulations, for example on the prevention of falls and the fencing of dangerous parts of machinery, mean that workers are not in most cases 'specially exposed' to risk if normal lighting fails. Emergency lighting is not therefore essential in most cases. Emergency lighting should, however, be provided in workrooms where sudden loss of light would present a serious risk, for example if process plant needs to be shut down under manual control or a potentially hazardous process needs to be made safe, and this cannot be done safely without lighting.

66 Emergency lighting should be powered by a source independent from that of normal lighting. It should be immediately effective in the event of failure of the normal lighting, without need for action by anyone. It should provide sufficient light to enable persons at work to take any action necessary to ensure their, and others', health and safety.

67 *(See Guidance Notes, p 368)*

Regulation 9

Cleanliness and waste materials

68 The standard of cleanliness required will depend on the use to which the workplace is put. For example, an area in which workers take meals would be expected to be cleaner than a factory floor, and a factory floor would be expected to be cleaner than an animal house. However, regulation 12(3) (avoidance of slipping, tripping and falling hazards) should be complied with in all cases.

69 Floors and indoor traffic routes should be cleaned at least once a week. In factories and other workplaces of a type where

dirt and refuse accumulates, any dirt and refuse which is not in suitable receptacles should be removed at least daily. These tasks should be carried out more frequently where necessary to maintain a reasonable standard of cleanliness or to keep workplaces free of pests and decaying matter. This paragraph does not apply to parts of workplaces which are normally visited only for short periods, or to animal houses.

70 Interior walls, ceilings and work surfaces should be cleaned at suitable intervals. Except in parts which are normally visited only for short periods, or where any soiling is likely to be light, ceilings and interior walls should be painted, tiled or otherwise treated so that they can be kept clean, and the surface treatment should be renewed when it can no longer be cleaned properly. This paragraph does not apply to parts of workplaces which cannot be safely reached using a 5-metre ladder.

71 Apart from regular cleaning, cleaning should also be carried out when necessary in order to clear up spillages or to remove unexpected soiling of surfaces. Workplaces should be kept free from offensive waste matter or discharges, for example, leaks from drains or sanitary conveniences.

72 Cleaning should be carried out by an effective and suitable method and without creating, or exposing anyone to, a health or safety risk.

73–75 *(See Guidance Notes, p 368)*

Regulation 10

Room dimensions and space

Minimum space

76 Workrooms should have enough free space to allow people to get to and from workstations and to move within the room, with ease. The number of people who may work in any particular room at any one time will depend not only on the size of the room, but on the space taken up by furniture, fittings, equipment, and on the layout of the room. Workrooms, except those where people only work for short periods, should be of sufficient height (from floor to ceiling) over most of the room to enable safe access to workstations. In older buildings with obstructions such as low beams the obstruction should be clearly marked.

77 The total volume of the room, when empty, divided by the number of people normally working in it should be at least 11 cubic metres. In making this calculation a room or part of a room which is more than 3.0 m high should be counted as 3.0 m high. The figure of 11 cubic metres per person is a minimum and may be insufficient if, for example, much of the room is taken up by furniture etc.

78 The figure of 11 cubic metres referred to in paragraph 77 does not apply to:

(a) retail sales kiosks, attendants' shelters, machine control cabs or similar small structures, where space is necessarily limited; or

(b) rooms being used for lectures, meetings and similar purposes.

79–80 *(See Guidance Notes, pp 368–369)*

Regulation 11

Workstations and seating

81 Workstations should be arranged so that each task can be carried out safely and comfortably. The worker should be at a suitable height in relation to the work surface. Work materials and frequently used equipment or controls should be within easy reach, without undue bending or stretching.

82 Workstations including seating, and access to workstations, should be suitable for any special needs of the individual worker, including workers with disabilities.

83 Each workstation should allow any person who is likely to work there adequate freedom of movement and the ability to stand upright. Spells of work which unavoidably have to be carried out in cramped conditions should be kept as short as possible and there should be sufficient space nearby to relieve discomfort.

84 There should be sufficient clear and unobstructed space at each workstation to enable the work to be done safely. This should allow for the manoeuvring and positioning of materials, for example lengths of timber.

85 Seating provided in accordance with regulation 11(3) should where possible provide adequate support for the lower back, and

a footrest should be provided for any worker who cannot comfortably place his or her feet flat on the floor.

86–88 *(See Guidance Notes, p 369)*

Regulation 12

Condition of floors and traffic routes

89 Floor and traffic routes should be of sound construction and should have adequate strength and stability taking account of the loads placed on them and the traffic passing over them. Floors should not be overloaded.

90 The surfaces of floors and traffic routes should be free from any hole, slope, or uneven or slippery surface which is likely to:

(a) cause a person to slip, trip or fall;

(b) cause a person to drop or lose control of anything being lifted or carried; or

(c) cause instability or loss of control of vehicles and/or their loads.

91 Holes, bumps or uneven areas resulting from damage or wear and tear, which may cause a person to trip or fall, should be made good. Until they can be made good, adequate precautions should be taken against accidents, for example by barriers or conspicuous marking. Temporary holes, for example an area where floor boards have been removed, should be adequately guarded. Account should be taken of people with impaired or no sight. Surfaces with small holes (for example metal gratings) are acceptable provided they are not likely to be a hazard. Deep holes into which people may fall are subject to regulation 13 and the relevant section of this Code.

92 Slopes should not be steeper than necessary. Moderate and steep slopes, and ramps used by people with disabilities, should be provided with a secure handrail where necessary.

93 Surfaces of floors and traffic routes which are likely to get wet or to be subject to spillages should be of a type which does not become unduly slippery. A slip-resistant coating should be applied where necessary. Floors near to machinery which could cause injury if anyone were to fall against it (for example a woodworking or grinding machine) should be slip-resistant and be kept free from slippery substances or loose materials.

94 Where possible, processes and plant which may discharge or leak liquids should be enclosed (for example by bunding), and leaks from taps or discharge points on pipes, drums and tanks should be caught or drained away. Stop valves should be fitted to filling points on tank filling lines. Where work involves carrying or handling liquids or slippery substances, as in food processing and preparation, the workplace and work surfaces should be arranged in such a way as to minimise the likelihood of spillages.

95 Where a leak or spillage occurs and is likely to be a slipping hazard, immediate steps should be taken to fence it off, mop it up, or cover it with absorbent granules.

96 Arrangements should be made to minimise risks from snow and ice. This may involve gritting, snow clearing and closure of some routes, particularly outside stairs, ladders and walkways on roofs.

97 Floors and traffic routes should be kept free of obstructions which may present a hazard or impede access. This is particularly important on or near stairs, steps, escalators and moving walkways, on emergency routes, in or near doorways or gangways, and in any place where an obstruction is likely to cause an accident, for example near a corner or junction. Where a temporary obstruction is unavoidable and is likely to be a hazard, access should be prevented or steps should be taken to warn people or the drivers of vehicles of the obstruction by, for example, the use of hazard cones. Where furniture or equipment is being moved within a workplace, it should if possible be moved in a single operation and should not be left in a place where it is likely to be a hazard. Vehicles should not be parked where they are likely to be a hazard. Materials which fall onto traffic routes should be cleared as soon as possible.

98 Effective drainage should be provided where a floor is liable to get wet to the extent that the wet can be drained off. This is likely to be the case in, for example, laundries, textile manufacture (including dyeing, bleaching and finishing), work on hides and skins, potteries and food processing. Drains and channels should be positioned so as to minimise the area of wet floor, and the floor should slope slightly towards the drain. Where necessary to prevent tripping hazards, drains and channels should have covers which should be as near flush as possible with the floor surface.

99 Every open side of a staircase should be securely fenced. As a minimum the fencing should consist of an upper rail at 900 mm or higher and a lower rail.

100 A secure and substantial handrail should be provided and maintained on at least one side of every staircase, except at points where a handrail would obstruct access or egress, as in the case of steps in a theatre aisle. Handrails should be provided on both sides if there is a particular risk of falling, for example where stairs are heavily used, or are wide, or have narrow treads, or where they are liable to be subject to spillages. Additional handrails should be provided down the centre of particularly wide staircases where necessary.

101–105 *(See Guidance Notes, pp 369–370)*

Regulation 13

Falls or falling objects

106–107 *(See Guidance Notes, p 370)*

Provision of fencing or covers

108 Secure fencing should be provided wherever possible at any place where a person might fall 2 metres or more. Secure fencing should also be provided where a person might fall less than 2 metres, where there are factors which increase the likelihood of a fall or the risk of serious injury; for example where a traffic route passes close to an edge, where large numbers of people are present, or where a person might fall onto a sharp or dangerous surface or into the path of a vehicle. Tanks, pits or similar structures may be securely covered instead of being fenced.

109 Fencing should be sufficiently high, and filled in sufficiently, to prevent falls (of people or objects) over or through the fencing. As a minimum, fencing should consist of two guard-rails (a top rail and a lower rail) at suitable heights. In the case of fencing installed after 1 January 1993 (but not repairs or partial replacement) the top of the fencing should be at least 1100 mm above the surface from which a person might fall except in cases where lower fencing has been approved by a local authority under Building Regulations.

110 Fencing should be of adequate strength and stability to restrain any person or object liable to fall on to or against it.

Untensioned chains, ropes and other non-rigid materials should not be used.

111 Fencing should be designed to prevent objects falling from the edge including items used for cleaning or maintenance. Where necessary an adequate upstand or toeboard should be provided.

112 Covers should be capable of supporting all loads liable to be imposed upon them, and any traffic which is liable to pass over them. They should be of a type which cannot be readily detached and removed, and should not be capable of being easily displaced.

113 Paragraphs 108 to 111 do not apply to edges on roofs or to places to which there is no general access. Nevertheless, secure, adequate fencing should be provided wherever possible in such cases. Tanks, pits or similar structures containing dangerous substances should always be provided with secure fencing or a secure cover.

114–115 *(See Guidance Notes, pp 370–371)*

Temporary removal of fencing or covers

116 When an opening or an edge is being used to transfer goods or materials from one level to another, it should be fenced as far as possible. Secure handholds should be provided where workers have to position themselves at an unfenced opening or edge, such as a teagle opening or similar doorway used for the purpose of hoisting or lowering goods. Where the operation necessarily involves the use of an unguarded edge, as little fencing or rail as possible should be removed, and should be replaced as soon as possible.

117 *(See Guidance Notes, p 371)*

118 Covers should be kept securely in place except when they have to be removed for inspection purposes or in order to gain access. Covers should be replaced as soon as possible.

Fixed ladders

119 Fixed ladders should not be provided in circumstances where it would be practical to install a staircase (see paragraph 162 of this Code). Fixed ladders or other suitable means of access or egress should be provided in pits, tanks and similar structures into which workers need to descend. In this Code a 'fixed ladder'

includes a steep stairway (a staircase which a person normally descends facing the treads or rungs).

120 Fixed ladders should be of sound construction, properly maintained and securely fixed. Rungs of a ladder should be horizontal, give adequate foothold and not depend solely upon nails, screws or similar fixings for their support.

121 Unless some other adequate handhold exists, the stiles of the ladder should extend at least 1100 mm above any landing served by the ladder or the highest rung used to step or stand on except that in the case of chimneys the stiles should not project into the gas stream.

122 Fixed ladders installed after 31 December 1992 with a vertical distance of more than 6 m should normally have a landing or other adequate resting place at every 6 m point. Each run should, where possible, be out of line with the last run, to reduce the distance a person might fall. Where it is not possible to provide such landings, for example on a chimney, the ladders should only be used by specially trained and proficient people.

123 Where a ladder passes through a floor, the opening should be as small as possible. The opening should be fenced as far as possible, and a gate should be provided where necessary to prevent falls.

124 Fixed ladders at an angle of less than 15 degrees to the vertical (a pitch of more than 75 degrees) which are more than 2.5 m high should where possible be fitted with suitable safety hoops or permanently fixed fall arrest systems. Hoops should be at intervals of not more than 900 mm measured along the stiles, and should commence at a height of 2.5 m above the base of the ladder. The top hoop should be in line with the top of the fencing on the platform served by the ladder. Where a ladder rises less than 2.5 m, but is elevated so that it is possible to fall a distance of more than 2 m, a single hoop should be provided in line with the top of the fencing. Where the top of a ladder passes through a fenced hole in a floor, a hoop need not be provided at that point.

125–128 *(See Guidance Notes, p 371)*

129 Where regular access is needed to roofs (including internal roofs, for example a single-storey office within a larger building) suitable permanent access should be provided and there should be fixed physical safeguards to prevent falls from edges and through fragile roofs. Where occasional access is required, other

safeguards should be provided, for example crawling boards, temporary access equipment etc.

130 A fragile roof or surface is one which would be liable to fracture if a person's weight were to be applied to it, whether by walking, falling on to it or otherwise. All glazing and asbestos cement or similar sheeting should be treated as being fragile unless there is firm evidence to the contrary. Fragile roofs or surfaces should be clearly identified.

131 *(See Guidance Notes, p 371)*

Falls into dangerous substances

132 The tanks, pits and structures mentioned in regulation 13(5) are referred to here as 'vessels' and include sumps, silos, vats, and kiers which persons could fall into. (Kiers are fixed vessels which are used for boiling textile materials in workplaces where the printing, bleaching or dyeing of textile materials or waste is carried out).

133 Every vessel containing a dangerous substance should be adequately protected to prevent a person from falling into it. Vessels installed after 31 December 1992 should be securely covered, or fenced to a height of at least 1100 mm unless the sides extend to at least 1100 mm above the highest point from which people could fall into them. In the case of existing vessels the height should be at least 915 mm or, in the case of atmospheric or open kiers, 840 mm.

Changes of level

134 *(See Guidance Notes, p 372)*

Stacking and racking

135 Materials and objects should be stored and stacked in such a way that they are not likely to fall and cause injury. Racking should be of adequate strength and stability having regard to the loads placed on it and its vulnerability to damage, for example by vehicles.

136–137 *(See Guidance Notes, p 372)*

Loading or unloading vehicles

138 The need for people to climb on top of vehicles or their loads should be avoided as far as possible. Where it is unavoidable, effective measures should be taken to prevent falls.

139 Where a tanker is loaded from a fixed gantry and access is required on to the top of the tanker, fencing should be provided where possible. The fencing may be collapsible fencing on top of the tanker or may form part of the gantry. In the latter case if varying designs of tankers are loaded the fencing should be adjustable, where necessary. Similar fencing should also be provided wherever people regularly go on top of tankers at a particular location, for example for maintenance.

140 Where loaded lorries have to be sheeted before leaving a workplace, suitable precautions should be taken against falls. Where sheeting is done frequently it should be carried out in designated parts of the workplace which are equipped for safe sheeting. Where reasonably practicable, gantries should be provided which lorries can drive under or alongside, so that the load is sheeted from the gantry without any need to stand on the cargo. In other situations safety lines and harnesses should be provided for people on top of the vehicle.

Measures other than fencing, covers, etc

141 When fencing or covers cannot be provided, or have to be removed, effective measures should be taken to prevent falls. Access should be limited to specified people and others should be kept out by, for example, barriers; in high risk situations suitable formal written permit to work systems should be adopted. A safe system of work should be operated which may include the provision and use of a fall arrest system, or safety lines and harnesses, and secure anchorage points. Safety lines should be short enough to prevent injury should a fall occur and the safety line operate. Adequate information, instruction, training and supervision should be given.

142 People should not be allowed into an area where, despite safeguards, they would be in danger, for example from work going on overhead.

143–144 *(See Guidance Notes, p 372)*

Scaffolding

145 Scaffolding and other equipment used for temporary access may either follow the provisions of this Code or the requirements of Construction Regulations.

Other Regulations

146 *(See Guidance Notes, p 373)*

Regulation 14

Windows, and transparent or translucent doors, gates and walls

147 In assessing whether it is necessary, for reasons of health or safety for transparent or translucent surfaces in doors, gates, walls and partitions to be of a safety material or be adequately protected against breakage, particular attention should be paid to the following cases:

(a) in doors and gates, and door and gate side panels, where any part of the transparent or translucent surface is at shoulder level or below;

(b) in windows, walls and partitions, where any part of the transparent or translucent surface is at waist level or below, except in glasshouses where people there will be likely to be aware of the presence of glazing and avoid contact.

This paragraph does not apply to narrow panes up to 250 mm wide measured between glazing beads.

148 'Safety materials' are:

(a) materials which are inherently robust, such as polycarbonates or glass blocks; or

(b) glass which, if it breaks, breaks safely; or

(c) ordinary annealed glass which meets the thickness criteria in the following table:

Nominal thickness	*Maximum size*
8 mm	1.10 m x 1.10 m
10 mm	2.25 m x 2.25 m
12 mm	3.00 m x 4.50 m
15 mm	Any size

149 As an alternative to the use of safety materials, transparent or translucent surfaces may be adequately protected against breakage. This may be achieved by means of a screen or barrier which will prevent a person from coming into contact with the glass if he or she falls against it. If a person going through the glass would fall from a height, the screen or barrier should also be designed to be difficult to climb.

150 A transparent or translucent surface should be marked where necessary to make it apparent. The risk of collision is greatest in large uninterrupted surfaces where the floor is at a similar level on each side, so that people might reasonably think they can walk straight through. If features such as mullions, transoms, rails, door frames, large pull or push handles, or heavy tinting make the surface apparent, marking is not essential. Where it is needed, marking may take any form (for example coloured lines or patterns), provided it is conspicuous and at a conspicuous height.

151–152 *(See Guidance Notes, p 373)*

Regulation 15

Windows, skylights and ventilators

153 It should be possible to reach and operate the control of openable windows, skylights and ventilators in a safe manner. Where necessary, window poles or similar equipment should be kept available, or a stable platform or other safe means of access should be provided. Controls should be so placed that people are not likely to fall through or out of the window. Where there is a danger of falling from a height, devices should be provided to prevent the window opening too far.

154 Open windows, skylights or ventilators should not project into an area where persons are likely to collide with them. The bottom edge of opening windows should normally be at least 800 mm above floor level, unless there is a barrier to prevent falls.

155 *(See Guidance Notes, p 373)*

Regulation 16

Ability to clean windows etc safely

156 Suitable provision should be made so that windows and skylights can be cleaned safely if they cannot be cleaned from the ground or other suitable surface.

157 Suitable provision includes:

(a) fitting windows which can be cleaned safely from the inside, for example windows which pivot so that the outer surface is turned inwards;

(b) fitting access equipment such as suspended cradles, or travelling ladders with an attachment for a safety harness;

(c) providing suitable conditions for the future use of mobile access equipment, including ladders up to 9 metres long. Suitable conditions are adequate access for the equipment, and a firm level surface in a safe place on which to stand it. Where a ladder over 6 metres long will be needed, suitable points for tying or fixing the ladder should be provided;

(d) suitable and suitably placed anchorage points for safety harnesses.

158 *(See Guidance Notes, p 373)*

Regulation 17

Organisation etc of traffic routes

159 This section of the Code applies to both new and existing workplaces. In paragraphs 160, 165 and 171 special provision is made for traffic routes in existence before 1 January 1993. This is because it might, in a few cases, otherwise be difficult for existing routes to comply fully with the Code. These special provisions reflect regulation 17(5) which has the effect of requiring existing traffic routes to comply with regulation 17(2) and 17(3) only to the extent that it is reasonably practicable. 'Traffic route' is defined in regulation 2 as 'a route for pedestrian traffic, vehicles or both and includes any stairs, staircase, fixed ladder, doorway, gateway, loading bay or ramp'.

160 There should be sufficient traffic routes, of sufficient width and headroom, to allow people on foot or in vehicles to circulate safely and without difficulty. Features which obstruct routes should be avoided. On traffic routes in existence before 1 January 1993, obstructions such as limited headroom are acceptable provided they are indicated by, for example, the use of conspicuous tape. Consideration should be given to the safety of people with impaired or no sight.

161 In some situations people in wheelchairs may be at greater risk than people on foot, and special consideration should be given to their safety. Traffic routes used by people in wheelchairs should be wide enough to allow unimpeded access, and ramps should be provided where necessary. Regulation 12(4) and paragraph 92 of this Code also deal with ramps.

162 Access between floors should not normally be by way of ladders or steep stairs. Fixed ladders or steep stairs may be used where a conventional staircase cannot be accommodated, provided they are only used by people who are capable of using them safely and any loads to be carried can be safely carried.

163 Routes should not be used by vehicles for which they are inadequate or unsuitable. Any necessary restrictions should be clearly indicated. Uneven or soft ground should be made smooth and firm if vehicles might otherwise overturn or shed their loads. Sharp or blind bends on vehicle routes should be avoided as far as possible; where they are unavoidable, measures such as one-way systems or the use of mirrors to improve vision should be considered. On vehicle routes, prominent warning should be given of any limited headroom, both in advance and at the obstruction itself. Any potentially dangerous obstructions such as overhead electric cables or pipes containing, for example, flammable or hazardous chemicals should be shielded. Screens should be provided where necessary to protect people who have to work at a place where they would be at risk from exhaust fumes, or to protect people from materials which are likely to fall from vehicles.

164 Sensible speed limits should be set and clearly displayed on vehicle routes except those used only by slow vehicles. Where necessary, suitable speed retarders such as road humps should be provided. These should always be preceded by a warning sign or a mark on the road. Arrangements should be made where necessary to avoid fork lift trucks having to pass over road humps unless the truck is of a type which can negotiate them safely.

165 Traffic routes used by vehicles should be wide enough to allow vehicles to pass oncoming or parked vehicles without leaving the route. One-way systems or restrictions on parking should be introduced as necessary. On traffic routes in existence before

1 January 1993, where it is not practical to make the route wide enough, passing places or traffic management systems should be provided as necessary.

166 Traffic routes used by vehicles should not pass close to any edge, or to anything that is likely to collapse or be left in a dangerous state if hit (such as hollow cast-iron columns and storage racking), unless the edge or thing is fenced or adequately protected.

167 The need for vehicles with poor rear visibility to reverse should be eliminated as far as possible, for example by the use of one-way systems.

168–169 *(See Guidance Notes, p 374)*

170 Where a load has to be tipped into a hopper, waste pit, or similar place, and the vehicle is liable to fall into it, substantial barriers or portable wheel stops should be provided at the end of the traffic route to prevent this type of occurrence.

Separation of people and vehicles

171 Any traffic route which is used by both pedestrians and vehicles should be wide enough to enable any vehicle likely to use the route to pass pedestrians safely. On traffic routes in existence before 1 January 1993, where it is not practical to make the route wide enough, passing places or traffic management systems should be provided as necessary. In buildings, lines should be drawn on the floor to indicate routes followed by vehicles such as fork lift trucks.

172 On routes used by automatic, driverless vehicles which are also used by pedestrians, steps should be taken to ensure that pedestrians do not become trapped by vehicles. The vehicles should be fitted with safeguards to minimise the risk of injury, sufficient clearance should be provided between the vehicles and pedestrians, and care should be taken that fixtures along the route do not create trapping hazards.

173 In doorways, gateways, tunnels, bridges, or other enclosed routes, vehicles should be separated from pedestrians by a kerb or barrier. Where necessary, for safety, separate routes through should

be provided and pedestrians should be guided to use the correct route by clear marking. Such routes should be kept unobstructed. Similar measures should be taken where the speed or volume of vehicles would put pedestrians at risk.

174 Workstations should be adequately separated or shielded from vehicles.

Crossings

175 Where pedestrian and vehicle routes cross, appropriate crossing points should be provided and used. Where necessary, barriers or rails should be provided to prevent pedestrians crossing at particularly dangerous points and to guide them to designated crossing places. At crossing places where volumes of traffic are particularly heavy, the provision of suitable bridges or subways should be considered.

176 At crossing points there should be adequate visibility and open space for the pedestrian where the pedestrian route joins the vehicle route. For example, where an enclosed pedestrian route, or a doorway or staircase, joins a vehicle route there should be an open space of at least one metre from which pedestrians can see along the vehicle route in both directions (or in the case of a one-way route, in the direction of oncoming traffic). Where such a space cannot be provided, barriers or rails should be provided to prevent pedestrians walking directly onto the vehicular route.

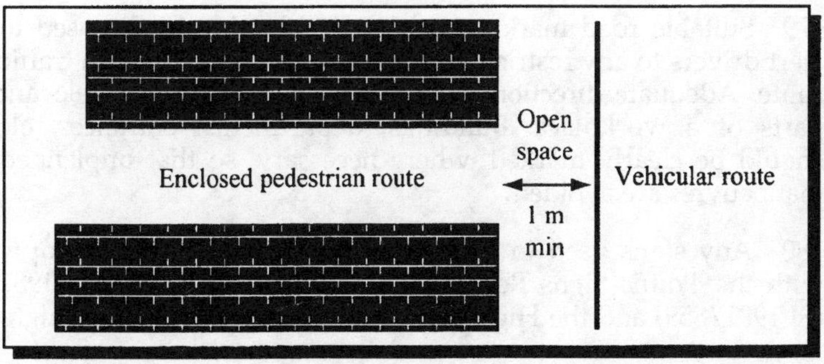

Loading bays

177 Loading bays should be provided with at least one exit point from the lower level. Wide loading bays should be provided with at least two exit points, one being at each end. Alternatively, a refuge should be provided which can be used to avoid being struck or crushed by a vehicle.

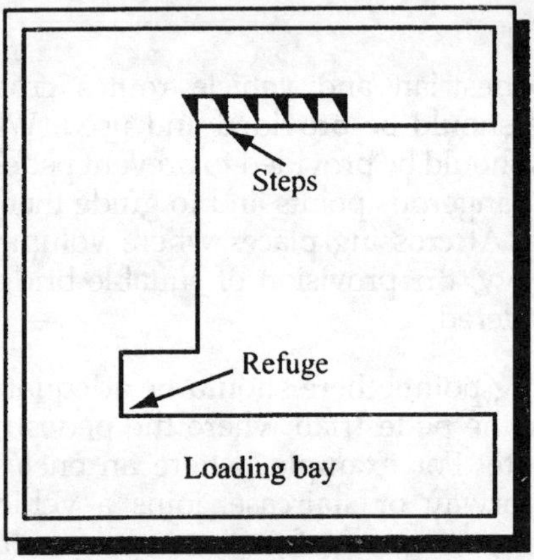

Signs

178 Potential hazards on traffic routes used by vehicles should be indicated by suitable warning signs. Such hazards may include: sharp bends, junctions, crossings, blind corners, steep gradients or roadworks.

179 Suitable road markings and signs should also be used to alert drivers to any restrictions which apply to the use of a traffic route. Adequate directions should also be provided to relevant parts of a workplace. Buildings, departments, entrances, etc should be clearly marked, where necessary, so that unplanned manoeuvres are avoided.

180 Any signs used in connection with traffic should comply with the Traffic Signs Regulations and General Directions 1981 (SI 1981/859) and the Highway Code for use on the public highway.

181–182 *(See Guidance Notes, p 374)*

Regulation 18

Doors and gates

183 Doors and gates which swing in both directions should have a transparent panel except if they are low enough to see over. Conventionally hinged doors on main traffic routes should also be fitted with such panels. Panels should be positioned to enable a person in a wheelchair to be seen from the other side.

184 Sliding doors should have a stop or other effective means to prevent the door coming off the end of the track. They should also have a retaining rail to prevent the door falling should the suspension system fail or the rollers leave the track.

185 Upward opening doors should be fitted with an effective device such as a counterbalance or ratchet mechanism to prevent them falling back in a manner likely to cause injury (see paragraph 190).

186 Power operated doors and gates should have safety features to prevent people being injured as a result of being struck or trapped. Safety features include:

(a) a sensitive edge, or other suitable detector, and associated trip device to stop, or reverse, the motion of the door or gate when obstructed;

(b) a device to limit the closing force so that it is insufficient to cause injury;

(c) an operating control which must be held in position during the whole of the closing motion. This will only be suitable where the risk of injury is low and the speed of closure is slow. Such a control, when released, should cause the door to stop or reopen immediately and should be positioned so that the operator has a clear view of the door throughout its movement.

187 Where necessary, power operated doors and gates should have a readily identifiable and accessible control switch or device so that they can be stopped quickly in an emergency. Normal on/off controls may be sufficient.

188 It should be possible to open a power operated door or gate if the power supply fails, unless it opens automatically in such circumstances, or there is an alternative way through. This does not apply to lift doors and other doors and gates which are there to prevent falls or access to areas of potential danger.

189 Where tools are necessary for manual opening they should be readily available at all times. If the power supply is restored while the door is being opened manually, the person opening it should not be put at risk.

190 *(See Guidance Notes, pp 374–375)*

Regulation 19

Escalators and moving walkways

191 *(See Guidance Notes, p 375)*

Regulation 21

Washing facilities

192 In paragraphs 193–211 'facilities' means sanitary and washing facilities, 'sanitary accommodation' means a room containing one or more sanitary conveniences and 'washing station' means a wash-basin or a section of a trough or fountain sufficient for one person.

193 Sufficient facilities should be provided to enable everyone at work to use them without undue delay. Minimum numbers of facilities are given in paragraphs 201–205 but more may be necessary if, for example, breaks are taken at set times or workers finish work together and need to wash before leaving.

194 Special provision should be made if necessary for any worker with a disability to have access to facilities which are suitable for his or her use.

195 The facilities do not have to be within the workplace, but they should if possible be within the building. Where arrangements are made for the use of facilities provided by someone else, for example the owner of the building, the facilities should still meet the provisions of this Code and they should be available at all material times. The use of public facilities is only acceptable as a last resort, where no other arrangement is possible.

196 Facilities should provide adequate protection from the weather.

197 Water closets should be connected to a suitable drainage system and be provided with an effective means for flushing with water. Toilet paper in a holder or dispenser and a coat hook should be provided. In the case of water closets used by women,

suitable means should be provided for the disposal of sanitary dressings.

198 Washing stations should have running hot and cold, or warm water, and be large enough to enable effective washing of face, hands and forearms. Showers or baths should also be provided where the work is:

(a) particularly strenuous;

(b) dirty; or

(c) results in contamination of the skin by harmful or offensive materials.

This includes, for example, work with molten metal in foundries and the manufacture of oil cake.

199 Showers which are fed by both hot and cold water should be fitted with a device such as a thermostatic mixer valve to prevent users being scalded.

200 The facilities should be arranged to ensure adequate privacy for the user. In particular:

(a) each water closet should be situated in a separate room or cubicle, with a door which can be secured from the inside;

(b) it should not be possible to see urinals, or into communal shower or bathing areas, from outside the facilities when any entrance or exit door opens;

(c) windows to sanitary accommodation, shower or bathrooms should be obscured by means of frosted glass, blinds or curtains unless it is not possible to see into them from outside; and

(d) the facilities should be fitted with doors at entrances and exits unless other measures are taken to ensure an equivalent degree of privacy.

Minimum numbers of facilities

201 Table 1 shows the minimum number of sanitary conveniences and washing stations which should be provided. The number of people at work shown in column 1 refers to the maximum number likely to be in the workplace at any one time. Where separate sanitary accommodation is provided for a group

of workers, for example men, women, office workers or manual workers, a separate calculation should be made for each group.

TABLE 1

1 *Number of people at work*	2 *Number of water closets*	3 *Number of washstations*
1 to 5	1	1
6 to 25	2	2
26 to 50	3	3
51 to 75	4	4
76 to 100	5	5

202 In the case of sanitary accommodation used only by men, Table 2 may be followed if desired, as an alternative to column 2 of Table 1. A urinal may either be an individual urinal or a section of urinal space which is at least 600 mm long.

TABLE 2

1 *Number of men at work*	2 *Number of water closets*	3 *Number of urinals*
1 to 15	1	1
16 to 30	2	1
31 to 45	2	2
46 to 60	3	2
61 to 75	3	3
76 to 90	4	3
91 to 100	4	4

203 An additional water closet, and one additional washing station, should be provided for every 25 people above 100 (or fraction of 25). In the case of water closets used only by men, an additional water closet for every 50 men (or fraction of 50) above 100 is sufficient provided at least an equal number of additional urinals are provided.

204 Where work activities result in heavy soiling of face, hands and forearms, the number of washing stations should be increased to one for every 10 people at work (or fraction of 10) up to 50 people; and one extra for every additional 20 people (or fraction of 20).

205 Where facilities provided for workers are also used by members of the public the number of conveniences and washing stations specified above should be increased as necessary to ensure that workers can use the facilities without undue delay.

Remote workplaces and temporary work sites

206 In the case of remote workplaces without running water or a nearby sewer, sufficient water in containers for washing, or other means of maintaining personal hygiene, and sufficient chemical closets should be provided. Chemical closets which have to be emptied manually should be avoided as far as possible. If they have to be used a suitable deodorising agent should be provided and they should be emptied and recharged at suitable intervals.

207 In the case of temporary work sites, which are referred to in paragraph 12 of this document, regulation 3(2) requires that suitable and sufficient sanitary conveniences and washing facilities should be provided so far as is reasonably practicable. As far as possible, water closets and washing stations which satisfy this Code should be provided. In other cases, mobile facilities should be provided wherever possible. These should if possible include flushing sanitary conveniences and running water for washing and meet the other requirements of this Code.

Ventilation, cleanliness and lighting

208 Any room containing a sanitary convenience should be well ventilated, so that offensive odours do not linger. Measures should also be taken to prevent odours entering other rooms. This may be achieved by, for example, providing a ventilated area

between the room containing the convenience and the other room. Alternatively it may be possible to achieve it by mechanical ventilation or, if the room containing the convenience is well sealed from the workroom and has a door with an automatic closer, by good natural ventilation. However, no room containing a sanitary convenience should communicate directly with a room where food is processed, prepared or eaten.

209 Arrangements should be made to ensure that rooms containing sanitary conveniences or washing facilities are kept clean. The frequency and thoroughness of cleaning should be adequate for this purpose. The surfaces of the internal walls and floors of the facilities should normally have a surface which permits wet cleaning, for example ceramic tiling or a plastic coated surface. The rooms should be well lit; this will also facilitate cleaning to the necessary standard and give workers confidence in the cleanliness of the facilities. Responsibility for cleaning should be clearly established, particularly where facilities are shared by more than one workplace.

Other Regulations and publications

210–211 *(See Guidance Notes, p 375)*

Regulation 22

Drinking water

212 Drinking water should normally be obtained from a public or private water supply by means of a tap on a pipe connected directly to the water main. Alternatively, drinking water may be derived from a tap on a pipe connected directly to a storage cistern which complies with the requirements of the UK Water Bye-laws. In particular, any cistern, tank or vessel used as a supply should be well covered, kept clean and tested and disinfected as necessary. Water should only be provided in refillable containers where it cannot be obtained directly from a mains supply. Such containers should be suitably enclosed to prevent contamination and should be refilled at least daily.

213 Drinking water taps should not be installed in places where contamination is likely, for example in a workshop where lead is handled or processed. As far as is reasonably practicable they should also not be installed in sanitary accommodation.

214 Drinking cups or beakers should be provided unless the supply is by means of a drinking fountain. In the case of non-disposable cups a facility for washing them should be provided nearby.

215 Drinking water supplies should be marked as such if people may otherwise drink from supplies which are not meant for drinking. Marking is not necessary if non-drinkable cold water supplies are clearly marked as such.

216 *(See Guidance Notes, p 375)*

Regulation 23

Accommodation for clothing

217 *(See Guidance Notes, p 376)*

218 Accommodation for work clothing and workers' own personal clothing should enable it to hang in a clean, warm, dry, well-ventilated place where it can dry out during the course of a working day if necessary. If the workroom is unsuitable for this purpose then accommodation should be provided in another convenient place. The accommodation should consist of, as a minimum, a separate hook or peg for each worker.

219 Where facilities to change clothing are required by regulation 24, effective measures should be taken to ensure the security of clothing. This may be achieved, for example, by providing a lockable locker for each worker.

220 Where work clothing (including personal protective equipment) which is not taken home becomes dirty, damp or contaminated due to the work it should be accommodated separately from the worker's own clothing. Where work clothing becomes wet, the facilities should enable it to be dried by the beginning of the following work period unless other dry clothing is provided.

221–222 *(See Guidance Notes, p 376)*

Regulation 24

Facilities for changing clothing

223 A changing room or rooms should be provided for workers who change into special work clothing (see paragraph 217) and

where they remove more than outer clothing. Changing rooms should also be provided where necessary to prevent workers' own clothing being contaminated by a harmful substance.

224 Changing facilities should be readily accessible from workrooms and eating facilities, if provided. They should be provided with adequate seating and should contain, or communicate directly with, clothing accommodation and showers or baths if provided. They should be constructed and arranged to ensure the privacy of the user.

225 The facilities should be large enough to enable the maximum number of persons at work expected to use them at any one time, to do so without overcrowding or unreasonable delay. Account should be taken of starting and finishing times of work and the time available to use the facilities.

226 *(See Guidance Notes, p 376)*

Regulation 25

Facilities for rest and to eat meals

227 For workers who have to stand to carry out their work, suitable seats should be provided for their use if the type of work gives them an opportunity to sit from time to time.

228 Suitable seats should be provided for workers to use during breaks. These should be in a suitable place where personal protective equipment (for example respirators or hearing protection) need not be worn. In offices and other reasonably clean workplaces, work seats or other seats in the work area will be sufficient, provided workers are not subject to excessive disturbance during breaks, for example, by contact with the public. In other cases one or more separate rest areas should be provided (which in the case of new workplaces, extensions and conversions should include a separate rest room).

229 Rest areas or rooms provided in accordance with regulation 25(2) should be large enough, and have sufficient seats with backrests and tables, for the number of workers likely to use them at any one time.

230 If workers frequently have to leave their work area, and to wait until they can return, there should be a suitable rest area where they can wait.

231 Where workers regularly eat meals at work suitable and sufficient facilities should be provided for the purpose. Such facilities should also be provided where food would otherwise be likely to be contaminated, including by dust or water, for example:

(a) cement works, clay works, foundries, potteries, tanneries, and laundries;

(b) the manufacture of glass bottles and pressed glass articles, sugar, oil cake, jute, and tin or terne plates; and

(c) glass bevelling, fruit preserving, gut scraping, tripe dressing, herring curing, and the cleaning and repairing of sacks.

232 Seats in work areas can be counted as eating facilities provided they are in a sufficiently clean place and there is a suitable surface on which to place food. Eating facilities should include a facility for preparing or obtaining a hot drink, such as an electric kettle, a vending machine or a canteen. Workers who work during hours or at places where hot food cannot be obtained in, or reasonably near to, the workplace should be provided with the means for heating their own food.

233 Eating facilities should be kept clean to a suitable hygiene standard. Responsibility for cleaning should be clearly allocated. Steps should be taken where necessary to ensure that the facilities do not become contaminated by substances brought in on footwear or clothing. If necessary, adequate washing and changing facilities should be provided in a conveniently accessible place.

234 Canteens or restaurants may be used as rest facilities, provided that there is no obligation to purchase food in order to use them.

235 Good hygiene standards should be maintained in those parts of rest facilities used for eating or preparing food and drinks.

236 *(See Guidance Notes, p 376)*

Facilities for pregnant women and nursing mothers

237 Facilities for pregnant women and nursing mothers to rest should be conveniently situated in relation to sanitary facilities and, where necessary, include the facility to lie down.

238 *(See Guidance Notes, p 376)*

Prevention of discomfort caused by tobacco smoke

239 Rest areas and rest rooms should be arranged to enable employees to use them without experiencing discomfort from tobacco smoke. Methods of achieving this include:

(a) the provision of separate areas or rooms for smokers and non-smokers; or

(b) the prohibition of smoking in rest areas and rest rooms.

240 *(See Guidance Notes, p 377)*

GUIDANCE NOTES

Regulation 1
Citation and commencement

1 The Regulations come into effect in four stages. Workplaces which are used for the first time after 31 December 1992, and modifications, extensions and conversions started after that date, should comply as soon as they are in use. In existing workplaces (apart from any modifications) the Regulations take effect on 1 January 1996 and the laws in Schedule 2 will continue to apply until that date. Any workplaces or parts of workplaces located at a quarry or above ground at a mine used for the first time after 26 October 1995, and modifications, extensions and conversions started after that date, should comply as soon as they are in use. In existing workplaces at a quarry or above ground at a mine (apart from any modifications) the Regulations take effect on 26 July 1998.

Regulation 2
Interpretation

2 These Regulations apply to a very wide range of workplaces, not only factories, shops and offices but, for example, schools, hospitals, hotels and places of entertainment. The term workplace also includes the common parts of shared buildings, private roads and paths on industrial estates and business parks, and temporary work sites (but not construction sites).

3 'Workplace' is defined in regulation 2(1). Certain words in the definition are themselves defined in sections 52 and 53 of the Health and Safety at Work etc Act 1974. In brief:

— 'Work' means work as an employee or self-employed person, and also:

(a) work experience on certain training schemes (Health and Safety (Training for Employment) Regulations 1990, SI 1990/138, regulation 3);

(b) training which includes operations involving ionising radiations (Ionising Radiations Regulations 1985, SI 1985/1333, regulation 2(1));

(c) activities involving genetic manipulation (Genetic Manipulation Regulations 1989, SI 1989/1810, regulation 3); and

(d) work involving the keeping and handling of a listed pathogen (Health and Safety (Dangerous Pathogens) Regulations 1981, SI 1981/1011, regulation 9).

— 'Premises' means any place (including an outdoor place).

— 'Domestic Premises' means a private dwelling. These Regulations do not apply to domestic premises, and do not therefore cover homeworkers. They do, however, apply to hotels, nursing homes and the like, and to parts of workplaces where 'domestic' staff are employed such as the kitchens of hostels or sheltered accommodation.

4 These Regulations aim to ensure that workplaces meet the health, safety and welfare needs of each member of the workforce which may include people with disabilities. Several of the Regulations require things to be 'suitable' as defined in regulation 2(3) in a way which makes it clear that traffic routes, facilities and workstations which are used by people with disabilities should be suitable for them to use.

5 Building Regulations contain requirements which are intended to make new buildings accessible to people with limited mobility, or impaired sight or hearing. There is also a British Standard on access to buildings for people with disabilities.[1]

New workplaces

6 A 'new workplace' is one that is taken into use for the first time after 31 December 1992, or July 1995 in the case of quarries and workplaces above ground at mines. Therefore if a building was a workplace at any time in the past it is not a new workplace (although it may be a conversion).

Modifications, extensions and conversions

7 Any modification or extension started after 31 December 1992, or 26 July 1995 in the case of quarries and workplaces above ground at mines, should comply with any relevant requirements of these Regulations as soon as it is in use. This applies only to the actual modification or extension. The rest of the workplace should comply as from 1 January 1996, or 26 July 1998 in the case of quarries and workplaces above ground at mines. A 'modification' includes any alteration but not a simple replacement.

8 The whole of any conversion started after 31 December 1992 should comply as soon as it is in use. 'Conversion' is not defined and is therefore any workplace which would ordinarily be considered to be a 'conversion'. Examples of conversions include:

(a) a large building converted into smaller industrial units. Each unit is a 'conversion';

(b) a private house, or part of a house, converted into a workplace;

(c) workplaces which undergo a radical change of use involving structural alterations.

Note: certain modifications, extensions and conversions will also be subject to Building Regulations and may need planning consent. Advice can be obtained from the local authority.

Regulation 3

Application of these Regulations

Means of transport

9 All operational ships, boats, hovercraft, aircraft, trains and road vehicles are excluded from these Regulations, except that regulation 13 applies to aircraft, trains and road vehicles when stationary in a workplace (but not when on a public road). Non-operational means of transport used as, for example, restaurants or tourist attractions, are subject to these Regulations.

Extractive industries (mines, quarries etc)

10 These Regulations apply to workplaces or parts of workplaces located at a quarry or above ground at a mine. They do not apply to underground workplaces at mines, quarries or other mineral

extraction sites, including those off-shore. Nor do they apply to any related workplace on the same site. Other legislation applies to this sector.

Construction sites

11 Construction sites (including site offices) are excluded from these Regulations. Where construction work is in progress within a workplace, it can be treated as a construction site and so excluded from these Regulations, if it is fenced off; otherwise, these Regulations and Construction Regulations will both apply.

Temporary work sites

12 At temporary work sites the requirements of these Regulations for sanitary conveniences, washing facilities, drinking water, clothing accommodation, changing facilities and facilities for rest and eating meals (regulations 20–25) apply 'so far as is reasonably practicable'. Temporary work sites include:

(a) work sites used only infrequently or for short periods; and

(b) fairs and other structures which occupy a site for a short period.

Farming and forestry

13 Agricultural or forestry workplaces which are outdoors and away from the undertaking's main buildings are excluded from these Regulations, except for the requirements on sanitary conveniences, washing facilities and drinking water (regulations 20–22) which apply 'so far as is reasonably practicable'.

Regulation 4

Requirements under these Regulations

14 Employers have a general duty under section 2 of the Health and Safety at Work etc Act 1974 to ensure, so far as is reasonably practicable, the health, safety and welfare of their employees at work. Persons in control of non-domestic premises also have a duty under section 4 of the Act towards people who are not their employees but use their premises. (These sections are reproduced on pages 382–383). These Regulations expand on these duties.

They are intended to protect the health and safety of everyone in the workplace, and to ensure that adequate welfare facilities are provided for people at work.

15 Employers have a duty to ensure that workplaces under their control comply with these Regulations. Tenant employers are responsible for ensuring that the workplace which they control complies with the Regulations, and that the facilities required by the Regulations are provided, for example that sanitary conveniences are sufficient and suitable, adequately ventilated and lit and kept in a clean and orderly condition. Facilities should be readily accessible but it is not essential that they are within the employer's own workplace; arrangements can be made to use facilities provided by, for example, a landlord or a neighbouring business but the employer is responsible for ensuring that they comply with the Regulations.

16 People other than employers also have duties under these Regulations if they have control, to any extent, of a workplace. For example, owners and landlords (of business premises) should ensure that common parts, common facilities, common services and means of access within their control, comply with the Regulations. Their duties are limited to matters which are within their control. For example, an owner who is responsible for the general condition of a lobby, staircase and landings, for shared toilets provided for tenants' use, and for maintaining ventilation plant, should ensure that those parts and plant comply with these Regulations. However, the owner is not responsible under these Regulations for matters outside his control, for example a spillage caused by a tenant or shortcomings in the day-to-day cleaning of sanitary facilities where this is the tenants' responsibility. Tenants should co-operate with each other, and with the landlord, to the extent necessary to ensure that the requirements of the Regulations are fully met.

17 In some cases, measures additional to those indicated in the Regulations and the Approved Code of Practice may be necessary in order to fully comply with general duties under the Health and Safety at Work etc Act. The Management of Health and Safety at Work Regulations 1992[2–3] require employers and self-employed people to assess risks; an associated Approved Code of Practice states that it is always best if possible to avoid a risk altogether, and that work should, where possible, be adapted to the individual. A risk assessment may show that the workplace or the

work should be reorganised so that the need for people to work, for example, at an unguarded edge or to work in temperatures which may induce stress does not arise in the first place.

18 It is often useful to seek the views of workers before and after changes are introduced, for example on the design of workstations, the choice of work chairs, and traffic management systems such as one-way vehicle routes or traffic lights. As well as promoting good relations, consultation can result in better decisions and in some cases help employers avoid making expensive mistakes. The Management of Health and Safety at Work Regulations[2–3] extend the law which requires employers to consult employees' safety representatives on matters affecting health and safety.

19 Where employees work at a workplace which is not under their employer's control, their employer has no duty under these Regulations, but should (as part of his or her general duties under the Health and Safety at Work etc Act 1974) take any steps necessary to ensure that sanitary conveniences and washing facilities will be available. It may be necessary to make arrangements for the use of facilities already provided on site, or to provide temporary facilities. This applies, for example, to those who employ seasonal agricultural workers to work on someone else's land.

Regulation 5

Maintenance of workplace, and of equipment, devices and systems

20–23 *(See Approved Code of Practice, pp 329–330)*

24 The frequency of regular maintenance, and precisely what it involves, will depend on the equipment or device concerned. The likelihood of defects developing, and the foreseeable consequences, are highly relevant. The age and condition of equipment, how it is used and how often it is used should also be taken into account. Sources of advice include published HSE guidance, British and EC standards and other authoritative guidance, manufacturers' information and instructions, and trade literature.

25 The Management of Health and Safety at Work Regulations 1992 include requirements on the competence of people whom employers appoint to assist them in matters affecting health and safety and on employees' duties to report serious dangers and shortcomings in health and safety precautions.[2–3]

26 There are separate HSE publications covering maintenance of escalators and window access equipment.[4–7]

27 Advice on systems of maintenance for buildings can be found in a British Standard[8] and in publications by the Chartered Institution of Building Services Engineers (CIBSE).[9–10] The maintenance of work equipment; personal protective equipment; and electrical systems, equipment and conductors is addressed in other Regulations.[11–14]

Regulation 6

Ventilation

28–34 *(See Approved Code of Practice, pp 330–331)*

35 Regulation 6 covers general workplace ventilation, not local exhaust ventilation for controlling employees' exposure to asbestos, lead, ionising radiations or other substances hazardous to health. There are other health and safety regulations and approved codes of practice on the control of such substances.[15–22]

36 It may not always be possible to remove smells coming in from outside, but reasonable steps should be taken to minimise them. Where livestock is kept, smells may be unavoidable, but they should be controlled by good ventilation and regular cleaning.

37 Where a close, humid atmosphere is necessary, for example in mushroom growing, workers should be allowed adequate breaks in a well-ventilated place.

38 The fresh air supply rate should not normally fall below 5 to 8 litres per second, per occupant. Factors to be considered include the floor area per person, the processes and equipment involved, and whether the work is strenuous.

39 More detailed guidance on ventilation is contained in HSE publications[23–24] and in publications by the Chartered Institution of Building Services Engineers.[25–27]

40 Guidance on the measures necessary to avoid legionnaires' disease, caused by bacteria which can grow in water cooling towers and elsewhere, is covered in separate HSE publications[28–29] and in a CIBSE publication.[30]

41 The legislation referred to in regulation 6(3) deals with what are known as 'confined spaces' where breathing apparatus may be necessary.

Regulation 7

Temperature in indoor workplaces

42–50 *(See Approved Code of Practice, pp 331–332)*

51 Detailed guidance on thermal comfort is expected to be published by HSE shortly.

52 The Regulations do not prevent the use of proprietary unflued heating systems designed and installed to be used without a conventional flue. Care needs to be taken when siting temporary heaters so as to prevent burns from contact with hot surfaces. The Provision and Use of Work Equipment Regulations[11–12] require protection from hot surfaces.

53 The Personal Protective Equipment at Work Regulations 1992[13–14] apply to the protective clothing provided for workers' use.

54 Information about Food Hygiene Regulations can be obtained from the Environmental Health Departments of local authorities.

55 Design data relevant to workplace temperatures are published by the Chartered Institution of Building Services Engineers.[25]

56–57 *(See Approved Code of Practice, p 333)*

Regulation 8

Lighting

58–60 *(See Approved Code of Practice, p 333)*

61 More detailed guidance is given in a separate HSE publication.[31] There are also a number of publications on lighting by the Chartered Institution of Building Services Engineers.[32–36]

62 Requirements on lighting are also contained in the Docks Regulations 1988,[37–38] the Provision and Use of Work Equipment Regulations 1992[11–12] and the Health and Safety (Display Screen Equipment) Regulations 1992.[39–42] The electrical safety of lighting installations is subject to the Electricity at Work Regulations 1989.[43–44]

63 *(See Approved Code of Practice, p 334)*

64 People generally prefer to work in natural rather than artificial light. In both new and existing workplaces workstations should be

sited to take advantage of the available natural light. Natural lighting may not be feasible where windows have to be covered for security reasons or where process requirements necessitate particular lighting conditions.

65–66 *(See Approved Code of Practice, p 334)*

67 Fire precautions legislation may require the lighting of escape routes. Advice can be obtained from local fire authorities.

Regulation 9

Cleanliness and waste materials

68–72 *(See Approved Code of Practice, pp 334–335)*

73 Care should be taken that methods of cleaning do not expose anyone to substantial amounts of dust, including flammable or explosive concentrations of dusts, or to health or safety risks arising from the use of cleaning agents. The Control of Substances Hazardous to Health Regulations 1988 are relevant.

74 Absorbent floors, such as untreated concrete or timber, which are likely to be contaminated by oil or other substances which are difficult to remove, should preferably be sealed or coated, for example with a suitable non-slip floor paint. Carpet should also be avoided in such situations.

75 Washable surfaces, and high standards of cleanliness, may be essential for the purposes of infection control (as in the case of post-mortem rooms and pathology laboratories), for the control of exposure to substances hazardous to health or for the purposes of hygiene in the processing or handling of food. In such cases, steps should be taken to eliminate traps for dirt or germs by, for example, sealing joints between surfaces and fitting curved strips or coving along joins between walls and floors and between walls and work surfaces. Further information about food hygiene can be obtained from Environmental Health Departments of local authorities.

Regulation 10

Room dimensions and space

76–78 *(See Approved Code of Practice, pp 335–336)*

79 In a typical room, where the ceiling is 2.4 m high, a floor area of 4.6 m^2 (for example 2.0 x 2.3 m) will be needed to provide a

space of 11m³. Where the ceiling is 3.0 m high or higher the minimum floor area will be 3.7 m² (for example 2.0 x 1.85 m). (These floor areas are only for illustrative purposes and are approximate.)

80 The floor space per person indicated in paragraphs 77 and 79 will not always give sufficient unoccupied space, as required by the Regulation. Rooms may need to be larger, or to have fewer people working in them, than indicated in those paragraphs, depending on such factors as the contents and layout of the room and the nature of the work. Where space is limited careful planning of the workplace is particularly important.

Regulation 11

Workstations and seating

81–85 *(See Approved Code of Practice, pp 336–337)*

86 More detailed guidance on seating is given in an HSE publication.[45] There are other HSE publications on visual display units and ergonomics.[41–42, 46–47]

87 Static and awkward posture at the workstation, the use of undesirable force and an uncomfortable hand grip, often coupled with continuous repetitive work without sufficient rest and recovery, may lead to chronic injury. Guidance is contained in an HSE publication.[48]

88 This Regulation covers all workstations. Workstations where visual display units, process control screens, microfiche readers and similar display units are used are subject to the Health and Safety (Display Screen Equipment) Regulations 1992.[39–40]

Regulation 12

Condition of floors and traffic routes

89–100 *(See Approved Code of Practice, pp 337–339)*

101 At workplaces at quarries and above ground at mines, regulation 12 only applies to floors and traffic routes inside buildings.

102 Methods of draining and containing toxic, corrosive or highly flammable liquids should not result in the contamination of drains, sewers, watercourses, or groundwater supplies, or put people or the environment at risk. Maximum concentration levels

are specified in the Environmental Protection (Prescribed Processes and Substances) Regulations 1991, and the Surface Waters (Dangerous Substances) (Classification) Regulations 1989 and 1992. Consent for discharges may be required under the Environmental Protection Act 1990, the Water Resources Act 1991 and the Water Industry Act 1991.

103 Consideration should be given to providing slip resistant footwear in workplaces where slipping hazards arise despite the precautions set out in paragraph 93.

104 Further guidance on slips, trips and falls, and on the containment of pesticides in storage, is contained in separate HSE publications.[49–50] Building Regulations also have requirements on floors and stairs. Advice may be obtained from the local authority. There is also a British Standard on the construction and maximum loading of floors.[51]

105 Steep stairways are classed as fixed ladders and are dealt with under regulation 13.

Regulation 13

Falls or falling objects

106 The consequences of falling from heights or into dangerous substances are so serious that a high standard of protection is required. Secure fencing should normally be provided to prevent people falling from edges, and the fencing should also be adequate to prevent objects falling onto people. Where fencing cannot be provided or has to be removed temporarily, other measures should be taken to prevent falls. Dangerous substances in tanks, pits or other structures should be securely fenced or covered.

107 The guarding of temporary holes, such as an area where floorboards have been removed, is dealt with in paragraph 91 of this Code.

Provision of fencing or covers

108–113 *(See Approved Code of Practice, pp 339–340)*

114 Additional safeguards may be necessary in places where unauthorised entry is foreseeable. A separate HSE publication gives guidance on safeguards for effluent storage in farms.[52]

115 Building Regulations also have requirements on fencing. Advice can be obtained from local authorities. There is a British Standard on the construction of fencing.[53]

Temporary removal of fencing or covers

116 *(See Approved Code of Practice, p 340)*

117 One method of fencing an opening or edge where articles are raised or lowered by means of a lift truck is to provide a special type of fence or barrier which the worker can raise without having to approach the edge, for example by operating a lever, to give the lift truck access to the edge.

118–124 *(See Approved Code of Practice, pp 340–341)*

125 Stairs are much safer than ladders, especially when loads are to be carried. A sloping ladder is generally easier and safer to use than a vertical ladder (see regulation 17 and paragraph 162 of the Code.)

126 British Standards deal with ladders for permanent access.[54–55]

Roof work

127 Slips and trips which may be trivial at ground level may result in fatal accidents when on a roof. It is therefore vital that precautions are taken, even when access is only occasional, for example for maintenance or cleaning.

128 As well as falling from the roof edge, there may be a risk of falling through a fragile material. Care should be taken of old materials which may have become fragile because of corrosion. The risks may be increased by moss, lichen, ice, etc. Surfaces may also be deceptive.

129–130 *(See Approved Code of Practice, pp 341–342)*

131 Construction Regulations contain specific requirements on roof work. An HSE publication gives more detailed advice on roof work.[56] There is also a British Standard on imposed roof loads.[57]

Falls into dangerous substances

132–133 *(See Approved Code of Practice, p 342)*

Changes of level

134 Changes of level, such as a step between floors, which are not obvious should be marked to make them conspicuous.

Stacking and racking

135 *(See Approved Code of Practice, p 342)*

136 Appropriate precautions in stacking and storage include:

(a) safe palletisation;

(b) banding or wrapping to prevent individual articles falling out;

(c) setting limits for the height of stacks to maintain stability;

(d) regular inspection of stacks to detect and remedy any unsafe stacks; and

(e) particular instruction and arrangements for irregularly shaped objects.

137 Further guidance on stacking materials is given in HSE publications.[58–59]

Loading or unloading vehicles

138–142 *(See Approved Code of Practice, pp 342–343)*

143 Systems which do not require disconnection and reconnection of safety harnesses from safety lines, when at risk of falling, should be used in preference to those that do. Where there is no need to approach the edge the length of the line and the position of the anchorage should be such as to prevent the edge being approached.

144 The provision and use of safety harnesses etc are also subject to the Personal Protective Equipment at Work Regulations 1992.[13–14] There are also relevant British Standards.[60–61]

Scaffolding

145 *(See Approved Code of Practice, p 343)*

Other Regulations

146 Other Regulations concerning shipyards, docks and agricultural workplaces also contain specific requirements for preventing injury from falls.[37–38, 62–63] Those specific requirements stand. Regulation 13 of these Regulations and relevant parts of this Code will also apply to such premises (subject to regulation 3(4) which partially excludes open farmland). However, it is not intended that regulation 13 or this Code should be interpreted as overriding or increasing those specific requirements of other Regulations.

Regulations 14

Windows, and transparent or translucent doors, gates and walls

147–150 *(See Approved Code of Practice, pp 344–345)*

151 The term 'safety glass' is used in a British Standard[64] which is concerned with the breakage of flat glass or flat plastic sheet. Materials meeting that Standard, for example laminated or toughened glass, will break in a way that does not result in large sharp pieces and will fulfil paragraph 148(b) above. 'Safety materials' as used in these Regulations includes safety glass, but also other materials as described in paragraphs 148(a) and (c) above. There is also a British Standard which contains a code of practice for the glazing for buildings.[65]

152 Building Regulations also have similar requirements. Advice may be obtained from local authorities.

Regulation 15

Windows, skylights and ventilators

153–154 *(See Approved Code of Practice, p 345)*

155 There is a British Standard on windows and skylights.[66]

Regulation 16

Ability to clean windows etc safely

156–157 *(See Approved Code of Practice, pp 345–346)*

158 Further guidance on safe window cleaning and access equipment is given in other HSE publications.[6–7] There is also a relevant British Standard.[66]

Regulation 17

Organisation etc of traffic routes

159–167 *(See Approved Code of Practice, pp 346–348)*

168 Where large vehicles have to reverse, measures for reducing risks to pedestrians and any people in wheelchairs should be considered, such as:

(a) restricting reversing to places where it can be carried out safely;

(b) keeping people on foot or in wheelchairs away;

(c) providing suitable high visibility clothing for people who are permitted in the area;

(d) fitting reversing alarms to alert, or with a detection device to warn the driver of an obstruction or apply the brakes automatically; and

(e) employing banksmen to supervise the safe movement of vehicles.

Whatever measures are adopted, a safe system of work should operate at all times. Account should be taken of people with impaired sight or hearing.

169 If crowds of people are likely to overflow on to roadways, for example at the end of a shift, consideration should be given to stopping vehicles from using the routes at such times.

170–180 *(See Approved Code of Practice, pp 348–350)*

181 Further guidance on workplace transport is given in separate HSE publications.[67–71]

182 There are also separate Regulations on dock work which have requirements on traffic routes.[37–38]

Regulation 18

Doors and gates

183–189 *(See Approved Code of Practice, pp 351–352)*

190 Where the device referred to in paragraph 185 already forms part of the door mechanism, additional devices are not required. The fire resistance of doors is dealt with in Building Regulations

and in fire precautions legislation. Advice can be obtained from local authorities and fire authorities.

Regulation 19

Escalators and moving walkways

191 There are HSE publications on the safe use and periodic thorough examination of escalators.[4-5, 72] There is also a relevant British Standard.[73]

Regulation 21

Washing facilities

192–209 *(See Approved Code of Practice, pp 352–356)*

Other Regulations and publications

210 Legionnaires' disease is caused by bacteria which may be found where water stands for long periods at lukewarm or warm temperatures in, for example, tanks or little-used pipes. Separate HSE publications are available.[28-29]

211 Other Regulations and Approved Codes of Practice on the control of substances hazardous to health also deal with washing facilities.[15-22] Information about the requirements of food hygiene legislation can be obtained from the Environmental Health Department of local authorities. The requirement in the last sentence of paragraph 208 is not intended to apply to rest rooms in which workers may eat food they have brought into the workplace.

Regulation 22

Drinking water

212–215 *(See Approved Code of Practice, pp 356–357)*

216 Water supplies likely to be grossly contaminated, such as in supplies meant for process use only, should be clearly marked by a suitable sign. Bottled water/water dispensing systems may still be provided as a secondary source of drinking water.

Regulation 23

Accommodation for clothing

217 Special work clothing includes all clothing which is only worn at work such as overalls, uniforms, thermal clothing and hats worn for food hygiene purposes.

218–220 *(See Approved Code of Practice, p 357)*

221 Separate Regulations deal with personal protective equipment at work in greater detail.[13–14]

222 Other Regulations and Approved Codes of Practice on the control of substances hazardous to health also deal with accommodation for clothing.[15–22] Information about the requirements for food hygiene legislation can be obtained from the Environmental Health Department of local authorities.

Regulation 24

Facilities for changing clothing

223–225 *(See Approved Code of Practice, pp 357–358)*

226 Other Regulations and Approved Codes of Practice on the control of substances hazardous to health also deal with changing facilities.[15–22]

Regulation 25

Facilities for rest and to eat meals

227–235 *(See Approved Code of Practice, pp 358–359)*

236 The subject of eating in the workplace is also dealt with in other Regulations concerning asbestos, lead, and ionising radiations, and in Approved Codes of Practice on the control of substances hazardous to health and on work in potteries.[15–22, 74]

Facilities for pregnant women and nursing mothers

237 *(See Approved Code of Practice, p 359)*

238 There is an HSE guidance sheet on health aspects of pregnancy.[75]

Prevention of discomfort caused by tobacco smoke

239 *(See Approved Code of Practice, p 360)*

240 Passive smoking in the workplace is dealt with in a separate HSE publication.[76]

APPENDIX 1

REFERENCES

Where reference is made to a British Standard, there may also be an equivalent European Standard.

1 BS 5810: 1979 *Code of practice for access for the disabled to buildings*

2 *Management of Health and Safety at Work Regulations* 1992 SI 1992/2051 HMSO ISBN 0 11 25051 6

3 HSE *Management of health and safety at work. Management of Health and Safety at Work Regulations: Approved code of practice* L2 1 HSE Books 2000 ISBN 0 7176 2488 9

4 HSE *Safety in the use of escalators* PM 34 HSE Books 1983 ISBN 0 1188 3572 6

5 HSE *Escalators: periodic thorough examination* PM 45 HSE Books 1984 ISBN 0 1188 3593 3 withdrawn

6 HSE *Prevention of falls to window cleaners* GS 25 HSE Books 1991 ISBN 0 1188 5682 0

7 HSE *Suspended access equipment* PM 30 HSE Books 1983 ISBN 0 1188 3577 7

8 BS 8210: 1986 *Guide to building maintenance management* ISBN 0 580 15241 3

9 Chartered Institution of Building Services Engineers *Maintenance management for building services* TM 17 1994 ISBN 0 90 095368 3

10 Building Services Research and Information Association *Operating and maintenance manuals for building services installations* AG1 1/87.1 ISBN 0 86 022255 1

11 *Provision and Use of Work Equipment Regulations* 1998 SI 1998/2306 ISBN 0 110 79599 7

12 HSE *Work equipment. Provision and Use of Work Equipment Regulations 1992: Guidance on Regulations* L22 HSE Books 1992 ISBN 0 7176 0414 4

13 *Personal Protective Equipment at Work Regulations 1992* SI 1992/2966 ISBN 0 110 25832 0

14 HSE *Personal protective equipment at work. Personal Protective Equipment at Work Regulations 1992: Guidance on Regulations* L25 HSE Books 1992 ISBN 0 7176 0415 2

15 *The Control of Substances Hazardous to Health Regulations 1999* SI 1999/437 Stationery Office 1999 ISBN 0 11 082087 8

16 *General COSHH ACOP (Control of substances hazardous to health), Carcinogens ACOP (Control of carcinogenic substances) and Biological agents ACOP (Control of biological agents). Control of Substances Hazardous to Health Regulations 1994. Approved Codes of Practice* HSE Books 1995 ISBN 0 7176 0819 0

17 *The Control of Asbestos at Work Regulations 1987* SI 1987/2115 HMSO ISBN 0 11 078115 5

18 HSE *Work with asbestos insulation, asbestos coating and asbestos insulating board. The Control of Asbestos at Work Regulations 1987. Approved Code of Practice* L28 HSE Books 1999 ISBN 0 7176 1674 6 *Control of asbestos at work: The Control of Asbestos at Work Regulations 1987. Approved Code of Practice* L27 HSE Books 1999 ISBN 0 7176 1673 8

19 *The Control of Lead at Work Regulations 1998* SI 1998/543 Stationery Office ISBN 0 11 065646 6

20 HSE *Control of lead at work approved code of practice.* COP 2 HSE Books 1998 ISBN 0 7176 1506 5

21 *Ionising Radiation Regulations 1999* SI 1999/3232

22 HSE *Protection of persons against ionising radiation arising from any work activity: the Ionising Radiations Regulations 1985: approved code of practice* L58 1985 HSE Books ISBN 0 7176 0508 6

23 HSE *Ventilation of the workplace* EH 22(Rev) HSE Books 1988 ISBN 0 7176 0551 5

24 HSE *Measurement of air change rates in factories and offices* MDHS 73 HSE Books 1992 ISBN 0 1188 5693 6

25 Chartered Institution of Building Services Engineers *CIBSE guide: Volume A. Design data GVA* 1986 ISBN 0 90 095329 2

26 Chartered Institution of Building Services Engineers *Air filtration and natural ventilation* GS A4 1986

27 Building Services Research and Information Association *Ventilation effectiveness in mechanical ventilation systems* TN 1/88 ISBN 0 86 022189 X

28 HSE *Approved Code of Practice for the prevention or control of legionellosis (including legionnaires' disease)* (Rev) L8 HSE Books 1995 ISBN 0 7176 0732 1

29 HSE *The control of legionellosis including legionnaires' disease* HSG70 HSE Books 1993 ISBN 0 7176 1451 9

30 Chartered Institution of Building Services Engineers *Minimising the risk of legionnaires' disease* TM 13:1991 ISBN 0 90 095352 7

31 HSE *Lighting at work* HSG38 HSE Books 1997 ISBN 0 7176 1232 5

32 Chartered Institution of Building Services Engineers *Code for interior lighting* CIL 1994 ISBN 0 90 095364 0

33 Chartered Institution of Building Services Engineers *Lighting guide: The industrial environment* LG1 1989 ISBN 0 90 095338 1

34 Chartered Institution of Building Services Engineers *Hospitals and health care buildings* 1989 ISBN 0 90 095337 3

35 Chartered Institution of Building Services Engineers *Lighting guide: Areas for visual display terminals* LG3 1989 ISBN 0 90 095341 1

36 Chartered Institution of Building Services Engineers *Lighting in hostile and hazardous environments* AG HHE 1983 ISBN 0 90 095326 8

37 *Docks Regulations* 1988 SI 1988/1655 HMSO ISBN 0 11 087655 5

38 HSE *Safety in docks: Docks Regulations 1988: Approved code of practice with regulations and guidance* COP 25 HSE Books 1988 ISBN 0 1188 5456 9

39 *Health and Safety (Display Screen Equipment) Regulations 1992* SI 1992/2792 ISBN 0 11 025919 X

40 HSE *Display screen equipment work. Guidance on the Health and Safety (Display Screen Equipment) Regulations 1992* L26 HSE Books 1992 ISBN 0 7176 0410 1

41 HSE *Visual display units* HSE Books 1983 ISBN 0 1188 3685 4

42 HSE *Working with VDUs* INDG36 1998

43 *Electricity at Work Regulations 1989* SI 1989/635 HMSO ISBN 0 11 096635 X

44 HSE *Memorandum of guidance on the Electricity at Work Regulations 1989* HSR25 HSE Books 2000 ISBN 0 7176 1602 9

45 HSE *Seating at work* HSG57 HSE Books 1997 ISBN 0 7176 1231 7

46 HSE *Ergonomics at work* INDG90L 1994

47 HSE *Reducing error and influencing behaviour* HSG48 HSE Books 1999 ISBN 0 7176 2452 8

48 HSE *Work related upper limb disorders: a guide to prevention* HSG60 HSE Books 1990 ISBN 0 7176 0475 6

49 HSE *Watch you step: prevention of slipping, tripping and falling accidents at work* HSE Books 1985 ISBN 0 7176 1145 0

50 HSE *Guidance on storing pesticides for farmers and other professional users* AS16 HSE Books 1996

51 BS 6399: Part 1: 1984 *Design loading for buildings: code of practice for dead and imposed loads*

52 HSE *Effluent storage on farms* GS12 HSE Books 1981 ISBN 0 1188 3386 3 (withdrawn)

53 BS 6180: 1995 *Code of practice for protective barriers in and about buildings*

54 BS 5395: 1985 *Code of practice for the design of industrial type stairs, permanent ladders and walkways*

55 BS 4211: 1994 *Specification for ladders for permanent access to chimneys, other high structures, silos and bins*

56 HSE *Safety in roofwork* HSG33 HSE Books 1998 ISBN 0 7176 1425 5

57 BS 6399: Part 3: 1997 *Design loading for buildings: code of practice for imposed roof loads*

58 HSE *Handling and stacking bales in agriculture* HSE Books INDG125L 1998

59 HSE *Health and safety in retail and wholesale warehouses* HSG76 HSE Books 1992 ISBN 0 1188 5731 2

60 BS EN 365: 1993 *Personal protective equipment against falls from a height. General requirements for instructions for use and for marking*

61 BS 5845: 1991 *Specification for permanent anchors for industrial safety belts and harnesses*

62 *Shipbuilding and Ship-repairing Regulations 1960* SI 1960/1932 HMSO 1960

63 *Agriculture (Safeguarding of Workplaces) Regulations 1959* SI 1959/428 (revoked by SI 1996/3022)

64 BS 6206: 1981(1994) *Specification for impact performance requirements for flat safety glass and safety plastics for use in buildings*

65 BS 6262: 1994 *Code of practice for glazing for buildings*

66 BS 8213: Part 1: 1991 *Windows, doors and rooflights: code of practice for safety in use and during cleaning of windows (including guidance on cleaning materials and methods)*

67 HSE *Road transport in factories and similar workplaces* GS9(Rev) HSE Books 1992 ISBN 0 1188 5732 0

68 HSE *Safety in working with lift trucks* HSG6(Rev) HSE Books 1992 ISBN 0 1188 6395 9

69 HSE *Container terminals: safe working practice* HSG7 HSE Books 1980 ISBN 0 1188 3302 2 (withdrawn)

70 HSE *Danger! Transport at work* INDG22L HSE Books 1985 (withdrawn)

71 HSE *Transport kills: a study of fatal accidents in industry* 1978–80 HSE Books 1982 ISBN 0 1188 3659 5 (withdrawn)

72 HSE *Ergonomic aspects of escalators used in retail organisations* CRR12/1989 USE Books 1989 ISBN 0 1188 5938 2

73 BS 5656: 1997 *Safety rules for the construction and installation of escalators and passenger conveyors*

74 HSE *Control of substances hazardous to health in the production of pottery. The Control of Substances Hazardous to Health Regulations 1994. The Control of Lead at Work Regulations 1998. Approved Code of Practice* L60 HSE Books 1998 ISBN 0 7176 0849 2

75 HSE *Occupational health aspects of pregnancy* MA 6 1989 HSE Books (available free from local HSE offices)

76 HSE *Passive smoking at work* INDG63 HSE Books (Rev) 1992

APPENDIX 2

EXTRACTS FROM RELEVANT HEALTH AND SAFETY LEGISLATION

Health and Safety at Work etc Act 1974—section 2

'(1) It shall be the duty of every employer to ensure, so far as is reasonably practicable, the health, safety and welfare at work of all his employees.

(2) Without prejudice to the generality of an employer's duty under the preceding subsection, the matters to which that duty extends include in particular—

(a) the provision and maintenance of plant and systems of work that are, so far as is reasonably practicable, safe and without risks to health;

(b) arrangements for ensuring, so far as is reasonably practicable, safety and absence of risks to health in connection with the use, handling, storage and transport of articles and substances;

(c) the provision of such information, instruction, training and supervision as is necessary to ensure, so far as is reasonably practicable, the health and safety at work of his employees;

(d) so far as is reasonably practicable as regards any place of work under the employer's control, the maintenance of it in a condition that is safe and without risks to health and the provision and maintenance of means of access to and egress from it that are safe and without such risks;

(e) the provision and maintenance of a working environment for his employees that is, so far as is reasonably practicable, safe, without risks to health, and adequate as regards facilities and arrangements for their welfare at work.'

Health and Safety at Work etc Act 1974—section 4

'(1) This section has effect for imposing on persons duties in relation to those who—

(a) are not their employees; but

(b) use non-domestic premises made available to them as a place of work or as a place where they may use plant or substances provided for their use there,

and applies to premises so made available and other non-domestic premises used in connection with them.

(2) It shall be the duty of each person who has, to any extent, control of premises to which this section applies or of the means of access thereto or egress therefrom or of any plant or substance in such premises to take such measures as it is reasonable for a person in his position to take to ensure, so far as is reasonably practicable, that the premises, all means of access thereto or egress therefrom available for use by persons using the premises, and any plant or substance in the premises or, as the case may be, provided for use there, is or are safe and without risks to health.

(3) Where a person has, by virtue of any contract or tenancy, an obligation of any extent in relation to—

(a) the maintenance or repair of any premises to which this section applies or any means of access thereto or egress therefrom; or

(b) the safety of or the absence of risks to health arising from plant or substances in any such premises;

that person shall be treated, for the purposes of subsection (2) above, as being a person who has control of the matters to which his obligations extends.

(4) Any reference in this section to a person having control of any premises or matter is a reference to a person having control of the premises or matter in connection with the carrying on by him of a trade, business or other undertaking (whether for profit or not).'

Factories Act 1961—section 175(5)

'Any workplace in which, with the permission of or under agreement with the owner or occupier, two or more persons carry on any work which would constitute a factory if the persons working therein were in the employment of the owner or occupier, shall be deemed to be a factory for the purposes of this Act, and, in the case of any such workplace not being a tenement factory or part of a tenement factory, the provisions of this Act shall apply as if the owner or occupier of the workplace were the occupier of the factory and the persons working therein were persons employed in the factory.'

Appendix 3

Provision and Use of Work Equipment Regulations 1998

THE REGULATIONS (SI 1998/2306)

Made *15 September 1998*

Laid before Parliament *25 September 1998*

Initial Commencement *5 December 1998*

The Secretary of State, in the exercise of the powers conferred on him by sections 15(1), (2), (3)(a), (5) and (6)(a), 49 and 82(3)(a) of, and paragraphs 1(1), (2) and (3), 9, 14, 15(1) and 16 of Schedule 3 to, the Health and Safety at Work etc Act 1974 ('the 1974 Act') and of all other powers enabling him in that behalf and for the purpose of giving effect without modifications to proposals submitted to him by the Health and Safety Commission under section 11(2)(d) of the 1974 Act, after the carrying out by the said Commission of consultations in accordance with section 50(3) of that Act, hereby makes the following Regulations:

PART I
INTRODUCTION

Regulation 1 *(See Guidance Notes, pp 437–438)*

Citation and commencement

These Regulations may be cited as the Provision and Use of Work Equipment Regulations 1998 and shall come into force on 5 December 1998.

Regulation 2 *(See Guidance Notes, pp 438–440)*

Interpretation

(1) In these Regulations, unless the context otherwise requires—

— 'the 1974 Act' means the Health and Safety at Work etc Act 1974;

— 'employer' except in regulation 3(2) and (3) includes a person to whom the requirements imposed by these Regulations apply by virtue of regulation 3(3)(a) and (b);

— 'essential requirements' means requirements described in regulation 10(1);

— 'the Executive' means the Health and Safety Executive;

— 'inspection' in relation to an inspection under paragraph (1) or (2) of regulation 6—

(a) means such visual or more rigorous inspection by a competent person as is appropriate for the purpose described in the paragraph;

(b) where it is appropriate to carry out testing for the purpose, includes testing the nature and extent of which are appropriate for the purpose;

— 'power press' means a press or press brake for the working of metal by means of tools, or for die proving, which is power driven and which embodies a flywheel and clutch;

— 'thorough examination' in relation to a thorough examination under paragraph (1), (2), (3) or (4) of regulation 32—

(a) means a thorough examination by a competent person;

(b) includes testing the nature and extent of which are appropriate for the purpose described in the paragraph;

— 'use' in relation to work equipment means any activity involving work equipment and includes starting, stopping, programming, setting, transporting, repairing, modifying, maintaining, servicing and cleaning;

— 'work equipment' means any machinery, appliance, apparatus, tool or installation for use at work (whether exclusively or not);

and related expressions shall be construed accordingly.

(2) Any reference in regulations 32 to 34 or Schedule 3 to a guard or protection device is a reference to a guard or protection device provided for the tools of a power press.

(3) Any reference in regulation 32 or 33 to a guard or protection device being on a power press shall, in the case of a guard or protection device designed to operate while adjacent to a power press, be construed as a reference to its being adjacent to it.

(4) Any reference in these Regulations to—

(a) a numbered regulation or Schedule is a reference to the regulation or Schedule in these Regulations so numbered; and

(b) a numbered paragraph is a reference to the paragraph so numbered in the regulation in which the reference appears.

Regulation 3 *(See Guidance Notes, pp 440–446)*

Application

(1) These Regulations shall apply—

(a) in Great Britain; and

(b) outside Great Britain as sections 1 to 59 and 80 to 82 of the 1974 Act apply by virtue of the Health and Safety at Work etc Act 1974 (Application outside Great Britain) Order 1995 ('the 1995 Order').

(2) The requirements imposed by these Regulations on an employer in respect of work equipment shall apply to such equipment provided for use or used by an employee of his at work.

(3) The requirements imposed by these Regulations on an employer shall also apply—

(a) to a self-employed person, in respect of work equipment he uses at work;

(b) subject to paragraph (5), to a person who has control to any extent of—

(i) work equipment;

(ii) a person at work who uses or supervises or manages the use of work equipment; or

(iii) the way in which work equipment is used at work,

and to the extent of his control.

(4) Any reference in paragraph (3)(b) to a person having control is a reference to a person having control in connection with the carrying on by him of a trade, business or other undertaking (whether for profit or not).

(5) The requirements imposed by these Regulations shall not apply to a person in respect of work equipment supplied by him by way of sale, agreement for sale or hire-purchase agreement.

(6) Subject to paragraphs (7) to (10), these Regulations shall not impose any obligation in relation to a ship's work equipment (whether that equipment is used on or off the ship).

(7) Where merchant shipping requirements are applicable to a ship's work equipment, paragraph (6) shall relieve the shore employer of his obligations under these Regulations in respect of that equipment only where he has taken all reasonable steps to satisfy himself that the merchant shipping requirements are being complied with in respect of that equipment.

(8) In a case where the merchant shipping requirements are not applicable to the ship's work equipment by reason only that for the time being there is no master, crew or watchman on the ship, those requirements shall nevertheless be treated for the purpose of paragraph (7) as if they were applicable.

(9) Where the ship's work equipment is used in a specified operation paragraph (6) shall not apply to regulations 7 to 9, 11 to 13, 20 to 22 and 30 (each as applied by regulation 3).

(10) Paragraph (6) does not apply to a ship's work equipment provided for use or used in an activity (whether carried on in or outside Great Britain) specified in the 1995 Order save that it does apply to—

(a) the loading, unloading, fuelling or provisioning of the ship; or

(b) the construction, reconstruction, finishing, refitting, repair, maintenance, cleaning or breaking up of the ship.

(11) In this regulation—

— 'master' has the meaning assigned to it by section 313(1) of the Merchant Shipping Act 1995;

— 'merchant shipping requirements' means the requirements of regulations 3 and 4 of the Merchant Shipping (Guarding of Machinery and Safety of Electrical Equipment) Regulations

1988 and regulations 5 to 10 of the Merchant Shipping (Hatches and Lifting Plant) Regulations 1988;

— 'ship' has the meaning assigned to it by section 313(1) of the Merchant Shipping Act 1995 save that it does not include an offshore installation;

— 'shore employer' means an employer of persons (other than the master and crew of any ship) who are engaged in a specified operation;

— 'specified operation' means an operation in which the ship's work equipment is used—

(a) by persons other than the master and crew; or

(b) where persons other than the master and crew are liable to be exposed to a risk to their health or safety from its use.

PART II
GENERAL

Regulation 4 *(See Approved Code of Practice, p 416; Guidance Notes, pp 446–451)*

Suitability of work equipment

(1) Every employer shall ensure that work equipment is so constructed or adapted as to be suitable for the purpose for which it is used or provided.

(2) In selecting work equipment, every employer shall have regard to the working conditions and to the risks to the health and safety of persons which exist in the premises or undertaking in which that work equipment is to be used and any additional risk posed by the use of that work equipment.

(3) Every employer shall ensure that work equipment is used only for operations for which, and under conditions for which, it is suitable.

[(4) In this regulation 'suitable'—

(a) subject to sub-paragraph (b), means suitable in any respect which it is reasonably foreseeable will affect the health or safety of any person;

(b) in relation to—

(i) an offensive weapon within the meaning of section 1(4) of the Prevention of Crime Act 1953 provided for use as self-defence or as deterrent equipment; and

(ii) work equipment provided for use for arrest or restraint,

by a person who holds the office of constable or an appointment as police cadet, means suitable in any respect which it is reasonably foreseeable will affect the health or safety of such person.][1]

Regulation 5 *(See Guidance Notes, pp 451–454)*

Maintenance

(1) Every employer shall ensure that work equipment is maintained in an efficient state, in efficient working order and in good repair.

(2) Every employer shall ensure that where any machinery has a maintenance log, the log is kept up to date.

Regulation 6 *(See Approved Code of Practice, pp 417–418; Guidance Notes, pp 454–460)*

Inspection

(1) Every employer shall ensure that, where the safety of work equipment depends on the installation conditions, it is inspected—

(a) after installation and before being put into service for the first time; or

(b) after assembly at a new site or in a new location,

to ensure that it has been installed correctly and is safe to operate.

(2) Every employer shall ensure that work equipment exposed to conditions causing deterioration which is liable to result in dangerous situations is inspected—

(a) at suitable intervals; and

1 Substituted by SI 1999/860, reg 5(1). Date in force: 14 April 1999: see SI 1999/860, reg 1.

(b) each time that exceptional circumstances which are liable to jeopardise the safety of the work equipment have occurred,

to ensure that health and safety conditions are maintained and that any deterioration can be detected and remedied in good time.

(3) Every employer shall ensure that the result of an inspection made under this regulation is recorded and kept until the next inspection under this regulation is recorded.

(4) Every employer shall ensure that no work equipment—

(a) leaves his undertaking; or

(b) if obtained from the undertaking of another person, is used in his undertaking,

unless it is accompanied by physical evidence that the last inspection required to be carried out under this regulation has been carried out.

(5) This regulation does not apply to—

(a) a power press to which regulations 32 to 35 apply;

(b) a guard or protection device for the tools of such power press;

(c) work equipment for lifting loads including persons;

(d) winding apparatus to which the Mines (Shafts and Winding) Regulations 1993 apply;

(e) work equipment required to be inspected by regulation 29 of the Construction (Health, Safety and Welfare) Regulations 1996.

Regulation 7 *(See Approved Code of Practice, pp 418–419; Guidance Notes, p 460)*

Specific risks

(1) Where the use of work equipment is likely to involve a specific risk to health or safety, every employer shall ensure that—

(a) the use of that work equipment is restricted to those persons given the task of using it; and

(b) repairs, modifications, maintenance or servicing of that work equipment is restricted to those persons who have been specifically designated to perform operations of that description (whether or not also authorised to perform other operations).

(2) The employer shall ensure that the persons designated for the purposes of sub-paragraph (b) of paragraph (1) have received adequate training related to any operations in respect of which they have been so designated.

Regulation 8 *(See Guidance Notes, pp 461–463)*

Information and instructions

(1) Every employer shall ensure that all persons who use work equipment have available to them adequate health and safety information and, where appropriate, written instructions pertaining to the use of the work equipment.

(2) Every employer shall ensure that any of his employees who supervises or manages the use of work equipment has available to him adequate health and safety information and, where appropriate, written instructions pertaining to the use of the work equipment.

(3) Without prejudice to the generality of paragraphs (1) or (2), the information and instructions required by either of those paragraphs shall include information and, where appropriate, written instructions on—

(a) the conditions in which and the methods by which the work equipment may be used;

(b) foreseeable abnormal situations and the action to be taken if such a situation were to occur; and

(c) any conclusions to be drawn from experience in using the work equipment.

(4) Information and instructions required by this regulation shall be readily comprehensible to those concerned.

Regulation 9 *(See Approved Code of Practice, pp 419–420; Guidance Notes, pp 464–466)*

Training

(1) Every employer shall ensure that all persons who use work equipment have received adequate training for purposes of health

and safety, including training in the methods which may be adopted when using the work equipment, any risks which such use may entail and precautions to be taken.

(2) Every employer shall ensure that any of his employees who supervises or manages the use of work equipment has received adequate training for purposes of health and safety, including training in the methods which may be adopted when using the work equipment, any risks which such use may entail and precautions to be taken.

Regulation 10 *(See Guidance Notes, pp 466–468)*

Conformity with Community requirements

(1) Every employer shall ensure that an item of work equipment has been designed and constructed in compliance with any essential requirements, that is to say requirements relating to its design or construction in any of the instruments listed in Schedule 1 (being instruments which give effect to Community directives concerning the safety of products).

(2) Where an essential requirement applied to the design or construction of an item of work equipment, the requirements of regulations 11 to 19 and 22 to 29 shall apply in respect of that item only to the extent that the essential requirement did not apply to it.

(3) This regulation applies to items of work equipment provided for use in the premises or undertaking of the employer for the first time after 31 December 1992.

Regulation 11 *(See Guidance Notes, pp 468–471)*

Dangerous parts of machinery

(1) Every employer shall ensure that measures are taken in accordance with paragraph (2) which are effective—

(a) to prevent access to any dangerous part of machinery or to any rotating stock-bar; or

(b) to stop the movement of any dangerous part of machinery or rotating stock-bar before any part of a person enters a danger zone.

(2) The measures required by paragraph (1) shall consist of—

(a) the provision of fixed guards enclosing every dangerous part or rotating stock-bar where and to the extent that it is practicable to do so, but where or to the extent that it is not, then

(b) the provision of other guards or protection devices where and to the extent that it is practicable to do so, but where or to the extent that it is not, then

(c) the provision of jigs, holders, push-sticks or similar protection appliances used in conjunction with the machinery where and to the extent that it is practicable to do so, but where or to the extent that it is not, then

(d) the provision of information, instruction, training and supervision.

(3) All guards and protection devices provided under sub-paragraphs (a) or (b) of paragraph (2) shall—

(a) be suitable for the purpose for which they are provided;

(b) be of good construction, sound material and adequate strength;

(c) be maintained in an efficient state, in efficient working order and in good repair;

(d) not give rise to any increased risk to health or safety;

(e) not be easily bypassed or disabled;

(f) be situated at sufficient distance from the danger zone;

(g) not unduly restrict the view of the operating cycle of the machinery, where such a view is necessary;

(h) be so constructed or adapted that they allow operations necessary to fit or replace parts and for maintenance work, restricting access so that it is allowed only to the area where the work is to be carried out and, if possible, without having to dismantle the guard or protection device.

(4) All protection appliances provided under sub-paragraph (c) of paragraph (2) shall comply with sub-paragraphs (a) to (d) and (g) of paragraph (3).

(5) In this regulation—

— 'danger zone' means any zone in or around machinery in which a person is exposed to a risk to health or safety from contact with a dangerous part of machinery or a rotating stock-bar;

— 'stock-bar' means any part of a stock-bar which projects beyond the head-stock of a lathe.

Regulation 12 *(See Guidance Notes, pp 471–474)*

Protection against specified hazards

(1) Every employer shall take measures to ensure that the exposure of a person using work equipment to any risk to his health or safety from any hazard specified in paragraph (3) is either prevented, or, where that is not reasonably practicable, adequately controlled.

(2) The measures required by paragraph (1) shall—

(a) be measures other than the provision of personal protective equipment or of information, instruction, training and supervision, so far as is reasonably practicable; and

(b) include, where appropriate, measures to minimise the effects of the hazard as well as to reduce the likelihood of the hazard occurring.

(3) The hazards referred to in paragraph (1) are—

(a) any article or substance falling or being ejected from work equipment;

(b) rupture or disintegration of parts of work equipment;

(c) work equipment catching fire or overheating;

(d) the unintended or premature discharge of any article or of any gas, dust, liquid, vapour or other substance which, in each case, is produced, used or stored in the work equipment;

(e) the unintended or premature explosion of the work equipment or any article or substance produced, used or stored in it.

(4) For the purposes of this regulation 'adequately' means adequately having regard only to the nature of the hazard and the nature and degree of exposure to the risk.

(5) This regulation shall not apply where any of the following Regulations apply in respect of any risk to a person's health or

safety for which such Regulations require measures to be taken to prevent or control such risk, namely—

(a) the Ionising Radiations Regulations 1985;

(b) the Control of Asbestos at Work Regulations 1987;

(c) the Control of Substances Hazardous to Health Regulations 1994;

(d) the Noise at Work Regulations 1989;

(e) the Construction (Head Protection) Regulations 1989;

(f) the Control of Lead at Work Regulations 1998.

Regulation 13 *(See Guidance Notes, pp 474–475)*

High or very low temperature

Every employer shall ensure that work equipment, parts of work equipment and any article or substance produced, used or stored in work equipment which, in each case, is at a high or very low temperature shall have protection where appropriate so as to prevent injury to any person by burn, scald or sear.

Regulation 14 *(See Guidance Notes, pp 475–477)*

Controls for starting or making a significant change in operating conditions

(1) Every employer shall ensure that, where appropriate, work equipment is provided with one or more controls for the purposes of—

(a) starting the work equipment (including re-starting after a stoppage for any reason); or

(b) controlling any change in the speed, pressure or other operating conditions of the work equipment where such conditions after the change result in risk to health and safety which is greater than or of a different nature from such risks before the change.

(2) Subject to paragraph (3), every employer shall ensure that, where a control is required by paragraph (1), it shall not be possible to perform any operation mentioned in sub-paragraph (a) or (b) of that paragraph except by a deliberate action on such control.

(3) Paragraph (1) shall not apply to re-starting or changing operating conditions as a result of the normal operating cycle of an automatic device.

Regulation 15 *(See Guidance Notes, pp 475–476, 478–479)*

Stop controls

(1) Every employer shall ensure that, where appropriate, work equipment is provided with one or more readily accessible controls the operation of which will bring the work equipment to a safe condition in a safe manner.

(2) Any control required by paragraph (1) shall bring the work equipment to a complete stop where necessary for reasons of health and safety.

(3) Any control required by paragraph (1) shall, if necessary for reasons of health and safety, switch off all sources of energy after stopping the functioning of the work equipment.

(4) Any control required by paragraph (1) shall operate in priority to any control which starts or changes the operating conditions of the work equipment.

Regulation 16 *(See Guidance Notes, pp 475–476, 479)*

Emergency stop controls

(1) Every employer shall ensure that, where appropriate, work equipment is provided with one or more readily accessible emergency stop controls unless it is not necessary by reason of the nature of the hazards and the time taken for the work equipment to come to a complete stop as a result of the action of any control provided by virtue of regulation 15(1).

(2) Any control required by paragraph (1) shall operate in priority to any control required by regulation 15(1).

Regulation 17 *(See Guidance Notes, pp 475–476, 479–482)*

Controls

(1) Every employer shall ensure that all controls for work equipment are clearly visible and identifiable, including by appropriate marking where necessary.

(2) Except where necessary, the employer shall ensure that no control for work equipment is in a position where any person operating the control is exposed to a risk to his health or safety.

(3) Every employer shall ensure where appropriate—

(a) that, so far as is reasonably practicable, the operator of any control is able to ensure from the position of that control that no person is in a place where he would be exposed to any risk to his health or safety as a result of the operation of that control, but where or to the extent that it is not reasonably practicable;

(b) that, so far as is reasonably practicable, systems of work are effective to ensure that, when work equipment is about to start, no person is in a place where he would be exposed to a risk to his health or safety as a result of the work equipment starting, but where neither of these is reasonably practicable;

(c) that an audible, visible or other suitable warning is given by virtue of regulation 24 whenever work equipment is about to start.

(4) Every employer shall take appropriate measures to ensure that any person who is in a place where he would be exposed to a risk to his health or safety as a result of the starting or stopping of work equipment has sufficient time and suitable means to avoid that risk.

Regulation 18 *(See Guidance Notes, pp 475–476, 482–483)*

Control systems

(1) Every employer shall—

(a) ensure, so far as is reasonably practicable, that all control systems of work equipment are safe; and

(b) are chosen making due allowance for the failures, faults and constraints to be expected in the planned circumstances of use.

(2) Without prejudice to the generality of paragraph (1), a control system shall not be safe unless—

(a) its operation does not create any increased risk to health or safety;

(b) it ensures, so far as is reasonably practicable, that any fault in or damage to any part of the control system or the loss of supply of any source of energy used by the work equipment cannot result in additional or increased risk to health or safety;

(c) it does not impede the operation of any control required by regulation 15 or 16.

Regulation 19 *(See Guidance Notes, pp 483–485)*

Isolation from sources of energy

(1) Every employer shall ensure that, where appropriate, work equipment is provided with suitable means to isolate it from all its sources of energy.

(2) Without prejudice to the generality of paragraph (1), the means mentioned in that paragraph shall not be suitable unless they are clearly identifiable and readily accessible.

(3) Every employer shall take appropriate measures to ensure that re-connection of any energy source to work equipment does not expose any person using the work equipment to any risk to his health or safety.

Regulation 20 *(See Guidance Notes, pp 485–486)*

Stability

Every employer shall ensure that work equipment or any part of work equipment is stabilised by clamping or otherwise where necessary for purposes of health or safety.

Regulation 21 *(See Guidance Notes, pp 486–487)*

Lighting

Every employer shall ensure that suitable and sufficient lighting, which takes account of the operations to be carried out, is provided at any place where a person uses work equipment.

Regulation 22 *(See Guidance Notes, pp 487–488)*

Maintenance operations

Every employer shall take appropriate measures to ensure that work equipment is so constructed or adapted that, so far as is reasonably practicable, maintenance operations which involve a risk to health or safety can be carried out while the work equipment is shut down, or in other cases—

(a) maintenance operations can be carried out without exposing the person carrying them out to a risk to his health or safety; or

(b) appropriate measures can be taken for the protection of any person carrying out maintenance operations which involve a risk to his health or safety.

Regulation 23 *(See Guidance Notes, pp 488–489)*

Markings

Every employer shall ensure that work equipment is marked in a clearly visible manner with any marking appropriate for reasons of health and safety.

Regulation 24 *(See Guidance Notes, pp 489–491)*

Warnings

(1) Every employer shall ensure that work equipment incorporates any warnings or warning devices which are appropriate for reasons of health and safety.

(2) Without prejudice to the generality of paragraph (1), warnings given by warning devices on work equipment shall not be appropriate unless they are unambiguous, easily perceived and easily understood.

PART III
MOBILE WORK EQUIPMENT

Regulation 25 *(See Approved Code of Practice, p 422; Guidance Notes, pp 493–495)*

Employees carried on mobile work equipment

Every employer shall ensure that no employee is carried by mobile work equipment unless—

(a) it is suitable for carrying persons; and

(b) it incorporates features for reducing to as low as is reasonably practicable risks to their safety, including risks from wheels or tracks.

Regulation 26 *(See Approved Code of Practice, p 423; Guidance Notes, pp 495–499)*

Rolling over of mobile work equipment

(1) Every employer shall ensure that where there is a risk to an employee riding on mobile work equipment from its rolling over, it is minimised by—

(a) stabilising the work equipment;

(b) a structure which ensures that the work equipment does no more than fall on its side;

(c) a structure giving sufficient clearance to anyone being carried if it overturns further than that; or

(d) a device giving comparable protection.

(2) Where there is a risk of anyone being carried by mobile work equipment being crushed by its rolling over, the employer shall ensure that it has a suitable restraining system for him.

(3) This regulation shall not apply to a fork-lift truck having a structure described in sub-paragraph (b) or (c) of paragraph (1).

(4) Compliance with this regulation is not required where—

(a) it would increase the overall risk to safety;

(b) it would not be reasonably practicable to operate the mobile work equipment in consequence; or

(c) in relation to an item of work equipment provided for use in the undertaking or establishment before 5 December 1998 it would not be reasonably practicable.

Regulation 27 *(See Approved Code of Practice, p 423; Guidance Notes, pp 499–500)*

Overturning of fork-lift trucks

Every employer shall ensure that a fork-lift truck to which regulation 26(3) refers and which carries an employee is adapted or equipped to reduce to as low as is reasonably practicable the risk to safety from its overturning.

Regulation 28 *(See Approved Code of Practice, p 424; Guidance Notes, pp 500–503)*

Self-propelled work equipment

Every employer shall ensure that, where self-propelled work equipment may, while in motion, involve risk to the safety of persons—

(a) it has facilities for preventing its being started by an unauthorised person;

(b) it has appropriate facilities for minimising the consequences of a collision where there is more than one item of rail-mounted work equipment in motion at the same time;

(c) it has a device for braking and stopping;

(d) where safety constraints so require, emergency facilities operated by readily accessible controls or automatic systems are available for braking and stopping the work equipment in the event of failure of the main facility;

(e) where the driver's direct field of vision is inadequate to ensure safety, there are adequate devices for improving his vision so far as is reasonably practicable;

(f) if provided for use at night or in dark places—

(i) it is equipped with lighting appropriate to the work to be carried out; and

(ii) is otherwise sufficiently safe for such use;

(g) if it, or anything carried or towed by it, constitutes a fire hazard and is liable to endanger employees, it carries appropriate fire-fighting equipment, unless such equipment is kept sufficiently close to it.

Regulation 29 *(See Guidance Notes, pp 503–504)*

Remote-controlled self-propelled work equipment

Every employer shall ensure that where remote-controlled self-propelled work equipment involves a risk to safety while in motion—

(a) it stops automatically once it leaves its control range; and

(b) where the risk is of crushing or impact it incorporates features to guard against such risk unless other appropriate devices are able to do so.

Regulation 30 *(See Guidance Notes, pp 504–505)*

Drive shafts

(1) Where the seizure of the drive shaft between mobile work equipment and its accessories or anything towed is likely to involve a risk to safety every employer shall—

(a) ensure that the work equipment has a means of preventing such seizure; or

(b) where such seizure cannot be avoided, take every possible measure to avoid an adverse effect on the safety of an employee.

(2) Every employer shall ensure that—

(a) where mobile work equipment has a shaft for the transmission of energy between it and other mobile work equipment; and

(b) the shaft could become soiled or damaged by contact with the ground while uncoupled,

the work equipment has a system for safeguarding the shaft.

PART IV

POWER PRESSES

Regulation 31

Power presses to which Part IV does not apply

Regulations 32 to 35 shall not apply to a power press of a kind which is described in Schedule 2.

Regulation 32

Thorough examination of power presses, guards and protection devices

(1) Every employer shall ensure that a power press is not put into service for the first time after installation, or after assembly at a new site or in a new location unless—

(a) it has been thoroughly examined to ensure that it—

(i) has been installed correctly; and

(ii) would be safe to operate; and

(b) any defect has been remedied.

(2) Every employer shall ensure that a guard, other than one to which paragraph (3) relates, or protection device is not put into service for the first time on a power press unless—

(a) it has been thoroughly examined when in position on that power press to ensure that it is effective for its purpose; and

(b) any defect has been remedied.

(3) Every employer shall ensure that that part of a closed tool which acts as a fixed guard is not used on a power press unless—

(a) it has been thoroughly examined when in position on any power press in the premises to ensure that it is effective for its purpose; and

(b) any defect has been remedied.

(4) For the purpose of ensuring that health and safety conditions are maintained, and that any deterioration can be detected and remedied in good time, every employer shall ensure that—

(a) every power press is thoroughly examined, and its guards and protection devices are thoroughly examined when in position on that power press—

(i) at least every 12 months, where it has fixed guards only; or

(ii) at least every 6 months, in other cases; and

(iii) each time that exceptional circumstances have occurred which are liable to jeopardise the safety of the power press or its guards or protection devices; and

(b) any defect is remedied before the power press is used again.

(5) Where a power press, guard or protection device was before the coming into force of these Regulations required to be thoroughly examined by regulation 5(2) of the Power Presses Regulations 1965 the first thorough examination under paragraph (4) shall be made before the date by which a thorough examination would have been required by regulation 5(2) had it remained in force.

(6) Paragraph (4) shall not apply to that part of a closed tool which acts as a fixed guard.

(7) In this regulation 'defect' means a defect notified under regulation 34 other than a defect which has not yet become a danger to persons.

Regulation 33
Inspection of guards and protection devices

(1) Every employer shall ensure that a power press is not used after the setting, re-setting or adjustment of its tools, save in trying out its tools or save in die proving, unless—

(a) its every guard and protection device has been inspected and tested while in position on the power press by a person appointed in writing by the employer who is—

(i) competent; or

(ii) undergoing training for that purpose and acting under the immediate supervision of a competent person,

and who has signed a certificate which complies with paragraph (3); or

(b) the guards and protection devices have not been altered or disturbed in the course of the adjustment of its tools.

(2) Every employer shall ensure that a power press is not used after the expiration of the fourth hour of a working period unless its every guard and protection device has been inspected and tested while in position on the power press by a person appointed in writing by the employer who is—

(a) competent; or

(b) undergoing training for that purpose and acting under the immediate supervision of a competent person,

and who has signed a certificate which complies with paragraph (3).

(3) A certificate referred to in this regulation shall—

(a) contain sufficient particulars to identify every guard and protection device inspected and tested and the power press on which it was positioned at the time of the inspection and test;

(b) state the date and time of the inspection and test; and

(c) state that every guard and protection device on the power press is in position and effective for its purpose.

(4) In this regulation 'working period', in relation to a power press, means—

(a) the period in which the day's or night's work is done; or

(b) in premises where a shift system is in operation, a shift.

Regulation 34

Reports

(1) A person making a thorough examination for an employer under regulation 32 shall—

(a) notify the employer forthwith of any defect in a power press or its guard or protection device which in his opinion is or could become a danger to persons;

(b) as soon as is practicable make a report of the thorough examination to the employer in writing authenticated by him or on his behalf by signature or equally secure means and containing the information specified in Schedule 3; and

(c) where there is in his opinion a defect in a power press or its guard or protection device which is or could become a danger to persons, send a copy of the report as soon as is practicable to the enforcing authority for the premises in which the power press is situated.

(2) A person making an inspection and test for an employer under regulation 33 shall forthwith notify the employer of any defect in a guard or protection device which in his opinion is or could become a danger to persons and the reason for his opinion.

Regulation 35

Keeping of information

(1) Every employer shall ensure that the information in every report made pursuant to regulation 34(1) is kept available for inspection for 2 years after it is made.

(2) Every employer shall ensure that a certificate under regulation 33(1)(a)(ii) or (2)(b) is kept available for inspection—

(a) at or near the power press to which it relates until superseded by a later certificate; and

(b) after that, until 6 months have passed since it was signed.

PART V
MISCELLANEOUS

Regulation 36

Exemption for the armed forces

(1) The Secretary of State for Defence may, in the interests of national security, by a certificate in writing exempt any of the home forces, any visiting force or any headquarters from any requirement or prohibition imposed by these Regulations and any such exemption may be granted subject to conditions and to a limit of time and may be revoked by the said Secretary of State by a certificate in writing at any time.

(2) In this regulation—

(a) 'the home forces' has the same meaning as in section 12(1) of the Visiting Forces Act 1952;

(b) 'headquarters' has the same meaning as in article 3(2) of the Visiting Forces and International Headquarters (Application of Law) Order 1965;

(c) 'visiting force' has the same meaning as it does for the purposes of any provision of Part I of the Visiting Forces Act 1952.

Regulation 37

Transitional provision

The requirements in regulations 25 to 30 shall not apply to work equipment provided for use in the undertaking or establishment before 5 December 1998 until 5 December 2002.

Regulation 38

Repeal of enactment

Section 19 of the Offices, Shops and Railway Premises Act 1963 is repealed.

Regulation 39

Revocation of instruments

The instruments specified in column 1 of Schedule 4 are revoked to the extent specified in column 3 of that Schedule.

Signed by authority of the Secretary of State

Alan Meale

Parliamentary Under Secretary of State,
Department of the Environment, Transport and the Regions

15 September 1998

SCHEDULE 1
INSTRUMENTS WHICH GIVE EFFECT TO COMMUNITY DIRECTIVES CONCERNING THE SAFETY OF PRODUCTS

Regulation 10

(1) *Title*	*(2)* *Reference*
The Construction Plant and Equipment (Harmonisation of Noise Emission Standards) Regulations 1985	SI 1985/1968, amended by SI 1989/1127
The Construction Plant and Equipment (Harmonisation of Noise Emission Standards) Regulations 1988	SI 1988/361, amended by SI 1992/488, 1995/2357
The Electro-medical Equipment (EEC Requirements) Regulations 1988	SI 1988/1586, amended by SI 1994/3017
The Low Voltage Electrical Equipment (Safety) Regulations 1989	SI 1989/728, amended by SI 1994/3260
The Construction Products Regulations 1991	SI 1991/1620, amended by SI 1994/3051

The Simple Pressure Vessels (Safety) Regulations 1991	SI 1991/2749, amended by SI 1994/3098
The Lawnmowers (Harmonisation of Noise Emission Standards) Regulations 1992	SI 1992/168
The Gas Appliances (Safety) Regulations 1992	SI 1992/711
The Electromagnetic Compatibility Regulations 1992	SI 1992/2372, amended by SI 1994/3080
The Supply of Machinery (Safety) Regulations 1992	SI 1992/3073, amended by SI 1994/2063
The Personal Protective Equipment (EC Directive) Regulations 1992	SI 1992/3139, amended by SI 1993/3074, 1994/2326, 1996/3039
The Active Implantable Medical Devices Regulations 1992	SI 1992/3146, amended by SI 1995/1671
The Medical Devices Regulations 1994	SI 1994/3017
The Electrical Equipment (Safety) Regulations 1994	SI 1994/3260
The Gas Appliances (Safety) Regulations 1995	SI 1995/1629
The Equipment and Protective Systems Intended for Use in Potentially Explosive Atmospheres Regulations 1996	SI 1996/192

The Lifts Regulations 1997	SI 1997/831
[The Pressure Equipment Regulations 1999][2]	[SI 1999/2001]

SCHEDULE 2
POWER PRESSES TO WHICH REGULATIONS 32 TO 35 DO NOT APPLY

Regulation 31

1

A power press for the working of hot metal.

2

A power press not capable of a stroke greater than 6 millimetres.

3

A guillotine.

4

A combination punching and shearing machine, turret punch press or similar machine for punching, shearing or cropping.

5

A machine, other than a press brake, for bending steel sections.

6

A straightening machine.

7

An upsetting machine.

2 Entry relating to The Pressure Equipment Regulations 1999 inserted by SI 1999/2001, reg 29(1). Date in force: 29 November 1999: see SI 1999/2001, reg 1(3).

8

A heading machine.

9

A riveting machine.

10

An eyeletting machine.

11

A press-stud attaching machine.

12

A zip fastener bottom stop attaching machine.

13

A stapling machine.

14

A wire stitching machine.

15

A power press for the compacting of metal powders.

SCHEDULE 3
INFORMATION TO BE CONTAINED IN A REPORT OF A THOROUGH EXAMINATION OF A POWER PRESS, GUARD OR PROTECTION DEVICE

Regulation 34(1)(b)

1

The name of the employer for whom the thorough examination was made.

2

The address of the premises at which the thorough examination was made.

3

In relation to each item examined—

(a) that it is a power press, interlocking guard, fixed guard or other type of guard or protection device;

(b) where known its make, type and year of manufacture;

(c) the identifying mark of—

(i) the manufacture,

(ii) the employer.

4

In relation to the first thorough examination of a power press after installation or after assembly at a new site or in a new location—

(a) that it is such thorough examination;

(b) either that it has been installed correctly and would be safe to operate or the respects in which it has not been installed correctly or would not be safe to operate;

(c) identification of any part found to have a defect, and a description of the defect.

5

In relation to a thorough examination of a power press other than one to which paragraph 4 relates—

(a) that it is such other thorough examination;

(b) either that the power press would be safe to operate or the respects in which it would not be safe to operate;

(c) identification of any part found to have a defect which is or could become a danger to persons, and a description of the defect.

6

In relation to a thorough examination of a guard or protection device—

(a) either that it is effective for its purpose or the respects in which it is not effective for its purpose;

(b) identification of any part found to have a defect which is or could become a danger to persons, and a description of the defect.

7

Any repair, renewal or alteration required to remedy a defect found to be a danger to persons.

8

In the case of a defect which is not yet but could become a danger to persons—

(a) the time by which it could become such danger;

(b) any repair, renewal or alteration required to remedy it.

9

Any other defect which requires remedy.

10

Any repair, renewal or alteration referred to in paragraph 7 which has already been effected.

11

The date on which any defect referred to in paragraph 8 was notified to the employer under regulation 34(1)(a).

12

The qualification and address of the person making the report; that he is self-employed or if employed, the name and address of his employer.

13

The date of the thorough examination.

14

The date of the report.

15

The name of the person making the report and where different the name of the person signing or otherwise authenticating it.

SCHEDULE 4
REVOCATION OF INSTRUMENTS

Regulation 39

(1) *Title*	(2) *Reference*	(3) *Extent of revocation*
The Operations at Unfenced Machinery (Amended Schedule) Regulations 1946	SR & O 1946/156	The whole Regulations
The Agriculture (Circular Saws) Regulations 1959	SI 1959/427	The whole Regulations
The Prescribed Dangerous Machines Order 1964	SI 1964/971	The whole Order
The Power Presses Regulations 1965	SI 1965/1441	The whole Regulations
The Abrasive Wheels Regulations 1970	SI 1970/535	The whole Regulations
The Power Presses (Amendment) Regulations 1972	SI 1972/1512	The whole Regulations
The Woodworking Machines Regulations 1974	SI 1974/903	The whole Regulations

The Operations at Unfenced Machinery (Amendment) Regulations 1976	SI 1976/955	The whole Regulations
The Factories (Standards of Lighting) (Revocation) Regulations 1978	SI 1978/1126	The whole Regulations
The Offshore Installations (Application of Statutory Instruments) Regulations 1984	SI 1984/419	The whole Regulations
The Offshore Installations (Operational Safety, Health and Welfare and Life-Saving Appliances) (Revocations) Regulations 1989	SI 1989/1672	The whole Regulations
The Provision and Use of Work Equipment Regulations 1992	SI 1992/2932	The whole Regulations
The Construction (Health, Safety and Welfare) Regulations 1996	SI 1996/1592	Regulation 27

APPROVED CODE OF PRACTICE

Introduction

1–8 *(See Guidance Notes, pp 424–426)*

Background

9–31 *(See Guidance Notes, pp 426–432)*

Erecting or dismantling work equipment

32 You should ensure that work equipment is erected or dismantled in a safe way, in particular observing any manufacturers' or suppliers' instructions where they exist.[3]

33–44 *(See Guidance Notes, pp 433–435)*

Risks to pedestrians

45 You should take measures, where appropriate, to prevent pedestrians coming within the area of operation of self-propelled work equipment. Where this is not reasonably practicable, appropriate measures should be taken to reduce the risks involved, including the operation of appropriate traffic rules.[4]

46–49 *(See Guidance Notes, pp 435–436)*

50 You should ensure that, where there is a risk to workers arising from lightning strikes to work equipment when it is being used, appropriate safety precautions are followed.[5]

51–54 *(See Guidance Notes, pp 436–437)*

PART I: INTRODUCTION

Regulation 1

Citation and commencement

55–59 *(See Guidance Notes, pp 437–438)*

3 This implements point 1.2 of Annex II of AUWED (see Guidance Notes, paras 9–12 on pp 426–427): 'Work equipment must be erected or dismantled under safe conditions, in particular observing any instructions which may have been furnished by the manufacturer.'

4 This implements AUWED, Annex II, point 2.2: 'if work equipment is moving around in a work area, appropriate traffic rules must be drawn up and followed'; point 2.3: 'Organisational measures must be taken to prevent workers on foot coming within the area of operation of self-propelled work equipment. If work can be done properly, only if workers on foot are present, appropriate measures should be taken to prevent them from being injured by the equipment'; and point 2.4: 'The transport of workers on mechanically driven mobile work equipment is authorised only where safe facilities are provided to this effect. If work needs to be carried out during the journey, speeds must be adjusted as necessary.'

5 This implements AUWED, Annex II, point 1.3: 'Work equipment which may be struck by lightning whilst being used must be protected by devices or appropriate means against the effects of lightning.'

Regulation 2

Interpretation

60–67 *(See Guidance Notes, pp 438–440)*

Regulation 3

Application

68–91 *(See Guidance Notes, pp 440–446)*

PART II: GENERAL

Regulation 4

Suitability of work equipment

92–99 *(See Guidance Notes, pp 446–447)*

100 When selecting work equipment, employers should take account of ergonomic risks.

101–105 *(See Guidance Notes, pp 447–448)*

106 You should ensure that work equipment is installed, located and used in such a way as to reduce risks to users of work equipment and for other workers, such as ensuring that there is sufficient space between the moving parts of work equipment and fixed or moving parts in its environment.[6]

107 When determining the suitability of work equipment, you should ensure that where appropriate:

(a) all forms of energy used or produced; and

(b) all substances used or produced can be supplied and/or removed in a safe manner.[6]

108 You should ensure that where mobile work equipment with a combustion engine is in use there is sufficient air of good quality.[7]

6 This implements AUWED, Annex II, point 1.1: 'Work equipment must be installed, located and used in such a way as to reduce risks to users of the work equipment and for other workers, for example by ensuring that there is sufficient space between the moving parts of work equipment and fixed or moving parts in its environment and that all forms of energy and substances used or produced can be supplied and/or removed in a safe manner.'

7 This implements AUWED, Annex II, point 2.5: 'Mobile work equipment with a combustion engine may not be used in working areas unless sufficient quantities of air presenting no health or safety risk to workers can be guaranteed.'

109–117 *(See Guidance Notes, pp 448–451)*

Regulation 5

Maintenance

118–133 *(See Guidance Notes, pp 451–454)*

Regulation 6

Inspection

134–135 *(See Guidance Notes, p 454)*

Identifying what needs to be inspected

136 Where the risk assessment under regulation 3 of the Management of Health and Safety at Work Regulations 1992 has identified a significant risk to the operator or other workers from the installation or use of the work equipment, a suitable inspection should be carried out.

137–140 *(See Guidance Notes, pp 454–455)*

What should be included in the inspection

141 The extent of the inspection required will depend on the potential risks from the work equipment. Inspection should include, where appropriate, visual checks, functional checks and testing.

142–147 *(See Guidance Notes, pp 455–456)*

Competent persons

148 You should ensure that persons who determine the nature of the inspections required and who carry out inspections are competent to do so.

149 The competent person should have the necessary knowledge and experience.

150–152 *(See Guidance Notes, pp 456–457)*

Installation

153 Where work equipment is of a type where the safe operation is critically dependent on it being properly installed (or

re-installed), and where failure to carry this out would lead to a significant risk to the operator, or other worker, you should arrange for a suitable inspection to be carried out before it is put into service.

154 *(See Guidance Notes, p 457)*

'Conditions causing deterioration' and 'Dangerous situations'

155 Where work equipment is of a type where the safe operation is critically dependent on its condition in use and deterioration would lead to a significant risk to the operator or other worker, you should arrange for suitable inspections to be carried out.

156–157 *(See Guidance Notes, pp 457–458)*

Frequency of inspection

158 The frequency of inspections should be based on how quickly the work equipment or parts of it are likely to deteriorate and therefore give rise to a significant risk. This should take into account the type of equipment, how it is used and the conditions to which it is exposed.

159–163 *(See Guidance Notes, pp 458–459)*

164 The physical evidence should be appropriate to the type of work equipment being inspected.

165–166 *(See Guidance Notes, pp 459–460)*

Regulation 7

Specific risks

167 You should ensure that, wherever possible, risks are always controlled by (in the order given):

(a) eliminating the risks, or if that is not possible;

(b) taking 'hardware' (physical) measures to control the risks such as the provision of guards; but if the risks cannot be adequately controlled;

(c) taking appropriate 'software' measures to deal with the residual (remaining) risk, such as following safe systems of work and the provision of information, instruction and training.

Normal operation

168 Where the risks from the use of work equipment cannot be adequately controlled by hardware measures, such as guards or protection devices, during its normal operation, it is particularly important that only the persons whose task it is should be allowed to use such equipment. They should have received sufficient information, instruction and training to enable them to carry out the work safely.

Repairs, modifications etc

169 Where the risks from the use of work equipment cannot be adequately controlled by hardware measures such as guards or protection devices during repair, maintenance, or other similar work, only persons who have received sufficient information, instruction and training to enable them to carry out the work safely should do the work. They shall be the designated person for the purpose of this regulation.

170–172 *(See Guidance Notes, p 460)*

Regulation 8

Information and instructions

173–184 *(See Guidance Notes, pp 461–463)*

Regulation 9

Training

185–193 *(See Guidance Notes, pp 464–465)*

Driver training

194 You should ensure that self-propelled work equipment, including any attachments or towed equipment, is only driven by workers who have received appropriate training in the safe driving of such work equipment.[8]

8 This implements AUWED, Annex II, point 2.1: 'Self-propelled work equipment shall be driven only by workers who have been appropriately trained in the safe driving of such equipment.'

Chainsaw operators

195 All workers who use a chainsaw should be competent to do so. Before using a chainsaw to carry out work on or in a tree, a worker should have received appropriate training and obtained a relevant certificate of competence or national competence award, unless they are undergoing such training and are adequately supervised. However, in the agricultural sector, this requirement only applies to first-time users of a chainsaw.

196–198 *(See Guidance Notes, pp 465–466)*

Regulation 10

Conformity with Community requirements

199–208 *(See Guidance Notes, pp 466–468)*

Regulation 11

Dangerous parts of machinery

209–220 *(See Guidance Notes, pp 468–471)*

Regulation 12

Protection against specified hazards

221–230 *(See Guidance Notes, pp 471–474)*

Regulation 13

High or very low temperature

231–243 *(See Guidance Notes, pp 474–476)*

Regulation 14

Controls for starting or making a significant change in operating conditions

244–250 *(See Guidance Notes, pp 476–477)*

Regulation 15

Stop controls

251–255 *(See Guidance Notes, pp 478–479)*

Regulation 16

Emergency stop controls

256–259 *(See Guidance Notes, p 479)*

Regulation 17

Controls

260–271 *(See Guidance Notes, pp 479–482)*

Regulation 18

Control systems

272–275 *(See Guidance Notes, pp 482–483)*

Regulation 19

Isolation from sources of energy

276–287 *(See Guidance Notes, pp 483–485)*

Regulation 20

Stability

288–291 *(See Guidance Notes, pp 485–486)*

Regulation 21

Lighting

292–296 *(See Guidance Notes, pp 486–487)*

Regulation 22

Maintenance operations

297–303 *(See Guidance Notes, pp 487–488)*

Regulation 23

Markings

304–307 *(See Guidance Notes, pp 488–489)*

Regulation 24

Warnings

308–313 *(See Guidance Notes, pp 489–491)*

PART III: MOBILE WORK EQUIPMENT

Introduction

314–323 *(See Guidance Notes, pp 491–493)*

Regulation 25

Employees carried on mobile work equipment

324 You should ensure that risks to the operator and other workers due to the mobile work equipment travelling are controlled. Workers should be protected against falling out of the equipment and from unexpected movement.

325–332 *(See Guidance Notes, pp 493–495)*

Speed adjustment

333 If work needs to be carried out during the journey, speeds should be adjusted as necessary.

334 *(See Guidance Notes, p 495)*

Guards and barriers

335 You should ensure that guards and/or barriers fitted to mobile work equipment, which are designed to prevent contact with wheels and tracks, are suitable and effective.

336 *(See Guidance Notes, p 495)*

Regulation 26

Rolling over of mobile work equipment

337–346 *(See Guidance Notes, pp 495–497)*

Roll-over protective structures (ROPS)

347 You should fit suitable roll-over protective structures to mobile work equipment where necessary to minimise the risks to workers carried, should roll-over occur.

348–351 *(See Guidance Notes, pp 497–498)*

Restraining systems

352 You should provide restraining systems on mobile work equipment, where appropriate, if they can be fitted to the equipment, to prevent workers carried from being crushed between any part of the work equipment and the ground, should roll-over occur.

353–355 *(See Guidance Notes, pp 498–499)*

Tractors

356 If a tractor is fitted with a ROP rather than a cab, a restraining system will be needed.

357 *(See Guidance Notes, p 499)*

Regulation 27

Overturning of fork-lift trucks

358–359 *(See Guidance Notes, pp 499–500)*

Restraining systems

360 For fork-lift trucks fitted with either a mast or a roll-over protective structure, you should provide restraining systems where appropriate, if such systems can be fitted to the equipment, to prevent workers carried from being crushed between any part of the truck and the ground, should it overturn.

361–363 *(See Guidance Notes, p 500)*

Regulation 28

Self-propelled work equipment

364–373 *(See Guidance Notes, pp 500–503)*

Regulation 28(g)

The carriage of appropriate fire-fighting appliances

374 Where escape from self-propelled work equipment in the event of a fire could not be achieved easily, you should ensure that fire-fighting appliances are carried on that equipment.

375–376 *(See Guidance Notes, p 503)*

Regulation 29

Remote-controlled self-propelled work equipment

377–380 *(See Guidance Notes, pp 503–504)*

Regulation 30

Drive shafts

381–385 *(See Guidance Notes, pp 504–505)*

PART IV: POWER PRESSES

Regulations 31 to 35 and related Schedules 2 and 3 refer to power presses and are not included here. They can be found in the power presses ACOP.[9]

GUIDANCE NOTES

INTRODUCTION

The Provision and Use of Work Equipment Regulations 1998

1 The Provision and Use of Work Equipment Regulations 1998 (PUWER 98) are made under the Health and Safety at Work etc Act 1974 (HSW Act) and come into force on 5 December 1998. PUWER 98 brings into effect the non-lifting aspects of the Amending Directive to the Use of Work Equipment Directive (AUWED). The primary objective of PUWER 98 is to ensure that

9 *Safe use of power presses. Provision and Use of Work Equipment Regulations 1998 as applied to power presses. Approved Code of Practice and Guidance on Regulations* L112 HSE Books ISBN 0 7176 1627 4.

work equipment should not result in health and safety risks, regardless of its age, condition or origin.

What does PUWER 98 apply to?

2 PUWER 98 applies to the provision and use of all work equipment, including mobile and lifting equipment.

Where does PUWER 98 apply?

3 It applies to all workplaces and work situations where the HSW Act applies and extends outside Great Britain to certain offshore activities in British territorial waters and on the UK Continental Shelf.

Who needs to read this?

4 Anyone with responsibility directly or indirectly for work equipment and its use, for example employers, employees, the self-employed and those who hire work equipment. Throughout this document we have referred to the employer and self-employed people who have duties as 'you'. Where the guidance is addressed to some other duty holder, for example a competent person, the text makes it clear who it is intended for.

What is in the document?

5 This document contains:

(a) the PUWER 98 Regulations;

(b) the Approved Code of Practice (ACOP); and

(c) guidance material that has been written to help people use these Regulations.

HSE is publishing separate guidance specific to particular industry sectors. These link the requirements of PUWER 98 to the specialised work equipment used in industry sectors such as agriculture and construction.

What is an Approved Code of Practice (ACOP)?

6 (The formal status of ACOP material has not been set out in this book). ACOP material gives practical guidance on how to

comply with the law. If you follow the advice in the ACOP you will be doing enough to ensure compliance with the law on the matters that it covers. ACOP material has special legal status. If you are prosecuted for a breach of health and safety law, and it is proved that you did not follow the relevant provisions of the ACOP, you will need to show that you have complied with the law in some other way or a court will find you at fault.

What is guidance?

7 Guidance material describes practical means of complying with the Regulations. It does not have special status in law, but is seen as best practice. Following the guidance is not compulsory and you are free to take other action. But if you do follow the guidance you will normally be doing enough to comply with the law. Health and safety inspectors seek to secure compliance with the law and may refer to this guidance as illustrating good practice.

Other HSC/E information

8 You should also take account of any relevant HSC/HSE publications giving guidance on other regulations, industries or equipment. Up-to-date information on these publications can be obtained from HSE's InfoLine which deals with public telephone requests (0541 545500).

BACKGROUND

The Use of Work Equipment Directive and PUWER 92

9 PUWER 98 replaces PUWER 92 which implemented the first European Community (EC) Directive about work equipment, the Use of Work Equipment Directive (UWED). This required all Member States of the EC to have the same minimum requirements for the selection and use of work equipment.

The Amending Directive to the Use of Work Equipment Directive and PUWER 98

10 The European Council of Ministers agreed a further Directive about work equipment in 1995. This was the Amending Directive

to the Use of Work Equipment Directive (AUWED). AUWED extended the requirements of UWED. PUWER 98 brought the requirements of AUWED into British law.

11 PUWER 98, while replacing PUWER 92, continues its requirements in addition to introducing the new measures required by AUWED. This was done to ensure that the law relating to work equipment was consolidated into one set of legislation.

AUWED's new requirements

12 AUWED's main new requirements concern mobile work equipment, lifting equipment and the inspection of work equipment. The 'hardware' requirements of AUWED outline physical features that relate to the equipment itself and the 'software' requirements consist of the management provisions. The specific requirements for mobile work equipment have been implemented through PUWER 98. The lifting requirements have been implemented through the Lifting Operations and Lifting Equipment Regulations 1998 (LOLER). The inspection requirements are incorporated in both PUWER 98 and LOLER.

Other changes—PUWER 92 to PUWER 98

13 PUWER 98 differs from PUWER 92 in a number of other ways. They are:

(a) an extension of the definition of 'work equipment' to include installations;

(b) an extension of the duty holder application to include a duty on people who have control of work equipment such as plant hire;

(c) guidance and ACOP material about regulation 7 (specific risks);

(d) changes to regulation 10 (conformity with Community requirements);

(e) minor changes to regulation 18 (control systems) as a result of a change required by AUWED;

(f) new regulations to replace the previous regulations dealing with power presses.

Lifting operations and lifting equipment

14 The lifting requirements of AUWED are brought into effect through a set of regulations called LOLER. LOLER replaces most of the existing legislation on lifting, for example lifting law that applies to factories, offices, shops, railway premises and construction sites. In doing so, it creates a single set of regulations that apply to all sectors.

15 Though PUWER 98 applies to all lifting equipment, LOLER applies over and above the general requirements of PUWER 98, in dealing with specific hazards/risks associated with lifting equipment and lifting operations.

16 LOLER, its supporting ACOP and guidance material are not included in this document. They are found in document L113.[10]

Power presses

17 PUWER 98 revokes the Power Presses Regulations 1965 and the Power Presses (Amendment) Regulations 1972 (the Power Presses Regulations). To ensure that safety levels for power presses are maintained, Part IV of PUWER 98 contains specific regulations applying to power presses. This part of PUWER 98 and its supporting ACOP and guidance material is available separately in L112.[11]

Woodworking

18 PUWER 98 revokes the remaining requirements of the Woodworking Machines Regulations 1974 which concern the training of operators, removal of material cut by circular sawing machines and requirements for machines with two-speed motors. The other requirements of the Woodworking Machines Regulations 1974 were revoked by PUWER 92. Because of the high accident rate and inherent risks associated with woodworking machines, a separate ACOP with supporting guidance material has been produced concerning the safeguarding of woodworking

10 *Safe use of lifting equipment. Lifting Operations and Lifting Equipment Regulations 1998. Approved Code of Practice and Guidance on Regulations* L113 HSE Books ISBN 0 7176 1628 2.

11 *Safe use of power presses. Provision and Use of Work Equipment Regulations 1998 as applied to power presses. Approved Code of Practice and Guidance on Regulations* L112 HSE Books ISBN 0 7176 1627 4.

machines and the training of people who use them. The ACOP and guidance material relating to woodworking machines is published in L114.[12]

What does PUWER 98 require?

19 PUWER 98 sets out important health and safety requirements for the provision and safe use of work equipment. PUWER 98 is set out as follows:

(a) Regulations 1 to 3 deal with the date PUWER 98 comes into force, interpretation of the Regulations (definition of terms used in the Regulations), where PUWER 98 applies and who has duties under the Regulations.

(b) Regulations 4 to 10 are the 'management' duties of PUWER 98 covering selection of suitable equipment, maintenance, inspection, specific risks, information, instructions and training. It also covers the conformity of work equipment with legislation which brings into effect the requirements of EC Directives on product safety.

(c) Regulations 11 to 24 deal with the physical aspects of PUWER 98. They cover the guarding of dangerous parts of work equipment, the provision of appropriate stop and emergency stop controls, stability, suitable and sufficient lighting and suitable warning markings or devices.[13]

(d) Regulations 25 to 30 deal with certain risks from mobile work equipment.[13]

(e) Regulations 31 to 35 deal with the management requirements for the safe use of power presses.

(f) Regulations 36 to 39 cover transitional provisions, repeal of Acts and revocation of instruments.

(g) Schedule 1 lists the regulations implementing relevant EC 'product' Directives referred to in regulation 10.

(h) Schedule 2 lists the power presses to which regulations 32 to 35 do not apply.

12 *Safe use of woodworking machinery. Provision and Use of Work Equipment Regulations 1998 as applied to woodworking. Approved Code of Practice and Guidance Regulations* L114 HSE Books ISBN 0 7176 1630 4.

13 Regulations 11–19 and 22–29 will not apply in certain circumstances—see regulation 10(2).

(i) Schedule 3 lists the information to be contained in a report of a thorough examination of a power press, guard or protection device.

(j) Schedule 4 lists the regulations revoked by PUWER 98.

How PUWER 98 relates to other health and safety legislation

20 PUWER 1998 cannot be considered in isolation from other health and safety legislation. In particular, it needs to be considered with the requirements of the HSW Act.

21 Some of AUWED's provisions are not implemented by regulation in PUWER 98 as they are already covered adequately by the requirements of section 2 of the HSW Act. Supporting ACOP and guidance on these provisions is contained in this publication in paragraphs 31 to 54.

22 There is also some overlap between PUWER 98 and other sets of regulations, for example the Workplace (Health, Safety and Welfare) Regulations 1992 (workplace risks to pedestrians from vehicles) and the Health and Safety (Display Screen Equipment) Regulations 1992 (for example, on lighting) and the Personal Protective Equipment at Work Regulations 1992 (for example, on maintenance), the Construction (Health, Safety and Welfare) Regulations 1996 (for example, on standards for work equipment such as scaffolding) and the Road Vehicles (Construction and Use) Regulations 1986. If you comply with the more specific regulations, it will normally be sufficient to comply with the more general requirements in PUWER 98.

23 The Management of Health and Safety at Work Regulations 1992 (the Management Regulations) also have important general provisions relating to the safety of work equipment, including the requirement to carry out a risk assessment. This is discussed in the following paragraphs.

Risk assessment

24 Because of the general risk assessment requirements in the Management Regulations, there is no specific regulation requiring a risk assessment in PUWER. HSE has produced guidance in a booklet called *5 steps to risk assessment*.[14]

14 *5 steps to risk assessment* INDG163(rev) HSE Books 1998 (single copies free; ISBN 0 7176 1565 0 for priced packs).

25 Risks to health and safety should be assessed taking into account matters such as the type of work equipment, substances and electrical or mechanical hazards to which people may be exposed.

26 Action to eliminate/control any risk might include, for example, during maintenance:

(a) disconnecting the power supply to the work equipment;

(b) supporting parts of the work equipment which could fall;

(c) securing mobile work equipment so that it cannot move;

(d) removing or isolating flammable or hazardous substances; and

(e) depressurising pressurised equipment.

27 Other matters to consider include environmental conditions such as:

(a) lighting;

(b) problems caused by weather conditions;

(c) other work being carried out which may affect the operation; or

(d) the activities of people who are not at work.

How does PUWER 98 affect older health and safety legislation about work equipment?

28 PUWER 98 repeals or revokes most of the remaining legislation about work equipment. This is listed at regulations 38 and 39 (referring to Schedule 4) of PUWER 98.

Structure of the ACOP

29 This publication has ACOP material for PUWER 98. Regulations that have been carried forward unchanged from PUWER 92 do not have any ACOP material except for regulations 4, 7 and 9, where it is necessary to bring into effect AUWED (regulations 4 and 9) or where needed to clarify the meaning of a regulation (regulation 7). We have also included some ACOP and guidance material on aspects of the HSW Act, where relevant.

Links with other legislation and other health and safety principles

30 The requirements of ACOP and guidance material which follows link with other health and safety legislation, for example the Workplace (Health and Safety) Regulations. These aspects are described, where relevant, in the following guidance. You may need to obtain further information on specific points and there is a list of useful references in the footnotes to this text.

The Health and Safety at Work etc Act 1974, section 2

Introduction

31 AUWED contains some requirements for the management of work equipment. These requirements are fulfilled through section 2 of the HSW Act and the new ACOP to support it which is reproduced in the following paragraphs.

'*The Health and Safety at Work etc Act 1974*

Section 2

(1) It shall be the duty of every employer to ensure, so far as is reasonably practicable, the health, safety and welfare at work of all his employees.

(2) Without prejudice to the generality of an employer's duty under the preceding subsection, the matters to which that duty extends include in particular—

(a) the provision and maintenance of plant and systems of work that are, so far as is reasonably practicable, safe and without risks to health;

(b) arrangements for ensuring, so far as is reasonably practicable, safety and absence of risks to health in connection with the use, handling, storage and transport of articles and substances;

(c) the provision of such information, instruction, training and supervision as is necessary to ensure, so far as is reasonably practicable, the health and safety at work of his employees;

(d) so far as is reasonably practicable as regards any place of work under the employer's control, the maintenance of it in a condition that is safe and without risks to health and the provision and maintenance of means of access and egress from it that are safe and without risks;

(e) the provision and maintenance of a working environment for his employees that is, so far as is reasonably practicable, safe, without risks to health, and adequate as regards facilities and arrangements for their welfare at work.'

Erecting or dismantling work equipment

32 *(See Approved Code of Practice, p 415)*

33 The assembly and dismantling of some items of work equipment, for example airbridges at airports, may be subject to the requirements of the Construction (Design and Management) Regulations 1994.

Systems of work

34 Work equipment should be erected, assembled or dismantled safely and without risk to health. Safe systems of work and safe working practice should be followed to achieve this. A safe system of work is a formal procedure which should be followed to ensure that work is carried out safely and is necessary where risks cannot be adequately controlled by other means.

35 The work should be planned and potential hazards identified. You should ensure that the systems of work to be followed are properly implemented and monitored and that details have been communicated to those at risk.

36 HSE's leaflet *Five steps to successful health and safety management*[15] gives further detailed information about safe systems of work.

Competence of workers

37 The knowledge, experience and abilities of the people carrying out the work should be considered. They should be competent to carry out the work safely.

Groups at risk

38 You have a duty under health and safety law to ensure, as far as is reasonably practicable, the health, safety and welfare of your employees. When carrying out an assessment of the risk to their health and safety, you should identify groups of workers that might be particularly at risk such as young or disabled people. The outcome of your risk assessment will be helpful in meeting your duty to provide information, instruction, training and supervision

15 *5 steps to successful health and safety management* INDG132 HSE Books 1992 (free).

necessary to ensure the health and safety of your employees. You will want to take account of factors such as their competence, experience, maturity, etc. Formal qualifications, training certificates, aptitude tests, etc might be used to help identify competence.

Mobile work equipment and the HSW Act

39 Sections 2(1) and 2(2) of the HSW Act cover the risks to pedestrians from the movement of mobile work equipment.

Risks from self-propelled work equipment

40 This section of ACOP and guidance deals with some of the risks from the movement of mobile work equipment. This includes risks of pedestrians being struck by vehicles, or their loads.

Assessing risks from self-propelled work equipment

41 The Management Regulations require you to carry out a risk assessment. This risk assessment will identify the hazards, help you evaluate the risks and also decide how to control the risks. Appropriate preventive and protective measures should be taken in the light of the risks identified. The risks which the use of mobile work equipment can create where pedestrians are present include the danger of people being struck, crushed or run over by self-propelled work equipment or being struck by an object falling from a vehicle.

Dealing with the risks from self-propelled mobile equipment

42 The precautions covered here that may be required to control the risks which the use of self-propelled work equipment can create for pedestrians are:

(a) separation of pedestrians and self-propelled mobile work equipment;

(b) traffic rules;

(c) traffic signs;

(d) planning traffic routes;

(e) traffic speed.

43 Regulation 17 of the Workplace Regulations and its supporting ACOP material also deal with the organisation, etc of traffic

routes and the organisation of workplaces so that pedestrians and vehicles can circulate in a safe manner. This requirement should be considered together with this publication to achieve compliance with both PUWER 98 and the Workplace Regulations.

44 The explanations describing mobile work equipment are in Part III of PUWER 98.

Risks to pedestrians

45 *(See Approved Code of Practice, p 415)*

46 Wherever possible you should keep pedestrians away from self-propelled work equipment. Where this is not possible you are required, so far as is reasonably practicable, to provide and maintain a safe system of work.

Traffic rules

47 Appropriate traffic rules should limit the risks to pedestrians and operators when mobile work equipment is in use, for example, fork-lift trucks operating in a loading bay where there are pedestrians and other vehicles. Traffic rules should be established as part of a safe system of work (see paragraph 35) following risk assessment.

Further guidance

48 Workplace transport safety is covered in detail in *Workplace transport safety*.[16] It covers the following topics:

(a) accidents—numbers, costs and causes;

(b) legal duties;

(c) managing risks;

(d) risk assessment;

(e) organising for safety including control, communication, co-operation and competence;

(f) a safe workplace;

(g) vehicle safety;

16 *Workplace transport safety* HSG136 HSE Books 1995 ISBN 0 7176 0935 9.

(h) maintenance;

(i) selection and training of drivers and other employees;

(j) contractors, visiting drivers and shared workplaces;

(k) safe working practices;

(l) reversing of vehicles;

(m) parking of vehicles;

(n) access on to vehicles;

(o) loading and unloading;

(p) tipping of loads;

(q) sheeting and unsheeting of loads.

Lightning

49 AUWED requires that where work equipment may be struck by lightning while being used it must be protected as appropriate against the effects of the lightning. This requirement is implemented through the following ACOP and guidance material.

50 *(See Approved Code of Practice, p 415)*

Assessing the risks from lightning

51 When assessing whether lightning protection is required for work equipment, you should consider whether:

(a) the area is one in which lightning generally occurs;

(b) the work equipment is tall or isolated;

(c) the work equipment contains flammable or explosive substances;

(d) large numbers of people will be affected by a lightning strike.

52 Types of work equipment which you should consider include:

(a) cranes being used in isolated areas;

(b) fairground equipment operating in similar conditions or in the open where large numbers of people are likely to be affected by a lightning strike;

(c) employees engaged in field work—surveyors etc, where there is a possible risk of lightning strike if the surveying staff/prism is used during a lightning strike.

Lightning protection

53 Protection from lightning can be provided by conductors or insulation. There are circumstances where the best way of reducing the risk is to cease working during a lightning storm. For example, a golf professional should avoid playing or teaching during a storm as a metal golf club is an excellent conductor of lightning and golf is often played in open areas or near trees where lightning strikes are likely to occur. Likewise a surveying team may need to stop work during a storm as would certain utility workers.

Other information about lightning and lightning protection

54 Detailed information about lightning, risk assessment, the likelihood of lightning strike and suitable protection is contained in BS 6651:1992, the British Standard Code of Practice for protection of structures against lightning.

PART I: INTRODUCTION

Regulation 1

Citation and commencement

When does PUWER 98 come into force?

55 PUWER 98 comes into force on 5 December 1998. Some of the Regulations dealing with mobile work equipment do not come into effect until 5 December 2002. The date the Regulations will apply will depend on whether the work equipment is new, existing, or second-hand on 5 December 1998. These transitional arrangements can be found in regulation 37.

New work equipment

56 Items of work equipment first provided for use from 5 December 1998 (NEW WORK EQUIPMENT) will need to meet ALL the requirements of PUWER 98 (see paragraphs on regulation 10).

Existing work equipment

57 If work equipment is first provided for use before 5 December 1998 (EXISTING WORK EQUIPMENT), PUWER 98 regulations 1–24 and 31–39 come into force on 5 December 1998. These Regulations are essentially the same as the requirements of PUWER 92, except for regulation 6 and those relating to power presses. Regulations 26–30 which for existing mobile work equipment come into force from December 2002 form Part III of PUWER 98 and cover mobile work equipment. However, Parts I and II of PUWER 98 will apply to ALL mobile work equipment from 5 December 1998.

Second-hand work equipment

58 When existing work equipment is sold by one company to another and brought into use by the purchasing company from 5 December 1998, it becomes new work equipment in the sense of paragraph 56 even though it is second-hand. This means that the purchasing company will need to ensure that the work equipment meets the provisions of PUWER 98 before it is put into use.

'Provided for use'

59 The phrase 'provided for use' refers to the date on which work equipment is first supplied in the premises or undertaking. This is not the same as first brought into use. Provided for use does not necessarily mean that it has actually been put into use. For example, equipment delivered to a company before 5 December 1998 and put into storage would be considered to be 'existing equipment' even though it might remain in store and not be put into use until after that date. This is set out in regulation 37 which deals with the transitional provisions for these Regulations.

Regulation 2

Interpretation

Inspection

60 The term 'inspection' is used in PUWER 98. The purpose of the inspection is to identify whether the equipment can be operated, adjusted and maintained safely and that any deterioration (for example, defect, damage, wear) can be detected and remedied before it results in unacceptable risks.

Use

61 The definition of 'use' is wide and includes all activities involving work equipment such as stopping or starting the equipment, repair, modification, maintenance and servicing. In addition to operations normally considered as use, cleaning and transport of the equipment are also included. In this context 'transport' means, for example, using a lift truck to carry goods around a warehouse.

Work equipment

62 The scope of 'work equipment' is extremely wide. It covers almost any equipment used at work, including:

(a) 'tool box tools' such as hammers, knives, handsaws, meat cleavers etc;

(b) single machines such as drilling machines, circular saws, photocopiers, combine harvesters, dumper trucks, etc;

(c) apparatus such as laboratory apparatus (Bunsen burners etc);

(d) lifting equipment such as hoists, lift trucks, elevating work platforms, lifting slings, etc;

(e) other equipment such as ladders, pressure water cleaners etc;

(f) an installation such as a series of machines connected together, for example a paper-making line or enclosure for providing sound insulation or scaffolding or similar access equipment (except where the Construction (Health, Safety and Welfare) Regulations 1996 impose more detailed requirements).

63 'Installation' does not include an offshore installation but would include any equipment attached or connected to it. The word 'installation' has replaced the phrase used in PUWER 92 'any assembly of components which in order to achieve a common end are arranged and controlled so that they function as a whole'.

64 The following are not classified as work equipment:

(a) livestock;

(b) substances (for example, acids, alkalis, slurry, cement, water);

(c) structural items (for example, walls, stairs, roof, fences);

(d) private car.

Motor vehicles

65 Motor vehicles which are not privately owned fall within the scope of PUWER 98. However, the more specific road traffic legislation will take precedence when these vehicles are used on public roads or in a public place. When such vehicles are used off the public highway and the road traffic law does not apply, for example on a dock road, PUWER 98 and the HSW Act would normally take precedence unless relevant local by-laws are in operation—for example, road traffic by-laws at some airports. Car drivers should hold a Department of Transport driving licence and cars should be maintained to the normal standards required for use on the public highway, ie they should have an MOT certificate, where necessary, or maintained to equivalent standard where statutory testing is not a legal requirement

Aircraft

66 The design, operation and maintenance of aircraft airworthiness is subject to other specific legislation, such as the Air Navigation (No 2) Order 1995 and Civil Aviation Authority (CAA) standards, such as JAR25 'Design requirements for large aeroplanes' which is an annex to EC Regulation 3922/91. This legislation takes precedence over PUWER 98.

'For use at work'

67 The phrase 'for use at work' has also been added to the end of the definition of 'work equipment'. Section 52(1)(b) and (c) of the HSW Act define this as 'an employee is at work throughout the time when he is in the course of his employment, but not otherwise and a self-employed person is at work throughout such time as he devotes to work as a self-employed person.'

Regulation 3
Application

Where PUWER 98 applies

68 PUWER 98 applies:

(a) to all work equipment used where the HSW Act applies, ie to all sectors, not only factories, offices and shops but also, for example, schools, universities, hospitals, hotels, places of entertainment and offshore oil and gas installations;

(b) to work equipment used in the common parts of shared buildings (such as lifts), private roads and paths on industrial estates and business parks and temporary work sites, including construction sites;

(c) throughout Great Britain and has effect wherever work is done by the employed or the self-employed except for domestic work in a private household;

(d) to homeworkers and will also apply to hotels, nursing homes and similar establishments and to parts of workplaces where 'domestic' staff are employed such as the kitchens of hostels or sheltered accommodation.

Application offshore

69 PUWER applies offshore as the HSW Act applies by virtue of the Health and Safety at Work etc Act 1974 (Application outside Great Britain) Order 1995 (SI 1995/263). This Order applies the Act to offshore installations, wells, pipelines and pipeline works, and to connected activities within the territorial waters of Great Britain or in designated areas of the United Kingdom Continental Shelf, plus certain other activities within territorial waters.

How does PUWER apply to marine activities?

70 Ships are subject to merchant shipping legislation which is dealt with by the Maritime and Coastguard Agency. Apart from certain regulations and in certain circumstances, PUWER 98 does not apply to ships' work equipment, no matter where it is used. However, regulations 7–9, 11–13, 20–22 and 30 of PUWER 98 will apply in what are called 'specified operations'. Specified operations are where the ships' equipment is used by those other than the master and crew of the vessel or where only the master and crew are involved in the work, but other people are put at risk by the work being carried out.

71 Where shore-based workers are to use ship's equipment, and their employers wish to take advantage of this disapplication from PUWER 98, they are required by the Regulations to take reasonable steps to satisfy themselves that the appropriate merchant shipping requirements have been met. The ship's records should normally contain sufficient information to satisfy reasonable enquiries.

72 PUWER 98 may apply to other work equipment not belonging to but used on board a ship, for example where a shore-based contractor carries out work on a ship within territorial waters. Work equipment used in such circumstances would be subject to PUWER 98 but PUWER 98 does not apply to foreign registered vessels on passage.

73 Most mobile offshore installations are also ships. PUWER will apply without qualification to mobile installations while at or near their working stations and when in transit to their working stations. PUWER will also apply without qualification to work equipment used on ships for the purposes of carrying out activities in connection with offshore installations or wells and for pipeline works.

Who has duties under PUWER 98?

74 PUWER 98 places duties on:

(a) employers;

(b) the self-employed; and

(c) people who have control of work equipment.

The duty on people who have control of work equipment reflects the way that work equipment is used in industry where there may not necessarily be a direct 'employment' relationship between the user and the person who controls the work equipment. For example, where a subcontractor carries out work at another person's premises with work equipment provided by that person or someone else who controls the equipment but not its use, such as a plant hire company. This approach is in line with that taken in the Construction (Health, Safety and Welfare) Regulations 1996 and in LOLER.

Who does not have duties under PUWER 98?

75 If you provide work equipment as part of a work activity for use by members of the public, you do not have duties under PUWER 98. Examples are compressed air equipment on a garage forecourt or lifts provided for use by the public in a shopping mall. In such circumstances members of the public will continue to be protected by the requirements of the HSW Act, principally sections 3 and 4.

What do you need to do if you have duties under PUWER 98?

76 If you have duties under PUWER 98 you need to ensure that work equipment you provide for use at work complies with PUWER 98.

Employer's duties

77 If you are an employer (whether as an individual, partnership or company) you have a duty to ensure that items of work equipment provided for your employees and the self-employed working for you comply with PUWER 98.

Self-employed people's duties

78 If you are self-employed you have a duty to ensure that work equipment you provide for work or use at work complies with PUWER 98.

The duties of 'those in control of work equipment'

79 If you provide work equipment for use at work where you do not control its use or the premises where it is to be used, you should still ensure that the work equipment complies with PUWER 98. People in control of non-domestic premises who provide work equipment which is used by other people at work should also comply with PUWER 98. PUWER places duties on employers and the self-employed; offshore this includes owners, operators and contractors. Their duties cover both their own employees and, as people having control of work equipment, other workers who may be affected. Meeting these duties where a number of employers and their employees are involved requires co-operation and co-ordination of activities. For example, the owner of a multi-occupied building has a legal responsibility to ensure that a lift complies with the Regulations, and the main contractor of a construction site would be responsible for a scaffold.

Where employees provide their own work equipment for use at work

80 PUWER 98 also covers situations where employers allow their employees to provide their own work equipment. For example, where builders use their own trowels or hammers.

Employees' duties

81 If you are an employee you do not have any specific duties under PUWER 98. These are covered in other legislation, in particular in section 7 of the HSW Act and regulation 12 of the Management Regulations.

The Trade Union Reform and Employment Rights Act 1993

82 This Act implements the employment protection requirements of the EC Health and Safety Framework Directive. The Act applies to all employees including those working offshore. It gives rights to all employees, regardless of their length of service, hours of work or age. The Act entitles employees to complain to an industrial tribunal about dismissal, selection for redundancy or any other matter for taking or proposing to take specified types of action on health and safety grounds, such as leaving the workplace in circumstances of danger or taking appropriate steps to protect themselves or others from the danger.

Multi-occupancy or multi-occupier sites

83 On multi-occupancy or multi-contractor sites where several duty holders may share the use of equipment, you can agree amongst yourselves that one of you takes responsibility for ensuring that the equipment complies with PUWER 98 (and any other relevant legislation), particularly regulation 9 of the Management Regulations. The following paragraphs examine such situations in detail in the construction and offshore sectors. Similar principles apply in other sectors.

Application to the construction industry

84 In the construction industry items of work equipment on sites are often used by a number of different contractors. Regulation 3 places a duty on each contractor to ensure that any work equipment used by their employees (or themselves in the case of self-employed contractors) conforms to, and is used in accordance with, these Regulations.

85 It also requires, to the extent that their control allows, the same duty from those who exercise control over the equipment or the way that it is used. For example, those hiring out equipment for others to use will often play the leading role in inspecting and

maintaining the equipment since they determine the maintenance schedules and availability of their machines. On the other hand, the users may be more directly concerned, for example, with organising, instructing and training their employees to use it safely since the conduct of their own employees is clearly a matter for them rather than the hirer.

86 The actions of others such as hirers may assist employers and the self-employed to meet their duties. However, that does not reduce the employer's or the self-employed's duty to ensure their own compliance with the Regulations. Effective co-ordination between the parties involved can be essential in order to satisfy their own legal duties.

87 The arrangements required by regulation 9 of the Management Regulations have been strengthened by the Construction (Design and Management) Regulations 1994 (CDM).

88 The CDM Regulations require the appointment of a single person or firm ('the principal contractor') to co-ordinate health and safety matters on site. The principal contractor also has a duty to ensure that all contractors co-operate on health and safety matters. Where the use of equipment by a wide range of personnel from a number of different employers requires particular attention or co-ordination, this should be addressed in the construction phase health and safety plan. Co-operation and exchanging information is vital when equipment is shared. All users need to know:

(a) who is responsible for the co-ordination of the equipment;

(b) that changes in conditions of use need to be reported to that person;

(c) whether there are any limitations on the use of the equipment; and

(d) how the equipment can be used safely.

Application to the offshore industry

89 PUWER places duties on employers and the self-employed; offshore this includes owners, operators and contractors. Their duties cover both their own employees and, as people having control of work equipment, other workers who may be affected. Meeting these duties where a number of employers and their employees are involved requires co-operation and co-ordination

of activities. The person in control of an operation should ensure that adequate arrangements are in place to ensure that work equipment provided for use at work is suitable, properly used and maintained, etc. This will often be an installation owner or operator, for example, but contractors who take equipment offshore are primarily responsible for risks arising from that equipment.

90 Legal requirements for co-operation between offshore duty holders are set out in the Offshore Installations (Safety Case) Regulations 1992 and the Offshore Installations and Pipeline Works (Management and Administration) Regulations 1995.

91 Equipment for use on offshore installations that is safety-critical, as defined by regulation 2(1) of the Offshore Installations (Safety Case) Regulations 1992, will be subject to the verification arrangements required elsewhere in those Regulations.

PART II: GENERAL

Regulation 4

Suitability of work equipment

92 This Regulation deals with the safety of work equipment from three aspects:

(a) its initial integrity;

(b) the place where it will be used, and

(c) the purpose for which it will be used.

93 The selection of suitable work equipment for particular tasks and processes makes it possible to reduce or eliminate many risks to the health and safety of people at the workplace. This applies both to the normal use of the equipment as well as to other operations such as maintenance.

94 The risk assessment carried out under regulation 3(1) of the Management Regulations will help to select work equipment and assess its suitability for particular tasks.

95 Most dutyholders will be capable of making the risk assessment themselves using expertise within their own organisations to identify the measures which need to be taken regarding their work equipment. In a few cases, for example where there are

complex hazards or equipment, it may need to be done in conjunction with the help of external health and safety advisers, appointed under regulation 6 of the Management Regulations.

96 For many items of work equipment, particularly machinery, you will know from experience what measures need to be taken to comply with previous legal requirements. Generally, these measures will ensure compliance with PUWER 98. Where this is not the case, there is usually a straightforward method of identifying the measures that need to be taken, because these are described in either general guidance or guidance specific to a particular industry or piece of equipment. However, the user will need to decide whether these are appropriate.

97 Where guidance does not exist, or is not appropriate, the main factors that need to be taken into account are the severity of any likely injury or ill health likely to result from any hazard present, the likelihood of that happening and the numbers of people exposed. You can then identify the measures that need to be taken to eliminate or reduce the risk to an acceptable level.

98 Further guidance on risk assessment is to be found in the ACOP on Management Regulations (see paragraph 24) which includes advice on the selection of preventive and protective measures and HSE's free leaflet entitled *5 steps to risk assessment*.[17]

Ergonomics

99 One of the factors to be considered is ergonomics.

100 *(See Approved Code of Practice, p 416)*

101 Ergonomic design takes account of the size and shape of the human body and should ensure that the design is compatible with human dimensions. Operating positions, working heights, reach distances, etc can be adapted to accommodate the intended operator. Operation of the equipment should not place undue strain on the user. Operators should not be expected to exert undue force or stretch or reach beyond their normal strength or physical reach limitations to carry out tasks. This is particularly important for highly repetitive work such as working on supermarket checkouts or high speed 'pick and place' operations.

17 *5 steps to risk assessment* INDG163(rev) HSE Books 1998 (single copies free; ISBN 0 7176 1565 0 for priced packs).

102 Risks could arise as a result of using mobile work equipment, for which the measures in Part III are relevant. However, for existing equipment (that which was in use before 5 December 1998) the requirements in Part III do not come into force until 5 December 2002. Therefore, until this date, you do not need to comply with Part III when considering the selection of suitable mobile work equipment. However, where the risks are significant, you may wish to select alternative equipment anyway.

Regulation 4(1)

103 Equipment must be suitable, by design, construction or adaptation, for the actual work it is provided to do. This means in practice that when you provide work equipment you should ensure that it is suitable for the work to be undertaken and that it is used in accordance with the manufacturer's specifications and instructions. If work equipment is adapted it must still be suitable for its intended purpose.

104 This requirement provides the focal point for the other Regulations—for example, compliance with regulation 10 should ensure the initial integrity of equipment in many cases, and compliance as appropriate with the specific requirements of regulations 11 to 24.

Regulation 4(2)

105 This requires you to assess the location in which the work equipment is to be used and to take account of any risks that may arise from the particular circumstances. Such factors can invalidate the use of work equipment in a particular place. For example, electrically powered equipment is not suitable for use in wet or flammable atmospheres unless it is designed for this purpose. In such circumstances you should consider selecting suitably protected electrical equipment or alternative pneumatically or hydraulically powered equipment.

106–108 *(See Approved Code of Practice, p 416)*

109 You should also take account of the fact that work equipment itself can sometimes cause risks to health and safety in particular locations which would otherwise be safe. Such an example is a petrol engine generator discharging exhaust fumes into an enclosed space.

Why ventilation may be necessary

110 Exhaust gases from mobile work equipment with a combustion engine may contribute significantly to airborne pollution in workplaces. For example, in motor vehicle workshops, underground car parks, in buildings where forklift trucks are used and in tunnels. In such circumstances a high standard of ventilation and/or extraction may be necessary to allow the combustion process to take place and to dilute toxic combustion products (such as carbon monoxide, carbon dioxide and oxides of nitrogen) to an acceptable level. Combustion products can be harmful to health if there is insufficient fresh air for people to breathe.

When to ventilate workplaces

111 Ventilation requirements will vary depending on the type of fuel, condition of the engine and pattern of use. If mobile work equipment is fitted with pollution control services, lower ventilation rates may be necessary. The method of ventilation will depend on where the work equipment is used, for example in a warehouse or a tunnel and if sufficient air of good quality is not naturally available it may need to be supplied.

How to ensure there is sufficient clean air

112 Examples of how to ensure there is sufficient clean air include:

(a) the exhausts of stationary vehicles under test or repair should be connected to exhaust removal systems;

(b) flexible exhaust systems or box filters should be used where necessary;

(c) natural and/or mechanical ventilation should be used where necessary;

(d) air quality should be monitored on a regular basis to ensure that the control systems in place are working adequately.

Ventilation requirements of the Workplace (Health, Safety and Welfare) Regulations 1992 (the Workplace Regulations)

113 Regulation 6 of the Workplace Regulations and its supporting ACOP contain general requirements about ventilation of the workplace and equipment used to ventilate workplaces.

The Confined Spaces Regulations 1997

114 The ACOP to the Confined Spaces Regulations 1997 does not allow the use of petrol-fuelled internal combustion engines in a confined space unless special precautions are taken. A confined space is one which is substantially or entirely closed and there will be a reasonably foreseeable risk of serious injury from hazardous substances or conditions within the space or nearby. Other forms of fuel such as diesel or gas are nearly as dangerous and are inappropriate unless adequate precautions are taken. Where their use is unavoidable, adequate ventilation needs to be provided to prevent a build-up of harmful gases, for example from leakage of the fuel or the exhaust system and to allow the engine to operate properly.

115 The exhaust from engines must be vented to a safe place well away from the confined space and downwind of any ventilator intakes for the confined space. You should check for the build-up of harmful gases within the confined space. Fuelling of portable engine-driven equipment should be conducted outside the confined space except in rare cases such as tunnelling. Using such equipment within the confined space requires constant atmospheric monitoring for gas. Full guidance and ACOP material on the Confined Spaces Regulations 1997 is contained in HSE's publication *Safe work in confined spaces.*[18]

Control of Substances Hazardous to Health Regulations 1998 (COSHH)

116 Regulation 7 of COSHH requires employers to prevent or control the exposure of employees to substances hazardous to health. ACOP material on the COSHH Regulations is contained in HSE's publication *General COSHH ACOP.*[19]

Regulation 4(3)

117 This requirement concerns each particular process for which the work equipment is to be used and the conditions under which

18 *Safe work in confined spaces* INDG258 HSE Books 1997.

19 *General COSHH ACOP (Control of substances hazardous to health) and Carcinogens ACOP (Control of carcinogenic substances) and Biological agents ACOP (Control of biological agents). Control of Substances Hazardous to Health Regulations 1994. Approved Codes of Practice* L5 HSE Books 1997 ISBN 0 7176 1308 9.

it will be used. You must ensure that the equipment is suitable for the process and conditions of use. Examples include:

(a) a circular saw is generally not suitable for cutting a rebate whereas a spindle moulding machine would be suitable because it can be guarded to a high standard;

(b) knives with unprotected blades are often used for cutting operations where scissors or other cutting tools could be used, reducing both the probability and severity of injury.

Regulation 5
Maintenance

Application of the Regulation

118 This Regulation builds on the general duty in the HSW Act, which requires work equipment to be maintained so that it is safe. It does not cover the maintenance process (that is covered by the general duties of the HSW Act) or the construction of work equipment so that maintenance can be carried out without risk to health or safety (these are the subject of regulation 10 and regulation 22).

119 It is important that equipment is maintained so that its performance does not deteriorate to the extent that it puts people at risk. In regulation 5, 'efficient' relates to how the condition of the equipment might affect health and safety. It is not concerned with productivity. Some parts of equipment such as guards, ventilation equipment, emergency shutdown systems and pressure relief devices have to be maintained to do their job at all times. The need to maintain other parts may not be as obvious, for example failure to lubricate bearings or replace clogged filters might lead to danger because of seized parts or overheating. Some maintenance routines affect both the way the equipment works and its safety. Checking and replacing worn or damaged friction linings in the clutch on a guillotine will ensure it operates correctly, but could also prevent the drive mechanism jamming, so reducing the risk of repeat uncovenanted strokes.

Frequency of maintenance

120 Equipment may need to be checked frequently to ensure that safety-related features are functioning correctly. A fault which affects production is normally apparent within a short time;

however, a fault in a safety-critical system could remain undetected unless appropriate safety checks are included in maintenance activities.

121 The frequency at which maintenance activities are carried out should also take into account the:

(a) intensity of use—frequency and maximum working limits;

(b) operating environment, for example marine, outdoors;

(c) variety of operations—is the equipment performing the same task all the time or does this change?

(d) risk to health and safety from malfunction or failure.

Maintenance management

122 The extent and complexity of maintenance can vary substantially from simple checks on basic equipment to integrated programmes for complex plant. In all circumstances, for maintenance to be effective it needs to be targeted at the parts of work equipment where failure or deterioration could lead to health and safety risks. Maintenance should address those parts which have failed or are likely to deteriorate and lead to health and safety risks.

123 A number of maintenance management techniques could be used:

(a) planned preventive;

(b) condition-based;

(c) breakdown.

Appropriate techniques should be selected through risk assessment and used independently or in combination to address the risks involved.

124 Simple hand tools usually require minimal maintenance, but could require repair or replacement at intervals. More complex powered equipment will normally be accompanied by a manufacturer's maintenance manual, which specifies routine and special maintenance procedures to be carried out at particular intervals. Some of the procedures will be necessary to keep the equipment in working order, others will be required for safety reasons.

125 It should be remembered that different maintenance management techniques have different benefits.

(a) Planned preventive maintenance involves replacing parts and consumables or making necessary adjustments at preset intervals so that risks do not occur as a result of the deterioration or failure of the equipment.

(b) Condition-based maintenance involves monitoring the condition of safety-critical parts and carrying out maintenance whenever necessary to avoid hazards which could otherwise occur.

(c) Breakdown maintenance involves carrying out maintenance only after faults or failures have occurred. It is appropriate only if the failure does not present an immediate risk and can be corrected before risk occurs, for example through effective fault reporting and maintenance schemes.

126 Where safety-critical parts could fail and cause the equipment, guards or other protection devices to fail and lead to immediate or hidden potential risks, a formal system of planned preventative or condition-based maintenance is likely to be needed.

127 Some equipment may not be owned by the user. Many items of plant and equipment are hired. It is important for both the hire company and the person responsible for hiring equipment to establish which party will carry out safety-related maintenance. This is particularly important for equipment on long-term hire and the terms of the agreement set out or recorded in writing.

128 In many cases, safety-related maintenance work is not carried out by the person with ultimate responsibility for the work equipment, in the mistaken belief that the other party will do it. If the hire company is some distance from the user site, it would be uneconomical for their staff to carry out simple checks and make minor adjustments, so the user may agree to carry them out. However, both parties should agree exactly what they are responsible for.

Maintenance log

129 There is no requirement for you to keep a maintenance log. However, it is recommended that you keep a record of maintenance for high-risk equipment. A detailed maintenance log can provide information for future planning of maintenance activities and inform maintenance personnel and others of previous action taken.

130 If you have a maintenance log, you should keep it up to date.

131 Maintenance procedures should be carried out in accordance with any manufacturer's recommendations which relate to the equipment, for example periodic lubrication, replacement and adjustment of parts.

132 However, additional maintenance measures may be required if particularly arduous conditions of use are foreseen or have been experienced in use. There may be times when these additional measures need to be reviewed and revised in the light of ongoing operating experiences.

Maintenance workers

133 Maintenance work should only be done by those who are competent to do the work. For details of the information, instructions and training required, see also regulations 8 and 9.

Regulation 6

Inspection

134 This requirement for the inspection of work equipment builds on the current but often informal practice of regular in-house inspection of work equipment, some of which is already recommended in other HSE guidance.

135 Inspection does not normally include the checks that are a part of the maintenance activity although certain aspects may well be common. Nor, for the purposes of this Regulation, does inspection include a pre-use check that an operator may make before using the work equipment. Additionally, while inspections need to be recorded, such checks do not.

Identifying what needs to be inspected

136 *(See Approved Code of Practice, p 417)*

Significant risk

137 A significant risk is one which could foreseeably result in a major injury or worse.

138 Inspection is only necessary where there is a significant risk resulting from:

(a) incorrect installation or re-installation;

(b) deterioration; or

(c) as a result of exceptional circumstances which could affect the safe operation of the work equipment.

139 The types of major injury are listed in Schedule 1 of the Reporting of Injuries, Diseases and Dangerous Occurrences Regulations 1995. They are reproduced on pages 505–506.

Purpose of an inspection

140 The purpose of an inspection is to identify whether the equipment can be operated, adjusted and maintained safely and that any deterioration (for example defect, damage, wear) can be detected and remedied before it results in unacceptable risks.

What should be included in the inspection

141 *(See Approved Code of Practice, p 417)*

142 The extent of the inspection that is needed will depend upon:

(a) the type of equipment;

(b) where it is used; and

(c) how it is used.

143 An inspection will vary from a simple visual external inspection to a detailed comprehensive inspection, which may include some dismantling and/or testing.

144 An inspection should always include those safety-related parts which are necessary for safe operation of equipment, for example overload warning devices and limit switches.

145 The level of inspection required would normally be less detailed and less intrusive than the thorough examination required for the purposes of regulation 33 on power presses and regulation 9 of LOLER for certain items of lifting equipment.

146 Some work equipment will need examinations and thorough examinations under other legislation such as the Pressure Systems Regulations, COSHH, Control of Lead at Work Regulations (CLAW), Control of Asbestos at Work Regulations (CAW). Inspections will only be needed for such work equipment if these

other examinations do not fully cover all the significant health and safety risks which are likely to arise from the use of the equipment, in a way that satisfies the requirements of PUWER.

Testing

147 As part of an inspection, a functional or other test may be necessary to check that the safety-related parts, for example interlocks, protection devices, controls, etc are working as they should be and that the work equipment and relevant parts are structurally sound, for example non-destructive testing of safety-critical parts. The need for any testing (for example non-destructive testing of safety-critical parts) should be decided by the competent person who determines the nature of the inspection.

Competent persons

148–149 *(See Approved Code of Practice, p 417)*

150 *'Determining the nature of the inspection'*—the knowledge and experience required by a person to determine the nature of the inspection needs to be sufficient for them to be able to decide what the inspection should include, how it should be done and when it should be carried out. Experienced, in-house employees such as a department manager or supervisor may be able to do this. They should know what will need to be inspected to detect damage or faults resulting from deterioration. They should also be able to determine whether any tests are needed during the inspection to see if the equipment is working safely or is structurally sound.

151 *'Carrying out the inspection'*—the person who actually carries out the inspection may not necessarily be the same person who determines the nature of the inspections. The actual inspection will normally be done by an in-house employee with an adequate knowledge of the equipment to:

(a) enable them to know what to look at (know the key components);

(b) know what to look for (fault-finding); and

(c) know what to do (reporting faults, making a record, who to report to).

Where necessary, you should give them appropriate information, instruction and training so they can carry out the inspection properly and avoid danger. They should also be aware of and able to avoid danger to themselves and others.

152 The necessary level of competence will vary according to the type of equipment and where and how it is used. For some equipment, the level of competence to determine the nature of the inspections or even to carry them out may not be available in-house, in which case the help of another body with relevant competence may be necessary. An example of this will be the person who carries out the annual inspection under these Regulations of some fairground rides.

Installation

153 *(See Approved Code of Practice, pp 417–418)*

154 Re-installation includes assembling the equipment at a new site or in a new location. Equipment that has been installed or re-installed is normally in a permanent or long-term location and is usually fixed in position. Installation or re-installation does not normally include re-positioning or moving equipment, particularly where there is no element of dismantling, re-assembling and/or fixing the equipment in position, or if its location is transitory. Examples of work equipment where safety is critically dependent on the installation conditions include those where guarding is provided by presence-sensing devices (such as light curtains used for paper-cutting guillotines or pressure sensitive mats used with tube-bending machines). These devices allow free access to the danger zone but should be positioned so that if anyone approaches the danger zone they will be detected and the hazardous functions stopped before injury can occur.

'Conditions causing deterioration' and 'Dangerous situations'

155 *(See Approved Code of Practice, p 418)*

Equipment that should receive an inspection

156 The types of equipment whose use could result in significant risk as a result of deterioration and which may therefore need to be inspected include:

(a) most fairground equipment;

(b) machines where there is a need to approach the danger zone during normal operation such as horizontal injection moulding machines, paper-cutting guillotines, die-casting machines, shell-moulding machines;

(c) complex automated equipment;

(d) integrated production lines.

Equipment for which an inspection is not required

157 If failure or fault of the equipment cannot lead to significant risk or if safety is guaranteed through appropriate maintenance regimes (under regulation 5), inspection may not be necessary. Equipment unlikely to need an inspection includes office furniture, hand tools, non-powered machinery and also powered machinery such as a reciprocating fixed blade metal cutting saw.

Frequency of inspection

158 *(See Approved Code of Practice, p 418)*

159 The inspection frequency may be different for the same type of equipment because the rate of deterioration can vary in different situations. Where equipment is subject to frequent use in a harsh outdoor environment (for example, at a coastal site or on a construction site), it is likely to need more frequent inspection than if it is used occasionally in an indoor environment such as a warehouse.

160 To ensure that appropriate inspection intervals and procedures are in place, they should be reviewed in the light of experience. Intervals between inspections can be lengthened if an inspection history has shown that deterioration is negligible or the interval between inspections should be shortened if substantial amounts of deterioration are detected at each inspection.

Exceptional circumstances

161 Regulation 6(2) states that an inspection is necessary 'each time that exceptional circumstances which are liable to jeopardise the safety of the work equipment have occurred.'

162 Exceptional circumstances which may result in the need for inspection include:

(a) major modifications, refurbishment or major repair work;

(b) known or suspected serious damage;

(c) substantial change in the nature of use, for example from an extended period of inactivity.

Records

163 Records do not have to be kept in a particular form. They can be handwritten or stored electronically from a pre-printed form to an entry in a diary. There are no legal requirements stating what they contain. However, a record should normally include:

(a) information on the type and model of equipment;

(b) any identification mark or number that it has;

(c) its normal location;

(d) the date that the inspection was carried out;

(e) who carried out the inspection;

(f) any faults; and/or

(g) any action taken;

(h) to whom the faults have been reported;

(i) the date when repairs or other necessary action were carried out.

164 *(See Approved Code of Practice, p 418)*

165 For large items of equipment for which inspection is necessary, the physical evidence can be in the form of a copy of the record of the last inspection carried out. For smaller items of equipment, a tagging, colour coding or labelling system can be used. The purpose of the physical evidence is to enable a user to check if an inspection has been carried out and whether or not it is current, where required, and also to determine the results of that inspection, by being able to link back from the physical evidence to the records.

166 These inspection requirements do not cover the following work equipment as set out in regulation 6(5):

(a) *Power presses covered by regulations 31–39 of PUWER.* These include mechanically driven presses or press brakes (called 'power press(es)' in this booklet) which are power driven, have a flywheel and clutch, and which are wholly or partly used to work metal. A clutch, in relation to a power press, is a device to impart the movement of the flywheel to any tool when required.

(b) *Work equipment for lifting loads including people.* This is defined as work equipment for lifting or lowering loads and includes its attachments used for anchoring, fixing or supporting it. A load does include a person.

(c) Under the Mines (Shafts and Winding) Regulations 1993, *winding apparatus* means 'mechanically operated apparatus for lowering and raising loads through a (mine) shaft and includes a conveyance or counterweight attached to such apparatus and all ancillary apparatus.'

(d) Work equipment required to be inspected by regulation 29 of the Construction (Health, Safety and Welfare) Regulations 1996, such as *scaffolding* and *excavation supports.*

Regulation 7

Specific risks

167–169 *(See Approved Code of Practice, pp 418–419)*

170 Specific risks can be common to a particular class of work equipment, for example the risks from a platten printing machine or from a drop forging machine. There can also be a specific risk associated with the way a particular item of work equipment is repaired, set or adjusted as well as with the way it is used.

171 The person whose normal work includes the use of a piece of work equipment will have been given 'the task of using it' and the instruction and training provided should be appropriate to that work. For someone using a grinding machine for example, training should cover the proper methods of dressing the abrasive wheels. For someone carrying out a turning operation on a lathe, the training should cover the devices which should be used if working with emery cloth to obtain the required finish on a workpiece.

172 The designated person to carry out repairs, etc will be the person whose work includes these activities. This person could be the operator of the equipment, provided that they have received relevant instruction and training. For example, the training for a person who has to change the knives on guillotines should include any devices which could be used, such as knife handles, as well as the system of work.

Regulation 8

Information and instructions

How regulation 8 links with other health and safety law

173 This Regulation builds on the general duty in the HSW Act to provide employees with the information and instructions that are necessary to ensure, so far as is reasonably practicable, their health and safety. It also links with the general requirement in the Management Regulations to provide information to employees relating to their health and safety. The Health and Safety (Consultation with Employees) Regulations 1996 (HSCER) require employers to consult their employees on the information required under other regulations, including PUWER 98, about risks to their health and safety and preventative measures in place.

What does regulation 8 require?

174 Regulation 8 places a duty on employers to make available all relevant health and safety information and, where appropriate, written instructions on the use of work equipment to their workforce. Workers should have easy access to such information and instructions and be able to understand them.

Written instructions

175 Regulation 8 refers to written instructions. This can include the information provided by manufacturers or suppliers of work equipment such as instruction sheets or manuals, instruction placards, warning labels and training manuals. It can also include in-house instructions and instructions from training courses. There are duties on manufacturers and suppliers to provide sufficient information, including drawings, to enable the correct installation, safe operation and maintenance of the work equipment. You should ask or check that they are provided.

Consultation with employees

176 Under the Consultation with Employees Regulations, employers must consult their employees about these matters before decisions and changes are made. HSE's booklet[20] *A guide to*

20 *A guide to the Health and Safety (Consultation with Employees) Regulations 1996* L95 Guidance on Regulations HSE Books 1995 ISBN 0 7176 1234 1.

the Health and Safety (Consultation with Employees) Regulations 1996 contains the HSCER Regulations and supporting guidance.

177 There are requirements for consultation within the Safety Representatives and Safety Committees Regulations 1977 which provide for the appointment of safety representatives by recognised trade unions. The Health and Safety (Consultation with Employees) Regulations 1996 apply to employees that are not covered by the 1977 Regulations. The 1996 Regulations require employers to consult their employees on matters which link with the requirements of PUWER 98 and particularly its requirements for information and instruction:

(a) any measures which may affect their health and safety;

(b) information they must have about risks to health and safety and preventive measures;

(c) any arrangements for getting a competent person to help comply with health and safety requirements;

(d) planning and organising of any health and safety training; and

(e) the health and safety consequences of any new equipment or technology.

These requirements for consultation link with several of PUWER 98's requirements, for example information, instruction and training on new work equipment.

178 Similar provisions apply offshore under the Offshore Installations (Safety Representatives and Safety Committees) Regulations 1989, which take in additional consultation requirements specific to the offshore sector.

To whom should the information and instructions be made available?

179 You should ensure that any written instructions are available to the people directly using the work equipment. You should also ensure that instructions are made available to other appropriate people, for example maintenance instructions are made available or passed to the people involved in maintaining your work equipment.

180 Supervisors and managers also need access to the information and written instructions. The amount of detailed health

and safety information they will need to have immediately available for day-to-day running of production lines will vary but it is important that they know what information is available and where it can be found.

How the information and instructions should be made available

181 Information can be made available in writing, or verbally where it is considered sufficient. It is your responsibility to decide what is appropriate, taking into consideration the individual circumstances. Where there are complicated or unusual circumstances, the information should be in writing. Other factors need to be taken into consideration, such as the degree of skill of the workers involved, their experience and training, the degree of supervision and the complexity and length of the particular job.

182 The information and written instructions should be easy to understand. They should be in clear English and/or other languages if appropriate for the people using them. They should be set out in logical order with illustrations where appropriate. Standard symbols should be used where appropriate.

183 You should give special consideration to any employees with language difficulties or with disabilities which could make it difficult for them to receive or understand the information or instructions. You may need to make special arrangements in these cases.

What the information and instructions should cover

184 Any information and written instructions you provide should cover:

(a) all health and safety aspects arising from the use of the work equipment;

(b) any limitations on these uses;

(c) any foreseeable difficulties that could arise;

(d) the methods to deal with them; and

(e) using any conclusions drawn from experience using the work equipment, you should either record them or take steps to ensure that all appropriate members of the workforce are aware of them.

Regulation 9

Training

What is 'adequate training'?

185 It is not possible to detail here what constitutes 'adequate training' as requirements will vary according to the job or activity and work equipment etc. In general, you will need to:

(a) evaluate the existing competence of employees to operate the full range of work equipment that they will use;

(b) evaluate the competence they need to manage or supervise the use of work equipment; and

(c) train the employee to make up any shortfall between their competence and that required to carry out the work with due regard to health and safety.

186 Account should be taken of the circumstances in which the employee works. For example, do they work alone or under close supervision of a competent person?

When is training necessary?

187 Training needs are likely to be greatest on recruitment. However, training needs are also required:

(a) if the risks to which people are exposed change due to a change in their working tasks; or

(b) because new technology or equipment is introduced; or

(c) if the system of work changes.

188 Also, you should provide refresher training if necessary. Skills decline if they are not used regularly. Pay particular attention to people who deputise for others on occasions—as they may need more frequent refresher training than those who do the work regularly.

Training for young people

189 Training and proper supervision of young people is particularly important because of their relative immaturity and unfamiliarity with the working environment. Induction training is of particular importance. There are no general age restrictions in

legislation relating to the use of work equipment although there is some ACOP material in the relevant publications dealing with lifting, power presses and wood working; all employees should be competent to use work equipment with due regard to health and safety regardless of their age.

190 The Management Regulations contain specific requirements relating to the employment of young people under the age of 18. These require employers to assess risks to young people before they start work, taking into account their inexperience, lack of awareness of potential risks and their immaturity. Employers must provide information to parents of school-age children (for example when they are on work experience) about the risks and the control measures introduced and take account of the risk assessment in determining whether the young person should undertake certain work activities.

Other health and safety legislation relating to training

191 PUWER 98 revokes the remaining sector specific law on training in the use of work equipment. The laws that have been removed are listed on pages 406 and 413 of this document.

192 This regulation covers health and safety training related to the provision and use of work equipment. Other health and safety legislation contains general requirements relating to training such as the HSW Act and regulation 11 of the Management Regulations which requires employers to provide employees with general health and safety training. Regulation 9 of PUWER 98 is concerned more specifically with what such training should comprise, ie the precautions to be taken during the use of work equipment.

When should training take place?

193 The Management Regulations specify that health and safety training should take place within working hours.

194–195 *(See Approved Code of Practice, pp 419–420)*

196 Portable hand-held chainsaws are dangerous machines which need to be handled with the greatest care. Everyone who uses a chainsaw at work for whatever task must have received adequate training under this regulation. The training should cover:

(a) dangers arising from the chainsaw itself;

(b) dangers arising from the task for which the chainsaw is to be used; and

(c) the precautions to control these dangers, including relevant legal requirements.

197 Over and above this, due to the significant risks involved, if a chainsaw is to be used on or in a tree, the operator will be expected to hold a certificate of competence or national competence award relevant to the work they undertake.

198 The requirement for a certificate or award applies to people working with chainsaws on or in trees on agricultural holdings *unless* it is done as part of agricultural operations (for example, hedging, clearing fallen branches, or pruning trees to maintain clearance for machines, etc) by the occupier or his employees and they have used a chainsaw before 5 December 1998.

Regulation 10

Conformity with Community requirements

What regulation 10 requires

199 This regulation aims to ensure that when work equipment is first provided for use in the workplace after 5 December 1998, it meets certain health and safety requirements. It contains duties that complement those on manufacturers and suppliers in other legislation regarding the initial integrity of equipment.

200 There are legal requirements covering all those involved in the chain of supply of work equipment which are designed to ensure that new work equipment is safe. For example, section 6 of the HSW Act places general duties on designers, manufacturers, importers and suppliers to ensure this so far as is reasonably practicable.

201 Existing legislation on the manufacture and supply of new work equipment is increasingly being supplemented by new and more detailed Regulations implementing EC Directives made under Article 100a of the Treaty of Rome. These new Regulations place duties on the manufacturer and supplier of new work equipment.

Article 100a Directives

202 The aim of Article 100a Directives is to achieve the free movement of goods in the Community Single Market by removing

different national controls and harmonising essential health and safety requirements. Examples important to safety at work include the Machinery, the Personal Protective Equipment and the Simple Pressure Vessels Directives.

Essential health and safety requirements

203 Article 100a Directives set out 'essential health and safety requirements' which must be satisfied before products may be sold in the European Economic Area. Products which comply with the Directives must be given free circulation within the European Economic Area. These Directives also apply to equipment made and put into service in-house. Suppliers must ensure that their products, when placed on the market, comply with the legal requirements implementing the Directives applicable to their product. It is a common feature of these Directives that compliance is claimed by the manufacturer affixing a mark—'CE Marking'—to the equipment.

204 At present, not all work equipment is covered by a product Directive; nor are product Directives retrospective. However, equipment which was provided for use in the European Economic Area before compliance with the relevant product Directives was required, may need to be modified to comply immediately with regulations 11–24 of PUWER 98.

205 One of the most significant relevant Directives is the Machinery Directive. The Machinery Directive was brought into UK law by the Supply of Machinery (Safety) Regulations 1992 as amended. These Regulations apply to machinery that was first placed on the market after 1 January 1993, though there was a transitional period until 31 December 1994.

206 In practice, regulation 10 means that you will need to check, for example, that adequate operating instructions have been provided with the equipment and that there is information about residual hazards such as noise and vibration. More importantly, you should also check the equipment for obvious faults. Products should also carry a CE marking and be accompanied by relevant certificates or declarations, as required by relevant product Directives. Your supplier should be able to give further advice about what the equipment is designed for and what it can or cannot be used for, or, alternatively make further enquiries about such

matters with the manufacturer. Further advice is given in HSE's leaflet *Buying new machinery.*[21]

Regulation 10(1)

207 Regulation 10(1) places a duty on you as users of work equipment. When first providing work equipment for use in the workplace you should ensure that it has been made to the requirements of the legislation implementing any product Directive which is relevant to the equipment. (For interpretation of 'provided for use' see the guidance on regulation 1.) This means that in addition to specifying that work equipment should comply with current health and safety legislation, you should also specify that it should comply with the legislation implementing any relevant EC Directive. Where appropriate, you can check to see that the equipment bears a CE marking and ask for a copy of the EC Declaration of Conformity.

Regulation 10(2)

208 As mentioned in paragraph 19, which sets out the requirements of PUWER, there are certain circumstances where regulations 11–19 and 22–29 will not apply. This can happen when work equipment, first supplied, complies with the applicable parts of the legislation that implements any relevant product Directive. An example of this is the Supply of Machinery (Safety) Regulations 1992 which implement the machinery Directive. Regulations 11–19 and 22–29 do not apply here as this would duplicate the requirements. They *will* apply, however, if they include relevant requirements that are not included in the corresponding product legislation or if that product legislation has not been complied with; for example, if a new machine has been supplied without suitable guards, regulation 11 of PUWER 98 could be used to ensure the user does provide adequate guards.

Regulation 11

Dangerous parts of machinery

What regulation 11(1) requires

209 Regulation 11(1) requires employers to take effective measures to prevent access to dangerous parts of machinery or

21 *Buying new machinery* INDG271 HSE Books (single copies free; ISBN 0 7176 1559 6 for priced packs).

stop their movement before any part of a person enters a danger zone. This Regulation also applies to contact with a rotating stock-bar which projects beyond the headstock of a lathe.

210 The term 'dangerous part' has been established in health and safety law through judicial decisions. In practice, this means that if a piece of work equipment could cause injury if it is being used in a foreseeable way, it can be considered a dangerous part.

211 Protection against other hazards associated with machinery is dealt with in regulations 12 and 13. However, the measures required by this regulation may also protect or help to protect against those other hazards such as ejected particles and heat.

Preventing contact with dangerous parts of machinery

212 There are many HSC/HSE publications which are specific to a machine or industry. They describe the measures that can be taken to protect against risks associated with dangerous parts of machinery. Current national, European and international standards may also be used for guidance, where appropriate.

213 Pages 506–515 give more detailed information about the available methods of safeguarding which may be used to conform with regulation 11.

Risk assessment

214 Your risk assessment carried out under regulation 3 of the Management Regulations should identify hazards presented by machinery. Your risk assessment should evaluate the nature of the injury, its severity and likelihood of occurrence for each hazard identified. This will enable you to decide whether the level of risk is acceptable or if risk reduction measures are needed. In most cases the objective of risk reduction measures is to prevent contact of part of the body or clothing with any dangerous part of the machine, for example guarding.

What regulation 11(2) requires

215 Regulation 11(2) specifies the measures which you should take to prevent access to the dangerous parts of the machinery and achieve compliance with regulation 11(1). The measures are ranked in the order they should be implemented, where practicable, to achieve an adequate level of protection. The levels of protection are:

(a) fixed enclosing guards;

(b) other guards or protection devices such as interlocked guards and pressure mats;

(c) protection appliances such as jigs, holders and push-sticks etc; and

(d) the provision of information, instruction, training and supervision.

An explanation of the guarding and protection terms used is given on pages 506–508.

216 The hazards from machinery will be identified as part of your risk assessment. The purpose of the risk assessment is to identify measures that you can take to reduce the risks that the hazards present. When selecting measures you should consider each level of protection from the first level of the scale listed in paragraph 215 and you should use measures from that level so far as it is practicable to do so, provided that they contribute to the reduction of risk. It may be necessary to select a combination of measures. The selection process should continue down the scale until the combined measures are effective in reducing the risks to an acceptable level, thus meeting the requirements of regulation 11(1). In selecting the appropriate combination you will need to take account of the requirements of the work, your evaluation of the risks, and the technical features of possible safeguarding solutions.

217 Most machinery will present more than one mechanical hazard, and you need to deal with the risks associated with all of these. For example, at belt conveyors there is a risk of entanglement with the rotating shafts and of being trapped by the intake between drum and moving belt—so you should adopt appropriate safety measures.

218 Any risk assessment carried out under regulation 3 of the Management Regulations should not just deal with the machine when it is operating normally, but must also cover activities such as setting, maintenance, cleaning or repair. The assessment may indicate that these activities require a different combination of protective measures from those appropriate to the machine doing its normal work. In particular, parts of machinery that are not dangerous in normal use because they are not then accessible may become accessible and therefore dangerous while this type of work is being carried out.

219 Certain setting or adjustment operations which may have to be done with the machine running may require a greater reliance on the provision of information, instruction, training and supervision than for normal use.

Regulation 11(3) and regulation 11(4)

220 Regulation 11(3) sets out various requirements for guards and protection devices which are set out on pages 509–515. These are largely common sense, and in large part are detailed in relevant national, European and international standards. Ways of achieving satisfactory guarding and other protection are discussed in more detail in the standards and in other guidance. Regulation 11(4) sets out requirements for protection appliances.

Regulation 12

Protection against specified hazards

What this regulation covers

221 This regulation covers risks arising from hazards during the use of work equipment. The hazards are listed in paragraph (3) of the Regulation.

Examples of hazards that the Regulation covers are:

(a) material falling from equipment, for example a loose board falling from scaffolding, a straw bale falling from a tractor foreloader or molten metal spilling from a ladle;

(b) material held in the equipment being unexpectedly thrown out, for example swarf ejected from a machine tool;

(c) parts of the equipment breaking off and being thrown out, for example an abrasive wheel bursting;

(d) parts of the equipment coming apart, for example collapse of scaffolding or falsework;

(e) overheating or fire due, for example, to friction (bearings running hot, conveyor belt on jammed roller), electric motor burning out, thermostat failing, cooling system failure;

(f) explosion of the equipment due to pressure build-up, perhaps due to the failure of a pressure-relief valve or the unexpected blockage or sealing off of pipework;

(g) explosion of substances in the equipment, due, for example, to exothermic chemical reaction or unplanned ignition of a flammable gas or vapour or finely divided organic material (for example flour, coal dust), or welding work on a container with flammable residues.

222 Your risk assessment carried out under regulation 3 of the Management Regulations should identify these hazards and assess the risks associated with them. You will need to consider the likelihood of such events occurring and the consequent danger if they do occur, in order to identify measures you should take to comply with this regulation.

223 Regulation 12(1) sets the primary aim, which is to prevent any of the events in regulation 12(3) arising, if that event exposes a person to risk. Where possible, the equipment should be designed so that events presenting a risk cannot occur. If this is not reasonably practicable, you should take steps to reduce the risk. Examples of measures that may be taken are the monitoring of solvent concentrations at evaporating ovens to detect the build-up of explosive atmospheres, or the use of inert gas systems to control and suppress dust explosions.

224 Regulation 12(1) permits the discharge or ejection of material as an intentional or unavoidable part of the process (for example grit-blasting of castings, sawdust from woodworking), but any risks to people must be controlled. The regulation also allows the use of equipment designed to make use of explosive forces in a controlled manner (for example an internal combustion engine or a rail detonator signal).

225 Equipment may have been designed before manufacture to eliminate or reduce the likelihood of the type of event listed in regulation 12(3). But equipment suppliers cannot control the materials used in equipment, or the environment in which it is used, and it is up to you to ensure that the equipment is suitable for their application, as required by regulation 4(2). Therefore risks associated with high temperature, vibration or a flammable atmosphere must be controlled.

226 Regulation 12(2)(b) requires that in addition to reducing the likelihood of the event occurring, measures must be taken to reduce the effect of any event which does give rise to risks. An example might be a blast wall or where there is a risk from a pressure-relief panel or vent bursting, ensuring that any gases or

liquids discharged are directed to a safe place, contained, or made safe as appropriate.

227 Regulation 12(2)(a) requires that risk-controlling measures should be provided as part of the equipment, so far as is reasonably practicable. Personal protective equipment may be appropriate where a risk remains that cannot be eliminated in some other way.

228 Training, supervision and provision of information will often have an important role to play. Firstly, they can help to ensure that equipment is operated in the correct way to prevent dangers occurring. Secondly, they can help to ensure that the appropriate safeguards are taken to prevent people being exposed to risk.

Abrasive wheels

229 One particular example of the application of these principles is the use of abrasive wheels.

(a) To minimise the risk of bursting, abrasive wheels should always be run within the specified maximum rotation speed.

(b) If they are large enough this will be marked on the wheel (in accordance with regulation 23).

(c) Smaller wheels should have a notice fixed in the workroom, giving the individual or class maximum permissible rotation speed.

(d) The power-driven spindle should be governed so that its rotation speed does not exceed this.

(e) Guarding must be provided to contain fragments of the wheel that might fly off if it did burst, so as to prevent them injuring anyone in the workplace. The guarding has an additional role in helping to meet the requirements of regulation 11; it should be designed, constructed and maintained to fulfil both functions.

(f) Providing information and training of workers in the correct handling and mounting of abrasive wheels (including pre-mounting and storing procedures) is also necessary to reduce the risk of bursting.

Relationship with other legislation

230 Regulation 12(5) refers to other regulations which cover specific hazards. For example, the Control of Substances Hazardous

to Health Regulations (COSHH) would apply to leakage of a toxic substance, whereas regulation 12 would apply to leakage of steam or cooling water from the same equipment. Similarly, COSHH would apply to the discharge of coolant mist from a machine tool, but regulation 12 would apply in the case of ejected swarf.

Regulation 13

High or very low temperature

231 Regulation 13 deals with the risk of injury from contact with hot or very cold work equipment, parts of work equipment or articles or substances in the work equipment. It does not cover any related risk such as radiant heat or glare.

232 Accessible surfaces of equipment or machinery, when hot or very cold, represent sources of risk of burn or other injury such as frostbite. Examples of relevant equipment might include a flat-iron, foundry equipment, drop forging, hot pressing, liquid nitrogen tank, gas cooker, blast furnace, snow-making machine, cold store, steam pipe, etc and employees may have to work close to the equipment. Touching such surfaces may take place intentionally, for example to operate a handle of the equipment, or unintentionally, when someone is near the equipment. In many cases, surfaces of equipment or devices have to be hot and accessible to operate, for example cooker hotplates, soldering iron bit, heated rolls. In such cases no engineering protective measures can be taken.

233 You should reduce the risk from contact with hot surfaces by engineering methods, ie reduction of surface temperature, insulation, shielding, barricading and guarding. The risk from hot process materials—contact, splashing, spilling, etc—should likewise be reduced by limiting maximum temperature, limiting liquor level, indirect steam heating methods, provision of doors, lids or covers, temperature interlocking of doors or lids and deflection systems for hot liquor (catch pan, spillway, etc).

234 In cases in which engineering protective measures can be applied, for example by reducing surface temperatures, you should adopt these in preference to personal protective measures. You will need to decide the choice of protective measures in each particular case and according to the particular circumstances.

235 While engineering measures should always be applied where appropriate, alternative or complementary forms of

protection may also be necessary. For example, these could include the use of personal protective equipment (see the Personal Protective Equipment at Work Regulations 1992) and/or organisational measures such as warning signs (warning signals, visual and noise alarm signals), instructions, training, supervision, technical documentation, operating instructions, instructions for use.

Regulations 14 to 18—Controls and control systems

Controls for starting or making a significant change in operating conditions

236 Regulations 14 to 18 require the provision of controls 'where appropriate'. This qualification relates both to the features and functioning of the work equipment itself and to whether there is a risk associated with its use.

237 The Regulations on controls and control systems apply not only to equipment with moving parts such as machinery, but may also apply to other equipment which might create a risk, such as ovens, X-ray generators, and lasers.

238 Start, stop and emergency stop controls are not generally appropriate where work equipment has no moving parts. Similarly, they are not appropriate where the risk of injury is negligible such as, for example, battery powered clocks or solar-powered calculators.

239 Some types of work equipment are manually powered and although their use involves risk of injury, human control makes the provision of controls inappropriate. Examples include the following when they use only manual power:

(a) guillotines;

(b) hand-drills;

(c) lawn-mowers.

240 Some types of manually-powered work equipment may need stop controls, if it does not come to a halt when the human effort stops.

241 A control is the manual actuator that the operator touches, for example a button, foot-pedal, knob, or lever. Normally it is part of a control device such as a brake, clutch, switch, or relay. The

control and control device are parts of the control system which may be considered as all the components which act together to monitor and control the functions of the work equipment. Control systems may operate using mechanical linkages, electricity, pneumatics, hydraulics etc, or combinations of these.

242 In practice, most individual items of equipment are likely to be provided with appropriate controls when supplied.

243 For complex items of equipment, installations or assemblies comprising several different items of equipment, it may be necessary to carry out a more detailed assessment of the risks and make special provisions to ensure that controls are provided that comply fully with these Regulations.

Regulation 14

Controls for starting or making a significant change in operating conditions

244 It should only be possible to start the equipment by using appropriate controls. Operating the control need not necessarily immediately start the equipment as control systems may require certain conditions (for example, those relating to operation or protection devices) to be met before starting can be achieved.

245 Restarting the equipment after any stoppage is subject to the same requirements. The stoppage may have been deliberate or may have happened, for example, by the activation of a protection device. You should not normally be able to restart the equipment simply by re-setting a protection device such as, for example, an interlock or a person's withdrawal from an area covered by a sensing device—operation of the start control should also be required.

246 Any change in the operating conditions of the equipment should only be possible by the use of a control unless the change does not increase risks to health and safety. Examples of operating conditions include speed, pressure, temperature and power. For example, certain multifunctional machines are used in the metal working industry for punching or shearing metal using different tools located on different parts of the machines. Safety in the use of these machines is achieved by means of a combination of safe systems of work and physical safeguards which match the

characteristics of the workpiece. It is essential that the function of the machine (for example punching or shearing) is changed by a conscious, positive action by the operator and that unused parts of the machine cannot start up unintentionally.

Regulation 14(1)(b) and 14(2)

247 The purpose of regulation 14(1)(b) and 14(2) is to ensure that users or other people are not caught unawares by any changes in the operating conditions or modes of the equipment in use.

Regulation 14(3)

248 Regulation 14(3) acknowledges that in the case of automatic machinery such as those controlled by programmable electronic systems, it is not appropriate to require separate controls for changing operating conditions when such changes are part of the normal operating cycle. (Nevertheless, these machines should be safeguarded as required by regulations 11 and 12.) However, where you need to make interventions outside the normal sequence, such as clearing blockages, setting or cleaning, you should provide proper controls in accordance with regulations 14(1) and (2).

249 The start control can be separate, combined with controls for operating conditions, or more than one of each type of control can be provided. The controls can be combined with stop controls as required by regulation 15 but not with an emergency stop control provided in accordance with regulation 16. 'Hold-to-run' devices are examples of combined stop and start controls. These should be designed so that the stop function has priority following the release of the control.

250 The controls provided should be designed and positioned so as to prevent, so far as possible, inadvertent or accidental operation. Buttons or levers, for example, should have an appropriate shrouding or locking facility. It should not be possible for the control to 'operate itself', for example due to the effects of gravity, vibration or failure of a spring mechanism. Starting that is initiated from a keyboard or other multifunction device should require some form of confirmation in addition to the start command. Furthermore, the results of the actuation should he displayed.

Regulation 15

Stop controls

251 Regulation 15(1) requires that the action of the stop control should bring the equipment to a safe condition in a safe manner. This acknowledges that it is not always desirable to bring all items of work equipment immediately to a complete stop if this could result in other risks. For example, stopping the mixing mechanism of a reactor during certain chemical reactions could lead to a dangerous exothermic reaction.

252 The stop control does not have to be instantaneous in its action and can bring the equipment to rest in sequence or at the end of an operating cycle if this is required for safety. This may be necessary in some processes, for example to prevent the unsafe build-up of heat or pressure or to allow a controlled run-down of large rotating parts with high inertia.

253 Regulation 15(2) is qualified by 'where necessary for reasons of health and safety'. Therefore all accessible dangerous parts must be rendered stationary which may mean they need to be locked into position and may be allowed to idle. However, parts of equipment which do not present a risk, such as suitably guarded cooling fans, do not need to be positively stopped.

254 Regulation 15(3) requires that the control should switch off all sources of energy from the equipment, after it has stopped, if this is necessary to prevent or minimise risk to health or safety. Where it is necessary to retain power for production reasons and a hazard could arise due to unexpected movement giving rise to risk of injury, control systems should be designed so as to immediately remove the power, should such an event occur. Where internally stored energy could lead to risk, it should be cut off by the action of the stop control. For example, horizontal plastic injection moulding machines may store hydraulic energy in internal hydraulic reservoirs which, under certain fault conditions, may cause uncovenanted movements which could cause injury. In this case, the stop control should effectively isolate or dissipate the stored energy so as to ensure safety.

255 The stop control should take priority over any operating or start control. Where possible, it should not require anything other than a short manual action to activate it, even though the stop and disconnection sequence so initiated may take some time to complete. Further information on the categories of stop function

can be found in BS EN 60204-1. Although this standard (which deals with specifications for general requirements for an individual machine) applies to new machinery, it gives valuable guidance which may be useful for any equipment—new or used.

Regulation 16

Emergency stop controls

256 An emergency stop control should be provided where the other safeguards in place are not adequate to prevent risk when an irregular event occurs. However, an emergency stop control should not be considered as a substitute for safeguarding.

257 Where it is appropriate to have one, based on the risk assessment, you should provide an emergency stop at every control point and at other appropriate locations around the equipment so that action can be taken quickly. The location of emergency stop controls should be determined as a follow-up to the risk assessment required under the Management Regulations. Although it is desirable that emergency stops rapidly bring work equipment to a halt, this must be achieved under control so as not to create any additional hazards.

258 As emergency stops are intended to effect a rapid response to potentially dangerous situations, they should not be used as functional stops during normal operation.

259 Emergency stop controls should be easily reached and actuated. Common types are mushroom-headed buttons, bars, levers, kick-plates, or pressure-sensitive cables. Guidance on specific features of emergency stops is given in national, European and international standards.

Regulation 17

Controls

Regulation 17(1)

260 It should be possible to identify easily what each control does and on which equipment it takes effect. Both the controls and their markings should be clearly visible. As well as having legible wording or symbols, factors such as the colour, shape and position of controls are important; a combination of these can often be used to reduce ambiguity. Some controls may need to be

distinguishable by touch, for example inching buttons on printing machines. Few controls will be adequately identifiable without marking of some sort.

261 The marking and form of many controls is covered by national, European and international standards either generic or specific to the type of equipment (BS 3641, prEN 50099). However, additional marking may often be desirable.

Regulation 17(2)

262 Controls used in the normal running of the equipment should normally not be placed where anybody using them might be exposed to risk. However, controls used for setting-up and fault-finding procedures may have to be positioned where people are at some risk, for example on a robot-teaching pendant. In such cases particular precautions should be employed to ensure safety; examples include using hold-to-run controls, enabling controls, emergency stop controls. Further precautions include the selection of reduced/limited capability of the work equipment during such operations.

Regulation 17(3)(a)

263 The provisions of regulation 17(3)(a) apply where physical safeguarding methods employed in accordance with regulation 11(2)(a) and (b) do not completely prevent access to dangerous parts of work equipment, or where people are at risk from other aspects of the operation, eg noise, or harmful radiation. The preferred aim is to position controls so that operators of equipment are able to see from the control position that no one is at risk from anything they set going. To be able to do this, operators need to have a view of any part of the equipment that may put anyone at risk. A direct view is best, but supplementing by mirrors or more sophisticated visual or sensing facilities may be necessary.

264 There will normally be little difficulty in meeting this requirement in the case of small and compact equipment. With larger equipment there is normally some latitude in the positioning of controls, and the safety aspect should be considered in deciding their location; this would apply, for example, on large process plant such as newspaper printing machinery or chemical plant.

265 Where people are at risk from dangerous parts of machinery, normal safeguarding procedures should restrict the need for

surveillance to vulnerable areas; an example would be on large newspaper printing machines. However, where regular intervention is necessary, which involves entry into, removal of, or opening of safeguards, for example for maintenance purposes, interlocks or similar devices may be necessary as appropriate to prevent start-up while people are at risk. You may need to employ additional measures to ensure that people do not remain inside safeguards at start-up. Similarly, where sensing devices are employed to aid surveillance, they may be interlocked with the controls so as to prevent start-up when people are at risk.

266 If anyone other than the operator is also working on the equipment, they may use permissive start controls. Such controls can indicate to the operator that everyone is clear and permit a start. These can be located at a position of safety from where they can ascertain that no one is at risk.

267 Where there is a risk other than from dangerous parts of machinery (for example noise, radiation), people at some distance from the work equipment may be affected. In such circumstances, it may not always be reasonably practicable for operators to have sight of all parts of the work equipment, so it may be necessary to employ systems of work or warning devices. Warning devices only provide limited protection and additional measures may be required if the risks are high. For example, it would not be acceptable to rely on audible or visible alarms where the risk is of an imminent potentially fatal dose of ionising radiation, but they may be adequate where the risk is from noisy plant.

Regulation 17(3)(b)

268 If the nature of the installation is such that it is not reasonably practicable for the operator at the control position to ensure that no one is at risk, then a system of work must be devised and used to achieve that aim. This should implement procedures to eliminate or reduce the probability of any workers being at risk as a result of starting-up. An example is the use of systems using signallers; these are often used to assist crane drivers, or tractor drivers setting a manned harvester in motion.

Regulation 17(3)(c)

269 The warning should comply with regulation 24, ie it should be unambiguous, easily perceived and easily understood. Signals

may be visual, audible, tactile or a combination of the three, as appropriate.

Regulation 17(4)

270 Warnings given in accordance with regulation 17(3)(c) should be given sufficiently in advance of the equipment actually starting to give those at risk time to get clear or take suitable actions to prevent risks. This may take the form of a device by means of which the person at risk can prevent start-up or warn the operator of their presence.

271 The provisions of regulation 17 do not preclude people from remaining in positions where they are at risk. Their aim is to prevent an operator unintentionally placing people at risk. Regulation 11, in its hierarchical approach to safeguarding, recognises that in exceptional circumstances people may have to approach dangerous parts of machinery, such as for maintenance purposes. Access to such positions should only be allowed under strictly controlled conditions and in accordance with regulation 11.

Regulation 18

Control systems

272 Another way of defining a control system is:

> 'a control system is a system or device which responds to input signals and generates an output signal which causes the equipment under control to operate in a particular manner.'

273 The input signals may be made by an operator via a manual control, or from the equipment itself, for example from automatic sensors or protection devices (photoelectric guards, guard interlock devices, speed limiters, etc). Signals from the equipment may also include information (feedback) on the condition of the equipment and its response (position, whether it is running, speed).

274 Failure of any part of the control system or its power supply should lead to a 'fail-safe' condition. Fail-safe can also be more correctly and realistically called 'minimised failure to danger'. This should not impede the operation of the 'stop' or 'emergency stop' controls. The measures which should be taken in the design and application of a control system to mitigate the effects of its failure will need to be balanced against the consequences of any failure. The greater the risk, the more resistant the control system

should be to the effects of failure. Bringing a machine to a safe halt may achieve the objective. Halting a chemical process, however, could create further hazards. Care should be taken to fully assess the consequences of such events and provide further protection, ie standby power plant or diverting chemicals to a place of safety. It should always be possible to recover to a safe condition.

275 There are national, European and international standards both current and in preparation (BS EN 60204-1, BS EN 954-1) which provide guidance on design of control systems so as to achieve high levels of performance related to safety. Though they are aimed at new machinery, they may be used as guidance for existing work equipment.

Regulation 19

Isolation from sources of energy

276 The main aim of this regulation is to allow equipment to be made safe under particular circumstances, such as when maintenance is to be carried out, when an unsafe condition develops (failure of a component, overheating, or pressure build-up), or where a temporarily adverse environment would render the equipment unsafe, for example electrical equipment in wet conditions or in a flammable or explosive atmosphere.

277 Isolation means establishing a break in the energy supply in a secure manner, ie by ensuring that inadvertent reconnection is not possible. You should identify the possibilities and risks of reconnection as part of your risk assessment, which should then establish how secure isolation can be achieved. For some equipment, this can be done by simply removing the plug from the electrical supply socket. For other equipment, an isolating switch or valve may have to be locked in the off or closed position to avoid unsafe reconnection. The closed position is not always the safe position: for example, drain or vent outlets may need to be secured in the open position.

278 If work on isolated equipment is being done by more than one person, it may be necessary to provide a locking device with multiple locks and keys. Each will have their own lock or key, and all locks have to be taken off before the isolating device can be removed. Keys should not be passed to anyone other than the nominated personnel and should not be interchanged between nominated people.

279 For safety reasons in some circumstances, sources of energy may need to be maintained when the equipment is stopped, for example when the power supply is helping to keep the equipment or parts of it safe. In such cases, isolation could lead to consequent danger, so it will be necessary to take appropriate measures to eliminate any risk before attempting to isolate the equipment.

280 It is appropriate to provide means of isolation where the work equipment is dependent upon external energy sources such as electricity, pressure (hydraulic or pneumatic) or heat. Where possible, means of dissipating stored energy should be provided. Other sources of energy such as its potential energy, chemical or radiological energy, cannot be isolated from the equipment. Nevertheless, there should be a means of preventing such energy from adversely affecting workers, by shielding, barriers or restraint.

281 Isolation of electrical equipment is dealt with by regulation 12 of the Electricity at Work Regulations 1989. Guidance to those Regulations expands on the means of isolating electrical equipment. Note that those Regulations are only concerned with electrical danger (electric shock or burn, arcing and fire or explosion caused by electricity), and do not deal with other risks (such as mechanical) that may arise from failure to isolate electrical equipment.

282 Thermal energy may be supplied by circulation of preheated fluid such as water or steam. In such cases, isolating valves should be fitted to the supply pipework.

283 Similar provision should be made for energy supplies in the form of liquids or gases under pressure. A planned preventive maintenance programme should therefore be instigated which assures effective means of isolation. It may be necessary to isolate pipework by physically disconnecting it or fitting spades in the line to provide the necessary level of protection. Redundancy in the form of more than one isolation valve fitted in series may also be used, but care should be exercised to check the efficacy of each valve function periodically. The performance of such valves may deteriorate over time, and their effectiveness often cannot be judged visually.

284 The energy source of some equipment is held in the substances contained within it; examples are the use of gases or liquids as fuel, electrical accumulators (batteries) and radionuclides. In such cases, isolation may mean removing the energy-containing material, although this may not always be necessary.

285 Also, it is clearly not appropriate to isolate the terminals of a battery from the chemical cells within it, since that could not be done without destroying the whole unit.

286 Some equipment makes use of natural sources of energy such as light or flowing water. In such cases, suitable means of isolation include screening from light, and the means of diverting water flow, respectively. Another natural energy source, wind power, is less easily diverted, so sail mechanisms should be designed and constructed so as to permit minimal energy transfer when necessary. Effective restraint should be provided to prevent uncovenanted movement when taken out of use for repair or maintenance.

287 Regulation 19(3) requires precautions to ensure that people are not put at risk following reconnection of the energy source. So, reconnection of the energy source should not put people at risk by itself initiating movement or other hazard. Measures are also required to ensure that guards and other protection devices are functioning correctly before operation begins.

Regulation 20

Stability

288 There are many types of work equipment that might fall over, collapse or overturn unless suitable precautions are taken to fix them to the ground by bolting, tying, fastening or clamping them in some way and/or to stabilise them by ballasting or counter-balancing. Where ballasting or counterbalancing is employed for portable equipment, you should reappraise the stabilising method each time the equipment is repositioned.

289 Most machines used in a fixed position should be bolted or otherwise fastened down so that they do not move or rock during use. This can be done by fastening the equipment to an appropriate foundation or supporting structure. Other means could include lashing or tying to a supporting structure or platform.

290 Where the stability of the work equipment is not inherent in its design or operation or where it is mounted in a position where its stability could be compromised, for example by severe weather conditions, additional measures should be taken to ensure its stability. Ladders should be at the correct angle, and tied or footed.

291 Certain types of mobile work equipment, for example access platforms, while inherently stable, can have their stability increased during use by means of outriggers or similar devices. While this equipment cannot be 'clamped' or 'fixed', steps must be taken to ensure that the equipment is always used within the limits of its stability at any given time.

Regulation 21

Lighting

292 Any place where a person uses work equipment should be suitably and sufficiently lit. If the ambient lighting provided in the workplace is suitable and sufficient for the tasks involved in the use of the equipment, special lighting need not be provided. But if the task involves the perception of detail, for example precision measurements, you would need to provide additional lighting to comply with the regulation. The lighting should be adequate for the needs of the task.

293 You should provide local lighting on the machine for the illumination of the work area when the construction of the machine and/or its guards render the normal lighting inadequate for the safe and efficient operation of the machine, for example on sewing machines. Local lighting may be needed to give sufficient view of a dangerous process or to reduce visual fatigue. Travelling cranes may obscure overhead lighting for the driver and others, particularly when there is no natural light available (nightworking), and supplementary lighting may be necessary.

294 You should also provide additional lighting in areas not covered by general lighting when work, such as maintenance or repairs, for example, is carried out in them. The arrangements for the provision of lighting could be temporary, by means of hand or other portable lights, for example by fixed lighting inside enclosures, such as lift shafts. The standard of lighting required will be related to the purpose for which the work equipment is used or to the work being carried out. Lighting levels should be checked periodically to ensure that the intensity is not diminished by dust and grime deposits. Where necessary, you should clean luminaires and reflectors at regular intervals to maintain lighting efficiency.

295 Where access is foreseeable on an intermittent but regular basis, you should always consider providing permanent lighting.

296 This regulation complements the requirement for sufficient and suitable workplace lighting in the Workplace (Health, Safety and Welfare) Regulations 1992 and the Electricity at Work Regulations 1989. Guidance is contained in HSE's guidance *Lighting at work.*[22]

Regulation 22

Maintenance operations

297 Regulation 5 requires that equipment is maintained. Regulation 22 requires that equipment is constructed or adapted in a way that takes account of the risks associated with carrying out maintenance work, such as routine and planned preventive maintenance, as described in the guidance to regulation 5. Compliance with this Regulation will help to ensure that when maintenance work is carried out, it is possible to do it safely and without risk to health, as required by section 2 of the HSW Act; it will also help to comply with regulation 5(1), since 'used' includes maintained. Regulation 11(3)(h) contains a requirement linked to regulation 22, but focusing on the narrower aspect of the design of guards for such work. Many accidents have occurred during maintenance work, often as a result of failure to adapt the equipment to reduce the risk.

298 In most cases the need for safe maintenance will have been considered at the design stage and attended to by the manufacturer, and you will need to do little other than review the measures provided. In other cases, particularly when a range of interconnecting components may be put together, for example in a research laboratory or a production line, you will need to consider when carrying out your risk assessment whether any extra features need to be incorporated so that maintenance can be done safely and without risks to health.

299 Ideally, there is no risk associated with the maintenance operation. For example, lubrication points on machines may be designed so that they can be accessed safely even while the machine is in motion, or adjustment points positioned so that they can be used without opening guards.

300 If, however, the maintenance work might involve a risk, this regulation requires that the installation should be designed so that

22 *Lighting at work* HSG38 HSE Books 1998 ISBN 0 7176 1232 5.

the work can, so far as is reasonably practicable, be carried out with the equipment stopped or inactive. This will probably be the case for most equipment.

301 If equipment will have to be running or working during a maintenance operation and this presents risks, you should take measures to enable the operation of the equipment in a way that reduces the risk. These measures include further safeguards or functions designed into the equipment, such as limiting the power, speed or range of movement that is available to dangerous parts or providing protection during maintenance operations. Examples are:

(a) providing temporary guards;

(b) limited movement controls;

(c) crawl speed operated by hold-to-run controls;

(d) using a second low-powered visible laser beam to align a powerful invisible one.

302 Other measures that can be taken to protect against any residual risk include wearing personal protective equipment and provision of instruction and supervision. Although the actual use of these measures falls outside the scope of this regulation, the work equipment should as far as possible be installed to be compatible with their use.

303 The design of equipment in relation to maintenance work on it may also be affected by other legislation. In particular, electrically powered equipment is subject to the Electricity at Work Regulations 1989 as regards risks of injury from electric shock or burn, or from explosion or ignition initiated by electricity. Guidance on those Regulations includes details of relevant equipment requirements.

Regulation 23

Markings

304 This regulation is closely related to the following one which deals with warnings; some markings may also serve as the warning required by regulation 24. There are many circumstances in which marking of equipment is appropriate for health or safety reasons. Stop and start controls for equipment need to be identified. The maximum rotation speed of an abrasive wheel

should be marked upon it. The maximum safe working load (rated capacity) should be marked on lifting equipment. Gas cylinders should indicate (normally by colour) the gas in them. Storage and feed vessels containing hazardous substances should be marked to show their contents, and any hazard associated with them. Pipework for water and compressed air and other mains services should be colour-coded to indicate contents.

305 Some legislation lays down specific circumstances in which markings are needed, and what form they should take. Examples of Regulations requiring particular markings are the Ionising Radiation Regulations 1985, and the Highly Flammable Liquids and Liquefied Petroleum Gases Regulations 1972 (regulations 6 and 7). Pressure vessels are subject to various Regulations, which include requirements for marking the vessel with specific information.

306 You should consider any other marking that might be appropriate for your own purposes, for example numbering machines to aid identification, particularly if the controls or isolators for the machines are not directly attached to them and there could otherwise be confusion.

307 Markings may use words, letters, numbers, or symbols, and the use of colour or shape may be significant. There are nationally or internationally agreed markings relating to some hazards, for example the international symbols for radiation and lasers. Markings should as far as possible conform to such published standards as BS 5378 or as required by any appropriate legislation such as the Health and Safety (Safety Signs and Signals) Regulations 1996.

Regulation 24

Warnings

308 Warnings or warning devices may be appropriate where risks to health or safety remain after other hardware measures have been taken. They may be incorporated into systems of work (including permit-to-work systems), and can enforce measures of information, instruction and training. A warning is normally in the form of a notice or similar. Examples are positive instructions ('hard hats must be worn'), prohibitions ('not to be operated by people under 18 years'), restrictions ('do not heat above 60°C'). A warning device is an active unit giving a signal; the signal may

typically be visible or audible, and is often connected into equipment so that it is active only when a hazard exists.

309 In some cases, warnings and warning devices will be specified in other legislation, for example automatic safe load indicators on mobile cranes on construction sites, or 'X-rays on' lights.

Warnings

310 Warnings can be permanent printed ones; these may be attached to or incorporated into the equipment or positioned close to it. There may also be a need for portable warnings to be posted during temporary operations such as maintenance; these may form part of a permit-to-work system.

311 In some cases words can be augmented or replaced by appropriate graphical signs. So as to be readily understood, such signs will normally need to be from a nationally or internationally agreed standard set. The Health and Safety (Safety Signs and Signals) Regulations 1996 are relevant here.

312 Warning devices can be:

(a) audible, for example reversing alarms on construction vehicles;

(b) visible, for example a light on a control panel that a fan on a microbiological cabinet has broken down or a blockage has occurred on a particular machine;

(c) an indication of imminent danger, for example machine about to start, or development of a fault condition (ie pump failure or conveyor blockage indicator on a control panel); or

(d) the continued presence of a potential hazard (for example, hotplate or laser on).

A particular warning may use both types of device simultaneously, for example, some automatic safe load indicators on mobile cranes.

313 Warnings must be easily perceived and understood, and unambiguous. It is important to consider factors which affect people's perception of such devices, especially for warnings of imminent danger. Visual warnings will be effective only if a person frequently looks in a particular direction, and therefore may not be as widely applicable as audible signals. Appropriate choice of colour and flashing can catch attention, and also

reinforce the warning nature of a visual signal. The sound given by an audible signal should be of such a type that people unambiguously perceive it as a warning. This means that it must be possible to distinguish between the warnings given by separate warning devices and between the warnings and any other, unrelated, signals which may be in operation at the time. It may not be possible to rely on audible signals in a noisy environment, nor in circumstances where many such signals are expected to be active at one time.

PART III: MOBILE WORK EQUIPMENT

Introduction

314 Part III is additional to the other requirements of PUWER 98 (such as training, guarding and inspection) which apply to all work equipment including mobile work equipment. The regulations in Part III of PUWER 98 implement additional requirements for mobile work equipment, which relate to the equipment when it is travelling. Except for the specific requirements of regulation 30 which deals with drive shafts, they are not intended to apply to moving parts of mobile work equipment which is carrying out work in a static position, for example an excavator involved in digging operations.

315 Some of these requirements build on the requirements of the Health and Safety at Work etc Act 1974 (HSW Act) and its supporting ACOP and guidance material that appears at paragraphs 31 to 54 of this publication.

316 Where vehicles are designed primarily for travel on public roads, compliance with the Road Vehicles (Construction and Use) Regulations 1986 will normally be sufficient to comply with the Regulations in Part III of PUWER 98.

317 New mobile work equipment taken into use from 5 December 1998 should comply with all the requirements of PUWER 98. Existing work equipment does not need to comply with Part III of PUWER 98 until 5 December 2002 (see paragraph 57).

What is mobile work equipment?

318 For the purposes of PUWER 98 (Part III), mobile work equipment is any work equipment which carries out work while

it is travelling or which travels between different locations where it is used to carry out work. Such equipment would normally be moved on, for example wheels, tracks, rollers, skids, etc. Mobile work equipment may be self-propelled, towed or remote controlled and may incorporate attachments.

Self-propelled mobile work equipment

319 Self-propelled mobile work equipment is work equipment which is propelled by its own motor or mechanism. The motor or mechanism may be powered by energy generated on the mobile work equipment itself, for example by an internal combustion engine, or through connection to a remote power source, such as an electric cable, electric induction or hydraulic line.

Attachment

320 Attachments are work equipment which may be mounted on self-propelled mobile work equipment to alter its characteristics. For example, a load rotator fitted to a fork-lift truck will alter its load-handling capabilities and may alter its safety characteristics, such as stability. Attachments are not considered to be mobile work equipment in their own right but if they can affect the safety of the self-propelled mobile work equipment when they are attached, they are considered to be part of the self-propelled work equipment. Attachments may be non-powered, powered by an independent power source or powered by the self-propelled work equipment to which they are attached.

Towed mobile work equipment

321 Towed mobile work equipment includes work equipment such as towed machines and trailers which are primarily self-supporting on, for example, their own wheels. They may have moving parts which are:

(a) powered by the vehicle (for example, a power harrow);

(b) an integral power source (for example, a powered crop sprayer); or

(c) they may have no moving parts and function as a result of the movement of the mobile work equipment (for example, a plough or trailer).

Remote controlled mobile work equipment

322 For the purposes of PUWER 98, remote controlled mobile work equipment is operated by controls which are not physically connected to it, for example radio control.

Pedestrian-controlled work equipment

323 You should note that pedestrian-controlled work equipment, for example a lawn-mower, is not likely to be covered by the Regulations in Part III of PUWER 98 irrespective of whether some functions are powered or not.

Regulation 25

Employees carried on mobile work equipment

324 *(See Approved Code of Practice, p 422)*

When regulation 25 applies

325 Regulation 25 contains general requirements which cover the risks to people (drivers, operators and passengers) carried by mobile work equipment when it is travelling. This includes risks associated with people falling from the equipment or from unexpected movement while it is in motion or stopping. It also covers risks associated with the environment and the place in which the mobile work equipment is used (for example, falling objects, low roofs and the surfaces on which it operates). Regulations 26–30 deal with particular risks.

326 Regulation 25(b) also specifically covers the risks from wheels and tracks when the equipment is travelling but it does not cover the risks from other moving parts, which are covered by regulation 11. In addition, it does not cover the risks associated with mounting or dismounting from the equipment which is covered by the HSW Act.

Suitable for carrying people

327 Operator stations with seats or work platforms normally provide a secure place on which the drivers and other people can travel on mobile work equipment.

Seating

328 Seats should be provided wherever necessary. They can provide security for:

(a) drivers who need to be seated when operating mobile work equipment, for example the seat on a site dumper;

(b) people who need to be seated while being transported by the mobile work equipment, for example bench seats in mine locomotive manriding carriages; and

(c) people who are involved in on-board work activities which are best carried out in a seated position.

Cabs, operators' stations and work platforms

329 Cabs, operators' stations and work platforms, with suitable side, front and rear barriers or guard rails can prevent people from falling from mobile work equipment when it is travelling. Where provided, they should be properly designed and constructed. They can be fully enclosed or may be open to the environment.

Equipment not specifically designed for carrying people

330 Under exceptional circumstances mobile work equipment may be used to carry people although it is not specifically designed for this purpose, for example trailers used to carry farmworkers during harvest time. Under these circumstances the mobile work equipment must have features to prevent people falling from it and to allow them to stabilise themselves while it is travelling, for example trailers with sides of appropriate height or by providing a secure handhold. People would also need to be able to safely mount and dismount.

Falling object protective structures (FOPS)

331 If people carried on the mobile work equipment are at significant risk of injury from objects falling on them while it is in use, a FOPS should be provided. This may be achieved by a suitably strong safety cab or protective cage which provides adequate protection in the working environment in which the mobile equipment is used.

Restraining systems

332 The need for restraining systems on mobile work equipment is determined by the risks to workers operating and riding on the mobile work equipment and the practicability of fitting and using such restraints. Restraining systems can be full-body seat belts, lap belts or purpose-designed restraining systems. When assessing the need for restraining systems and the nature of seat restraint required, the risk of people being injured through contact with or being flung from the mobile work equipment if it comes to a sudden stop, or moves unexpectedly, should be taken into account. The need for protection against risks for rolling over and overturning (regulations 26 and 27) should also be taken into account when deciding whether restraining systems should be fitted.

Speed adjustment

333 *(See Approved Code of Practice, p 422)*

334 When carrying people, mobile work equipment should be driven within safe speed limits to ensure that the equipment is stable when cornering and on all the surfaces and gradients on which it is allowed to travel. In addition, the speeds at which the mobile machinery travels should be limited to avoid sudden movements which could put people being carried at risk. See guidance on motor vehicles in paragraph 65 and on driver training in paragraph 194.

Guards and barriers

335 *(See Approved Code of Practice, p 422)*

336 Where there is a foreseeable risk of contact with wheels or tracks when mobile equipment is travelling, adequate separation needs to be provided between people and the wheels and tracks. This can be achieved by positioning cabs, operator stations or work platforms and any suitable barriers, such as robust guard rails or fenders, in positions which prevent the wheels and tracks being reached.

Regulation 26

Rolling over of mobile work equipment

When regulation 26 applies

337 In addition to the more general requirements of regulation 25, regulation 26 covers the measures necessary to protect employees

carried on mobile work equipment where there are risks from roll-over while it is travelling, for example a moving dumper truck on a construction site or an agricultural tractor forwarding or manoeuvring on a slope.

338 It covers roll-over in which the mobile work equipment may only roll over onto its side or end (ie through 90 degrees) or turn over completely (ie through 180 degrees or more).

339 It does not apply to the risk of mobile work equipment, such as an excavator or a vehicle with a winch, overturning when operating in a stationary position. This is covered by regulation 20.

Risk assessment

340 To assess the likelihood and potential consequences of roll-over, you will need to take into account the following to determine what safety measures are needed:

(a) nature of the mobile work equipment and any attachments or accessories fitted to it;

(b) the effects of any work being carried out on or by the mobile work equipment; and

(c) the conditions in which it is used.

341 When mobile work equipment is travelling, roll-over may be encouraged by uneven surfaces, variable or slippery ground conditions, excessive gradients, inappropriate speeds, incorrect tyre pressures and sudden changes in direction. It can also occur due to the inertia transmitted to the mobile work equipment by attachments used with it, particularly if those attachments are not securely restrained from movement.

342 When mobile work equipment is under power but is restrained from movement, for example when a forestry tractor is being used to drag fallen trees or logs from one place to another and the tree or log snags, you will need to take account of the inherent stability of the mobile work equipment and the forces it can apply.

343 When carrying out a risk assessment it is important to remember that although drivers should be trained to minimise the risk of roll-over, this is not a substitute for hardware measures to prevent roll-over (for example counterbalance weights) or protective structures (for example roll-over protective structures (ROPS) to minimise the risk of injury in the event of a roll-over) *where they are necessary*.

Stabilisation

344 Measures that can be taken to stabilise mobile work equipment (ie measures to reduce the risk of roll-over) include fitting appropriate counterbalance weights or increasing its track width by fitting additional or wider wheels. Also, moveable parts which could otherwise create instability by moving around when the mobile work equipment is travelling, may be locked or lashed in stable positions, particularly where locking features are provided for such purposes, for example locking devices for excavator back hoes.

Structures which prevent rolling over by more than 90 degrees

345 Some types of mobile work equipment will only turn onto their sides if roll-over occurs (ie 90 degree roll-over). For example, the boom of a hydraulic excavator, when positioned in its recommended travelling position, can prevent more than 90 degree roll-over.

Regulation 26(1)(b)

346 If parts of the mobile work equipment prevent it rolling over by more than 90 degrees, the requirements of regulation 26(1) will be met.

Roll-over protective structures (ROPS)

347 *(See Approved Code of Practice, p 423)*

Regulation 26(1)(c)

348 ROPS are normally fitted on mobile work equipment which is at risk from 180 degree or more roll-over. They may be structures, frames or cabs which, in the event of roll-over, prevent the work equipment from crushing the people carried by it. ROPS should be capable of withstanding the forces that they would sustain if the mobile work equipment were to roll over through 180 degrees or more.

Limitations on fitting protective structures

349 A protective structure may not be appropriate where it could increase the overall risk of injury to people operating, driving or riding on mobile work equipment. In these circumstances, where

possible, the risks of roll-over should be addressed by other means. An example of where protective structures are not appropriate is when mobile work equipment is required to enter and leave buildings with low roofs and contact could increase the risks to workers.

350 In workplaces such as orchards or a glasshouse, it may not be reasonably practicable to operate mobile work equipment fitted with a ROP.

Regulation 26(4)(c)

351 Before fitting ROPS to older mobile work equipment, which has no anchorage points provided on it (in use before 5 December 1998), an engineering analysis would be necessary. The analysis would need to assess whether it is reasonably practicable to fit adequate anchorage points to the equipment and the structural integrity of any anchorage provided. Some mobile work equipment may not be capable of being fitted with protective structures because mounting points of sufficient strength cannot be provided. This will be true of some equipment in use before 5 December 1998. If the risks associated with the use of the equipment are sufficiently high and it is not reasonably practicable to fit mounting points to allow the fitting of a protective structure, you may need to use other equipment which has, or can have, a protective structure fitted to it.

Restraining systems

352 *(See Approved Code of Practice, p 423)*

Regulation 26(2)

353 Where the operator is at risk of falling out and being crushed by the mobile work equipment or its protective structure in the event of roll-over, you should provide a restraining system (for example, a seat belt) if it can be fitted. This restraining system may also be necessary under the more general requirements of regulation 25 to protect against other risks.

354 If the operator is in a fully enclosed protective structure and unable to fall out of the mobile work equipment, they will not be at risk of being crushed between the mobile work equipment and the ground. However, if the operator or people carried are likely

to be injured through contact with the inside of the structure during roll-over, a restraining system may be necessary.

Mounting points for restraining systems

355 Any restraining system needs to be fitted to appropriate anchorage points on the mobile work equipment to ensure its integrity and reliability in use. Substantial structural modification may need to be made on some older types of work equipment in use before 5 December 1998 to allow a restraining system to be fitted. Under these circumstances it would only be considered reasonably practicable to fit a restraining system if the risks involved were of a sufficiently high order to justify the necessary modifications. Alternatively, you may need to use other work equipment which has or can have a restraining system fitted to it.

Tractors

356 *(See Approved Code of Practice, p 423)*

357 Despite compliance with the Agriculture (Tractor Cabs) Regulations 1974, if the operator or people carried are likely to be injured through contact with the inside of the structure during roll-over, it is likely that you will need to provide a seat restraining system.

Regulation 27

Overturning of fork-lift trucks

What regulation 27 covers

358 This regulation applies to fork-lift trucks (FLTs) fitted with vertical masts, which effectively protect seated operators from being crushed between the FLT and the ground in the event of roll-over, and other FLTs fitted with a ROPS, for example rough terrain variable reach trucks when they are used with fork-lift attachments. Other types of FLT are covered by regulation 26.

Roll-over protection

359 The mast of a vertical-masted FLT will generally prevent an FLT overturning by more than 90 degrees, provided it has sufficient strength and dimensions for this purpose. A variable reach truck FLT, however, is capable of rolling over 180 degrees or

more and would need a ROPS to protect the operator if it is used in circumstances where there is a risk of it rolling over.

Restraining systems

360 *(See Approved Code of Practice, p 423)*

361 If risk assessment shows that an FLT with a seated ride-on operator can roll over in use and there is a risk of the operator leaving the operating position and being crushed between the FLT and the ground, a restraining system, such as a seat belt, will be required. Restraining systems are also required on any FLT which is fitted with a ROPS, for example a variable reach truck to protect operators from the risks of injury from 180 degrees or more roll-over. To be effective, the restraining system should prevent operators or others carried from falling out or being trapped by the FLT or its protective structure in the event of roll-over.

362 There is a history of accidents on counterbalanced, centre control, high lift trucks that have a sit-down operator. Restraining systems will normally be required on these trucks to protect operators from the risks of roll-over.

Where restraining systems are not required

363 Substantial structural modification may be necessary on some older FLTs provided for use before 5 December 1998 in order to allow seat belts or other types of restraining system to be fitted. Under these circumstances it would only be considered reasonably practicable to fit a restraining system if the risks involved were of a sufficiently high order to justify the necessary modifications. Where seat restraints cannot be fitted, and the risks are sufficiently high, you may need to use another FLT which has a restraining system.

Regulation 28

Self-propelled work equipment

Regulation 28(a)

Preventing unauthorised start-up

364 Self-propelled work equipment may be prevented from unauthorised start-up if it has a starter key or device which is issued or made accessible only to authorised people. This means that access to starter keys and starting devices, such as removable

dumper starting handles, should be controlled. Vehicles designed primarily for travel on public roads are dealt with in paragraph 316.

Regulation 28(b)

Minimising the consequences of a collision of rail-mounted work equipment

365 If more than one item of rail-mounted work equipment can travel on the same rails at the same time and collision may be foreseen, safety precautions are required to control the risks involved. Where necessary, safe methods of working will need to be followed to reduce the chances of rail-mounted work equipment colliding with each other. Where collision may be foreseen, safety precautions, such as buffers or automatic means of preventing contact, should be provided.

Regulation 28(c)

Devices for stopping and braking

366 All self-propelled mobile work equipment should have brakes to enable it to slow down and stop in a safe distance and park safely. To this end, mobile work equipment should have adequate braking capacity to enable it to be operated safely on the gradients on which it will be used and its parking brakes should be capable of holding it stationary (where appropriate, fully loaded) on the steepest incline that the mobile work equipment may be parked in use.

367 Other relevant product legislation exists which deals with braking systems on vehicles which may be used on the road as well as at work, such as the Road Vehicles (Construction and Use) Regulations and Directive 76/432/EEC dealing with tractor braking. Under normal circumstances, vehicles meeting these requirements would be suitable for use at work.

Regulation 28(d)

Emergency braking and stopping facilities

368 Where there are significant risks associated with failure of the main braking device, a secondary braking system is required. The secondary braking system may operate automatically through spring applied brakes or through a dual circuit system on the service brakes. It may also be operated through the parking brake

system or other controls which are easily accessible to the driver. Self-propelled mobile work equipment which will not stop in a safe distance, for example through transmission drag, if service brake failure or faults occur, are normally fitted with secondary braking systems.

Regulation 28(e)

Driver's field of vision

369 This regulation applies when mobile work equipment is about to move or while it is travelling (including manoeuvring). Under these circumstances, where the driver's direct field of vision is inadequate to ensure safety then visibility aids or other suitable devices should be provided so far as is reasonably practicable. Regulation 17 requires that operators of mobile equipment should be able to see anyone who may be put at risk when any control is operated. Therefore, if direct vision is impaired, then mirrors or more sophisticated visual or sensing facilities may be necessary. Regulation 28(e) requires, so far as is reasonably practicable, mobile work equipment to have adequate devices to improve the driver's field of vision where this is otherwise inadequate. Such devices may include mirrors or closed-circuit television (CCTV) and the provision of these devices can be used to meet the requirements of both regulations.

370 Examples of devices which can aid the driver's vision include:

(a) plane, angled and curved mirrors;

(b) Fresnel lenses;

(c) radar; and

(d) CCTV systems.

The selection of these devices for use on mobile work equipment is a matter for risk assessment, taking account of the purposes for which the devices are provided and their ability to improve driver visibility.

Regulation 28(f)

Equipping mobile work equipment with lighting for use in the dark

371 In terms of this Regulation, 'dark' means any situation where the light levels are not good enough for the driver to operate the

self-propelled work equipment safely without risks to themselves or other people in the vicinity.

372 In such situations the equipment needs to be equipped with 'appropriate' lighting. The level of lighting required will depend on the type of equipment being operated, how it is being operated and the area in which it is operating. Factors you will need to consider are the presence of other people and/or obstacles in the vicinity of the equipment and ground conditions which could lead to risk. In situations where there is a significant risk of an accident, the lighting will need to be at a sufficient level to help control this risk.

373 Regulation 28(f) only covers lighting on mobile work equipment. Lighting provided at the workplace for the use of all work equipment is covered by regulation 21.

Regulation 28(g)

The carriage of appropriate fire-fighting appliances

374 *(See Approved Code of Practice, p 424)*

375 This regulation covers the risks to the operators of self-propelled work equipment if the equipment itself or any load handled by it catches fire. If the operators cannot readily escape from the equipment (such as from a tower crane), you will need to provide appropriate equipment for extinguishing the fire. This will depend on the type of equipment and/or any load it is intended to handle but could include appropriate extinguishers and fire blankets.

376 For self-propelled work equipment that is used on the public highways carrying a dangerous load, it may need to carry suitable fire extinguishers under the requirements of the Carriage of Dangerous Goods by Road and Rail (Classification, Packaging and Labelling) Regulations 1994. Further guidance is contained in HSE's publication *The carriage of dangerous goods explained: Part 2.*[23]

Regulation 29

Remote-controlled self-propelled work equipment

377 For the purposes of regulation 29, 'remote-controlled self-propelled work equipment' is self-propelled work equipment that

23 *Carriage of dangerous goods explained: Part* 2 HSG161 HSE Books 1996 ISBN 0 7176 1253 8.

is operated by controls which have no physical link with it, for example radio control. It should be noted that pendant-controlled mobile work equipment is not covered by regulation 29.

378 As part of your risk assessment you need to consider risks, due to the movement of the equipment, to the person controlling it and also anyone else who may be in the vicinity. You may need to consider alarms or flashing lights so that other people in the area are aware of its movement, or presence, sensing or contact devices which will protect people from the risks associated with the equipment, ie if people may come close to or contact it.

379 When the equipment is switched off you must ensure that every part of the equipment which could present a risk comes to a safe stop. If the equipment is controlled manually, the controls for its operation should be of the hold-to-run type so that any hazardous movements can stop when the controls are released.

380 If the equipment leaves its control range, any part of it which could present a risk should be able to stop and remain in a safe state.

Regulation 30

Drive shafts

Regulation 30(1)

381 A 'drive shaft' is a device which conveys the power from the mobile work equipment to any work equipment connected to it. In agriculture these devices are known as power take-off shafts.

382 'Seizure' refers to stalling of the drive shaft as a result of the operating mechanism of any accessory or anything connected to it becoming incapable of movement due to blockage or some other reason. Under these circumstances regulation 30 applies if the power output of the mobile work equipment is sufficient to cause damage to the connected work equipment which could lead to risk. Regulation 30 does not apply to the risks associated with trapped energy resulting from stalling of the drive shaft if the power output of the mobile work equipment is insufficient to cause damage which could lead to risk. This situation is covered by regulation 19 which deals with the isolation of work equipment from sources of energy.

383 You should assess the risks associated with seizure of the drive shaft. If seizure could lead to risk, for example the ejection of

parts, measures should be taken to protect against such risks. For example, slip clutches on the power input connection of the connected work equipment can protect it from damage and guards fitted in accordance with regulation 12 can protect people from ejection risks in the event of equipment break-up.

384 To prevent damage to power take-off shafts in the event of seizure, it is important to use shafts of adequate length. There should be sufficient overlap between the two halves of the shaft to ensure that it is stable in use, to protect against damage when movements occur in the hitch and to ensure that it has sufficient strength. The shaft needs to be capable of sustaining the full power output of the mobile work equipment, taking account of any slip clutches, shear bolts or similar devices which are provided to limit the torque that the shaft would sustain.

Regulation 30(2)

385 To prevent damage to the drive shaft and its guard when the equipment is not in use, the drive shaft should be supported on a cradle wherever one is provided. If there is no cradle, it should be supported by other means to give equivalent protection against damage. You should not rest the drive shafts on draw bars, nor drop them on the ground, as this could lead to damage.

APPENDIX 1

SCHEDULE 1 OF THE REPORTING OF INJURIES, DISEASES AND DANGEROUS OCCURRENCES REGULATIONS 1995

1 Any fracture, other than to the fingers, thumbs or toes.

2 Any amputation.

3 Dislocation of the shoulder, hip, knee or spine.

4 Loss of sight (whether temporary or permanent).

5 A chemical or hot metal burn to the eye or any penetrating injury to the eye.

6 Any injury resulting from an electric shock or electrical burn (including any electrical burn caused by arcing or arcing products) leading to unconsciousness or requiring resuscitation or admittance to hospital for more than 24 hours.

7 Any other injury:

(a) leading to hypothermia, heat-induced illness or to unconsciousness;

(b) requiring resuscitation; or

(c) requiring admittance to hospital for more than 24 hours.

8 Loss of consciousness caused by asphyxia or by exposure to a harmful substance or biological agent.

9 Either of the following conditions which result from the absorption of any substance by inhalation, ingestion or through the skin:

(a) acute illness requiring medical treatment; or

(b) loss of consciousness.

10 Acute illness which requires medical treatment where there is reason to believe that this resulted from exposure to a biological agent or its toxins or infected material.

APPENDIX 2

FURTHER GUIDANCE ON REGULATION 11—DANGEROUS PARTS OF MACHINERY

Explanation of safeguarding terms, regulation 11(2)

1 *Guards* are physical barriers which prevent access to the danger zone. *Fixed guards* in regulation 11(2)(b) have no moving parts and are fastened in a constant position relative to the danger zone (see Figure 1). They are kept in place either permanently, by welding for example, or by means of fasteners (screws, nuts, etc) making removal/opening impossible without using tools. If by themselves, or in conjunction with the structure of the equipment, they *enclose* the dangerous parts, fixed guards meet the requirements of the first level of the hierarchy. Note that fixed enclosing guards, and other types of guard, can have openings provided that they comply with appropriate safe reach distances (see BS EN 294: 1992).

2 *Other guards* in regulation 11(2)(b) include movable guards which can be opened without the use of tools, and fixed guards that are not fully enclosing. These allow limited access through openings, gates, etc for feeding materials, making adjustments,

cleaning, etc (see Figure 2). *Movable guards* may be power-operated, self-closing, adjustable, etc and are likely to require an interlocking device so that:

(a) the hazardous machine functions covered by the guard cannot operate until the guard is closed;

(b) if the guard is opened while hazardous machine functions are operating, a stop instruction is given;

(c) when the guard is closed, the hazardous machine functions covered by the guard can operate, but the closure of the guard does not by itself initiate their operation.

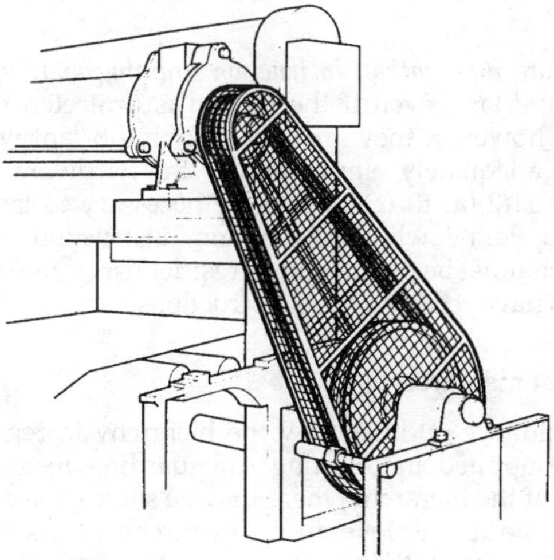

Figure 1 Fixed enclosing guard

Interlocking guards may be fitted with a locking device so that the guard remains closed and locked until any risk of injury from the hazardous machine functions has passed. A control guard (interlocking guard with a start function) is a particular type of interlocking guard which should be used only in certain situations where frequent access is required. It should also fulfil specific conditions, in particular, where there is no possibility of an operator or part of their body remaining in the danger zone or between the danger zone and the guard while the guard is closed (see BS EN 953: 1998).

3 *Protection devices* are devices which do not prevent access to the danger zone but stop the movement of the dangerous part before contact is made. They will normally be used in conjunction with a guard. Typical examples are mechanical trip devices, active opto-electronic devices such as light curtains (see Figure 3), pressure-sensitive mats and two-hand controls.

4 *Protection appliances* are used to hold or manipulate in a way which allows operators to control and feed a loose workpiece at a machine while keeping their body clear of the danger zone. They are commonly used in conjunction with manually fed wood-working machines (see Figure 4) and some other machines such as bandsaws for cuffing meat where it is not possible to fully guard the cuffing tool. These appliances will normally be used in addition to guards.

5 Adequate *information, instruction, training and supervision* are always important, even if the hazard is protected by hardware measures; however, they are especially important when the risk cannot be adequately eliminated by the hardware measures in regulation 11(2)(a) to (c). It may be necessary to lay down procedures to define what information, instruction, training and supervision must be given, and to restrict use of the equipment to those who have received such instructions etc.

Selection of measures

6 The guidance outlines how the hierarchy in regulation 11(2) should be applied in selecting safeguarding measures. Within each level of the hierarchy, there may be some choice available. In particular, the second level in the hierarchy allows a choice from among a number of different types of guard or protection device.

7 Regulation 11(2)(b) requires that when it is not practicable to use fixed enclosing guards, either at all or to the extent required for adequate protection, other guards and/or protection devices shall be used as far as practicable. Where, for example, frequent access is required, it may be necessary to choose between an interlocking movable guard or a protection device. The foreseeable probability and severity of injury will influence the choice of measures from among the range of guards and protection devices available. Regulation 11(3)(a) requires that these must be suitable for their purpose. It is likely that some fixed guarding will be required to ensure that access can only be made through the movable opening guard or protection device. The use of movable guards which are

interlocked is well established. Protection devices need to be carefully applied, taking into account the particular circumstances and the consequences of their failing to act as required.

8 Fixed distance guards, adjustable guards and other guards which do not completely enclose the dangerous parts should only be used in situations where it is not practicable to use fixed enclosing guards or protection devices which would give a greater level of protection.

Features of guards and protection devices, regulation 11(3)

Regulation 11(3)(a)

9 All guards and protection devices provided must be suitable for their purpose. In deciding what is suitable, employers should first establish the foreseeable risks from the machine and then follow guidance contained in national and international standards, guidance from HSC, HSE and industry associations, normal industrial practice and their own knowledge of the particular circumstances in which the machine is to be used.

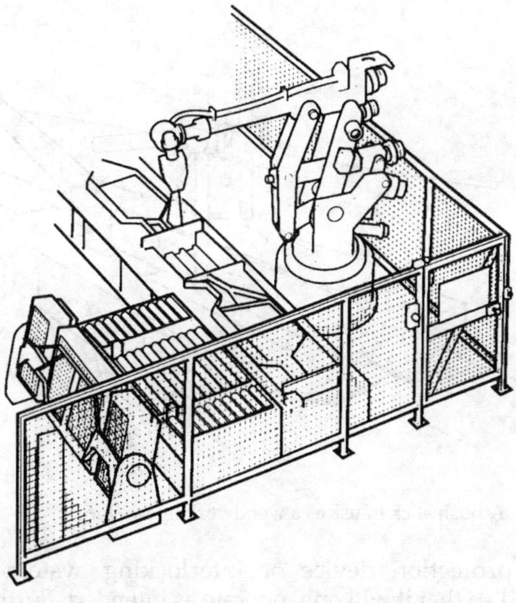

Figure 2 Perimeter fence guard with fixed panels and interlocking access door

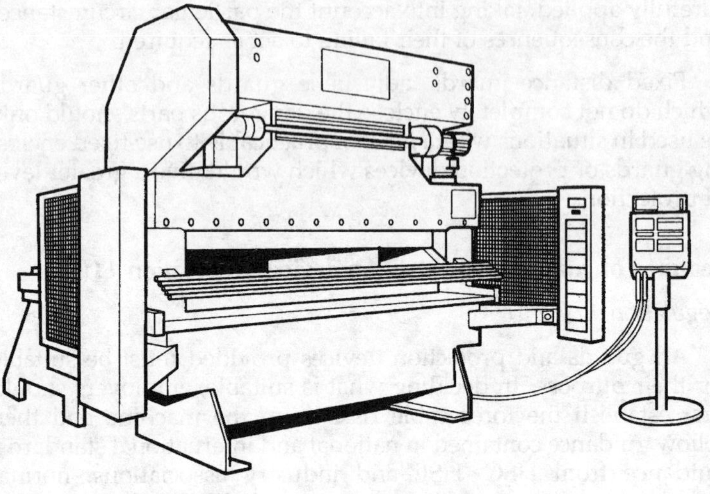

Figure 3 Photoelectric device fitted to a press brake

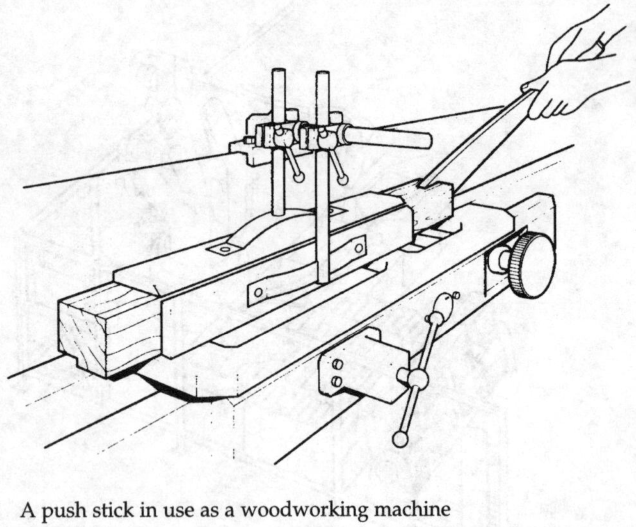

Figure 4 A push stick in use as a woodworking machine

10 A protection device or interlocking system should be designed so that it will only operate as intended. Furthermore, if a component deteriorates or fails, the device or system should as far

as possible fail in a safe manner by inhibiting the dangerous action of the machine. The force of this requirement depends on the combination of probability of failure and severity of the injury, should the system fail. If the overall risk is high, there should be adequate provision to counteract the effects of failure. Guidance on appropriate levels of protection is given in the publications referred to in paragraph 9.

Regulation 11(3)(b)

11 Guards and protection devices must be of good construction, sound material and adequate strength. They must be capable of doing the job they are intended to do. Several factors can be considered:

(a) material of construction (metal, plastic, laminated glass, etc);

(b) form of the material (sheet, open mesh, bars, etc);

(c) method of fixing.

12 Good construction involves design and layout as well as the mechanical nature and quality of the construction. Foreseeable use and misuse should be taken into account.

Regulation 11(3)(c)

13 Guards and protection devices must be maintained in an efficient state, in efficient working order and in good repair. This is an important requirement as many accidents have occurred when guards have not been maintained. It is a particular example of the general requirement under regulation 5 to maintain equipment. Compliance can be achieved by the use of an effective check procedure for guards and protection devices, together with any necessary follow-up action. In the case of protection devices or interlocks, some form of functional check or test is desirable.

14 For inspection and thorough examination of power presses and their guards and protection devices, see the Approved Code of Practice and guidance *Safe use of power presses.*[24]

24 *Safe use of power presses. Provision and Use of Work Equipment Regulations 1998 as applied to power presses. Approved Code of Practice and Guidance on Regulations* L112 HSE Books ISBN 0 7176 1627 4.

Regulation 11(3)(d)

15 Guards and protection devices must not themselves lead to any increased risk to health or safety. One effect of this sub-paragraph is to prevent use of inherently hazardous measures for guarding.

16 A second effect is that guards must be constructed so that they are not themselves dangerous parts. If a guard is power operated or assisted, the closing or opening action might create a potentially dangerous trap which needs secondary protection, for example a leading-edge trip bar or pressure-sensitive strip.

17 The main concern is the overall effect on risk. The fact that a guard may itself present a minor risk should not rule out its use if it can protect against the risk of major injury. For example, sweep-away guards or manually-actuated sliding access gates might be able to cause minor injury, but their use in guarding against more serious risks is justified.

Regulation 11(3)(e)

18 Guards and protection devices must be designed and installed so that they can not be easily bypassed or disabled. This refers to accidental or deliberate action that removes the protection offered. By regulation 11(3)(a), guards must be suitable for their purpose, and one consequence of this is that simple mechanical bypassing or disabling should not be possible.

19 Movable panels in guards giving access to dangerous parts or movable guards themselves will often need to be fitted with an interlocking device. This device must be designed and installed so that it is difficult or impossible to bypass or defeat. Guidance on the selection and design of interlocking devices is available from BS EN 1088: 1996 and the sources listed in paragraph 9.

20 In some cases, bypassing is needed for a particular purpose such as maintenance. The risks arising in such circumstances must be carefully assessed. As far as possible, the risks should be reduced or eliminated by appropriate design of the machinery—see regulation 22.

Regulation 11(3)(f)

21 Guards and protection devices must be situated at a sufficient distance from the danger zone they are protecting. In the case of

solid fixed enclosing guards, there is no minimum distance between guard and danger zone, except that required for good engineering design. However, the gap between a fence type guard or protection device and machine should normally be sufficiently small to prevent anybody remaining in it without being detected; alternatively, the space between guard or protection device and machine should be monitored by a suitable presence-sensing device.

22 Where guarding is provided with holes or gaps (for visibility, ventilation or weight reduction, for example), or is not fully enclosing, the holes must be positioned or sized so that it prevents foreseeable access to the danger zone. Published national and international standards (for example, BS EN 294: 1992) give guidance on suitable distances and opening sizes in different circumstances.

23 The positioning of protection devices which enable hazardous machine operation to stop before access can be gained to its danger zone will be affected by both the characteristics of the device itself (response time) and those of the machine to which it is fitted (time needed to stop). In these circumstances the device must be positioned so that it meets published criteria for the performance of such a system. Refer to the relevant standard (BS EN 999: 1998) and guidance (HSG180).[25]

24 Safeguarding is normally attached to the machine, but the Regulation does not preclude the use of free-standing guards or protection devices. In such cases, the guards or protection devices must be fixed in an appropriate position relative to the machine.

Regulation 11(3)(g)

25 Guards and protection devices must not unduly restrict the view of the operating cycle of the machinery, where such a view is necessary. It is not usually necessary to be able to see all the machine; the part that needs to be seen is normally that which is acting directly on material or a workpiece.

26 Operations for which it is necessary to provide a view include those where the operator controls and feeds a loose workpiece at a machine. Examples include manually fed woodworking machines

25 *The application of electro-sensitive protective equipment employing light curtains and light beam devices to machinery* HSG180 HSE Books 1998 ISBN 0 7176 1550 2.

and food slicers. Many of these operations involve the use of protection appliances.

27 If the machine process needs to be seen, but cannot be, there is a temptation for the operator to remove or disable guards or interlocks. In such cases, some view of the work may be considered necessary. Examples are a hopper feeding a screw conveyor, milling machines, and power presses.

28 In other cases it may be convenient but not absolutely necessary to see the entire operating cycle. The regulation does not prohibit providing a view in these cases, but does not require it; an example is an industrial tumble drier.

29 Where an operation protected by guards needs to be seen, the guard should be provided with viewing slits or properly constructed panels, perhaps backed up by internal lighting, enabling the operator to see the operation. The arrangements to ensure visibility should not prevent the guarding from carrying out its proper function; but any restriction of view should be the minimum compatible with that. An example of a guard providing necessary vision is viewing slits provided in the top guard of a circular saw.

Regulation 11(3)(h)

30 Guards and protection devices must be constructed or adapted so that they allow operations necessary to fit or replace parts and for maintenance work, restricting access so that it is allowed only to the area where the work is to be carried out and, if possible, without having to dismantle the guard or protection device.

31 This regulation applies to the design of guards or protection devices so as to reduce risks arising from some particular operations. Regulation 22 applies to the design of equipment as a whole so that maintenance and similar operations can be carried out safely; regulation 11 is restricted to machine safeguards.

32 The aim is to design the safeguards so that operations like fitting or changing parts or maintenance can be done with minimal risk. If risk assessment shows this is not already the case, it may be possible to adapt the safeguarding appropriately.

33 Ideally, the machine is designed so that operations can be done in an area without risk, for example by using remote adjustment or maintenance points. If the work has to be done in the

enclosed or protected area, the safeguarding should be designed to restrict access just to that part where the work is to be carried out. This may mean using a series of guards.

34 If possible, the guard or protection device should not have to be dismantled. This is because of the possibility that after reassembly, the guard or device may not work to its original performance standard.

Regulation 11(4)

35 Protection appliances also need to be suitable for their application. Factors for consideration come under the same headings as those for guards and protection devices as in regulation 11(3). Many of these are common sense matters. Their design, material, manufacturer and maintenance should all be adequate for the job they do. They should allow the person to use them without having to get too close to the danger zone, and they should not block the view of the workpiece.

REFERENCES

This section has been reproduced as footnotes 9–12 and 14–25 within pages 424–513.

FURTHER READING

This section is not reproduced in this book.

Appendix 4

Personal Protective Equipment at Work Regulations 1992

THE REGULATIONS (SI 1992/2966)

Made 25 November 1992

The Secretary of State, in exercise of the powers conferred upon her by sections 15(1), (2), (3)(a) and (b), (5)(b) and (9) of, and paragraphs 11 and 14 of Schedule 3 to, the Health and Safety at Work etc Act 1974, and of all other powers enabling her in that behalf and for the purpose of giving effect without modifications to proposals submitted to her by the Health and Safety Commission under section 11(2)(d) of the said Act after the carrying out by the said Commission of consultations in accordance with section 50(3) of that Act, hereby makes the following Regulations:

Regulation 1

Citation and commencement

These Regulations may be cited as the Personal Protective Equipment at Work Regulations 1992 and shall come into force on 1 January 1993.

Regulation 2

Interpretation

(1) In these Regulations, unless the context otherwise requires, 'personal protective equipment' means all equipment (including clothing affording protection against the weather) which is intended to be worn or held by a person at work and which protects him against one or more risks to his health or safety, and any addition or accessory designed to meet that objective.

(2) Any reference in these Regulations to—

(a) a numbered regulation or Schedule is a reference to the regulation or Schedule in these Regulations so numbered; and

(b) a numbered paragraph is a reference to the paragraph so numbered in the regulation in which the reference appears.

Regulation 3 *(See Guidance Notes, pp 531–535)*

Disapplication of these Regulations

(1) These Regulations shall not apply to or in relation to the master or crew of a sea-going ship or to the employer of such persons in respect of the normal ship-board activities of a ship's crew under the direction of the master.

(2) Regulations 4 to 12 shall not apply in respect of personal protective equipment which is—

(a) ordinary working clothes and uniforms which do not specifically protect the health and safety of the wearer;

(b) an offensive weapon within the meaning of section 1(4) of the Prevention of Crime Act 1953 used as self-defence or as deterrent equipment;

(c) portable devices for detecting and signalling risks and nuisances;

(d) personal protective equipment used for protection while travelling on a road within the meaning (in England and Wales) of section 192(1) of the Road Traffic Act 1988, and (in Scotland) of section 151 of the Roads (Scotland) Act 1984;

(e) equipment used during the playing of competitive sports.

(3) Regulations 4 and 6 to 12 shall not apply where any of the following Regulations apply and in respect of any risk to a person's health or safety for which any of them require the provision or use of personal protective equipment, namely—

(a) the Control of Lead at Work Regulations 1980;

(b) [the Ionising Radiations Regulations 1999 [SI 1999/3232]];[1]

1 The words 'the Ionising Radiations Regulations 1999 [SI 1999/3232]' in square brackets substituted by SI 1999/3232, reg 41(1), Sch 9, para 3. Date in force: 1 January 2000: see SI 1999/3232, reg 1(a).

(c) the Control of Asbestos at Work Regulations 1987;

(d) the Control of Substances Hazardous to Health Regulations 1988;

(e) the Noise at Work Regulations 1989;

(f) the Construction (Head Protection) Regulations 1989.

Regulation 4 *(See Guidance Notes, pp 535–539)*

Provision of personal protective equipment

(1) [Subject to paragraph (1A),][2] every employer shall ensure that suitable personal protective equipment is provided to his employees who may be exposed to a risk to their health or safety while at work except where and to the extent that such risk has been adequately controlled by other means which are equally or more effective.

[(1A) Where the characteristics of any policing activity are such that compliance by the relevant officer with the requirement in paragraph (1) would lead to an inevitable conflict with the exercise of police powers or performance of police duties, that requirement shall be complied with so far as is reasonably practicable.][3]

(2) Every self-employed person shall ensure that he is provided with suitable personal protective equipment where he may be exposed to a risk to his health or safety while at work except where and to the extent that such risk has been adequately controlled by other means which are equally or more effective.

(3) Without prejudice to the generality of paragraphs (1) and (2), personal protective equipment shall not be suitable unless—

(a) it is appropriate for the risk or risks involved and the conditions at the place where exposure to the risk may occur;

(b) it takes account of ergonomic requirements and the state of health of the person or persons who may wear it;

(c) it is capable of fitting the wearer correctly, if necessary, after adjustments within the range for which it is designed;

2 The words 'Subject to paragraph (1A),' in square brackets inserted by SI 1999/860, reg 4(1), (2). Date in force: 14 April 1999: see SI 1999/860, reg 1.

3 Inserted by SI 1999/860, reg 4(1), (3). Date in force: 14 April 1999: see SI 1999/860, reg 1.

(d) so far as is practicable, it is effective to prevent or adequately control the risk or risks involved without increasing overall risk;

(e) it complies with any enactment (whether in an Act or instrument) which implements in Great Britain any provision on design or manufacture with respect to health or safety in any relevant Community directive listed in Schedule 1 which is applicable to that item of personal protective equipment.

Regulation 5 *(See Guidance Notes, p 539)*

Compatibility of personal protective equipment

(1) Every employer shall ensure that where the presence of more than one risk to health or safety makes it necessary for his employee to wear or use simultaneously more than one item of personal protective equipment, such equipment is compatible and continues to be effective against the risk or risks in question.

(2) Every self-employed person shall ensure that where the presence of more than one risk to health or safety makes it necessary for him to wear or use simultaneously more than one item of personal protective equipment, such equipment is compatible and continues to be effective against the risk or risks in question.

Regulation 6 *(See Guidance Notes, pp 540–541)*

Assessment of personal protective equipment

(1) Before choosing any personal protective equipment which by virtue of regulation 4 he is required to ensure is provided, an employer or self-employed person shall ensure that an assessment is made to determine whether the personal protective equipment he intends will be provided is suitable.

(2) The assessment required by paragraph (1) shall include—

(a) an assessment of any risk or risks to health or safety which have not been avoided by other means;

(b) the definition of the characteristics which personal protective equipment must have in order to be effective against the risks referred to in sub-paragraph (a) of this paragraph, taking into account any risks which the equipment itself may create;

(c) comparison of the characteristics of the personal protective equipment available with the characteristics referred to in sub-paragraph (b) of this paragraph.

(3) Every employer or self-employed person who is required by paragraph (1) to ensure that any assessment is made shall ensure that any such assessment is reviewed if—

(a) there is reason to suspect that it is no longer valid; or

(b) there has been a significant change in the matters to which it relates,

and where as a result of any such review changes in the assessment are required, the relevant employer or self-employed person shall ensure that they are made.

Regulation 7 *(See Guidance Notes, pp 541–542)*

Maintenance and replacement of personal protective equipment

(1) Every employer shall ensure that any personal protective equipment provided to his employees is maintained (including replaced or cleaned as appropriate) in an efficient state, in efficient working order and in good repair.

(2) Every self-employed person shall ensure that any personal protective equipment provided to him is maintained (including replaced or cleaned as appropriate) in an efficient state, in efficient working order and in good repair.

Regulation 8 *(See Guidance Notes, pp 542–543)*

Accommodation for personal protective equipment

Where an employer or self-employed person is required, by virtue of regulation 4, to ensure personal protective equipment is provided, he shall also ensure that appropriate accommodation is provided for that personal protective equipment when it is not being used.

Regulation 9 *(See Guidance Notes, pp 543–545)*

Information, instruction and training

(1) Where an employer is required to ensure that personal protective equipment is provided to an employee, the employer shall also ensure that the employee is provided with such information, instruction and training as is adequate and appropriate to enable the employee to know—

(a) the risk or risks which the personal protective equipment will avoid or limit;

(b) the purpose for which and the manner in which personal protective equipment is to be used; and

(c) any action to be taken by the employee to ensure that the personal protective equipment remains in an efficient state, in efficient working order and in good repair as required by regulation 7(1).

(2) Without prejudice to the generality of paragraph (1), the information and instruction provided by virtue of that paragraph shall not be adequate and appropriate unless it is comprehensible to the persons to whom it is provided.

Regulation 10 *(See Guidance Notes, p 545)*

Use of personal protective equipment

(1) Every employer shall take all reasonable steps to ensure that any personal protective equipment provided to his employees by virtue of regulation 4(1) is properly used.

(2) Every employee shall use any personal protective equipment provided to him by virtue of these Regulations in accordance both with any training in the use of the personal protective equipment concerned which has been received by him and the instructions respecting that use which have been provided to him by virtue of regulation 9.

(3) Every self-employed person shall make full and proper use of any personal protective equipment provided to him by virtue of regulation 4(2).

(4) Every employee and self-employed person who has been provided with personal protective equipment by virtue of regulation 4 shall take all reasonable steps to ensure that it is returned to the accommodation provided for it after use.

Regulation 11 *(See Guidance Notes, p 545)*

Reporting loss or defect

Every employee who has been provided with personal protective equipment by virtue of regulation 4(1) shall forthwith report to his employer any loss of or obvious defect in that personal protective equipment.

Regulation 12

Exemption certificates

(1) The Secretary of State for Defence may, in the interests of national security, by a certificate in writing exempt—

(a) any of the home forces, any visiting force or any headquarters from those requirements of these Regulations which impose obligations on employers; or

(b) any member of the home forces, any member of a visiting force or any member of a headquarters from the requirements imposed by regulation 10 or 11;

and any exemption such as is specified in sub-paragraph (a) or (b) of this paragraph may be granted subject to conditions and to a limit of time and may be revoked by the said Secretary of State by a further certificate in writing at any time.

(2) In this regulation—

(a) 'the home forces' has the same meaning as in section 12(1) of the Visiting Forces Act 1952;

(b) 'headquarters' has the same meaning as in article 3(2) of the Visiting Forces and International Headquarters (Application of Law) Order 1965;

(c) 'member of a headquarters' has the same meaning as in paragraph 1(1) of the Schedule to the International Headquarters and Defence Organisations Act 1964; and

(d) 'visiting force' has the same meaning as it does for the purposes of any provision of Part I of the Visiting Forces Act 1952.

Regulation 13 *(See Guidance Notes, p 546)*

Extension outside Great Britain

These Regulations shall apply to and in relation to the premises and activities outside Great Britain to which sections 1 to 59 and 80 to 82 of the Health and Safety at Work etc Act 1974 apply by virtue of the Health and Safety at Work etc Act 1974 (Application

Outside Great Britain) Order 1989 as they apply within Great Britain.

Regulation 14 *(See Guidance Notes, p 546)*

Modifications, repeal and revocations

(1) The Act and Regulations specified in Schedule 2 shall be modified to the extent specified in the corresponding Part of that Schedule.

(2) Section 65 of the Factories Act 1961 is repealed.

(3) The instruments specified in column 1 of Schedule 3 are revoked to the extent specified in column 3 of that Schedule.

SCHEDULE 1
RELEVANT COMMUNITY DIRECTIVES

Regulation 4(3)(e)

1

Council Directive of 21 December 1989 on the approximation of the laws of the Member States relating to personal protective equipment (89/686/EEC), as amended by Council Directive 93/95/EEC of 29 October 1993 and Article 7 of Council Directive 93/63/EEC of 22 July 1993.

2

Council Directive 93/42/EEC concerning medical devices (OJ No L169, 12 July 1993, p 1).

SCHEDULE 2
MODIFICATIONS

Regulation 14(1)

Part I
The Factories Act 1961

1 In section 30(6), for 'breathing apparatus of a type approved by the chief inspector', substitute 'suitable breathing apparatus'.

Part II
The Coal and Other Mines (Fire and Rescue) Order 1956

2 In Schedule 1, in regulation 23(a), for 'breathing apparatus of a type approved by the Minister', substitute 'suitable breathing apparatus'.

3 In Schedule 1, in regulation 23(b), for 'smoke helmets or other apparatus serving the same purpose, being helmets or apparatus of a type approved by the Minister', substitute 'suitable smoke helmets or other suitable apparatus serving the same purpose'.

4 In Schedule 1, in regulation 24(a), for 'smoke helmet or other apparatus serving the same purpose, being a helmet or other apparatus of a type approved by the Minister', substitute 'suitable smoke helmet or other suitable apparatus serving the same purpose'.

Part III
The Shipbuilding and Ship-Repairing Regulations 1960

5 In each of regulations 50, 51(1) and 60(1), for 'breathing apparatus of a type approved for the purpose of this Regulation', substitute 'suitable breathing apparatus'.

Part IV
The Coal Mines (Respirable Dust) Regulations 1975

6 In regulation 10(a), for 'dust respirators of a type approved by the Executive for the purpose of this Regulation', substitute 'suitable dust respirators'.

Part V
The Control of Lead at Work Regulations 1980

7 In regulation 7—

(a) after 'respiratory protective equipment', insert 'which complies with regulation 8A or, where the requirements of that regulation do not apply, which is'; and

(b) after 'as will', insert ', in either case,'.

8 In regulation 8, for 'adequate protective clothing', substitute 'protective clothing which complies with regulation 8A or, where no requirement is imposed by virtue of that regulation, is adequate'.

9 After regulation 8, insert the following new regulations—

'Compliance with relevant Community directives

8A Any respiratory protective equipment or protective clothing shall comply with any enactment (whether in an Act or instrument) which implements any provision on design or manufacture with respect to health or safety in any relevant Community directive listed in Schedule 1 to the Personal Protective Equipment at Work Regulations 1992 which is applicable to that item of respiratory protective equipment or protective clothing.

Assessment of respiratory protective equipment or protective clothing

8B

(1) Before choosing respiratory protective equipment or protective clothing, an employer shall make an assessment to determine whether it will satisfy regulation 7 or 8, as appropriate.

(2) The assessment required by paragraph (1) shall involve—

(a) definition of the characteristics necessary to comply with regulation 7 or, as the case may be, 8, and

(b) comparison of the characteristics of respiratory protective equipment or protective clothing available with the characteristics referred to in sub-paragraph (a) of this paragraph.

(3) The assessment required by paragraph (1) shall be revised if—

(a) there is reason to suspect that it is no longer valid; or

(b) there has been a significant change in the work to which it relates,

and, where, as a result of the review, changes in the assessment are required, the employer shall make them.'

10 In regulation 9, for sub-paragraph (b), substitute the following sub-paragraph—

'(b) where he is required under regulations 7 or 8 to provide respiratory protective equipment or protective clothing, adequate changing facilities and adequate facilities for the storage of—

(i) the respiratory protective equipment or protective clothing, and

(ii) personal clothing not worn during working hours.'

11 At the end of regulation 13, add the following new paragraph—

'(3) Every employee shall take all reasonable steps to ensure that any respiratory protective equipment provided to him pursuant to regulation 7 and protective clothing provided to him pursuant to regulation 8 is returned to the accommodation provided for it after use.'

12 In regulation 18(2), omit the full stop and add 'and that any provision imposed by the European Communities in respect of the encouragement of improvements in the safety and health of workers at work will be satisfied.'

Part VI

13–15 . . .[4]

Part VII
The Control of Asbestos at Work Regulations 1987

16 In regulation 8(3), after 'shall' the first time that word appears, insert 'comply with paragraph (3A) or, where no requirement is imposed by that paragraph, shall'.

17 Insert the following new paragraph after regulation 8(3)—

'(3A) Any respiratory protective equipment provided in pursuance of paragraph (2) or protective clothing provided in pursuance of regulation 11(1) shall comply with this paragraph if it complies with any enactment (whether in an Act or instrument) which implements in Great Britain any provision on design or manufacture with respect to health or safety in any relevant Community directive listed in Schedule 1 to the Personal Protective Equipment at Work Regulations 1992 which is applicable to that item of respiratory protective equipment or protective clothing.'

18 In regulation 20(2), omit the full stop and add 'and that any provision imposed by the European Communities in respect of the encouragement of improvements in the safety and health of workers at work will be satisfied.'

4 Revoked by SI 1999/3232, reg 41(2)(c). Date in force: 1 January 2000: see SI 1999/3232, reg 1(a).

Part VIII
The Control of Substances Hazardous to Health Regulations 1988

19–21 . . .[5]

Part IX
The Noise at Work Regulations 1989

22 Add the following new paragraph at the end of regulation 8—

'(3) Any personal ear protectors provided by virtue of this regulation shall comply with any enactment (whether in an Act or instrument) which implements in Great Britain any provision on design or manufacture with respect to health or safety in any relevant Community directive listed in Schedule 1 to the Personal Protective Equipment at Work Regulations 1992 which is applicable to those ear protectors.'

Part X
The Construction (Head Protection) Regulations 1989

23 Add the following paragraphs at the end of regulation 3—

'(3) Any head protection provided by virtue of this regulation shall comply with any enactment (whether in an Act or instrument) which implements any provision on design or manufacture with respect to health or safety in any relevant Community directive listed in Schedule 1 to the Personal Protective Equipment at Work Regulations 1992 which is applicable to that head protection.

(4) Before choosing head protection, an employer or self-employed person shall make an assessment to determine whether it is suitable.

(5) The assessment required by paragraph (4) of this regulation shall involve—

(a) the definition of the characteristics which head protection must have in order to be suitable;

(b) comparison of the characteristics of the protection available with the characteristics referred to in sub-paragraph (a) of this paragraph.

5 The former paragraphs 19 to 21 referring to the 1988 Regulations were revoked by the Control of Substances Hazardous to Health Regulations 1994.

(6) The assessment required by paragraph (4) shall be reviewed if—

(a) there is reason to suspect that it is no longer valid; or

(b) there has been a significant change in the work to which it relates, and where as a result of the review changes in the assessment are required, the relevant employer or self-employed person shall make them.

(7) Every employer and every self-employed person shall ensure that appropriate accommodation is available for head protection provided by virtue of these Regulations when it is not being used.'

24 For regulation 6(4), substitute the following paragraph—

'(4) Every employee or self-employed person who is required to wear suitable head protection by or under these Regulations shall—

(a) make full and proper use of it; and

(b) take all reasonable steps to return it to the accommodation provided for it after use.'

25 In regulation 9(2), omit the full stop and add 'and that any provision imposed by the European Communities in respect of the encouragement of improvements in the safety and health of workers at work will be satisfied.'

SCHEDULE 3
REVOCATIONS

Regulation 14(3)

(1) *Title*	(2) *Reference*	(3) *Extent of revocation*
Regulations dated 26 February 1906 in respect of the processes of spinning and weaving of flax and tow and the processes incidental thereto (the Flax and Tow-Spinning and Weaving Regulations 1906)	SR & O 1906/177, amended by SI 1988/1657	In regulation 9, the words 'unless waterproof skirts, and bibs of suitable material, are provided by the occupier and worn by the workers' Regulation 13

Order dated 5 October 1917 (the Tin or Terne Plates Manufacture Welfare Order 1917)	SR & O 1917/1035	Paragraph 1
Order dated 15 August 1919 (the Fruit Preserving Welfare Order 1919)	SR & O 1919/1136, amended by SI 1988/1657	Paragraph 1
Order dated 23 April 1920 (the Laundries Welfare Order 1920)	SR & O 1920/654	Paragraph 1
Order dated 28 July 1920 (the Gut-Scraping, Tripe Dressing, etc Welfare Order 1920)	SR & O 1920/1437	Paragraph 1
Order dated 3 March 1921 (the Glass Bevelling Welfare Order 1921)	SR & O 1921/288	Paragraph 1
The Aerated Water Regulations 1921	SR & O 1921/1932; amended by SI 1981/686	The whole Regulations
The Sacks (Cleaning and Repairing) Welfare Order 1927	SR & O 1927/860	Paragraph 1
The Oil Cake Welfare Order 1929	SR & O 1929/534	Paragraph 1
The Cement Works Welfare Order 1930	SR & O 1930/94	Paragraph 1
The Tanning Welfare Order 1930	SR & O 1930/312	Paragraph 1 and the Schedule
The Magnesium (Grinding of Castings and Other Articles) Special Regulations 1946	SR & O 1946/2107	Regulation 12

The Clay Works (Welfare) Special Regulations 1948	SI 1948/1547	Regulation 5
The Iron and Steel Foundries Regulations 1953	SI 1953/1464; amended by SI 1974/1681 and SI 1981/1332	Regulation 8
The Shipbuilding and Ship-Repairing Regulations 1960	SI 1960/1932; amended by SI 1974/1681	Regulations 73 and 74
The Non-Ferrous Metals (Melting and Founding) Regulations 1962	SI 1962/1667; amended by SI 1974/1681	Regulation 13
The Abstract of Special Regulations (Aerated Water) Order 1963	SI 1963/2058	The whole Order
The Construction (Health and Welfare) Regulations 1966	SI 1966/95; to which there are amendments not relevant to these Regulations	Regulation 15
The Foundries (Protective Footwear and Gaiters) Regulations 1971	SI 1971/476	The whole Regulations
The Protection of Eyes Regulations 1974	SI 1974/1681; amended by SI 1975/303	The whole Regulations
The Aerated Water Regulations (Metrication) Regulations 1981	SI 1981/686	The whole Regulations

GUIDANCE NOTES

Introduction

1 The Personal Protective Equipment at Work Regulations came into force on 1 January 1993. The full text of the Regulations is available from HMSO. The Regulations, interspersed with respective guidance, are reproduced in Part 1 of this publication.

2 The Regulations are made under the Health and Safety at Work etc Act 1974 (HSW Act), and apply to all workers in Great Britain except the crews of sea-going ships.

3 The Regulations are based on a European Community (EC) Directive requiring similar basic laws throughout the Community on the use of personal protective equipment (PPE) in the workplace.

4 The guidance on the Regulations has been prepared by the Health and Safety Executive (HSE) for the Health and Safety Commission (HSC) after widespread consultation with industry. Part 1 deals with the main steps required under the Regulations.

5 Part 2 of this document contains advice on the selection of PPE. It considers the different types of PPE available and identifies some of the processes and activities which may require PPE to be worn. Hearing and respiratory protective equipment are not considered in this guidance as they are already dealt with by separate Regulations and guidance produced by the HSC/E.

6 Separate and more detailed guidance for specific industries is available as indicated in the text. Employers should also take into account any other HSC or HSE publications giving guidance on other Regulations, specific hazards or equipment, for example HSE booklet HS(G)53 *Respiratory protective equipment: a practical guide for users*. Up to date information on these publications can be obtained from the HSE Information Centre at Sheffield.

PART I: GUIDANCE ON THE PERSONAL PROTECTIVE EQUIPMENT AT WORK REGULATIONS 1992

Regulation 3

Disapplication of these Regulations

7 Personal protective equipment (PPE) includes both the following, when they are worn for protection of health and safety:

(a) protective clothing such as aprons, protective clothing for adverse weather conditions, gloves, safety footwear, safety helmets, high visibility waistcoats etc; and

(b) protective equipment such as eye protectors, life-jackets, respirators, underwater breathing apparatus and safety harnesses.

8 In practice, however, these Regulations will not apply to ear protectors, most respiratory protective equipment and some other types of PPE used at work. These types of PPE are specifically excluded from the scope of the PPE at Work Regulations because they are covered by existing Regulations such as the Noise at Work Regulations 1989 (see Table 1 and paragraph 16). Even if the PPE at Work Regulations do not apply, the advice given in this guidance may still be applicable, as the general principles of selecting and maintaining suitable PPE and training employees in its use are common to all Regulations which refer to PPE.

Table 1 Provisions on the use of personal protective equipment

Where there are other comprehensive Regulations which require PPE.	The PPE at Work Regulations will not apply.	For example: The Control of Substances Hazardous to Health Regulations 1994 require respirators to be used in certain circumstances.
Where there are no other comprehensive Regulations dealing with PPE.	The PPE at Work Regulations will apply.	For example: The PPE at Work Regulations will require that chainsaw operators are provided with and wear the appropriate PPE.
Where there are other but not comprehensive Regulations requiring the use of PPE.	The PPE at Work Regulations will apply, and will complement the requirements of the other Regulations.	For example: regulation 19 of the Docks Regulations 1988 requires the provision of high visibility clothing when an employee is working in specific areas of a dock. The PPE at Work Regulations complement this duty by laying down duties about the accommodation of PPE, training of employees in its use etc, and may also require the use of high visibility clothing in any other part of the dock where there is a risk from vehicle movements.

9 Items such as uniforms provided for the primary purpose of presenting a corporate image, and ordinary working clothes, are not subject to these Regulations. Likewise the Regulations will not apply to protective clothing provided in the food industry primarily for food hygiene purposes. However, where any uniform or clothing protects against a specific risk to health and safety, for example high visibility clothing worn by emergency services, it will be subject to the Regulations. Waterproof, weatherproof or insulated clothing is subject to the Regulations if it is worn to protect employees against risks to their health or safety, but not otherwise.

10 The Regulations do not cover the use of PPE such as cycle helmets, crash helmets or motor cycle leathers worn by employees on the public highway. Motor cycle crash helmets remain legally required for motor cyclists under road traffic legislation, and section 2 of the Health and Safety at Work etc Act 1974 (HSW Act 1974)—requiring employers to ensure the health and safety of employees, so far as is reasonably practicable—will still apply. The Regulations do apply to the use of such equipment at work elsewhere if there is a risk to health and safety, for example, farm workers riding motorcycles or all-terrain vehicles should use crash helmets.

11 The Regulations do not require professional sports people to use PPE such as shin guards or head protection during competition. However, they do apply to sports equipment used in other circumstances, for example, life-jackets worn by professional canoeing instructors, riding helmets worn by stable staff, or climbing helmets worn by steeplejacks.

12 The Regulations do not require employers to provide equipment for self defence or deterrent purposes, for example personal sirens/alarms or truncheons for use by security staff. However, they do apply to PPE (such as helmets or body armour) provided where staff are at risk from physical violence.

13 The Regulations do not cover personal gas detectors or radiation dosimeters. Although this equipment would come within the broad definition of PPE, the specific disapplication is included as many of the Regulations would not be appropriate to it (for example, the fitting and ergonomic requirements of regulation 4). However, employers will have a duty to provide such equipment under section 2 of the HSW Act 1974 if its use is necessary to ensure the health and safety of employees.

Application to merchant shipping

14 Sea-going ships are subject to separate merchant shipping legislation, administered by the Department of Transport, which gives protection to people on board. Regulation 3(1) disapplies the Regulations from these ships in respect of the normal ship-board activities of a ship's crew under the direction of the master. But it does not disapply them in respect of other work activities, for example, where a shore-based contractor goes on board to carry out work on the ship, that person's activities will be subject to the Regulations within territorial waters as provided for under regulation 14. This partial exemption applies to sea-going ships only. Therefore the Regulations will apply to PPE used on ships that only operate on inland waters.

Aircraft

15 Aircraft are subject to these Regulations while on the ground and in airspace for which the United Kingdom has jurisdiction.

Application of other Regulations

16 The sets of Regulations listed in regulation 3(3) require the provision and use of certain PPE against particular hazards, and the PPE at Work Regulations will not apply where these Regulations remain in force. For example, a person working with asbestos would, where necessary, have to use respiratory protective equipment and protective clothing under the Control of Asbestos at Work Regulations 1987, rather than the PPE at Work Regulations.

17 The PPE at Work Regulations revoked much older legislation on PPE, for example the Protection of Eyes Regulations 1974; a complete list is in Schedule 3 to the Regulations. However, because they provide for the particular circumstances of the relevant industry or risk, it was necessary to retain some provisions apart from those in regulation 3(3) (for example in the Diving Operations at Work Regulations 1981 and the Docks Regulations 1988). The more comprehensive PPE at Work Regulations will apply in addition to these Regulations (see Table 1). Where necessary, therefore, employers (and others with duties under the Regulations) will have to comply with both the specific Regulations and the PPE at Work Regulations. A list of these specific Regulations is on page 570.

18 From 1 July 1995, PPE provided by virtue of the Regulations listed in regulation 3(3) of the PPE at Work Regulations will have to comply with the requirements of the Personal Protective Equipment (EC Directive) Regulations 1992 (as amended), which implement the Personal Protective Equipment Product Directive (89/686/EEC), and be CE marked; as will PPE provided under these Regulations. PPE already in use can continue to be used without being CE marked for as long as it remains suitable for the use to which it is being put. See paragraphs 33 to 35 for a fuller explanation.

Application to non-employees

19 These Regulations do not apply to people who are not employees, for example voluntary workers, or school children while in school. However section 3 of the HSW Act 1974, which requires that 'It shall be the duty of every employer to conduct his undertaking in such a way as to ensure, so far as is reasonably practicable, that persons not in his employment who may be affected thereby are not exposed to risks to their health and safety', will still apply. If an employer needs to provide PPE to comply with this duty, then by following the requirements of these Regulations he will fully satisfy this duty. These Regulations do apply to trainees and children on work experience programmes.

Regulation 4

Provision of personal protective equipment

PPE as a 'last resort'

20 The Management of Health and Safety at Work Regulations (MHSWR) 1992 require employers to identify and assess the risks to health and safety present in the workplace, so enabling the most appropriate means of reducing those risks to an acceptable level to be determined. There is in effect a hierarchy of control measures, and PPE should always be regarded as the 'last resort' to protect against risks to safety and health; engineering controls and safe systems of work should always be considered first. It may be possible to do the job by another method which will not require the use of PPE or, if that is not possible, adopt other more effective safeguards: for example, fixed screens could be provided rather than individual eye protection to protect against swarf thrown off

a lathe. Employers' duties in this respect are contained in much of the legislation under the HSW Act 1974, including MHSWR. The practical guidance to MHSWR given in its Approved Code of Practice is also particularly relevant. However, in some circumstances PPE will still be needed to control the risk adequately, and the PPE at Work Regulations will then take effect.

21 There are a number of reasons for this approach. Firstly, PPE protects only the person wearing it, whereas measures controlling the risk at source can protect everyone at the workplace. Secondly, theoretical maximum levels of protection are seldom achieved with PPE in practice, and the actual level of protection is difficult to assess. Effective protection is only achieved by suitable PPE, correctly fitted and maintained and properly used. Thirdly, PPE may restrict the wearer to some extent by limiting mobility or visibility, or by requiring additional weight to be carried. Other means of protection should therefore be used whenever reasonably practicable.

22 Employers should, therefore, provide appropriate PPE and training in its usage to their employees wherever there is a risk to health and safety that cannot be adequately controlled by other means.

Providing personal protective equipment

23 In order to provide PPE for their employees, employers must do more than simply have the equipment on the premises. The employees must have the equipment readily available, or at the very least have clear instructions on where they can obtain it. Most PPE is provided on a personal basis, but in certain circumstances items of PPE may be shared by employees, for example where they are only required for limited periods—see the guidance on regulation 7 (maintenance).

24 By virtue of section 9 of the HSW Act 1974, no charge can be made to the worker for the provision of PPE which is used only at work.

25 Section 9 of the HSW Act 1974 states: 'No employer shall levy or permit to be levied on any employee of his any charge in respect of anything done or provided in pursuance of any specific requirement of the relevant statutory provisions'. Section 9 applies to these Regulations because they impose a 'specific requirement', ie to provide PPE.

26 Regulation 4 requires PPE to be provided where risks have not been adequately controlled by other means. Where risks are sufficiently low that they can be considered in effect to be adequately controlled, then PPE need not be provided. For example, in most workplaces there will be some risk of people dropping objects onto their feet, but it is only when there is manual handling of objects of sufficient weight that the risk will be sufficient to require the provision of safety footwear.

27 Adequate control of the risk is also in general the standard of protection which the PPE provided should achieve. However, there may be some circumstances where no PPE will provide adequate control of the risk (for example fire fighters' protective clothing can give only limited protection from radiant heat and flames). In these cases, the employer is required only to provide PPE offering the best protection practicable in the circumstances. Use of PPE must not increase the overall level of risk, ie PPE must not be worn if the risk caused by wearing it is greater than the risk against which it is meant to protect.

28 Regulation 4(3)(a) to (e) lists other factors which determine whether PPE is suitable. Further guidance on the suitability of PPE is given in paragraphs 29 to 35.

Ergonomic and other factors

29 When selecting PPE to be used while doing a job, the nature of the job and the demands it places on the worker should be taken into account. This will involve considering the physical effort required to do the job, the methods of work, how long the PPE needs to be worn, and requirements for visibility and communication. Those who do the job are usually best placed to know what is involved, and they should be consulted. Other factors may also influence selection: for example, PPE used in the food industry may need to be cleaned easily. The aim should always be to choose PPE which will give minimum discomfort to the wearer, as uncomfortable equipment is unlikely to be worn properly.

30 There will be considerable differences in the physical dimensions of different workers and therefore more than one type or size of PPE may be needed. The required range may not be available from a single supplier. Those having to use PPE should be consulted and involved in the selection and specification of the equipment as there is a better chance of PPE being used effectively if it is accepted by each wearer.

31 All PPE which is approved by HSE or bears the CE mark must pass basic performance requirements. These have usually been set following medical advice, and the use of such PPE should cause no problems to average healthy adults. Where problems occur, employers should seek medical advice as to whether the individual can tolerate wearing the PPE. Employers are able to take into account only those medical conditions of which they have been informed.

32 In some industries, particularly those where peripatetic workers (such as contract maintenance workers or building workers) are employed, the site operator will be better placed to provide the appropriate PPE than the peripatetic worker's employer. Although under these circumstances the employer does not have to repeat the provision of suitable PPE, it is still the employer's responsibility to ensure that suitable PPE is provided. Likewise, the site operator may in practice take the action necessary to meet the requirements of the Regulations which follow, but the employer still remains responsible for ensuring that this has been done.

The quality of personal protective equipment

33 Employers should ensure that any PPE they buy complies with the United Kingdom legislation implementing Community directives concerning the design or manufacture of PPE with regard to health and safety, listed in Schedule 1 of the Regulations, where that legislation is applicable. At 1 January 1995, the only provisions listed in Schedule 1 were:

(a) the PPE Product Directive (89/686/EEC) as amended by Directive 93/95/EEC (which extended the transition period before the PPE Product Directive takes full effect to 30 June 1995) and article 7 of Directive 93/68/EEC (which made a number of technical changes in connection with the CE-marking of PPE). The PPE (EC Directive) Regulations 1992 implemented the PPE Product Directive, and they have been amended by the PPE (EC Directive) (Amendment) Regulations 1993 and the PPE (EC Directive) (Amendment) Regulations 1994, whose primary purposes were to implement the two Amending Directives (93/95/EEC and 93/68/EEC) respectively;

(b) Directive 93/42/EEC on medical devices, which was implemented by the Medical Devices Regulations 1994.

There are a few types of PPE such as that used by law enforcers and the military, and escape equipment on ships and aircraft, that are not within the scope of the PPE Product Directive. Regulation 4(3)(e) does not apply to these types of PPE as they do not have to meet the PPE (EC Directive) Regulations. Gloves used by medical staff are likely to be the only PPE covered by the Medical Devices Regulations.

34 The PPE (EC Directive) Regulations will require almost all PPE supplied for use at work to be type examined by a body which is Approved and notified to the European Commission for that purpose. This body will, if the PPE meets the basic safety requirements, issue a certificate of conformity. For a few types of simple PPE protecting against low risks (eg gardening gloves) the manufacturer or importer can self-certify that the PPE meets the basic safety requirements. The manufacturer or importer is then able to display the CE mark on the product. From 1 July 1995 it will be illegal for manufacturers and importers to supply or sell PPE unless it meets these requirements and displays a CE mark. However, employers can continue to use any PPE without a CE mark which was legally supplied to them, provided that it continues to be suitable for the use to which it is put and is properly maintained.

35 In many cases, PPE will be made to harmonised European Standards or 'Norms' (ENs) which are systematically replacing existing British Standards. These standards are published in the Official Journal of the European Communities and equipment conforming with these standards will be considered to comply with the basic safety requirements of the PPE Product Directive. Part 2 and pages 571–575 refer to some of these standards, but many ENs on PPE are still in preparation.

Regulation 5

Compatibility of personal protective equipment

36 If more than one item of PPE is being worn, the different items of PPE must be compatible with each other. For example, certain types of respirators will not fit properly and give adequate protection if a safety helmet is worn. In such cases when selecting PPE it should be ensured that both items when used together will adequately control the risks against which they are provided to protect.

Regulation 6

Assessment of personal protective equipment

Assessment

37 The purpose of the assessment provision in regulation 6 is to ensure that the employer who needs to provide PPE chooses PPE which is correct for the particular risks involved and for the circumstances of its use. It follows on from, but does not duplicate, the risk assessment requirement of the Management of Health and Safety at Work Regulations (MHSWR) 1992, which involves identifying the hazards present in any undertaking and then evaluating the extent of the risks involved. Regulation 6(2) lays down the steps the employer should take to identify appropriate PPE.

38 Whatever PPE is chosen, it should be remembered that, although some types of equipment do provide very high levels of protection, none provides 100%. Some indication is needed of the level of risk so that the performance required of the PPE can be estimated. This information may have been gathered as part of the overall risk assessment required under MHSWR (described above), or more generalised data may be available from sources such as HSE guidance.

39 In the simplest and most obvious cases which can easily be repeated and explained at any time, the assessment to identify suitable PPE need not be recorded. In more complex cases, however, the assessment will need to be recorded and kept readily accessible to those who need to know the results.

Selection of suitable PPE

40 Once potential hazards are known there may be several types of PPE that would be suitable. The risks at the workplace and the parts of the body endangered are the two key elements to consider. A specimen risk survey table is produced on page 569 and is designed to help define the areas in which workers are at risk. Part 2 identifies types of PPE that may be suitable once the risks have been assessed.

41 For example, when assessing the need for eye protection, employers should first identify the types of hazard present, such as airborne dust, liquid splashes or projectiles, and then assess the degree of risk—for example the likely size and velocity of the projectiles. They can then select a suitable type of PPE from the

range of CE marked equipment available. In this case, eye protection is designed for dust or chemical protection, and to different levels of impact resistance.

42 Once a type of CE marked PPE has been selected for a given application, further advice and information may be necessary to ensure that the equipment can provide the protection needed. Manufacturers and suppliers have duties under the PPE (EC Directive) Regulations 1992 and under section 6 of the Health and Safety at Work etc Act 1974 to provide information of this type.

43 When selecting PPE to be used while doing a job, the nature of the job and the demands it places on the worker should be taken into account as explained in paragraphs 29 to 32. This will involve considering the physical effort required to do the job, the methods of work, how long the PPE needs to be worn, and requirements for visibility and communication.

44 Selection should be seen as only the first stage in a continuing programme which is also concerned with the proper use and maintenance of the equipment, and the training and supervision of employees.

Regulation 7

Maintenance and replacement of personal protective equipment

45 An effective system of maintenance of PPE is essential to make sure the equipment continues to provide the degree of protection for which it was designed. Maintenance is required under the Regulations and includes, where appropriate, cleaning, disinfection, examination, replacement, repair and testing. The responsibility for carrying out maintenance should be laid down, together with the details of the procedures to be followed and their frequency. Where appropriate, records of tests and examinations should also be kept. The maintenance programme will vary with the type of equipment and the use to which it is put. For example, mechanical fall-arrestors will require a regular planned preventative maintenance programme which will include examination, testing and overhaul. However, gloves may only require periodic inspection by the user, depending on what they are being used to protect against.

46 In general, PPE should be examined to ensure that it is in good working order, before being issued to the wearer. PPE should also

be examined before it is put on and should not be worn if it is found to be defective or has not been cleaned. Such examinations should be carried out by properly trained staff in accordance with the manufacturer's instructions. While most PPE will be provided on a personal basis, some may be used by a number of people. There should therefore be arrangements for cleaning and disinfecting if necessary before PPE is reissued.

47 A sufficient stock of spare parts, when appropriate, should be available to wearers. Only proper spare parts should be used in maintaining PPE, or the equipment may not provide the required degree of protection. The use of different parts may also be prohibited under regulation 4(3)(e)—some new PPE components also have to be CE marked.

48 Manufacturers' maintenance schedules and instructions (including recommended replacement periods and shelf lives) should normally be followed: any significant departure from them should be discussed beforehand with the manufacturers or their authorised agent. Some British or European Standards on PPE (many of which are listed in Part 2 and on pages 571–575) also contain useful information on maintenance.

49 Simple maintenance can be carried out by the trained wearer, but more intricate repairs should only be done by specialist personnel. With complex equipment, a high standard of training will be required. As an alternative to in-house maintenance, contract maintenance services are available from both manufacturers and suppliers of equipment and specialist maintenance firms.

50 In certain circumstances it may be more appropriate, instead of instituting a specific maintenance procedure, to provide a supply of disposable PPE (eg single use coveralls) which can simply be discarded after use. If disposable PPE is used, it is important that users know when it should be discarded and replaced.

Regulation 8

Accommodation for personal protective equipment

51 The employer needs to ensure that accommodation is provided for PPE so that it can be safely stored or kept when it is not in use. Accommodation may be simple, for example, pegs for weatherproof clothing or safety helmets. It need not be fixed, for example, safety spectacles could be kept by the user in a suitable

carrying case, and PPE used by mobile workers can be stored in suitable containers in their vehicle. The storage should be adequate to protect the PPE from contamination, loss, or damage by (for example) harmful substances, damp or sunlight. Where PPE becomes contaminated during use, the accommodation should be separate from any provided for ordinary clothing (accommodation for ordinary work clothing is dealt with in the Workplace (Health, Safety and Welfare) Regulations 1992), and where necessary be suitably labelled. If the PPE itself contains hazardous materials, for example asbestos, it may need special storage arrangements.

52 Where quantities of PPE are stored, equipment which is ready for use should be clearly segregated from that which is awaiting repair or maintenance.

Regulation 9

Information, instruction and training

53 The Regulations require employers to provide suitable information, instruction and training for their employees, to enable them to make effective use of the PPE provided to protect them against workplace hazards to their health and safety. A systematic approach to training is needed; this means that everyone who is involved in the use or maintenance of PPE should be trained appropriately.

54 Users must be trained in the proper use of PPE, how to correctly fit and wear it, and what its limitations are. Managers and supervisors must also be aware of why PPE is being used and how it is used properly. People involved in maintaining, repairing and testing the equipment and in its selection for use will also need training. Training should include elements of theory as well as practice in using the equipment, and should be carried out in accordance with the recommendations and instructions supplied by the PPE manufacturer.

55 The extent of the instruction and training will vary with the complexity and performance of the equipment. For PPE which is simple to use and maintain, safety helmets for example, some basic instructions to the users may be all that is required. On the other hand, the safe use of anti-static footwear or laser eye protection will depend on an adequate understanding of the principles behind them, and in the case of the former, regular maintenance and testing. The instruction and training should include:

Theoretical training

(a) an explanation of the risks present and why PPE is needed;

(b) the operation, performance and limitations of the equipment;

(c) instructions on the selection, use and storage of PPE related to the intended use. Written operating procedures such as permits to work involving PPE should be explained;

(d) factors which can affect the protection provided by the PPE such as: other protective equipment; personal factors; working conditions; inadequate fitting; and defects, damage and wear;

(e) recognising defects in PPE and arrangements for reporting loss or defects.

Practical training

(a) practice in putting on, wearing and removing the equipment;

(b) practice and instruction in inspection and, where appropriate, testing of the PPE before use;

(c) practice and instruction in the maintenance which can be done by the user, such as cleaning and the replacement of certain components;

(d) instruction in the safe storage of equipment.

Duration and frequency of training

56 The extent of the training that is required will depend on the type of equipment, how frequently it is used and the needs of the people being trained. Many manufacturers of PPE run training courses for users of their equipment and these courses may be of particular benefit to small users who do not have training facilities.

57 In addition to initial training, users of PPE and others involved with the equipment may need refresher training from time to time. Records of training details should be kept, to assist in the efficient administration of the training programme.

58 Employers must ensure not only that their employees undergo the appropriate training but that they understand what they are being taught. Employees may have difficulty in understanding their training for a number of reasons. For example, the risks (and

precautions) may be of a particularly complex nature, making it difficult for employees to understand the precise nature of the protective measures they must take. English may not be the first language of some employees, and in this case the instruction and training may have to be undertaken in the employee's mother tongue to ensure comprehensibility.

Regulation 10

Use of personal protective equipment

59 PPE should be used in accordance with the employer's instructions, which should in turn be based on the manufacturer's instructions for use. PPE should be used only after adequate training has been given to the user, and adequate levels of supervision should be provided to ensure that the training and instructions are being followed.

60 The self-employed user should ensure that he has been adequately trained to use PPE competently, to avoid creating risks to himself and others.

61 Most PPE should be returned after use to the storage place provided under regulation 8. However, there may be instances where the employee may take PPE away from the workplace, for example, some types of protective footwear or overalls. Equipment that is used or worn intermittently, welding visors for example, need only be returned at the end of the working period, shift or assignment.

Regulation 11

Reporting loss or defect

62 Employers should make arrangements to ensure that their employees can report to them (or their representative) the loss of or defects in PPE. These arrangements should also ensure that defective PPE is repaired or replaced before the employee concerned re-starts work.

63 Employees must take reasonable care of PPE provided and report to their employer any loss or obvious defect as soon as possible. If employees have any concerns about the serviceability of the PPE, they should immediately consult their employer or the employer's representative.

Regulation 13

Extension outside Great Britain

64 The Regulations apply to certain work activities carried out in British territorial waters and in designated areas of the UK Continental Shelf, except where disapplied by regulation 3(1). These activities are listed in the Health and Safety at Work etc Act (Application outside Great Britain) Order 1995 (which replaces the 1989 Order).

65 This applies to offshore installations, wells, pipelines, and activities carried out by vessels in connection with an offshore installation or a well such as construction, repair, dismantling, loading, unloading and diving. Also covered are other construction and similar activities carried out in territorial waters.

Regulation 14

Modifications, repeal and revocations

66 The Regulations specified in Schedule 2 have been amended to ensure that they are consistent with the requirements of the PPE at Work Regulations, particularly with regard to the assessment and provision of suitable PPE, and accommodation for PPE.

PART II: SELECTION, USE AND MAINTENANCE OF PERSONAL PROTECTIVE EQUIPMENT

67 This part aims to help employers to comply with their duties to select suitable PPE and maintain it. It contains information about the main types of PPE which are widely used in industry, but does not cover more specialised and less frequently used items (for example, safety harnesses). More detailed information about particular items of PPE can be obtained from suppliers. It is also wise to involve those who will wear the PPE in its selection. Where possible, more than one model satisfying the appropriate safety performance and other criteria of suitability should be made available.

HEAD PROTECTION

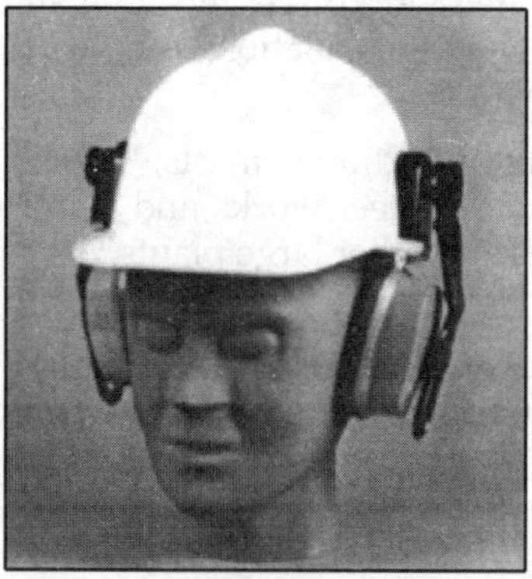

Figure 1 Safety helmet with hearing protection

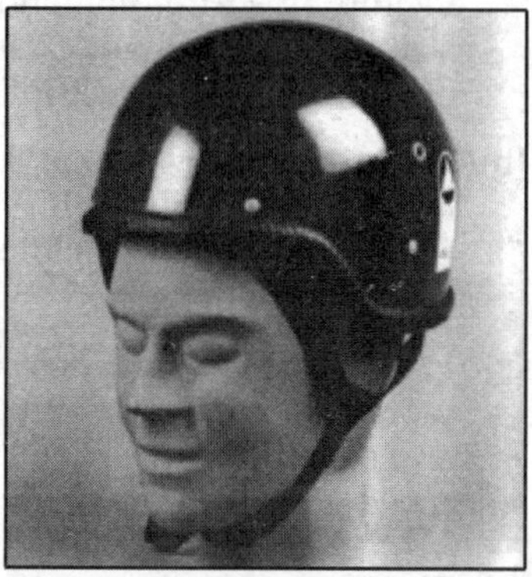

Figure 2 Climbing helmet

Types of protection

68 There are four widely used types of head protection:

(a) crash helmets, cycling helmets, riding helmets and climbing helmets which are intended to protect the user in falls;

(b) industrial safety helmets which can protect against falling objects or impact with fixed objects;

(c) industrial scalp protectors (bump caps) which can protect against striking fixed obstacles, scalping or entanglement; and

(d) caps, hairnets etc which can protect against scalping/entanglement.

69 The following guidance deals only with industrial safety helmets, scalp protectors and climbing helmets (ie it excludes caps and hairnets).

Processes and activities

70 The following are examples of activities and processes involving risks of falling objects or impacts, which may require the provision of head protection; it is not an exhaustive list. Some of these activities will also be subject to the Construction (Head Protection) Regulations 1989:

(a) Building work, particularly work on, underneath or in the vicinity of scaffolding and elevated workplaces, erection and stripping of formwork, assembly and installation work, work on scaffolding and demolition work.

(b) Construction work on bridges, buildings, masts, towers, hydraulic structures, blast furnaces, steel works and rolling mills, large containers, pipelines and other large plants, boiler plants and power stations.

(c) Work in pits, trenches, shafts and tunnels. Underground workings, quarries, opencast mining, minerals preparation and stocking.

(d) Work with bolt-driving tools.

(e) Blasting work.

(f) Work near hoists, lifting plant, cranes and conveyors.

(g) Work with blast furnaces, direct reduction plants, steelworks, rolling mills, metalworks, forging, drop forging and casting.

(h) Work with industrial furnaces, containers, machinery, silos, storage bunkers and pipelines.

(i) Building or repairing ships and offshore platforms.

(j) Railway shunting work, and other transport activities involving a risk of falling material.

(k) Slaughterhouses.

(1) Tree-felling and tree surgery.

(m) Work from suspended access systems, bosun's chairs etc.

71 Some relevant British and European standards:[1]

BS 5240 Part 1: 1987 *Industrial safety helmets—specification for construction and performance* (To be replaced by BS EN 397).

BS 4033: 1966 (1978) *Specification for industrial scalp protectors (light duty)* (To be replaced by BS EN 812).

1 Many British Standards will be replaced by harmonised European Standards, for example BS 3864: 1989 will be replaced by the European Standard EN 443. When the European Standard is introduced it will be prefixed by 'BS' so EN 443 will become BS EN 443 in the United Kingdom. Those with the prefix 'pr' are provisional at the time of going to print. See pages 571–575 for a more comprehensive list of appropriate standards.

BS 3864: 1989 *Specification for protective helmets for firefighters* (To be replaced by BS EN 443).

The selection of suitable head protection

72 To fit, head protection should:

(a) be of an appropriate shell size for the wearer; and

(b) have an easily adjustable headband, nape and chin strap.

The range of size adjustment should be large enough to accommodate thermal liners used in cold weather.

73 Head protection should be as comfortable as possible. Comfort is improved by the following:

(a) a flexible headband of adequate width and contoured both vertically and horizontally to fit the forehead;

(b) an absorbent, easily cleanable or replaceable sweat-band;

(c) textile cradle straps;

(d) chin straps (when fitted) which:

(i) do not cross the ears,

(ii) are compatible with any other PPE needed,

(iii) are fitted with smooth, quick-release buckles which do not dig into the skin,

(iv) are made from non-irritant materials,

(v) can be stowed on the helmet when not in use.

Compatibility with the work to be done

74 Whenever possible, the head protection should not hinder the work being done. For example, an industrial safety helmet with little or no peak is useful for a surveyor taking measurements using a theodolite or to allow unrestricted upward vision for a scaffold erector. If a job involves work in windy conditions, especially at heights, or repeated bending or constantly looking upwards, a secure retention system is required. Flexible headbands and Y-shaped chin straps can help to secure the helmet. Head protection worn in the food industry may need to be easily cleaned or compatible with other hygiene requirements.

75 If other PPE such as ear defenders or eye protectors are required, the design must allow them to be worn safely and in comfort. Check manufacturers instructions regarding the compatibility of head protection with other types of PPE.

Maintenance

76 Head protection must be maintained in good condition. It should:

(a) be stored, when not in use, in a safe place, for example, on a peg or in a cupboard. It should not be stored in direct sunlight or in excessively hot, humid conditions;

(b) be visually inspected regularly for signs of damage or deterioration;

(c) have defective harness components replaced (if the design or make allows this). Harnesses from one design or make of helmet cannot normally be interchanged with those from another;

(d) have the sweat-band regularly cleaned or replaced.

77 Before head protection is reissued to another person, it should be inspected to ensure it is serviceable and thoroughly cleaned in accordance with the manufacturer's instructions, eg using soap and water. The sweat-band should always be cleaned or replaced.

Damage to shell

78 Damage to the shell of a helmet can occur when:

(a) objects fall onto it;

(b) it strikes against a fixed object;

(c) it is dropped or thrown.

Deterioration in shock absorption or penetration resistance

79 Deterioration in shock absorption or penetration resistance of the shell can occur from:

(a) exposure to certain chemical agents;

(b) exposure to heat or sunlight;

(c) ageing due to heat, humidity, sunlight and rain.

80 Chemical agents which should be avoided include paint, adhesives or chemical cleaning agents. Where names or other markings need to be applied using adhesives, advice on how to do this safely should be sought from the helmet manufacturer.

81 Exposure to heat or sunlight can make the shell go brittle. Head protection should never be stored therefore near a window, eg the rear window of a motor vehicle, because excessive heat may build up.

Replacement

82 The head protection should normally be replaced at intervals recommended by the manufacturer. It will also need replacing when the harness is damaged and cannot be replaced, or when the shell is damaged or it is suspected that its shock absorption or penetration resistance has deteriorated—for example when:

(a) the shell has received a severe impact;

(b) deep scratches occur;

(c) the shell has any cracks visible to the naked eye.

EYE PROTECTION

Figure 3 Safety goggles

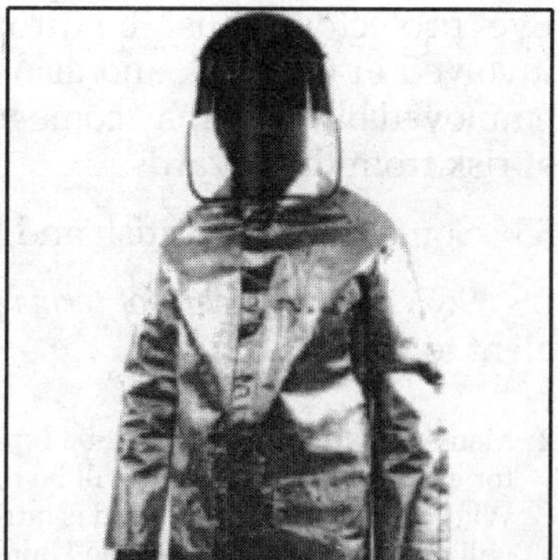

Figure 4 Face shield

Types of eye protection

83 Eye protection serves to guard against the hazards of impact, splashes from chemicals or molten metal, liquid droplets (chemical mists and sprays), dust, gases, welding arcs, non-ionising radiation and the light from lasers. Eye protectors include safety spectacles, eyeshields, goggles, welding filters, face-shields and hoods. Safety spectacles can be fitted with prescription lenses if required. Some types of eye protection can be worn over ordinary spectacles if necessary.

Processes and activities

84 The following are examples of activities and processes involving a risk to the face and eyes for which eye protectors should be used. It is not an exhaustive list.

(a) handling or coming into contact with acids, alkalis and corrosive or irritant substances;

(b) working with power-driven tools where chippings are likely to fly or abrasive materials be propelled;

(c) working with molten metal or other molten substances;

(d) during any welding operations where intense light or other optical radiation is emitted at levels liable to cause risk of injury;

(e) working on any process using instruments that produce light amplification or radiation; and

(f) using any gas or vapour under pressure.

Eye protectors must be provided both for persons directly involved in the work and also for others not directly involved or employed but who may come into contact with the process and be at risk from the hazards.

85 Some relevant British and European Standards:[2]

BS 6967: 1988 *Glossary of terms for personal eye protection* (Is equivalent to BS EN 165).

2 Many British Standards will be replaced by harmonised European Standards, for example BS 3864: 1989 will be replaced by the European Standard EN 443. When the European Standard is introduced it will be prefixed by 'BS' so EN 443 will become BS EN 443 in the United Kingdom. Those with the prefix 'pr' are provisional at the time of going to print. See pages 571–575 for a more comprehensive list of appropriate standards.

BS 2092: 1987 *Specification for eye protectors for industrial and non-industrial uses* (To be replaced by BS EN 166, 167 and 168).

BS 7028: 1988 *Guide for selection, use and maintenance of eye-protection for industrial and other uses.*

BS 1542: 1982 *Specification for equipment for eye, face and neck protection against non-ionising radiation arising during welding and similar operations.*

Selecting suitable eye protection

86 The selection of eye protection depends primarily on the hazard. However, comfort, style and durability should also be considered.

(a) *Safety spectacles* are similar in appearance to prescription spectacles but may incorporate optional sideshields to give lateral protection to the eyes. To protect against impact, the lenses are made from tough optical quality plastic such as polycarbonate. Safety spectacles are generally light in weight and are available in several styles with either plastic or metal frames. Most manufacturers offer a range of prescription safety spectacles which are individually matched to the wearer.

(b) *Eyeshields* are like safety spectacles but are heavier and designed with a frameless one-piece moulded lens. Vision correction is not possible as the lenses cannot be interchanged. Some eyeshields may be worn over prescription spectacles.

(c) *Safety goggles* are heavier and less convenient to use than spectacles or eyeshields. They are made with a flexible plastic frame and one-piece lens and have an elastic headband. They afford the eyes total protection from all angles as the whole periphery of the goggle is in contact with the face. Goggles may have toughened glass lenses or have wide vision plastic lenses. The lenses are usually replaceable. Safety goggles are more prone to misting than spectacles. Double glazed goggles or those treated with an anti-mist coating may be more effective where misting is a problem. Where strenuous work is done in hot conditions, direct ventilation goggles may be more suitable. However, these are unsuitable for protection against chemicals, gases and dust. Indirect ventilation goggles are not perforated, but are fitted with baffled ventilators to prevent liquids and dust from entering. Indirect ventilation goggles will not protect against gas or vapour.

(d) *Faceshields* are heavier and bulkier than other types of eye protector but are comfortable if fitted with an adjustable head harness. Faceshields protect the face but do not fully enclose the eyes and therefore do not protect against dusts, mist or gases. Visors on browguards or helmets are replaceable. They may be worn over standard prescription spectacles and are generally not prone to misting. Face shields with reflective metal screens permit good visibility while effectively deflecting heat and are useful in blast and open-hearth furnaces and other work involving radiant heat.

Maintenance

87 The lenses of eye protectors must be kept clean as dirty lenses restrict vision, which can cause eye fatigue and lead to accidents. There are two methods for cleaning eye protectors. Glass, polycarbonate and other plastic lenses can be cleaned by thoroughly wetting both sides of the lenses and drying them with a wet strength absorbent paper. Anti-static and anti-fog lens cleaning fluids may be used, daily if necessary, if static or misting is a problem. Alternatively lenses can be dry cleaned by removing grit with a brush and using a silicone treated non-woven cloth. However plastic or polycarbonate lenses should not be dry cleaned as the cloth used in this method can scratch them.

88 Eye protectors should be issued on a personal basis and used only by the person they are issued to. If eye protectors are re-issued they should be thoroughly cleaned and disinfected. Eye protectors should be protected by being placed in suitable cases when not in use. Eye protector headbands should be replaced when worn out or damaged.

89 Lenses that are scratched or pitted must be replaced as they may impair vision and their resistance to impact may be impaired. Transparent face shields must be replaced when warped, scratched or have become brittle with age.

FOOT PROTECTION

Figure 5 Foundry boots

Figure 6 Insulated safety boots

Types of safety footwear

90 The following are examples of types of safety footwear:

(a) *The safety boot or shoe* is the most common type of safety footwear. These normally have steel toe-caps. They may also have other safety features including slip resistant soles, steel midsoles and insulation against extremes of heat and cold.

(b) *Clogs* may also be used as safety footwear. They are traditionally made from beech wood which provides a good insulation against heat and absorbs shock. Clogs may be fitted with steel toe-caps and thin rubber soles for quieter tread and protection against slippage or chemicals.

(c) *Foundry boots* have steel toe-caps, are heat resistant and designed to keep out molten metal. They are without external features such as laces to avoid trapping molten metal blobs and should have velcro fasteners or elasticated sides for quick release.

(d) *Wellington boots* protect against water and wet conditions and can be useful in jobs where the footwear needs to be washed and disinfected for hygienic reasons, such as in the food industry. They are usually made from rubber but are available in polyurethane and PVC which are both warmer and have greater chemical resistance. Wellington boots can be obtained with corrosion resistant steel toe-caps, rot-proof insoles, steel midsoles, ankle bone padding and cotton linings. They range from ankle boots to chest-high waders.

(e) *Anti-static footwear* prevents the build up of static electricity on the wearer. It reduces the danger of igniting a flammable atmosphere and gives some protection against electric shock.

(f) *Conductive footwear* also prevents the build up of static electricity. It is particularly suitable for handling sensitive components or substances (eg explosive detonators). It gives no protection against electric shock.

Processes and activities

91 The following are examples of activities and processes involving risks to the feet. It is not an exhaustive list.

(a) *Construction:* Work on building and demolition sites will usually require safety footwear to protect the feet against a variety of hazards, particularly objects falling on them, or sharp objects (eg nails) on the ground piercing the shoe and injuring the sole of the foot.

(b) *Mechanical and manual handling:* There may be a risk of objects falling on or crushing the front of the foot. There may be a risk of a fall through slipping which could result in damage to the heel on impact. There is also a danger of treading on pointed or sharp objects which can penetrate the shoe and injure the sole of the foot

(c) *Electrical:* People who work where there are flammable atmospheres should wear anti-static footwear to help prevent ignitions due to static electricity. Such footwear is similar to conventional footwear in that the soles are sufficiently insulated to give some measure of protection against electric shock.

(d) *Thermal:* Working in cold conditions requires footwear with thermal insulation. Work in hot conditions requires footwear with heat-resistant and insulating soles.

(e) *Chemical:* Footwear provided when working with hazardous chemicals should be both impermeable and resistant to attack by chemicals.

(f) *Forestry:* Forestry chain-saw boots are water-resistant and are designed to offer protection against chain-saw contact.

(g) *Molten substances:* Foundry boots that are easily removed should be provided where there is a danger of splashing by molten substances.

92 Some relevant British and European Standards:[3]

BS 1870: Part 1: 1988 *Specification for safety footwear other than all rubber and all plastic moulded compounds.*

BS 1870: Part 2: 1986 *Specification for lined rubber safety boots.*

BS 1870: Part 3: 1981 *Specification for PVC moulded safety footwear.*

BS 4676: 1983 *Specification for gaiters and footwear for protection against burns and impact risks in foundries.*

BS 4972: 1973 *Specification for women's protective footwear.*

BS 5145: 1989 *Specification for lined industrial vulcanised rubber boots.*

BS 5462: *Footwear with midsole protection:*

BS 5462: Part 1: 1984 *Specification for lined vulcanised rubber footwear with penetration resistant midsoles;*

BS 5462: Part 2: 1984 *Specification for lined or unlined polyvinyl chloride (PVC) footwear with penetration resistant midsoles.*

BS 6159: *Polyvinyl chloride boots:*

BS 6159: Part 1: 1987 *Specification for general and industrial lined or unlined boots.*

Selecting suitable foot protection

93 The selection of foot protection depends primarily on the hazard. However, comfort, style and durability should also be considered. The choice should be made on the basis of suitability for protection, compatibility with the work and the requirements of the user.

94 Generally, safety footwear should be flexible, wet resistant and absorb perspiration. Inflexible or unnecessarily bulky footwear will

3 Many British Standards will be replaced by harmonised European Standards, for example BS 3864: 1989 will be replaced by the European Standard EN 443. When the European Standard is introduced it will be prefixed by 'BS' so EN 443 will become BS EN 443 in the United Kingdom. Those with the prefix 'pr' are provisional at the time of going to print. See pages 571–575 for a more comprehensive list of appropriate standards.

result in tired feet and legs. Boots and not shoes are required where ankles need protection. You should consider the ability of the footwear to resist corrosion, abrasion and industrial wear and tear. Always follow the manufacturer's instructions and markings for appropriate use and level of protection.

(a) *Soles:* Work shoes and boots should have treaded soles for slip-resistance. Soles can be heat and oil resistant, slip resistant, shock resistant, anti-static or conductive. Footwear intended to protect against oils, solvents or liquids need soles that are moulded or bonded to the upper. Soles that are stitched or glued may separate and expose the foot to hazard. Footwear with steel midsoles should be used where there is a risk that the sole could be pierced by nails and similar objects.

(b) *Steel toe-caps:* They should be capable of resisting a heavy sharp object falling from a considerable height. Footwear complying with BS 4676 will offer this resistance.

(c) *Heat resistance:* Leather or other heat resistant materials can be used in safety footwear to offer protection against heat, sparks and molten metal.

(d) *Waterproofing:* People working in wet places should wear safety footwear impervious to water. Rubber and PVC are suitable inexpensive water-proofing materials for footwear but they are not permeable. There are breathable materials which are water resistant, but which also allow air to get through and perspiration to get out, and may therefore be more comfortable and more hygienic. However, footwear manufactured from this type of material tends to be more expensive.

95 Electrical hazards: The following provide protection against electrical hazards.

(a) *Anti-static footwear:* Anti-static footwear offers suitable protection against the hazard of static electricity and will give some protection against mains electric shock. Anti-static footwear must be worn where there is both a hazard from static build up and the possibility of contact with mains electricity. The soles must have a resistance low enough to allow static electricity to leak slowly away while maintaining enough resistance to protect against a 240 volt mains electricity shock.

(b) *Conductive footwear* offers greater protection against static electricity and is used where the wearer handles very sensitive components or materials. *It must not be worn where there is a danger of electric shock.* The soles of conductive footwear must have an electrical resistance low enough to enable static electricity to be taken quickly away from the body to the earth.

96 Leg protection: The following are examples of leg protection.

(a) People working around molten metal need protection for their lower legs. For example this can be achieved by the use of foundry boots and gaiters, or a high foundry boot worn inside molten metal protective trousers.

(b) Hard fibre or metal guards should be used to protect shins against impact. The top of the foot up to the ankle can be protected by added-on metatarsal guards.

Maintenance

97 Safety footwear should be maintained in good condition, checked regularly and discarded if worn or deteriorated. Laces should be checked and replaced if necessary. Materials lodged into the tread should be removed. The stitching should be checked for loose, worn or cut seams. Spraying the upper layers of new footwear with a silicone spray or applying a protective wax will give extra protection against wet conditions.

HAND AND ARM PROTECTION

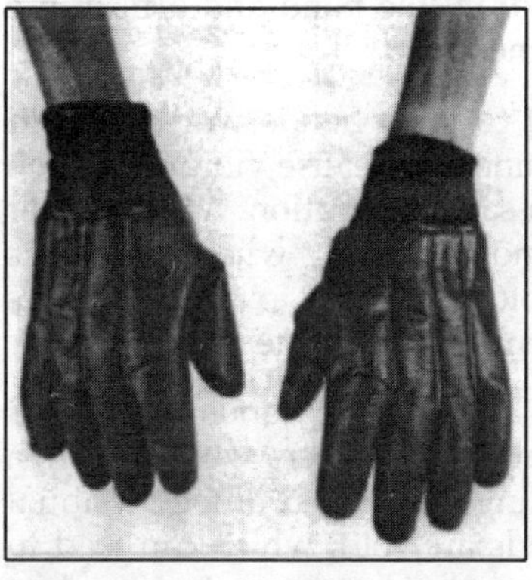

Figure 7 Vinyl-coated insulated gloves

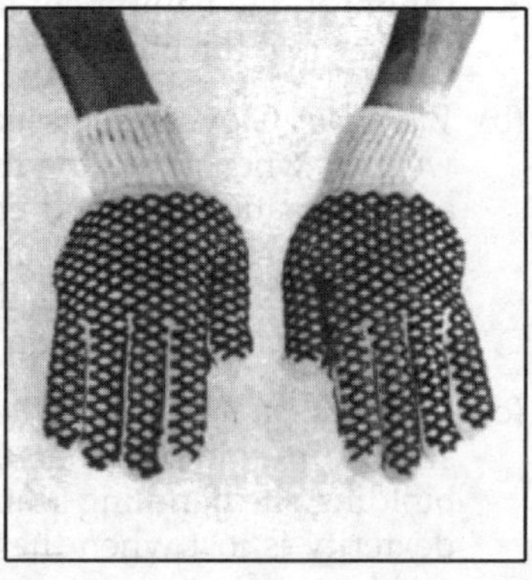

Figure 8 General purpose work gloves

Types of hand protection

98 Gloves of various designs provide protection against a range of industrial hazards, including:

(a) cuts and abrasions;

(b) extremes of temperature, hot and cold;

(c) skin irritation and dermatitis;

(d) contact with toxic or corrosive liquids.

99 The type and degree of protection depends on the glove material and the way in which it is constructed. Barrier creams may sometimes be used as an aid to skin hygiene in situations where gloves cannot be used. Experience shows, however, that barrier creams are less reliable than suitable gloves as a means of chemical protection.

Processes and activities

100 The following processes and activities involve risk of injury to the hands or hazards for which hand protection may be necessary. It is not an exhaustive list.

(a) *Manual handling:* Hands may be pierced by abrasive, sharp or pointed objects or damaged by impact when handling goods. However, gloves should not be worn when working near moving equipment and machinery parts as the glove may get caught in the equipment and draw the hand and arm of the worker into the moving machinery.

(b) *Vibration:* Gloves are essential to keep hands warm in cold weather when operating machines that cause vibrations such as pneumatic drills and chain-saws. Vibration White Finger occurs more frequently and more severely when the hands and fingers are cold as the blood supply to the fingers is reduced by the body in an attempt to conserve heat.

(c) *Construction and outdoor work:* Keeping the hands warm and supple in cold weather is important when working on a building site handling scaffolding, bricks and timber. Manual dexterity is lost when the hands are cold, which can lead to accidents if articles are dropped. Gloves protect against hazards in site clearance such as previous contamination of soil which may contain disease spores that may seriously infect small cuts and abrasions.

(d) *Hot and cold materials:* Gloves will also protect against hazards from handling hot or cold materials and work involving contact with naked flames or welding.

(e) *Electricity:* Danger from electric shock.

(f) *Chemical:* There are many tasks where the hands may come into contact with toxic or corrosive substances. Examples include maintenance of machinery, cleaning up chemical spillages and mixing and dispensing pesticide formulations. If correctly selected and used, gloves provide a barrier between the wearer's skin and the harmful substance, preventing local damage, or in some cases absorption through the skin.

(g) *Radioactivity:* Danger from contamination when handling radio-active materials.

101 Some relevant British and European Standards:[4]

BS EN 374: 1994 (Parts 1–3) *Protective gloves against chemicals and micro-organisms.*

BS EN 388: 1994 *Protective gloves against mechanical risks.*

BS EN 407: 1994 *Protective gloves against thermal risks (heat and/or fire).*

BS EN 420: 1994 *General requirements for gloves.*

BS 697: 1986 *Specification for rubber gloves for electrical purposes.*

Selecting suitable hand protection

102 Gloves or other hand protection should be capable of giving protection from hazards, be comfortable and fit the wearer. The choice should be made on the basis of suitability for protection, compatibility with the work and the requirements of the user. You should consider the ability of protective gloves to resist abrasion

4 Many British Standards will be replaced by harmonised European Standards, for example BS 3864: 1989 will be replaced by the European Standard EN 443. When the European Standard is introduced it will be prefixed by 'BS' so EN 443 will become BS EN 443 in the United Kingdom. Those with the prefix 'pr' are provisional at the time of going to print. See pages 571–575 for a more comprehensive list of appropriate standards.

and other industrial wear and tear. Always follow the manufacturer's instructions and markings for appropriate use and level of protection. When selecting gloves for chemical protection, reference should be made to chemical permeation and resistance data provided by manufacturers.

(a) *Penetration and abrasion:* Gloves made from chain-mail or leather protect against penetration and abrasion. Gloves knitted from special man-made fibres such as Kevlar will provide protection against cuts, and gloves manufactured from eg Kevlar needlefelt give good puncture resistance.

(b) *Thermal protection:* Depending upon their weight and construction, terrycloth gloves will provide protection against heat and cold. Gloves made from neoprene are good for handling oils in low temperatures. Gloves manufactured from other materials such as Kevlar, glass fibre and leather can be used to provide protection at higher temperatures.

(c) *Fire resistance:* Chromed leather gloves are fire retardant.

(d) *Chemicals protection:* Chemical protective gloves are available in a range of materials including natural rubber, neoprene, nitrile, butyl, PVA, PVC and viton. The degree of protection against chemical permeation depends on the glove material, its thickness and method of construction. As a general rule, gloves for use in handling toxic liquids should be chosen on the basis of breakthrough time. This means that the duration of use should not exceed the breakthrough time quoted by the manufacturer of the glove for the chemical substance concerned. Laboratory testing may be required in order to establish adequacy in some applications. When handling dry powders, any chemically resistant glove may be used. The durability of the gloves in the workplace should also be considered. Some glove materials may be adversely affected by abrasion.

(e) *General use gloves:* Rubber, plastic or knit fabric gloves are flexible, resist cuts and abrasions, repel liquids and offer a good grip. Rubber gloves allow a sensitive touch and give a firm grip in water or wet conditions. Leather, cotton knit or other general purpose gloves are suitable for most other jobs. General use gloves should only be used to protect against minimal risks to health and safety (eg for gardening and washing up and similar low risk tasks).

Maintenance

103 Care should be taken in the donning, use, removal and storage of protective gloves. They should be maintained in good condition, checked regularly and discarded if worn or deteriorated. Gloves should be free of holes or cuts and foreign materials and their shape should not be distorted. They should fit the wearer properly leaving no gap between the glove and the wearer's sleeve.

104 Gloves should always be cleaned according to the manufacturer's instructions as they may have particular finishes which may make the following general guidance inappropriate. For example, repeated washing may remove fungal and bacterial inhibitors from the lining of the glove which may ultimately lead to skin irritation. And there is also the risk of cross contamination as chemical residues can remain on the gloves even after washing.

105 Contact between the gloves and chemicals should be kept to a minimum as some chemicals can alter the physical characteristics of a glove and impair its protective properties. Gloves contaminated by chemicals should be washed as soon as possible and before their removal from the hands. Grossly contaminated gloves should be discarded. Gloves contaminated on the inside can be dangerous as the chemical contamination will be absorbed by the skin. Wear armlets if there is a danger of chemicals entering the glove at the cuff.

106 When wearing protective gloves do not touch other exposed parts of the body, equipment or furniture as contamination can be transferred to them. Cotton liners can be worn if hands sweat profusely.

Care for the hands when handling chemicals

107 Do not let chemicals come into contact with the skin. Wash hands frequently, dry them carefully and use a hand cream to keep the skin from becoming dry through loss of natural oils. Keep cuts and abrasions covered with waterproof plasters and change the dressing for a porous one after work. Handle and remove gloves carefully to avoid contamination of hands and the insides of the gloves.

PROTECTIVE CLOTHING FOR THE BODY

Figure 9 High visibility coat

Figure 10 Lamex apron

Types of protection

108 Types of clothing used for body protection include:

(a) coveralls, overalls and aprons to protect against chemicals and other hazardous substances;

(b) outfits to protect against cold, heat and bad weather;

(c) clothing to protect against machinery such as chain-saws.

109 Types of clothing worn on the body to protect the person include:

(a) high visibility clothing;

(b) life-jackets and buoyancy aids.

Processes and activities

110 The following are examples of the sorts of processes and activities that require protective clothing for the body. It is not an exhaustive list.

(a) Laboratory work or work with chemicals, dust or other hazardous substances;

(b) construction and outdoor work;

(c) work in cold-stores;

(d) forestry work using chainsaws;

(e) highway and road works;

(f) work on inland and inshore waters;

(g) spraying pesticides;

(h) food processing;

(i) welding;

(j) foundry work and molten metal processes;

(k) fire-fighting.

111 Some relevant British and European Standards:[5]

BS 1547: 1959 *Specification for flameproof industrial clothing (materials and design).*

BS 1771: Part 1: 1989 *Specification for fabrics of wool and wool blends.*

BS 1771: Part 2: 1990 *Specification for fabrics of cellulosic fibres, synthetic fibres and blends.*

BS 2653: 1955 *Specification for protective clothing for welders.*

BS 6249: 1982 *Materials and material assemblies used in clothing for protection against heat and flame.*

BS 6249: Part 1: 1982 *Specification for flammability testing and performance.*

BS EN 366: 1993 *Protective clothing. Protection against heat and fire. Method of test: evaluation of materials and material assemblies when exposed to a source of radiant heat.*

BS EN 367: 1992 *Protective clothing. Protection against heat and flames. Test methods. Determination of heat transmission on exposure to flame.*

BS EN 394: 1994 *Life-jackets and personal buoyancy aids: additional items.*

5 Many British Standards will be replaced by harmonised European Standards, for example BS 3864: 1989 will be replaced by the European Standard EN 443. When the European Standard is introduced it will be prefixed by 'BS' so EN 443 will become BS EN 443 in the United Kingdom. Those with the prefix 'pr' are provisional at the time of going to print. See pages 571–575 for a more comprehensive list of appropriate standards.

BS EN 396: 1994 *Life-jackets and personal buoyancy aids: life-jackets, 150N.*

BS EN 399: 1994 *Life-jackets and personal buoyancy aids: life-jackets, 275N.*

BS EN 471: 1994 *Specification for high-visibility warning clothing.*

Selection

112 Protection from chemicals and hazardous substances:

(a) *Low risk chemicals* can be protected against by wearing chemical-resistant clothing, coveralls and laboratory coats made from uncoated cotton or synthetic material such as nylon or Terylene with a water repellent finish.

(b) *Strong solvents, oils and greases* require heavier protection afforded by coats, overalls and aprons made from neoprene or polyurethane coated nylon, or Terylene or rubber aprons.

(c) *Chemical suits* protect against more potent chemicals. They are totally encapsulating suits which are either vapour-proof or liquid-splash proof and are fed with breathable air. They must be washed in warm water and a mild soap whenever they have come into contact with chemicals. The suit should be hung up to dry before being stored in cases or hung on hangers. Chemical suits have a life expectancy of three to four years and should be inspected every three months even if not in use. This entails an air test and looking at all of the seams.

(d) *Vapour suits* protect against hazardous vapours and are made of butyl, polyvinyl chloride (PVC), viton, a combination of viton and butyl or teflon. They should be air-tested with the manufacturer's test kit, before being stored in a protective case. Manufacturers of vapour proof suits generally provide a testing and repair service consisting of a visual inspection and air test.

(e) *Splash-resistant suits* are also made from the same polymers but may also be made of limited-use bonded olefin fabrics.

(f) *Fibres and dust:* Protection can be obtained by wearing suits made from bonded olefin that forms a dense shield which keeps out fibres and particles.

113 Thermal and weather protection:

(a) *Keeping dry:* Jackets, trousers and leggings made with PVC coated nylon or cotton will offer protection against rain. These materials are also resistant to abrasions, cracking and tearing and will protect against most oils, chemicals and acids. 'Breathable' water-proof fabrics will keep out water while allowing body perspiration to escape. Waxed cotton will also protect against rain.

(b) *Keeping warm:* Minus 25 and Minus 50 suits are available which are guaranteed to protect at these respective sub zero temperatures. More limited protection can be obtained from quilted and insulated coats and vests.

(c) *Keeping cool:*

(i) Aluminium-asbestos clothing made of dust-suppressed materials is heat-resistant. The outside is made of aluminium and the inside lining is cotton. This type of clothing is suitable for hot work, for example in foundries.

(ii) Welding and foundry clothing is flame retardant and is mainly of flame retardant cotton or wool materials. Chrome leather is used for aprons etc.

(iii) Molten metal splash clothing is heat resistant and should resist molten metal splash up to 1600 degrees centigrade.

(iv) Cotton or cotton and polyester coveralls with flame-retardant finishes are available to protect against sparks and flame.

114 *Food processing:* Food quality overalls and coveralls will protect against splashes from oils and fats. Butchers and slaughterhouse workers should wear lamex or chain-mail aprons if there is a risk of injury to the abdomen or chest, for example when using knives or choppers.

115 *Chainsaw protective clothing:* The front of the leg is most vulnerable to chainsaw accidents although the back of the leg is also at risk. Protective legwear incorporates layers of loosely woven long synthetic fibres. On contact with the saw chain, the fibres are drawn out and clog the chain saw sprocket, causing the chain to stop. Legwear is available with all-round protection or with protection only for the front of the legs. The legwear with all round protection offers the greatest protection for users. Jackets and gloves are also available with inserts of chainsaw resistant materials at vulnerable points. See paragraph 91(f) in the section on chainsaw boots.

Maintenance

116 Protective clothing should only be used for the purpose intended. It should be maintained in good condition and checked regularly. It should be repaired in accordance with the manufacturer's instructions, or discarded if damaged.

Personal protection worn on the body

High visibility clothing

117 This is made from materials impregnated with fluorescent pigments, and will include some areas of retroreflective material. It is designed to make the user conspicuous under any light conditions in the day and under illumination by vehicle headlights in the dark. It will need to be worn by workers on roadsides and other places where it is important to be seen to be safe. High visibility clothing to BS EN 471 is classified according to the area of its fluorescent material and the performance of its retroreflective material; in both cases the higher the class of material, the more conspicuous the wearer.

Personal buoyancy equipment

118 Life-jackets or buoyancy aids should be worn where there is a foreseeable risk of drowning when working on or near water.

(a) *A lifejacket* is a personal safety device which should be selected so that, when fully inflated (if inflatable), it provides sufficient buoyancy to turn and support even an unconscious person face upwards within five seconds (ten seconds if automatically inflated). The person's head should be supported with the mouth and nose well clear of the water.

Some people are reluctant to wear life-jackets as they find them bulky and restrictive. However, either an automatically inflatable life-jacket or a type which is inflated by a manual pull-cord should overcome these problems. These are usually compact and allow for a full range of movement. Automatically inflated life-jackets need to be carefully maintained and regularly inspected, as the automatic inflation mechanism can fail to operate if subject to rough handling or incorrect storage.

(b) *Buoyancy aids* are worn to provide extra buoyancy to assist a conscious person in keeping afloat. However, they will not turn over an unconscious person from a face down position.

APPENDIX 1: SPECIMEN RISK SURVEY TABLE FOR THE USE OF PERSONAL PROTECTIVE EQUIPMENT

			Risks																					
			The PPE at Work Regulations 1992 apply except where the Construction (Head Protection) Regulations 1989 apply										The CLW, IRR, CAW, COSHH and NAW Regulations[1] will each apply to the appropriate hazard											
			Mechanical					Thermal																
			Falls from a height	Blows, cuts, impact, crushing	Stabs, cuts, grazes	Vibration	Slipping, falling over	Scalds, heat, fire	Cold	Immersion	Non-ionising radiation	Electrical	Noise	Ionising radiation	Dust fibre	Fume	Vapours	Splashes, spurts	Gases, vapours	Harmful bacteria	Harmful viruses	Fungi	Non-micro bio-logical antigens
P	Head	Cranium																					
A		Ears																					
R		Eyes																					
T		Respiratory tract																					
S		Face																					
of		Whole head																					
the	Upper limbs	Hands																					
		Arms (parts)																					
	Lower limbs	Foot																					
B		Legs (parts)																					
O	Various	Skin																					
D		Trunk/abdomen																					
Y		Whole body																					

(1) The Control of Lead at Work Regulations 1980, The Ionising Radiations Regulations 1985, The Control of Asbestos at Work Regulations 1987, The Control of Substances Hazardous to Health Regulations 1994, The Noise at Work Regulations 1989.

APPENDIX 2

LEGISLATION ON PPE APPLYING IN ADDITION TO THE REGULATIONS

**The Coal and Other Mines (Fire and Rescue) Regulations 1956* SI 1956/1768 HMSO (regulations 23–25)

**The Construction (Working Places) Regulations 1966* SI 1966/94 HMSO ISBN 0 11 100264 8 (regulation 38)

**The Coal Mines (Respirable Dust) Regulations 1975* SI 1975/1433 HMSO ISBN 0 11 051433 5 (regulation 10)

The Dangerous Substances in Harbour Areas Regulations 1987 SI 1987/37 HMSO ISBN 0 11 076037 9 (regulation 17(1)(b) & (2)(b))

**The Diving Operations At Work Regulations 1981* SI 1981/399 HMSO ISBN 0 11 016399 0 (regulations 5(1), 7(1), 9(1), 12(1), (4) & (5), 13)

The Docks Regulations 1988 SI 1988/1655 HMSO ISBN 0 11 087655 5 (regulations 2(1), 19)

The Electricity at Work Regulations 1989 SI 1989/635 HMSO ISBN 0 11 096635 X (regulations 4(4), 14)

The Offshore Installations (Operational Safety, Health and Welfare) Regulations 1976 SI 1976/1019 HMSO ISBN 0 11 061019 9 (regulation 16 and Schedule 4)

The Offshore Installations (Life-saving Appliances) Regulations 1977 SI 1977/486 HMSO ISBN 0 11 070486 X (regulation 7)

Provisions dealing with entry into confined spaces

**The Factories Act 1961 Chapter 34* HMSO ISBN 0 10 850027 6 (Section 30)

**The Docks Regulations 1988* SI 1988/1655 HMSO ISBN 0 11 087655 5 (regulation 18)

**The Ship Building and Ship-repairing Regulations 1960* SI 1960/1932 (regulations 50, 51 and 60)

**The Breathing Apparatus etc (Report on Examination) Order 1961* SI 1961/1345 HMSO ISBN 0 11 100320 2 (The whole order)

* Regulations currently under review.

APPENDIX 3

BRITISH AND EUROPEAN STANDARDS

Head protection

BS 3864: 1989 *Specification for protective helmets for firefighters* (To be replaced by BS EN 443)

BS 4033: 1966 (1978) *Specification for industrial scalp protectors (light duty)* (To be replaced by BS EN 812)

BS 4423: 1969 *Specification for climbers' helmets*

BS 4472: 1988 *Specification for protective skull caps for jockeys*

BS 5240: Part 1: 1987 *Industrial safety helmets—specification for construction and performance* (To be replaced by BS EN 397)

BS 6473: 1984 *Specification for protective hats for horse and pony riders*

Eye protection

BS 1542: 1982 *Specification for equipment for eye, face and neck protection against non-ionising radiation arising during welding and similar operations*

BS 2092: 1987 *Specification for eye protection for industrial and non-industrial uses* (To be replaced by BS EN 166, 167 and 168)

BS 6967: 1988 *Glossary of terms for personal eye protection* (Is equivalent to BS EN 165)

BS 7028: 1988 *Guide for selection, use and maintenance of eye protection for industrial and other uses*

prEN 165 *Personal eye protection: Vocabulary*

prEN 166 *Personal eye protection: Specifications*

prEN 167 *Personal eye protection: Optical test methods*

prEN 168 *Personal eye protection: Non-optical test methods*

BS EN 169: 1992 *Specification for filters for personal eye protection equipment used in welding and similar operations*

BS EN 170: 1992 *Specification for ultraviolet filters used in personal eye protectors for industrial use*

BS EN 171: 1992 *Personal eye protection: Infrared filters: Transmittance requirements and recommended use*

BS EN 172: 1995 *Specification for sunglare filters used in personal eye protectors for industrial use*

BS EN 207: 1993 *Specification for filters and equipment used for personal eye protection against laser radiation*

BS EN 208: 1994 *Specification for personal eye protection used for adjustment work on lasers and laser systems*

BS EN 379: 1994 *Specification for filters with switchable or dual luminous transmittance for personal eye-protectors used in welding and similar operations*

Footwear

BS 953: 1979 *Methods of test for safety and protective footwear*

BS 1870: Part 1: 1988 *Specification for safety footwear other than all rubber and all plastic moulded compounds*

BS 1870: Part 2: 1976 (1986) *Specification for lined rubber safety boots*

BS 1870: Part 3: 1981 *Specification for polyvinyl chloride moulded safety footwear*

BS 2723: 1956 (1988) *Specification for fireman's leather boots*

BS 4676: 1983 *Specification for gaiters and footwear for protection against burns and impact risks in foundries*

BS 4972: 1973 *Specification for women's protective footwear*

BS 5145: 1989 *Specification for lined industrial vulcanised rubber boots*

BS 5462: *Footwear with midsole protection:*

BS 5462: Part 1: 1984 *Specification for lined vulcanised rubber footwear with penetration resistant midsoles*

BS 5462: Part 2: 1984 *Specification for lined or unlined polyvinyl chloride (PVC) footwear with penetration resistant midsoles*

BS 6159: *Polyvinyl chloride boots*:

BS 6159: Part 1: 1987 *Specification for general and industrial lined or unlined boots*

BS 7193: 1989 *Specification for lined lightweight rubber overshoes and overboots*

The following will probably replace BS 1870 and BS 953:

prEN 191 *Additional specifications for safety footwear for professional use*

BS EN 344: 1993 *Requirements and test methods for safety protective and occupational footwear for professional use*

BS EN 345: 1993 *Specification for safety footwear for professional use*

BS EN 346: 1993 *Specification for protective footwear for professional use*

BS EN 347 *Specification for occupational footwear for professional use*

BS EN 381: 1993 *Protective clothing for users of hand held chain saws:*

prEN 381: Part 5 *Requirements for leg protection*

prEN 381: Part 6 *Requirements for boots*

prEN 381: Part 9 *Requirements for chain saw protective gaiters*

Gloves

BS 697: 1986 *Specification for rubber gloves for electrical purposes*

BS EN 374 (Parts 1 to 3) *Protective gloves against chemicals and micro-organisms*

prEN 381 (Part 7) *Protective clothing for users of hand held chain saws: Requirements for chain saw protective gloves*

BS EN 388: 1994 *Protective gloves against mechanical risks*

BS EN 407: 1994 *Protective gloves against thermal risks (heat and/or fire)*

BS EN 420: 1994 *General requirements for gloves*

BS EN 421: 1994 *Protective gloves against ionising radiation to include irradiation and contamination*

BS EN 511: 1994 *Protective gloves against cold*

prEN 659 *Fire-fighters' gloves: Protection against heat and flame*

Protective clothing

BS 1547: 1959 *Specification for flameproof industrial clothing (materials and design)*

BS 1771: Part 1: 1989 *Specification for fabrics of wool and wool blends*

BS 1771: Part 2: 1990 *Specification for fabrics of cellulosic fibres, synthetic fibres and blends*

BS 2653: 1955 *Specification for protective clothing for welders*

BS 5426: 1993 *Specification for workwear and career wear*

BS 6249: 1982 *Materials and material assemblies used in clothing for protection against heat and flame*

BS 6249: Part 1: 1982 *Specification for flammability testing and performance*

BS EN 340: 1993 *General requirements for protective clothing*

prEN 342 *Protective clothing against cold weather*

prEN 343 *Protective clothing against foul weather*

BS EN 348: 1992 *Protective clothing: Determination of behaviour of materials on impact of small splashes of molten metal*

BS EN 366: 1993 *Protective clothing. Protection against heat and fire. Method of test: evaluation of materials and material assemblies when exposed to a source of radiant heat*

BS EN 367: 1992 *Protective clothing. Protection against heat and flames. Test methods. Determination of heat transmission on exposure to flame*

BS EN 368: 1993 *Protective clothing: Protection against liquid chemicals: Resistance of materials to penetration by liquids*

BS EN 373: 1993 *Protective clothing: Assessment of resistance of materials to molten metal splash*

BS EN 381: 1993 *Protective clothing for users of hand held chainsaws (Parts 1 to 9 will be produced progressively)*

BS EN 393: 1994 *Life-jackets and personal buoyancy aids: buoyancy aids, 50N*

BS EN 394: 1994 *Life-jackets and personal buoyancy aids: additional items*

prEN 395: 1994 *Life-jackets and personal buoyancy aids: life-jackets, 100 N*

BS EN 396: 1994 *Life-jackets and personal buoyancy aids: life-jackets, 150 N*

BS EN 399: 1994 *Life-jackets and personal buoyancy aids: life-jackets, 275 N*

BS EN 412: 1993 *Protective aprons for use with hand knives*

BS EN 463: 1995 *Protective clothing. Protection against liquid chemicals. Test method. Determination of resistance to penetration by a jet of liquid (Jet Test)*

BS EN 464: 1994 *Protective clothing for use against liquid and gaseous chemical including aerosols and solid particles. Test method. Determination of leak-tightness of gas-tight suits (Internal Pressure Test)*

prEN 465 *Protective clothing: protection against liquid chemicals: performance requirements: type 4 equipment: protective suits with spray-tight connections between different parts of the protective suit*

prEN 466 *Chemical protection clothing: protection against liquid chemicals (including liquid aerosols): performance requirements: type 3 equipment: chemical protective clothing with liquid-tight connections between different parts of the clothing*

prEN 467 *Protective clothing: protection against liquid chemicals: performance requirements: type 5 equipment garments providing chemical protection to parts of the body*

BS EN 468: 1995 *Protective clothing for use against liquid chemicals. Test method. Determination of resistance to penetration by spray (Spray Test)*

prEN 469 *Protective clothing for fire-fighters*

prEN 470 *Protective clothing for use in welding and similar activities*

BS EN 471: 1994 *Specification for high-visibility warning clothing*

BS EN 510: 1993 *Protective clothing against the risk of being caught up in moving parts*

BS EN 531: 1995 *Protective clothing for industrial workers exposed to heat (excluding fire-fighters' and welders' clothing)*

BS EN 532: 1995 *Protective clothing. Protection against heat and flame: Test method for limited flame spread*

prEN 533 *Clothing for protection against heat and flame: Performance specification for limited flame spread of materials*

BS EN 702: 1995 *Protective clothing: Protection against heat and flame—Test method: Determination of the contact heat transmission through protective clothing or its materials*

APPENDIX 4

FURTHER READING

This Appendix is not reproduced in this book.

Appendix 5

Manual Handling Operations Regulations 1992

THE REGULATIONS (SI 1992/2793)

Made *5 November 1992*

The Secretary of State, in exercise of the powers conferred on her by sections 15(1), (2), (3)(a), (5)(a) and (9) and 80(1), (2)(a) and (4) of, and paragraphs 1(1)(a) and (c) and 8 of Schedule 3 to, the Health and Safety at Work etc Act 1974 ('the 1974 Act') and of all other powers enabling her in that behalf and—

(a) for the purpose of giving effect without modifications to proposals submitted to her by the Health and Safety Commission under section 11(2)(d) of the 1974 Act after the carrying out by the said Commission of consultations in accordance with section 50(3) of that Act; and

(b) it appearing to her that the repeal of section 18(1)(f) of the Children and Young Persons Act 1933 and section 28(1)(f) of the Children and Young Persons (Scotland) Act 1937 except insofar as those provisions apply to such employment as is permitted under section 1(2) of the Employment of Women, Young Persons, and Children Act 1920 is expedient in consequence of the Regulations referred to below after the carrying out by her of consultations in accordance with section 80(4) of the 1974 Act,

hereby makes the following Regulations:

Regulation 1

Citation and commencement

These Regulations may be cited as the Manual Handling Operations Regulations 1992 and shall come into force on 1 January 1993.

Regulation 2 *(See Guidance Notes, pp 590–591)*

Interpretation

(1) In these Regulations, unless the context otherwise requires—

— 'injury' does not include injury caused by any toxic or corrosive substance which—

(a) has leaked or spilled from a load;

(b) is present on the surface of a load but has not leaked or spilled from it; or

(c) is a constituent part of a load;

and 'injured' shall be construed accordingly;

— 'load' includes any person and any animal;

— 'manual handling operations' means any transporting or supporting of a load (including the lifting, putting down, pushing, pulling, carrying or moving thereof) by hand or by bodily force.

(2) Any duty imposed by these Regulations on an employer in respect of his employees shall also be imposed on a self-employed person in respect of himself.

Regulation 3 *(See Guidance Notes, p 591)*

Disapplication of Regulations

These Regulations shall not apply to or in relation to the master or crew of a sea-going ship or to the employer of such persons in respect of the normal ship-board activities of a ship's crew under the direction of the master.

Regulation 4 *(See Guidance Notes, pp 591–630)*

Duties of employers

(1) Each employer shall—

(a) so far as is reasonably practicable, avoid the need for his employees to undertake any manual handling operations at work which involve a risk of their being injured; or

(b) where it is not reasonably practicable to avoid the need for his employees to undertake any manual handling operations at work which involve a risk of their being injured—

(i) make a suitable and sufficient assessment of all such manual handling operations to be undertaken by them, having regard to the factors which are specified in column 1 of Schedule 1 to these Regulations and considering the questions which are specified in the corresponding entry in column 2 of that Schedule,

(ii) take appropriate steps to reduce the risk of injury to those employees arising out of their undertaking any such manual handling operations to the lowest level reasonably practicable, and

(iii) take appropriate steps to provide any of those employees who are undertaking any such manual handling operations with general indications and, where it is reasonably practicable to do so, precise information on—

(aa) the weight of each load, and

(bb) the heaviest side of any load whose centre of gravity is not positioned centrally.

(2) Any assessment such as is referred to in paragraph (1)(b)(i) of this regulation shall be reviewed by the employer who made it if—

(a) there is reason to suspect that it is no longer valid; or

(b) there has been a significant change in the manual handling operations to which it relates;

and where as a result of any such review changes to an assessment are required, the relevant employer shall make them.

Regulation 5 *(See Guidance Notes, p 631)*

Duty of employees

Each employee while at work shall make full and proper use of any system of work provided for his use by his employer in compliance with regulation 4(1)(b)(ii) of these Regulations.

Regulation 6

Exemption certificates

(1) The Secretary of State for Defence may, in the interests of national security, by a certificate in writing exempt—

(a) any of the home forces, any visiting force or any headquarters from any requirement imposed by regulation 4 of these Regulations; or

(b) any member of the home forces, any member of a visiting force or any member of a headquarters from the requirement imposed by regulation 5 of these Regulations;

and any exemption such as is specified in sub-paragraph (a) or (b) of this paragraph may be granted subject to conditions and to a limit of time and may be revoked by the said Secretary of State by a further certificate in writing at any time.

(2) In this regulation—

(a) 'the home forces' has the same meaning as in section 12(1) of the Visiting Forces Act 1952;

(b) 'headquarters' has the same meaning as in article 3(2) of the Visiting Forces and International Headquarters (Application of Law) Order 1965;

(c) 'member of a headquarters' has the same meaning as in paragraph 1(1) of the Schedule to the International Headquarters and Defence Organisations Act 1964; and

(d) 'visiting force' has the same meaning as it does for the purposes of any provision of Part I of the Visiting Forces Act 1952.

Regulation 7 *(See Guidance Notes, p 631)*

Extension outside Great Britain

These Regulations shall, subject to regulation 3 hereof, apply to and in relation to the premises and activities outside Great Britain to which sections 1 to 59 and 80 to 82 of the Health and Safety at Work etc Act 1974 apply by virtue of the Health and Safety at Work etc Act 1974 (Application Outside Great Britain) Order 1989 as they apply within Great Britain.

Regulation 8 *(See Guidance Notes, pp 631–632)*

Repeals and revocations

(1) The enactments mentioned in column 1 of Part I of Schedule 2 to these Regulations are repealed to the extent specified in the corresponding entry in column 3 of that part.

(2) The Regulations mentioned in column 1 of Part II of Schedule 2 to these Regulations are revoked to the extent specified in the corresponding entry in column 3 of that part.

SCHEDULE 1
FACTORS TO WHICH THE EMPLOYER MUST HAVE REGARD AND QUESTIONS HE MUST CONSIDER WHEN MAKING AN ASSESSMENT OF MANUAL HANDLING OPERATIONS

Regulation 4(1)(b)(i)

	Column 1 *Factors*	*Column 2* *Questions*
1	The tasks	Do they involve: — holding or manipulating loads at distance from trunk? — unsatisfactory bodily movement or posture, especially: — twisting the trunk? — stooping? — reaching upwards? — excessive movement of loads, especially: — excessive lifting or lowering distances? — excessive carrying distances? — excessive pushing or pulling of loads? — risk of sudden movement of loads? — frequent or prolonged physical effort? — insufficient rest or recovery periods? — a rate of work imposed by a process?

2	The loads	Are they: — heavy? — bulky or unwieldy? — difficult to grasp? — unstable, or with contents likely to shift? — sharp, hot or otherwise potentially damaging?
3	The working environment	Are there: — space constraints preventing good posture? — uneven, slippery or unstable floors? — variations in level of floors or work surfaces? — extremes of temperature or humidity? — conditions causing ventilation problems or gusts of wind? — poor lighting conditions?
4	Individual capability	Does the job: — require unusual strength, height, etc? — create a hazard to those who might reasonably be considered to be pregnant or to have a health problem? — require special information or training for its safe performance?
5	Other factors	Is movement or posture hindered by personal protective equipment or by clothing?

SCHEDULE 2
REPEALS AND REVOCATIONS

Regulation 8

Part I
Repeals

Column 1 *Short title of enactment*	*Column 2* *Reference*	*Column 3* *Extent of repeal*
The Children and Young Persons Act 1933	1933 c 12	Section 18(1)(f) except insofar as that paragraph applies to such employment as is permitted under section 1(2) of the Employment of Women, Young Persons, and Children Act 1920 (1920 c 65)
The Children and Young Persons (Scotland) Act 1937	1937 c 37	Section 28(1)(f) except insofar as that paragraph applies to such employment as is permitted under section 1(2) of the Employment of Women, Young Persons, and Children Act 1920
The Mines and Quarries Act 1954	1954 c 70	Section 93; in section 115 the word 'ninety-three'
The Agriculture (Safety, Health and Welfare Provisions) Act 1956	1956 c 49	Section 2
The Factories Act 1961	1961 c 34	Section 72
The Offices, Shops and Railway Premises Act 1963	1963 c 41	Section 23 except insofar as the prohibition contained in that section applies to any person specified in section 90(4) of the same Act In section 83(1) the number '23'

Part II
Revocations

Column 1 *Title of instrument*	*Column 2* *Reference*	*Column 3* *Extent of revocation*
The Agriculture (Lifting of Heavy Weights) Regulations 1959	SI 1959/2120	The whole Regulations
The Construction (General Provisions) Regulations 1961	SI 1961/1580	In regulation 3(1)(a) the phrase 'and 55'; regulation 55

GUIDANCE NOTES

Introduction

1 This booklet gives general guidance on the Manual Handling Operations Regulations 1992 (the Regulations) which came into force on 1 January 1993. The Regulations are made under the Health and Safety at Work etc Act 1974 (the HSW Act). They implement European Directive 90/269/EEC on the manual handling of loads; supplement the general duties placed on employers and others by the HSW Act and the broad requirements of the Management of Health and Safety at Work Regulations 1992; and replace a number of earlier, outdated legal provisions.

2 More than a quarter of the accidents reported each year to the enforcing authorities are associated with manual handling—the transporting or supporting of loads by hand or by bodily force. While fatal manual handling accidents are rare, accidents resulting in a major injury such as a fractured arm are more common, accounting for nearly 10% of all major injuries reported in 1996/97. The vast majority of reported manual handling accidents result in over-three-day injury, most commonly a sprain or strain, often of the back. Figures 1 to 3 illustrate these patterns for over-three-day injuries reported in 1996/97.

3 Sprains and strains are caused by incorrect application and/or prolongation of bodily force. Poor posture and excessive repetition

of movement can be important factors in their onset. Many manual handling injuries are cumulative rather than being truly attributable to any single handling incident. The injured do not always make a full recovery; the result can be physical impairment or even permanent disability.

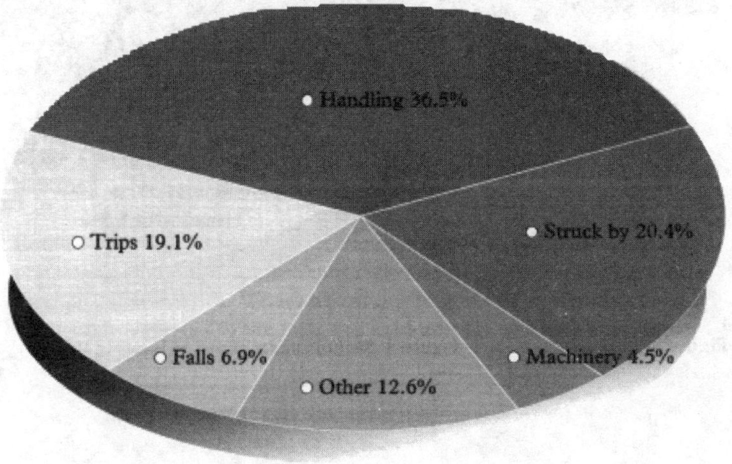

Figure 1 Kinds of accident causing injury 1996/97

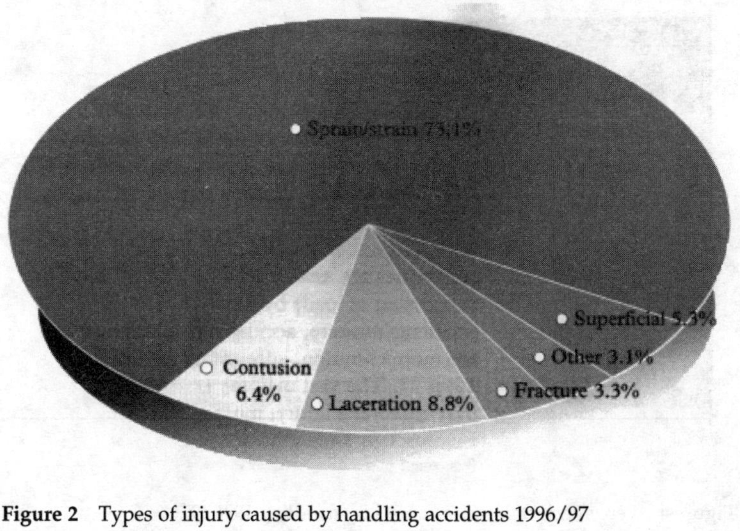

Figure 2 Types of injury caused by handling accidents 1996/97

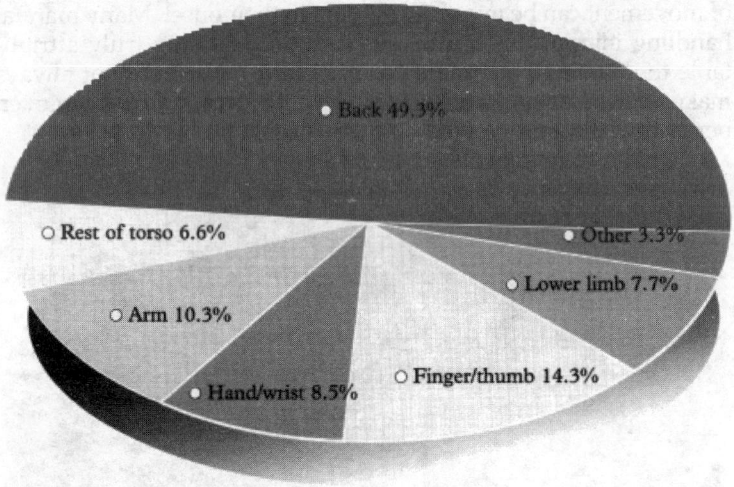

Figure 3 Sites of injuries caused by handling 1996/97

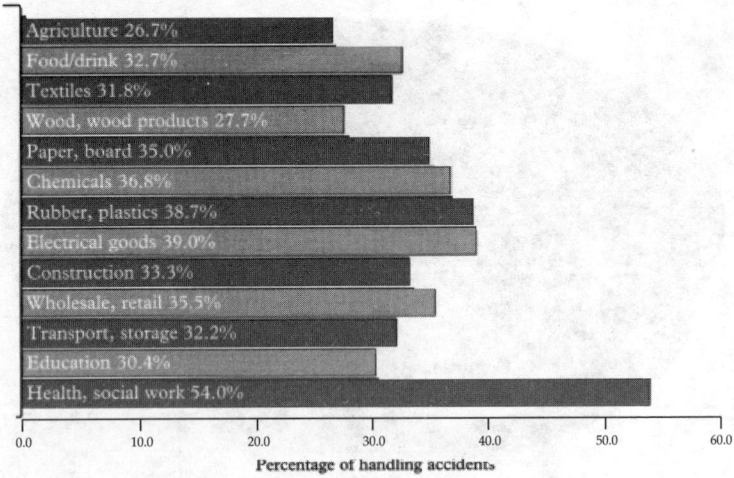

Figure 4 Percentage of injuries caused by handling 1996/97

4 Figure 4, also based on over-three-day injuries reported in 1996/97, shows that the problem of manual handling is not confined to a narrow range of industries but is widespread. Nor is the problem confined to industrial work: for example the comparable figure for wholesale and retail distribution is over 35% and for health and social work 54%.

5 There is now substantial international acceptance of both the scale of manual handling problems and methods of prevention. Modern medical and scientific knowledge stresses the importance of an ergonomic approach to remove or reduce the risk of manual handling injury. Ergonomics is sometimes described as fitting the job to the person, rather than the person to the job. The ergonomic approach, therefore, looks at manual handling as a whole. It takes into account a range of relevant factors, including the nature of the task, the load, the working environment and individual capability. This approach is central to the European Directive on manual handling, and to the Regulations. The flow chart in Figure 5 sets out the sequence of actions required when applying the Regulations to particular manual handling operations.

6 The Regulations should not be considered in isolation. Regulation 3(1) of the Management of Health and Safety at Work Regulations 1992 requires employers to make a suitable and sufficient assessment of the risks to the health and safety of their employees while at work. Where this general assessment indicates the possibility of risks to employees from the manual handling of loads, the requirements of the present Regulations should be followed.

7 The Regulations establish a clear hierarchy of measures:

(a) Avoid hazardous manual handling operations so far as is reasonably practicable. This may be done by redesigning the task to avoid moving the load or by automating or mechanising the process.

(b) Make a suitable and sufficient assessment of any hazardous manual handling operations that cannot be avoided.

(c) Reduce the risk of injury from those operations so far as is reasonably practicable. Particular consideration should be given to the provision of mechanical assistance. Where this is not reasonably practicable then other improvements to the task, the load and the working environment should be explored.

8 Although only the courts can give an authoritative interpretation of the law, in considering the application of these regulations and guidance to people working under another's direction, the following should be considered:

> If people working under the control and direction of others are treated as self-employed for tax/national insurance purposes, they are nevertheless treated as their employees for health and safety purposes. It may, therefore, be necessary to take appropriate action to protect them. If any doubt exists about who is responsible for the health and safety of a worker this could be clarified and included in the terms of the contract. However, a legal duty under the Health and Safety at Work Act (the HSW Act) cannot be passed on by means of a contract and there will still be duties towards others under section 3 of the HSW Act. If such workers are employed on the basis that they are responsible for their own health and safety, legal advice should be sought before doing so.

9 The Regulations set no specific requirements such as weight limits. The ergonomic approach shows clearly that such requirements are based on too simple a view of the problem and are likely to lead to incorrect conclusions. Instead, an ergonomic assessment based on a range of relevant factors is used to determine the risk of injury and point the way to remedial action.

10 However, a full assessment of every manual handling operation could be a major undertaking and might involve wasted effort. To enable assessment work to be concentrated where it is most needed, numerical guidelines are offered (pages 632–636) which can be used as an initial filter. This will help to identify those manual handling operations which warrant a more detailed examination. However, even operations lying within the guidelines boundaries should be avoided or made less demanding wherever it is reasonably practicable to do so. *Do not regard the guidelines as precise recommendations. Where there is doubt make a more detailed assessment.*

11 This booklet contains a general framework within which individual industries and sectors will be able to produce more specific guidance appropriate to their own circumstances.

12 Manual handling injuries are part of a wider family of musculoskeletal problems; the reader may also find it helpful to refer to the Health and Safety Executive (HSE) booklet *Work related upper limb disorders: A guide to prevention.*

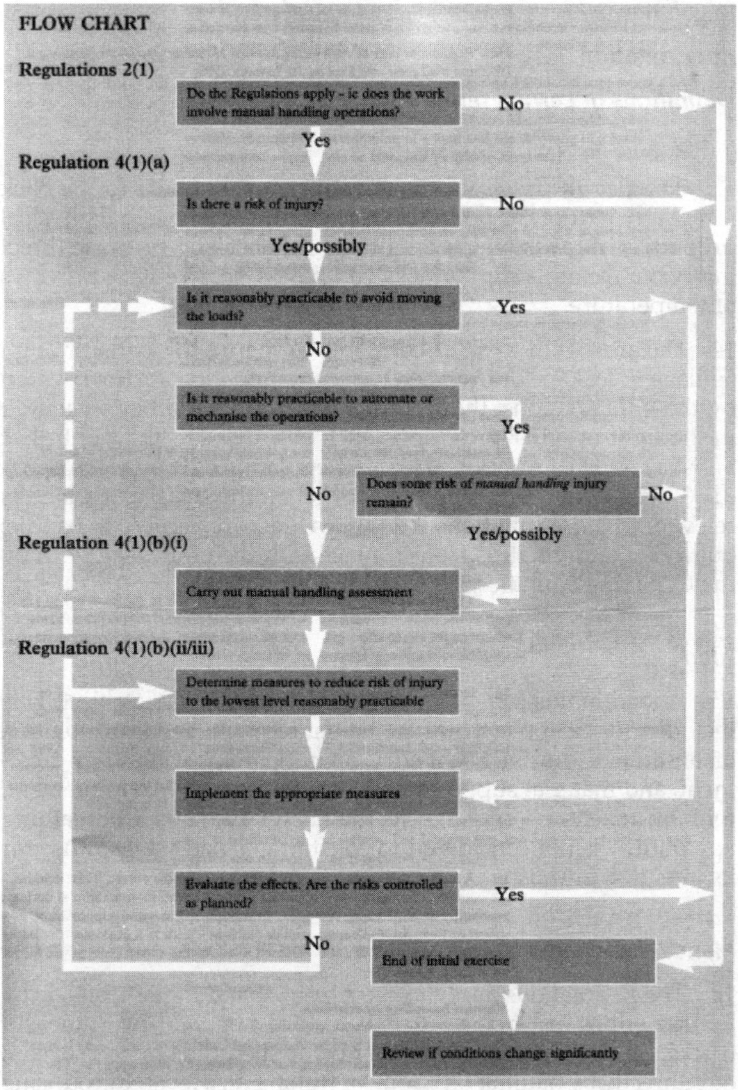
FLOW CHART
Regulations 2(1)
Do the Regulations apply - ie does the work involve manual handling operations?
No
Yes
Regulation 4(1)(a)
Is there a risk of injury?
No
Yes/possibly
Is it reasonably practicable to avoid moving the loads?
Yes
No
Is it reasonably practicable to automate or mechanise the operations?
Yes
Does some risk of manual handling injury remain?
No
No
Yes/possibly
Regulation 4(1)(b)(i)
Carry out manual handling assessment
Regulation 4(1)(b)(ii/iii)
Determine measures to reduce risk of injury to the lowest level reasonably practicable
Implement the appropriate measures
Evaluate the effects. Are the risks controlled as planned?
Yes
No
End of initial exercise
Review if conditions change significantly

Figure 5 How to follow the Manual Handling Operations Regulations 1992

Regulation 2

Interpretation

Definitions of certain terms

Injury

13 The Regulations seek to prevent injury not only to the back but to any part of the body. Take account of any external physical properties of loads which might either affect grip or cause direct injury, for example slipperiness, roughness, sharp edges, extremes of temperature.

14 Hazards from toxic or corrosive properties of loads through spillage or leakage or from external contamination are not covered by these Regulations, though such hazards should be considered in the light of other provisions such as COSHH—the Control of Substances Hazardous to Health Regulations 1994 (see Appendix 3). For example, the presence of oil on the surface of a load is relevant to the Regulations if it makes the load slippery to handle; but a risk of dermatitis from contact with the oil is dealt with by the COSHH Regulations.

Load

15 A load in this context must be a discrete movable object. This includes, for example, a human patient receiving medical attention or an animal during husbandry or undergoing veterinary treatment, and material supported on a shovel or fork. An implement, tool or machine—such as a chainsaw, fire hoses, breathing apparatus—is not considered to constitute a load while in use for its intended purpose.

Manual handling operations

16 The Regulations apply to the manual handling of loads, ie by human effort, as opposed to mechanical handling by crane, lift trucks, etc. The human effort may be applied directly to the load, or indirectly by hauling on a rope or pulling on a lever. Introducing mechanical assistance, for example a sack truck or a powered hoist, may reduce but not eliminate manual handling since human effort is still required to move, steady or position the load.

17 Manual handling includes both transporting a load and supporting a load in a static posture. The load may be moved or

supported by the hands or any other part of the body, for example the shoulder. Manual handling also includes the intentional dropping of a load and the throwing of a load, whether into a receptacle or from one person to another.

18 The application of human effort for a purpose other than transporting or supporting a load does not constitute a manual handling operation. For example, turning the starting handle of an engine or lifting a control lever on a machine is not manual handling, nor is the action of pulling on a rope while lashing down cargo on the back of a vehicle.

Duties of the self-employed

19 Regulation 2(2) makes the self-employed responsible for their own safety during manual handling. They should take the same steps to safeguard themselves as employers must to protect their employees in similar circumstances. Employers should remember, however, that they may be responsible for the health and safety of someone who is self-employed for tax and national insurance purposes but who works under their control and direction.

Regulation 3

Disapplication of Regulations

Sea-going ships

20 Sea-going ships are subject to separate Merchant Shipping legislation administered by the Department of the Environment, Transport and the Regions. The Regulations, therefore, do not apply to the normal ship-board activities of a ship's crew under the direction of the master. However, the Regulations may apply to other manual handling operations aboard a ship; for example where a shore-based contractor carries out the work, provided the ship is within territorial waters. The Regulations also apply to certain activities carried out offshore—see regulation 7.

Regulation 4

Duties of employers

Introduction

21 The Regulations should not be considered in isolation. Regulation 3(1) of the Management of Health and Safety at Work

Regulations 1992 (see Appendix 3) requires employers to make a suitable and sufficient assessment of the risks to the health and safety of their employees while at work. Where this general assessment indicates the possibility of risks to employees from the manual handling of loads the requirements of the Regulations should be observed, as follows.

Hierarchy of measures

22 The various parts of regulation 4(1) establish a clear hierarchy of measures:

(a) avoid hazardous manual handling operations so far as is reasonably practicable;

(b) assess any hazardous manual handling operations that cannot be avoided; and

(c) reduce the risk of injury so far as is reasonably practicable.

Extent of the employer's duties

23 The extent of the employer's duty to avoid manual handling or to reduce the risk of injury is determined by reference to what is 'reasonably practicable'. Such duties are satisfied if the employer can show that the cost of any further preventive steps would be grossly disproportionate to the further benefit from their introduction.

24 This approach is fully applicable to the work of the emergency services. Ultimately, the cost of prohibiting all potentially hazardous manual handling operations would be an inability to provide the general public with an adequate rescue service. A fire authority, for example may consider that it has discharged this duty when it can show that any further preventive steps would make the efficient discharge of its emergency functions unduly difficult.

A continuing duty

25 It is not sufficient simply to make changes and then hope that the problem has been dealt with. Monitor steps taken to avoid manual handling or reduce the risk of injury to check that they are having the desired effect in practice. If they are not, seek alternative steps.

26 Regulation 4(2) (discussed later) requires the assessment made under regulation 4(1) to be kept up to date.

Work away from the employer's premises

27 The Regulations impose duties upon the employer whose employees carry out the manual handling. However, manual handling operations may occur away from the employer's premises in situations over which the employer can exercise little direct control. Where possible the employer should seek close liaison with those in control of such premises. There will sometimes be a limit to employers' ability to influence the working environment; but the task and perhaps the load will often remain within their control, as will the provision of effective training, so it is still possible to establish a safe system of work.

28 Employers and others in charge of premises where visiting employees work also have duties towards those employees, particularly under sections 3 or 4 of the HSW Act, the Management of Health and Safety at Work Regulations 1992 and the Workplace (Health, Safety and Welfare) Regulations 1992. For example, they need to ensure that the premises and plant provided there are in a safe condition.

Those self-employed for tax/NI purposes

29 Those working under the control and direction of another may be regarded as employees for health and safety purposes even though they are treated as self-employed for tax/national insurance purposes. Those who employ workers on this basis, therefore, may need to take appropriate action to protect them. If any doubt exists about who is responsible for the health and safety of such workers, legal advice should be sought.

Avoidance of manual handling

Risk of injury

30 If the general assessment carried out under regulation 3(1) of the Management of Health and Safety at Work Regulations 1992 indicates a possibility of injury from manual handling operations, first consider avoiding the need for the operations in question. At this preliminary stage a judgement should be made as to the nature

and likelihood of injury. It may not be necessary to assess in great detail, particularly if the operations can readily be avoided or if the risk is clearly low. Pages 632–636 provide some simple numerical guidelines to assist with this initial judgement, at least in relatively straightforward cases.

Elimination of handling

31 In seeking to avoid manual handling the first question to ask is whether movement of the loads can be eliminated altogether: are the handling operations unnecessary; or could the desired result be achieved in some entirely different way? For example, can a process such as machining or wrapping be carried out in situ, without handling the loads? Can a treatment be brought to a patient rather than taking the patient to the treatment?

Automation or mechanisation

32 Secondly, if load handling operations, in some form, cannot be avoided entirely, ask further questions:

(a) Can the operations be automated?

(b) Can the operations be mechanised?

33 Remember that the introduction of automation or mechanisation may create other, different risks. Even an automated plant will require maintenance and repair. Mechanisation, for example by the introduction of lift trucks or powered conveyors, can introduce fresh risks requiring precautions of their own.

34 It is especially important to address these questions when plant or systems of work are being designed. For example, raw materials can be handled in the workplace in a way which eliminates or reduces the need for manual handling. Powders or liquids can be transferred from large containers and big bags by gravity feed or pneumatic transfer, avoiding bag or container handling. The layout of the process can often be designed to minimise transfer of materials or the distance over which containers have to be moved. Examination of existing activities may also reveal opportunities for avoiding manual handling operations that involve a risk of injury. Such improvements often bring additional benefits in terms of greater efficiency and productivity, and reduced damage to loads.

Assessment of risk

35 Where the general assessment carried out under regulation 3(1) of the Management of Health and Safety at Work Regulations 1992 indicates a possibility of injury from manual handling operations, but the conclusion reached under regulation 4(1)(a) is that avoidance of the operations is not reasonably practicable, a more specific assessment should be carried out as required by regulation 4(1)(b)(i). The extent to which this further assessment need be pursued will depend on the circumstances. Pages 632–636 offer some simple numerical guidelines to help with this decision. The guidelines are intended to be used as an initial filter, to help identify those operations deserving more detailed assessment.

36 Schedule 1 to the Regulations specifies factors which this assessment should take into account, including the *task*, the *load*, the *working environment* and *individual capability*. First, however, consider how the assessment is to be carried out—and by whom—and what other relevant information may be available to help.

Who should carry out the assessment?

37 In most cases employers should be able to carry out the assessment themselves or delegate it to others in their organisation, having regard to regulation 6 of the Management of Health and Safety at Work Regulations 1992. A meaningful assessment can only be based on a thorough practical understanding of the type of manual handling tasks to be performed, the loads to be handled and the working environment in which the tasks will be carried out. Employers and managers should be better placed to know about the manual handling taking place in their own organisation than someone from outside the organisation.

38 While one individual may be able to carry out a perfectly satisfactory assessment, at least in relatively straightforward cases, it can be helpful to draw on the knowledge and expertise of others. In some organisations this is done informally; others prefer to set up a small assessment team.

39 Areas of knowledge and expertise likely to be relevant to successful risk assessment of manual handling operations, and individuals who may be able to make a useful contribution, include:

(a) the requirements of the Regulations (manager, safety professional);

(b) the nature of the handling operations (supervisor, industrial engineer);

(c) a basic understanding of human capabilities (occupational health nurse, safety professional);

(d) identification of high risk activities (manager, supervisor, occupational health nurse, safety professional); and

(e) practical steps to reduce risk (manager, supervisor, industrial engineer, safety professional).

40 It may be appropriate to seek outside help, for example to give basic training to in-house assessors or where manual handling risks are novel or particularly difficult to assess. Outside specialist advice may also help solve unusual handling problems or contribute to ergonomic design. But employers will still wish to oversee the assessment as they have the final responsibility for it.

Employees' contribution

41 The views of staff can be particularly valuable in identifying manual handling problems and practical solutions to them. Encourage employees, their safety representatives and safety committees to play a positive part in the assessment process. They can assist the employer by highlighting difficulties from such things as the size or shape of loads, how often they are handled or the circumstances in which the handling operations are carried out. For example, staff can provide valuable information about when space constraints make it difficult to manoeuvre the load, and about the need to reorganise storage on shelving systems to minimise the risk to the handler.

Records of accidents and ill health

42 Well-kept records of accidents and ill health can play a useful part in the assessment process. They should identify accidents associated with manual handling. Careful analysis may also show evidence of links between manual handling and ill health, including injuries apparently unrelated to any specific event or accident. Other possible indicators of manual handling problems include high levels of absenteeism or staff turnover, poor productivity and morale, excessive product damage, and general dissatisfaction among the employees concerned. Investigate any regular occurrence of back disorders or other ailments possibly associated with

unsatisfactory manual handling practices. However, such indicators are not a complete guide and should be used only to augment other risk assessment methods.

How detailed should an assessment be?

43 Employers' assessments will be 'suitable and sufficient' if they look in a considered way at the totality of the manual handling operations their employees are required to perform. Properly based 'generic' assessments which draw together common threads from a range of broadly similar operations are quite acceptable. Indeed a more narrowly focused assessment may fail to reflect adequately the range of operations encountered.

44 An assessment made at the last minute is unlikely to be 'suitable and sufficient'. In conducting assessments employers should, therefore, use their experience of the type of work their employees perform, consulting the employees as appropriate. This approach will help with the assessment of work that is of a varied nature (such as construction or maintenance), peripatetic (such as making deliveries) or involves dealing with emergencies (such as fire-fighting and rescue).

45 In the case of delivery operations, for example, a useful technique is to list the various types of task, load and working environment concerned and then to review a selection of them. Aim to establish the range of manual handling risks to which employees are exposed and then to decide on appropriate preventive steps where these are shown to be necessary.

46 A distinction should be made between the employer's assessment required by regulation 4(1)(b)(i) and the everyday judgements which supervisors and others will have to make in dealing with manual handling operations. The assessment should identify in broad terms the problems likely to arise during the operations that can be foreseen and the measures needed to deal with them. These measures should include the provision of training to enable supervisors and, where appropriate, individual employees, to cope effectively with the operations they are likely to undertake.

47 This distinction is perhaps most clearly seen in the case of emergency work. Here it will be essential to provide training to enable fire officers, for example to make the rapid judgements that will inevitably be necessary in dealing satisfactorily with an emergency incident or in supervising realistic training.

Industry-specific data and assessments

48 Individual industries and sectors have a valuable role to play in identifying common manual handling problems and developing practical solutions. Industry associations and similar bodies can also act as a focus for the collection and analysis of accident and ill health data drawn from a far wider base than that available to the individual employer.

Recording the assessment

49 In general, the significant findings of the assessment should be recorded and the record kept, readily accessible, as long as it remains relevant. However, the assessment need not be recorded if:

(a) it could very easily be repeated and explained at any time because it is simple and obvious; or

(b) the manual handling operations are quite straightforward, of low risk, are going to last only a very short time, and the time taken to record them would be disproportionate.

Making a more detailed assessment

50 When a more detailed assessment is necessary it should follow the broad structure set out in Schedule 1 to the Regulations. The Schedule lists a number of questions in five categories including the *task*; the *load*; the *working environment*; and *individual capability*. Not all of these questions will be relevant in every case.

51 These categories are clearly interrelated: each may influence the others and, therefore, none can be considered in isolation. However, in order to carry out an assessment in a structured way it is often helpful to begin by breaking the operations down into separate, more manageable items.

Assessment checklist

52 It may be helpful to use a checklist during assessment as an aide-memoire. An example of such a checklist is provided on pages 636–642. This checklist addresses not only the analysis of risk required by regulation 4(1)(b)(i) but also the identification of steps to reduce the risk as required by regulation 4(1)(b)(ii), discussed later. The particular example given will not be suitable in all circumstances; it can be adapted or modified as appropriate.

53 *Remember*—assessment is not an end in itself, merely a structured way of analysing risks and pointing the way to practical solutions.

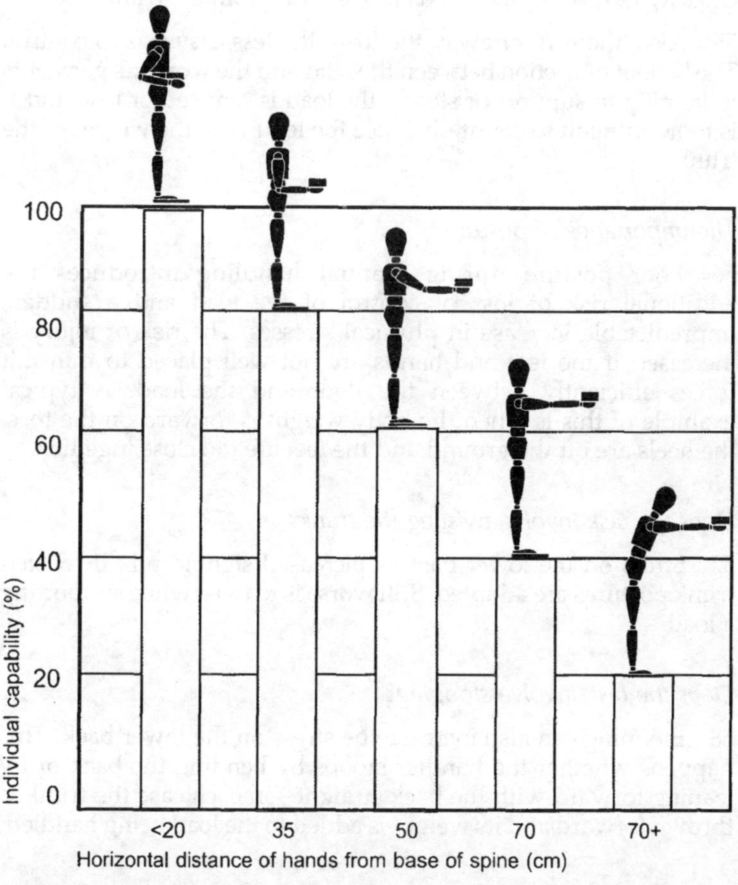

Figure 6 Reduction of individual handling capability as the hands move away from the trunk

The task—making an assessment

Is the load held or manipulated at a distance from the trunk?

54 As the load is moved away from the trunk the general level of stress on the lower back rises. Regardless of the handling technique used, not keeping the load close to the body will increase the stress.

As a rough guide, holding a load at arm's length imposes about five times the stress experienced when holding the same load very close to the trunk. Figure 6 shows how individual handling capacity reduces as the hands move away from the trunk.

55 Also, the further away the load, the less easy it is to control. The benefit of friction between the load and the worker's garments in helping to support or steady the load is reduced or lost, and it is more difficult to counterbalance the load with the weight of the trunk.

The importance of posture

56 Poor posture during manual handling introduces the additional risk of loss of control of the load and a sudden, unpredictable increase in physical stresses. The risk of injury is increased if the feet and hands are not well placed to transmit forces efficiently between the floor and the load. A typical example of this is when the body weight is forward on the toes, the heels are off the ground and the feet are too close together.

Does the task involve twisting the trunk?

57 Stress on the lower back is increased significantly if twisted trunk postures are adopted. Still worse is to twist while supporting a load.

Does the task involve stooping?

58 Stooping can also increase the stress on the lower back. This happens whether the handler stoops by bending the back or by leaning forward with the back straight—in each case the trunk is thrown forward and its weight is added to the load being handled.

Does the task involve reaching upwards?

59 Reaching upwards places additional stresses on the arms and back. Control of the load becomes more difficult and, because the arms are extended, they are more prone to injury.

The effect of combining risk factors

60 Individual capability can be reduced substantially if twisting is combined with stooping or stretching. Such combinations should

be avoided wherever possible, especially since their effect on individual capability can be worse than the simple addition of their individual effects might suggest.

Does the task involve excessive lifting or lowering distances?

61 The distance through which a load is lifted or lowered can also be important: large distances are considerably more demanding physically than small ones. Moreover lifting or lowering through a large distance is likely to necessitate a change of grip part way, further increasing the risk of injury. Lifts beginning at floor level should be avoided where possible; where unavoidable, preferably, they should finish no higher than waist height.

Does the task involve excessive carrying distances?

62 In general, if a load can safely be lifted and lowered, it can also be carried without endangering the back. However, if a load is carried for an excessive distance, physical stresses are prolonged, leading to fatigue and increased risk of injury. As a rough guide, if a load is carried further than about 10 m then the physical demands of carrying the load will tend to predominate over those of lifting and lowering and individual capability will be reduced.

Does the task involve excessive pushing or pulling of the load?

63 Like lifting, lowering and carrying, pushing or pulling a load can harm the handler. The risk of injury is increased if pushing or pulling is carried out with the hands much below knuckle height or above shoulder height.

64 Additionally, because of the way in which pushing and pulling forces have to be transmitted from the handler's feet to the floor, the risk of slipping and consequent injury is much greater. For this reason pushing or pulling a load in circumstances where the grip between foot and floor is poor—whether through the condition of the floor, footwear or both—is likely to increase significantly the risk of injury.

Does the task involve positioning the load precisely?

65 A requirement to position the load with precision can also add to the risk of injury.

Does the task involve a risk of sudden movement of the load?

66 If a load suddenly becomes free and the handler is unprepared or is not able to retain complete control of the load, unpredictable stresses can be imposed on the body, creating a risk of injury. For example, the freeing of a box jammed on a shelf or the release of a machine component during maintenance work can easily cause injury if handling conditions are not ideal. The risk is compounded if the handler's posture is unstable.

Does the task involve frequent or prolonged physical effort?

67 The frequency with which a load is handled can affect the risk of injury. A quite modest load, handled very frequently, can create as large a risk of injury as one-off handling of a more substantial load. The effect will be worsened by jerky, hurried movements which can multiply a load's effect on the body.

68 If physical stresses are prolonged then fatigue will occur, increasing the risk of injury. This effect will often be made worse by a relatively fixed posture, leading to a rapid increase in fatigue and a corresponding fall in muscular efficiency.

Does the task involve insufficient rest or recovery periods?

69 Research and experience in industry have shown that failure to counter fatigue during physically demanding work increases ill health and reduces output. Consider, therefore, whether there are adequate opportunities for rest (ie breaks from work) or recovery (ie changing to another task which uses a different set of muscles). The amount of work undertaken in fixed postures is also an important consideration since blood flow to the muscles is likely to be reduced, adding to fatigue. This problem is complicated by a large variation in individual susceptibility to fatigue.

Does the task involve a rate of work imposed by a process?

70 Particular care is necessary where the rate of work cannot be varied by the handler. Mild fatigue, which otherwise might quickly be relieved by a momentary pause or a brief spell doing another operation using different muscles, can soon become more pronounced, leading to an increased risk of injury.

Handling while seated

71 Handling loads while seated imposes considerable constraints. The relatively powerful leg muscles cannot be used. Nor can the weight of the handler's body be used as a counterbalance. Most of the work, therefore, has to be done by the weaker muscles of the arms and trunk.

72 Unless the load is presented close to the body the handler will have to reach and/or lean forward. Not only will handling in this position put the body under additional stress but the seat, unless firmly placed, will then tend to move as the handler attempts to maintain a stable posture.

73 Lifting from below the level of a work surface will almost inevitably result in twisting and stooping, the dangers of which were discussed in paragraphs 57 and 58.

Team handling

74 Handling by two or more people (see Figure 7) may make possible an operation that is beyond the capability of one person, or reduce the risk of injury to a solo handler. However, team handling may introduce additional problems which the assessment should consider. During the handling operation the proportion of the load that is borne by each member of the team will inevitably vary to some extent. Such variation is likely to be more pronounced on rough ground. Therefore, the load that a team can handle in safety is less than the sum of the loads that the individual team members could cope with when working alone.

75 As an approximate guide the capability of a two person team is two-thirds the sum of their individual capabilities; and for a

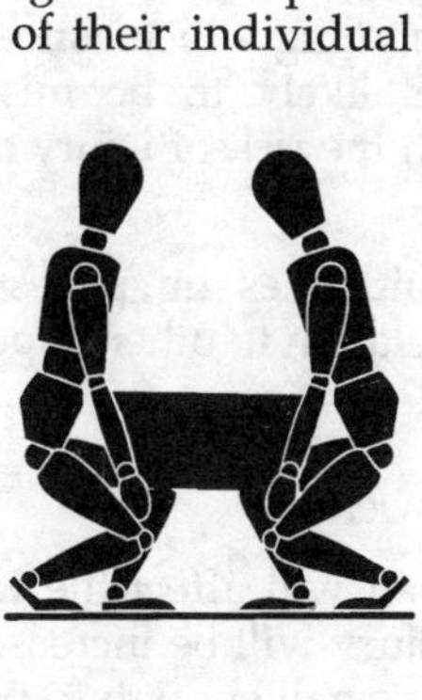

Figure 7 Team handling

three person team the capability is half the sum of their individual capabilities. If steps or slopes must be negotiated most of the weight may be borne by the handler or handlers at the lower end, further reducing the capability of the team as a whole.

76 Additional difficulties may arise if team members impede each others' vision or movement, or if the load offers insufficient good handholds. This can occur particularly with compact loads which force the handlers to work close together or where the space available for movement is limited.

The load—making an assessment

Is the load heavy?

77 For many years legislation and guidance on manual handling have concentrated on the weight of the load. It is now well established that the weight of the load is only one—and sometimes not the main—consideration affecting the risk of injury. Other features of the load such as its resistance to movement, its size, shape or rigidity must also be considered. Proper account must also be taken of the circumstances in which the load is handled; for example postural requirements, frequency and duration of handling, workplace design, and aspects of work organisation such as incentive schemes and piecework.

78 Moreover, traditional guidance, based on so-called 'acceptable' weights, has often considered only symmetrical, two-handed lifts, in front of and close to the body. In reality such lifting tasks are comparatively rare, since most will involve sideways movement, twisting of the trunk or some other asymmetry. For these reasons an approach to manual handling which concentrates solely upon the weight of the load is likely to be misleading, either failing adequately to deal with the risk of injury or imposing excessively cautious constraints.

79 The numerical guidelines on pages 632–636 consider the weight of the load in relation to other important factors.

Is the load bulky or unwieldy?

80 The shape of a load will affect the way it can be held. For example, the risk of injury will be increased if a load to be lifted from the ground is not small enough to pass between the knees,

since its bulk will hinder a close approach. Similarly, if the bottom front corners of a load are not within reach when carried at waist height it will be harder to get a good grip. And if handlers have to lean away from a load to keep it off the ground when carrying it at their side, they will be forced into unfavourable postures.

81 In general, if any dimension of the load exceeds about 75 cm, its handling is likely to pose an increased risk of injury, especially if this size is exceeded in more than one dimension. The risk will be further increased if the load does not provide convenient handholds.

82 The bulk of the load can also interfere with vision. Where restrictions of view by a bulky load cannot be avoided, take account of the increased risk of slipping, tripping, falling or colliding with obstructions.

83 The risk of injury will also be increased if the load is unwieldy and difficult to control. Well-balanced lifting may be difficult to achieve, the load may hit obstructions, or it may be affected by gusts of wind or other sudden air movements.

84 If the centre of gravity of the load is not positioned centrally within the load, inappropriate handling may increase the risk of injury. For example much of the weight of a typewriter is often at the back; therefore, an attempt to lift the typewriter from the front will place its centre of gravity further from the handler's body than if the typewriter is first turned around and lifted from the back.

85 Sometimes, as with a sealed and unmarked carton, an offset centre of gravity is not immediately apparent. In these circumstances there is a greater risk of injury since the handler may unwittingly hold the load with its centre of gravity further from the body than is necessary.

Is the load difficult to grasp?

86 If the load is difficult to grasp, for example because it is large, rounded, smooth, wet or greasy, its handling will call for extra grip strength—which is tiring—and will probably involve inadvertent changes of posture. There will also be a greater risk of dropping the load. Handling will be less sure and the risk of injury will be increased.

Is the load unstable, or are its contents likely to shift?

87 If the load is unstable, for example because it lacks rigidity or has contents that are liable to shift, the likelihood of injury is increased. The stresses arising during the manual handling of such a load are less predictable, and the instability may impose sudden additional stresses for which the handler is not prepared. This is particularly true if the handler is unfamiliar with a particular load and there is no cautionary marking on it.

88 Handling people or animals, for example hospital patients or livestock, can present additional problems. The load lacks rigidity, there is particular concern on the part of the handler to avoid damaging the load, and to complicate matters the load will often have a mind of its own, introducing an extra element of unpredictability. These factors are likely to increase the risk of injury to the handler as compared with the handling of an inanimate load of similar weight and shape.

Is the load sharp, hot or otherwise potentially damaging?

89 Risk of injury may also arise particularly in the event of a collision from the consistency or external state of the load. It may have sharp edges or rough surfaces, or be too hot or too cold to touch safely without protective clothing. In addition to the more obvious risk of direct injury, such characteristics may also impair grip, discourage good posture or otherwise interfere with safe handling.

The working environment—making an assessment

Are there space constraints preventing good posture?

90 If the working environment hinders working at a safe height or the adoption of good posture, the risk of injury from manual handling will be increased. Low work surfaces or restricted head room will enforce a stooping posture; furniture, fixtures or other obstructions may increase the need for twisting or leaning; constricted working areas and narrow gangways will hinder the manoeuvring of bulky loads.

Are there uneven, slippery or unstable floors?

91 In addition to increasing the likelihood of slips, trips and falls, uneven or slippery floors hinder smooth movement and create

additional unpredictability. Unstable footrests and floors susceptible to movement—for example on a boat, a moving train, or a mobile work platform—similarly increase the risk of injury through the imposition of sudden, unpredictable stresses.

Are there variations in level of floors or work surfaces?

92 The presence of steps, steep slopes, etc can increase the risk of injury by adding to the complexity of movement when handling loads. Carrying a load up or down a ladder, if it cannot be avoided, is likely to aggravate handling problems because of the additional need to maintain a proper hold on the ladder.

93 Excessive variation between the heights of working surfaces, storage shelving, etc will increase the range of movement and in consequence the scope for injury. This will be especially so if the variation is large and requires, for example, movement of the load from near floor level to shoulder height or beyond.

Are there extremes of temperature or humidity?

94 The risk of injury during manual handling will be increased by extreme thermal conditions. For example high temperatures or humidity can cause rapid fatigue; and perspiration on the hands may reduce grip. Work at low temperatures may impair dexterity. Gloves and other protective clothing which may be necessary in such circumstances may also hinder movement, impair dexterity and reduce grip. The influence of air movement on working temperatures—the wind chill factor—should not be overlooked.

Are there ventilation problems or gusts of wind?

95 Inadequate ventilation can hasten fatigue, increasing the risk of injury. Sudden air movements, whether caused by a ventilation system or the wind, can make large loads more difficult to manage safely.

Are there poor lighting conditions?

96 Poor lighting conditions can increase the risk of injury. Dimness or glare may cause poor posture, for example by encouraging stooping. Contrast between areas of bright light and deep shadow can aggravate tripping hazards and hinder the accurate judgement of height and distance.

Individual capability—making an assessment

Does the task require unusual strength, height, etc?

97 The ability to carry out manual handling in safety varies between individuals. However, the variations in individual capability are generally less important than the nature of the handling operations in causing manual handling injuries. Any assessment, therefore, which concentrates on individual capability at the expense of task or workplace design is likely to be misleading.

98 In general the lifting strength of women as a group is less than that of men. But for both men and women the range of individual strength and ability is large, and there is considerable overlap—some women can deal safely with greater loads than some men.

99 An individual's physical capability varies with age, typically climbing until the early 20s and declining gradually thereafter. This decline becomes more significant from the mid-40s and the risk of manual handling injury, therefore, may be somewhat higher for employees in their teens or in their 50s and 60s. Particular care is needed in the design of tasks for these groups who are more likely to be working close to their maximum capacity in manual handling. In addition, older workers may become fatigued sooner and will take longer to recover from musculoskeletal injury. However, the range of individual capability is large and the benefits of experience and maturity should not be overlooked.

100 Consider the nature of the work in deciding whether the physical demands imposed by manual handling operations should be regarded as unusual. For example, demands that would be considered unusual for a group of employees engaged in office work might not be out of the ordinary for those normally involved in heavy physical labour. It would also be unrealistic to ignore the element of self-selection that often occurs for jobs that are relatively demanding physically.

101 As a general rule, however, the risk of injury should be regarded as unacceptable if the manual handling operations cannot be performed satisfactorily by most reasonably fit, healthy employees.

Does the job put at risk those who might be pregnant or have a disability or a health problem?

102 Allowance should be made for pregnancy where the employer could reasonably be expected to be aware of it, ie where

the pregnancy is visibly apparent or the employee has informed her employer that she is pregnant. Manual handling has significant implications for the health of the pregnant worker (and the foetus), particularly if combined with long periods of standing and/or walking. Hormonal changes during pregnancy can affect the ligaments and joint laxity, thereby increasing the risk of injury during manual handling tasks. As pregnancy progresses, and particularly during the last three months, it becomes more difficult to achieve and maintain good postures and this further reduces manual handling capability. Particular care should also be taken for women who may handle loads during the three months following a return to work after childbirth.

103 When an employee informs her employer that she is pregnant, the risks to the health and safety of the worker and her unborn child must be assessed in accordance with the duties under the Management of Health and Safety at Work (Amendment) Regulations 1994. A useful way to secure compliance and ensure that workers can continue to work safely during pregnancy is to have a well-defined plan on how to respond when pregnancy is confirmed. Such a plan may include:

(a) re-assessment of the handling task (positioning of the load and feet, frequency of lifting) to consider what improvements might be made;

(b) training in recognising ways in which work organisation may be altered to accommodate changes in posture and physical capability, including the timing and frequency of rest periods;

(c) consideration of job-sharing, relocation or suspension on full pay where the risk cannot be reduced by a change to the working conditions;

(d) liaison with the GP to confirm that the pregnant worker is capable of performing work duties; and

(e) careful monitoring of the employees returning to work following childbirth to assess the need for changes to work organisation.

104 The Disability Discrimination Act 1995 places a duty on employers to make reasonable adjustments to the workplace or employment arrangements so that a disabled person is not at any substantial disadvantage compared to a non-disabled person. This might include, for example arranging to limit the number, size or

weight of loads handled by someone with a disability that limits their manual handling capacity. It might also mean providing suitable manual handling aids. Further guidance is given in the Department for Education and Employment *Code of practice for the elimination of discrimination against disabled persons or persons who have had a disability*.

105 Allowance should also be made for any health problem of which the employer could reasonably be expected to be aware and which might have a bearing on the ability to carry out manual handling operations in safety. If there is good reason to suspect that an individual's state of health might significantly increase the risk of injury from manual handling operations, seek medical advice.

Does the task require special information or training for its safe performance?

106 The risk of injury from a manual handling task will be increased where a worker does not have the information or training necessary for its safe performance. While section 2 of the HSW Act and regulation 11 of the Management of Health and Safety at Work Regulations 1992 require employers to provide health and safety training, this may need to be supplemented to enable employees to carry out manual handling operations safely.

107 For example, ignorance of any unusual characteristics of loads, or of a system of work designed to ensure safety during manual handling, may lead to injury. Remedial steps such as the provision of mechanical handling aids may themselves create a need for training, for example in the proper use of those aids.

Other factors—making an assessment

Personal protective equipment and other clothing

108 Personal protective equipment should be used only as a last resort, when engineering or other controls do not provide adequate protection. Where wearing personal protective equipment cannot be avoided, its implications for the risk of manual handling injury should be taken into consideration. For example, gloves may impair dexterity; the weight of gas cylinders used with breathing apparatus will increase the stresses on the body. Other clothing, such as a uniform required to be worn, may inhibit free movement during manual handling.

Reducing the risk of injury

Striking a balance

109 It will usually be convenient to continue with the same structured approach used during the assessment of risk, considering in turn the *task*, the *load*, the *working environment* and *individual capability*.

110 The emphasis given to each of these factors may depend in part on the nature and circumstances of the manual handling operations. Routine manual handling operations carried out in essentially unchanging circumstances, for example in manufacturing processes, may lend themselves particularly to improvement of the task and working environment.

111 However, manual handling operations carried out in circumstances which change continually, for example certain activities carried out in mines or on construction sites, may offer less scope for improvement of the working environment and perhaps the task. More interest may, therefore, focus on the load—for example, can it be made easier to handle?

112 For varied work of this kind, including of course much of the work of the emergency services, the provision of effective training will be especially important. It should enable employees to recognise potentially hazardous handling operations. It should also give them a clear understanding of why they should avoid or modify such operations where possible, make full use of appropriate equipment and apply good handling technique.

An ergonomic approach

113 Health, safety and productivity are most likely to be optimised if an ergonomic approach is used to design the manual handling operations as a whole. Wherever possible full consideration should be given to the *task*, the *load*, the *working environment*, *individual capability* and the relationship between them, with a view to fitting the operations to the individual rather than the other way around.

114 While better job or workplace design may not eliminate handling injuries, the evidence is that it can greatly reduce them. Particular consideration should be given to the provision of mechanical assistance where this is reasonably practicable.

Mechanical assistance

115 Mechanical assistance involves the use of handling aids—an element of manual handling is retained but bodily forces are applied more efficiently, reducing the risk of injury. There are many examples. A simple lever can reduce the risk of injury merely by lessening the bodily force required to move a load, or by removing fingers from a potentially damaging trap. A hoist, either powered or hand operated, can support the weight of a load and leave the handler free to control its positioning. A trolley, sack truck or roller conveyor can greatly reduce the effort required to move a load horizontally. Chutes are a convenient way of using gravity to move loads from one place to another. Handling devices such as hand-held hooks or suction pads can simplify the problem of handling a load that is difficult to grasp. Examples of some common handling aids are illustrated in Figures 8–15.

Figure 8 Small hand-powered hydraulic hoist

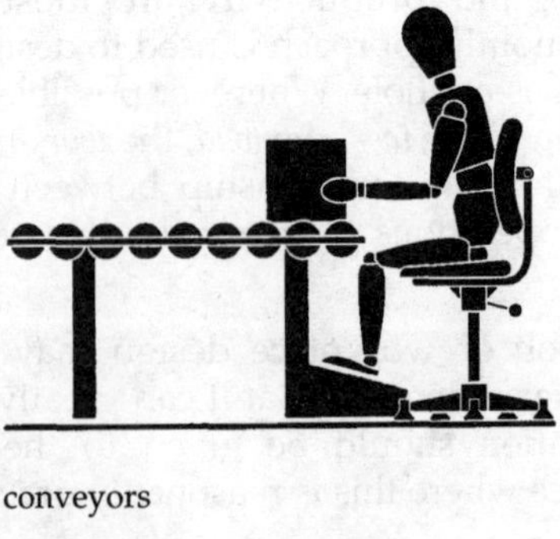

Figure 9 Roller conveyors

Figure 10 Moving large sheet material

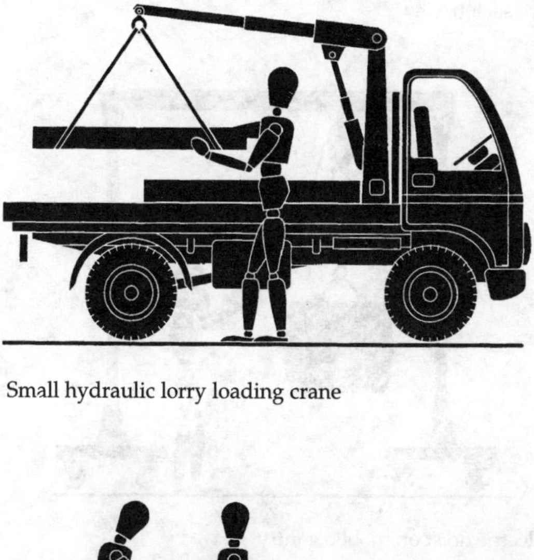

Figure 11 Small hydraulic lorry loading crane

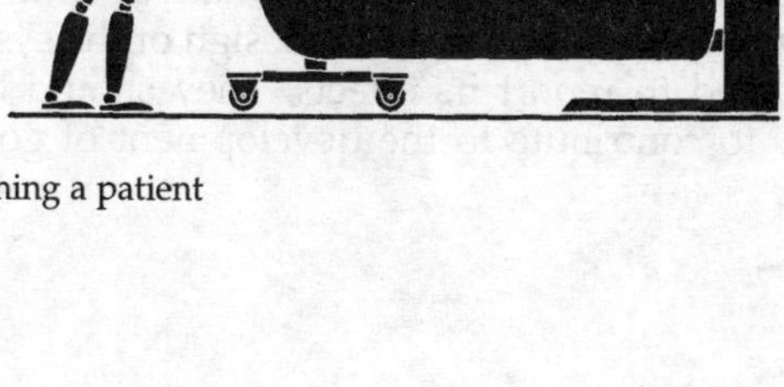

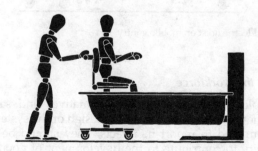

Figure 12 Bathing a patient

Figure 13 The simple low-tech sack trolley

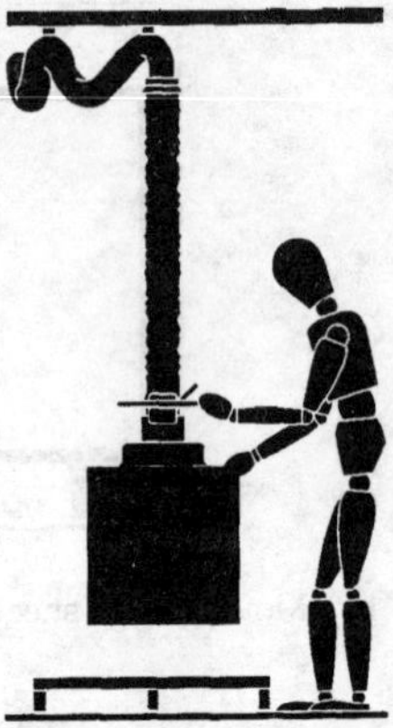

Figure 14 Powered vacuum lifter

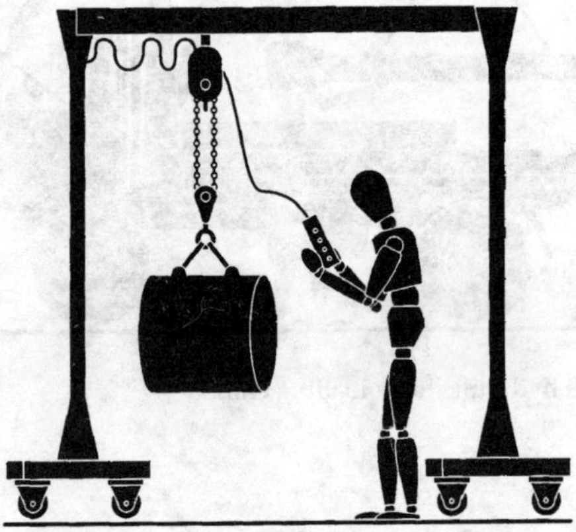

Figure 15 Electric hoist on mobile gantry

Involving the workforce

116 Employees, their safety representatives and safety committees should be involved in any redesign of the system of work and encouraged to report its effects. They should be given the opportunity to contribute to the development of good handling practice.

Industry-specific guidance

117 The development of industry-specific guidance within the framework established by the Regulations and this general guidance will provide a valuable source of information on preventive action that has been found effective for particular activities or types of work.

'Appropriate' steps

118 Above all, the steps taken to reduce the risk of injury should be 'appropriate'. They should address the problem in a practical and effective manner. Their effectiveness should be monitored. This can be done by observation of the effect of the changes made and discussions with the handlers or, less directly, through regular checks of accident statistics. If they do not have the desired effect the situation should be reappraised (see also paragraph 180 *'Reviewing the assessment'*).

Checklist

119 It may be helpful to use a checklist as an aide-memoire while seeking practical steps to reduce the risk of injury. An example of such a checklist is given on pages 636–642 which combines the assessment of risk required by regulation 4(1)(b)(i) with the identification of remedial steps required by regulation 4(1)(b)(ii). The particular example given will not be suitable in all circumstances; it can be adapted or modified as appropriate.

The task—reducing the risk of injury

Improving task layout

120 Changes to the layout of the task can reduce the risk of injury by, for example improving the flow of materials or products. Such changes will often bring the additional benefits of increased efficiency and productivity.

121 The optimum position for storage of loads, for example, is around waist height; storage much above or below this height should be reserved for loads that are lighter or more easily handled, or loads that are handled infrequently.

Using the body more efficiently

122 A closely related set of considerations concerns the way in which the handler's body is used. Changes to the task layout, the equipment used, or the sequence of operations can reduce or remove the need for twisting, stooping and stretching.

123 In general, any change that allows the load to be held closer to the body is likely to reduce the risk of injury. The level of stress in the lower back will be reduced; the weight of the load will be more easily counterbalanced by the weight of the body; and the load will be more stable and the handler less likely to lose control of it. Moreover, if the load is hugged to the body, friction with the handler's garments will steady it and may help to support its weight. The need for protective clothing should be considered (see paragraphs 136 and 137).

124 When lifting of loads at or near floor level is unavoidable, handling techniques which allow the use of the relatively strong leg muscles rather than those of the back are preferable, provided the load is small enough to be held close to the trunk. In addition, if the task includes lifting to shoulder height, an intermediate step to allow the handler to change hand grip (see Figure 16) will help to reduce risk. Bear in mind, however, that such techniques impose heavy forces on the knees and hip joints which must carry both the weight of the load and the weight of the rest of the body. If such techniques are needed for handling on a daily basis, take steps to eliminate or modify the task.

Figure 16 Use of midway stage to change grip

125 The closeness of the load to the body can also be influenced by foot placement. The elimination of obstacles which need to be

reached over or into—for example poorly placed pallets, excessively deep bins—will permit the handler's feet to be placed beneath or adjacent to the load (see Figure 17).

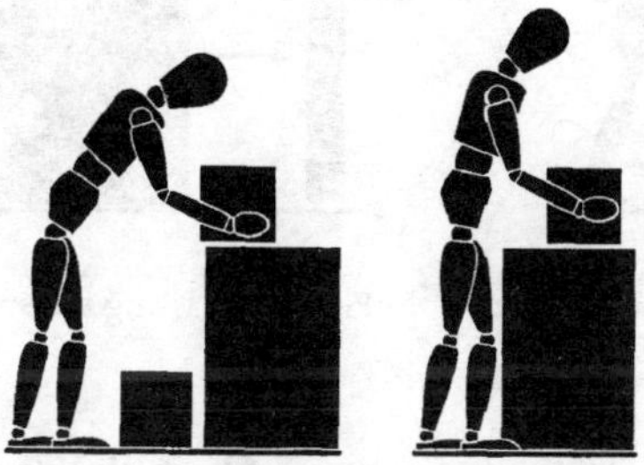

Figure 17 Avoiding an obstructed lift. Organise the workplace so that the handler can get as close to the load as possible

126 Where possible the handler should be able to move in close to the load before beginning the manual handling operation. The handler should also be able to address the load squarely, preferably facing in the direction of intended movement.

127 The risk of injury may also be reduced if lifting can be replaced by controlled pushing or pulling. For example, it may be possible to slide the load or roll it along (see Figure 18). However, uncontrolled sliding or rolling, particularly of large or heavy loads, may introduce fresh risks of injury.

128 For both pulling and pushing, a secure footing should be ensured, and the hands applied to the load at a height between waist and shoulder wherever possible. A further option, where other safety considerations allow, is to push with the handler's back against the load (see Figure 19), using the strong leg muscles to exert the force.

Improving the work routine

129 The risk of manual handling injury can also be reduced by careful attention to the work routine. Minimising the need for fixed postures dictated by sustained holding or supporting of a load will reduce fatigue and the associated fall-off in muscular

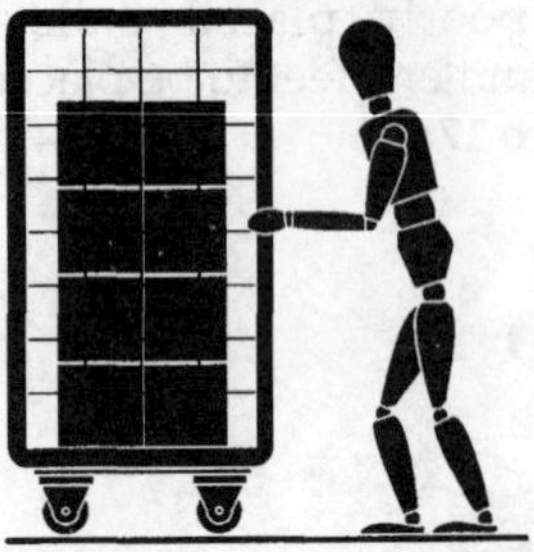

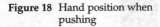

Figure 18 Hand position when pushing

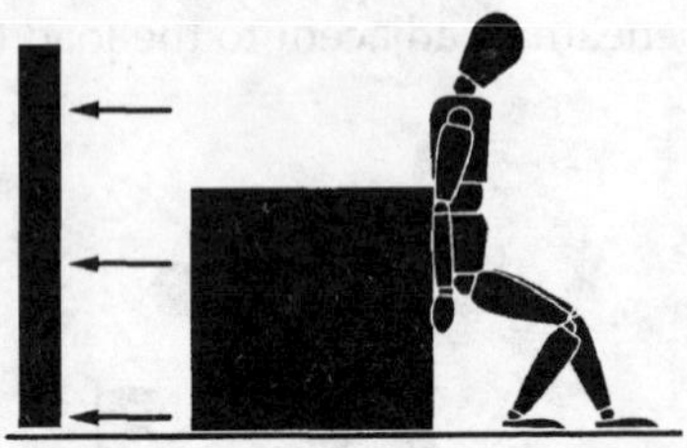

Figure 19 Using the strong leg muscles

efficiency. Attention to the frequency of handling loads, especially those that are heavy or awkward, can also reduce fatigue and the risk of injury. Where possible, tasks should be self-paced and employees trained to adjust their rate of working to optimise safety and productivity.

130 An inflexible provision of rest pauses may not be an efficient method of reducing the risk of injury. The large variation in individual susceptibility to muscular fatigue means that mandatory, fixed breaks are generally less effective than those taken voluntarily within the constraints of what is possible in terms of work organisation.

131 A better solution can often be found in job rotation where this allows one group of muscles to rest while others are being used. Periods of heavy work may be interspersed with lighter activities such as paperwork or the monitoring of instruments. Job rotation can also bring advantages in reduced monotony and increased attentiveness. However, where rotation merely repeats the use of the same group of muscles, albeit on a different task, it is generally ineffective in reducing the risk of manual handling injury.

Handling while seated

132 For the reasons given in paragraphs 71–73, the loads that can be handled in safety by a person who is seated are substantially less than can be dealt with while standing. This activity, therefore, demands particular care. Lifting loads from the floor while seated should be avoided where possible.

133 The possibility of accidental movement of the seat should be considered. Castors may be inadvisable, especially on hard floors. A swivel-action seat will help the handler to face the load without having to twist the trunk. The relative heights of seats and work surfaces should be well matched. Further advice on this is given in the HSE booklet *Seating at work*.

Team handling

134 Where a handling operation would be difficult or unsafe for one person, handling by a team of two or more may provide an answer. However, team handling can introduce additional hazards and caution should be exercised for the reasons given in paragraphs 74–76.

135 For safe team handling there should be enough space for the handlers to manoeuvre as a group. They should have adequate access to the load, and the load should provide sufficient handholds; if the load is particularly small or difficult to grasp then a handling aid such as a stretcher or slings should be used. One person should plan and then take charge of the operation, ensuring that movements are co-ordinated. Team members should preferably be of broadly similar build and physical capability. Where the weight of the load is unevenly distributed, the strongest members of the team should take the heavier end.

Personal protective equipment and other clothing

136 The nature of the load, or the environment in which it is handled, may necessitate the use of personal protective equipment (PPE) such as gloves, aprons, overalls, gaiters or safety footwear. In these cases the protection offered by PPE should not be compromised to facilitate the manual handling operations. Alternative methods of handling may need to be considered where manual handling is likely to lead to risks from the contents of the load or from external contamination.

137 PPE and indeed all work clothing should be well-fitting and restrict movement as little as possible. Fasteners, pockets and other features on which loads might snag should be concealed. Gloves should be close-fitting and supple, so that they interfere with manual dexterity as little as possible. Footwear should provide adequate support, a stable, non-slip base and proper protection. Restrictions on the handler's movement caused by

wearing protective clothing need to be recognised in the design of the task. Refer to the Personal Protective Equipment at Work Regulations 1992.

Maintenance and accessibility of equipment

138 All equipment provided for use during manual handling, including handling aids and PPE, should be well maintained and there should be a defect reporting and correction system. The siting of equipment can be important: handling aids and PPE that are not readily accessible are less likely to be used fully and effectively. Refer to The Provision and Use of Work Equipment Regulations 1992 and from 5 December 1998, The Provision and Use of Work Equipment Regulations 1998 and the Lifting Operations and Lifting Equipment Regulations 1998.

Safety of machinery—European standards

139 The Supply of Machinery (Safety) Regulations 1992 cover the essential health and safety requirements in the design of machinery and its component parts. The Regulations require machinery to be capable of being handled safely. If manual handling is involved, the machinery and component parts must be easily movable or equipped for picking up, for example with hand grips. Machinery and component parts unsuitable for manual handling must be fitted with attachments for lifting gear or designed so that standard lifting gear can be easily attached.

140 In support of the requirement that machinery be designed to facilitate its safe handling, CEN, one of the European standards making bodies, is preparing a harmonised standard entitled, *Safety of machinery—Human physical performance*. The purpose of the standard is to assist machinery designers and manufacturers. A parallel standard, *Ergonomics—Manual handling*, is being developed by ISO, the international standards setting body, to help job designers identify and eliminate risks to health from workplace manual handling activities on a more general basis. Neither standard will have direct relevance to the Manual Handling Operations Regulations 1992.

Duties of manufacturers—articles for use at work

141 The Health and Safety at Work etc Act 1974 requires designers and manufacturers to ensure the safety so far as is

reasonably practicable of any article for use at work and to provide adequate information about the conditions necessary to ensure that when put to use such articles will be safe and without risk to health (see paragraph 179).

142 To ensure that adequate information is available for articles which are likely to cause injury if manually handled, it may be helpful to provide information on the weight. The simplest way of doing this is to mark the article with its weight. Alternatively, mark its package with the total weight prominently in a place or places where the handler will see it easily. For asymmetric articles likely to cause injury when lifted manually, the centre of gravity should be marked on the article or package.

The load—reducing the risk of injury

Making it lighter

143 Where a risk of injury from the manual handling of a load is identified, consider reducing its weight. For example, liquids and powders may be packaged in smaller containers. Where loads are bought in it may be possible to specify lower package weights. However, the breaking down of loads will not always be the safest course: do not overlook the consequent increase in the frequency of handling. The effort associated with moving the handler's own body weight becomes more significant as the rate of handling rises. The result can be increased risk of fatigue and undue stresses on particular parts of the body, for example the shoulders.

144 If a great variety of weights is to be handled, it may be possible to sort the loads into weight categories so that additional precautions can be applied selectively, where most needed.

Making it smaller or easier to manage

145 Similarly, consider making loads less bulky so that they can be grasped more easily and the centre of gravity brought closer to the handler's body. Again, it may be possible to specify smaller or more manageable loads, or to redesign those produced in-house.

Making it easier to grasp

146 Where the size, surface texture or nature of a load makes it difficult to grasp, consider providing handles, hand grips, indents or any other feature designed to improve the handler's grasp.

Alternatively it may be possible to place the load securely in a container which is itself easier to grasp. Where a load is bulky rather than heavy it may be easier to carry it at the side of the body if it has suitable handholds or if slings or other carrying devices can be provided.

147 The positioning of handholds can play a part in reducing the risk of injury. For example, handholds at the top of a load may reduce the temptation to stoop when lifting it from a low level. However, depending upon the size of the load, this might also necessitate carriage with bent arms which could increase fatigue.

148 Handholds should be wide enough to clear the breadth of the palm, and deep enough to accommodate the knuckles and any gloves which may need to be worn.

Making it more stable

149 Where possible, packaging should be such that objects will not shift unexpectedly while being handled. Where the load as a whole lacks rigidity it may be preferable to use slings or other aids to maintain effective control during handling. Ideally, containers holding liquids or free-moving powders should be well filled, leaving only a small amount of free space; where this is not possible alternative means of handling should be considered.

Making it less damaging to hold

150 As far as possible, loads should be clean and free from dust, oil, corrosive deposits, etc. To prevent injury during the manual handling of hot or cold materials an adequately insulated container should be used; failing this, suitable handling aids or PPE will be necessary. Sharp corners, jagged edges, rough surfaces etc should be avoided where possible; again, where this cannot be achieved the use of handling aids or PPE will be necessary. In selecting personal protective equipment, note the advice in paragraphs 136–137.

The working environment—reducing the risk of injury

Removing space constraints

151 Gangways and other working areas should where possible allow adequate room to manoeuvre during manual handling operations. The provision of sufficient clear floor space and head room is important; constrictions caused by narrow doorways and

the positioning of fixtures, machines etc should be avoided as far as possible. In many cases, much can be achieved simply by improving the standard of housekeeping. Refer to The Workplace (Health, Safety and Welfare) Regulations 1992.

The nature and condition of floors

152 On permanent sites, both indoors and out, a flat, well maintained and properly drained surface should be provided. In construction, agriculture and other activities where manual handling may take place on temporary surfaces, the ground should be prepared if possible and kept even and firm; if possible, suitable coverings should be provided. Temporary work platforms should be firm and stable.

153 Spillages of water, oil, soap, food scraps and other substances likely to make the floor slippery should be cleared away promptly. Where necessary, and especially where floors can become wet, give attention to the choice of slip-resistant surfacing.

154 Particular care is needed where manual handling is carried out on a surface that is unstable or susceptible to movement, for example on a boat, a moving train, a mobile work platform or when using a footrest. In these conditions, the capability to handle loads in safety may be reduced significantly.

Working at different levels

155 Where possible, all manual handling activities should be carried out on a single level. Where more than one level is involved, the transition should preferably be made by a gentle slope or, failing that, by well-positioned and properly maintained steps. Manual handling on steep slopes should be avoided as far as possible. Working surfaces such as benches should, where possible, be at a uniform height to reduce the need for raising or lowering loads.

The thermal environment and ventilation

156 There is less risk of injury if manual handling is performed in a comfortable working environment. Extremes of temperature, excessive humidity and poor ventilation should be avoided where possible, either by improving environmental control or relocating the work.

157 Where these conditions cannot be changed, for example when manual handling is necessarily performed out of doors in extreme weather, or close to a very hot process, or in a refrigerated storage area, and the use of PPE is necessary, the advice given in paragraphs 136–137 should be noted.

Strong air movements

158 Particular care should be taken when handling bulky or unwieldy loads in circumstances in which high winds or powerful ventilation systems could catch a load and destabilise the handler. Possible improvements include relocating the handling operations or taking a different route, providing handling aids to give greater control of the load, or team handling.

Lighting

159 There should be sufficient well-directed light to enable handlers to see clearly what they are doing and the layout of the workplace, and to make accurate judgements of distance and position.

Individual capability—reducing the risk of injury

Personal considerations

160 Particular consideration should be given to employees who are or have recently been pregnant, or who are known to have a history of back, knee or hip trouble, hernia or other health problems which could affect their manual handling capability. However, beyond such specific pointers to increased risk of injury the scope for preventive action on an individual basis is limited.

161 Clearly an individual's state of health, fitness and strength can significantly affect the ability to perform a task safely. But even though these characteristics vary enormously, studies have shown no close correlation between any of them and injury incidence. There is, therefore, insufficient evidence for reliable selection of individuals for safe manual handling on the basis of such criteria. It is recognised, however, that there is often a degree of self-selection for work that is physically demanding.

162 It is also recognised that motivation and self-confidence in the ability to handle loads are important factors in reducing the

risk of injury. These are linked with fitness and familiarity. Unaccustomed exertion—whether in a new task or on return from holiday or sickness absence—can carry a significant risk of injury and requires particular care.

Abdominal and back support belts

163 There are many different types of abdominal and back support belts. Unfortunately, information is not always readily available on how each is designed to work. In general, however, most claim to be a lifting aid that helps reduce the physical demands of the task and so reduce the risk of injury to the handler.

164 Continuing claims that abdominal and back support belts produce impressive reductions in the number of manual handling injuries, however, are balanced by studies that show no effect on injury rates. There is also some evidence to suggest that wearing a belt may make particular individuals more susceptible to injury or to more severe injury. In addition, there is some concern about the possible long-term effects of prolonged use, such as a weakening of support muscles, but these have not yet been studied. The effectiveness of back belts to reduce risk, therefore, remains controversial.

165 The decision on whether to advocate the wearing of abdominal and back support belts as part of a risk control strategy rests with employers. Risk reduction measures, however, must be appropriate to the findings of the risk assessment. It will normally be possible to reduce the risk more directly and effectively, therefore, through safer systems of working. These could incorporate engineering, design or organisational changes to alter features concerned with the task, load or the working environment. Such measures will provide protection for the whole group of workers involved rather than to individual workers. Control measures which rely solely on the use of belts as lifting aids are unlikely to satisfy the requirements of the Regulations.

166 Employers who choose to advocate the use of abdominal and back support belts should select a design and fit appropriate to the work activities, for example some belts prohibit forward stooping movements and some belts, if worn tightly for prolonged periods, can cause breathlessness. It might also be helpful to introduce the belts cautiously on a limited scale until their effectiveness in particular circumstances has been evaluated. In all cases,

however, employees should be given full information on the way the belts are to be worn and the possible adverse effects in both the long and short term.

Information and training

167 Section 2 of the HSW Act and regulations 8 and 11 of the Management of Health and Safety at Work Regulations 1992 require employers to provide their employees with health and safety information and training. This should be supplemented as necessary with more specific information and training on manual handling injury risks and prevention, as part of the steps to reduce risk required by regulation 4(1)(b)(ii) of the Regulations.

168 It should not be assumed that the provision of information and training alone will ensure safe manual handling. The primary objective in reducing the risk of injury should always be to optimise the design of the manual handling operations, improving the task, the load and the working environment as appropriate. Where possible the manual handling operations should be designed to suit individuals, not the other way round. As a complement to a safe system of work, effective training has an important part to play in reducing the risk of manual handling injury. It should not be regarded as a substitute for it. The right kind of training will contribute to safe working methods, but it is not the whole answer.

169 Employers should ensure that their employees understand clearly how manual handling operations have been designed to ensure their safety. Employees, their safety representatives and safety committees should be involved in developing and implementing manual handling training, and monitoring its effectiveness. This might also involve checking whether behaviour patterns have improved and accident rates reduced.

170 In devising a training programme for safe manual handling, pay particular attention, therefore, to imparting a clear understanding of how to recognise potentially hazardous handling operations:

(a) how the hazard might be avoided;

(b) how to deal with unavoidable and unfamiliar handling operations;

(c) the proper use of handling aids;

(d) the proper use of personal protective equipment;

(e) features of the working environment that contribute to safety;

(f) the importance of appropriate housekeeping;

(g) factors affecting individual capability;

(h) good handling technique (see paragraphs 173–174).

171 Employees should be trained to recognise loads whose weight, in conjunction with their shape and other features, and the circumstances in which they are handled, might cause injury. Simple methods for estimating weight on the basis of volume may be taught. Where volume is less important than the density of the contents, as for example in the case of a dustbin containing refuse, an alternative technique for assessing the safety of handling should be taught, such as rocking the load from side to side before attempting to lift it (see Figure 20).

Figure 20 Rocking a load to assess its ease of handling

172 In general, unfamiliar loads should be treated with caution. For example it should not be assumed that apparently empty drums or other closed containers are in fact empty. The load may first be tested, for example by attempting to raise one end. Employees should be taught to apply force gradually until either undue strain is felt, in which case the task should be reconsidered, or it is apparent that the task is within the handler's capability.

Good handling technique

173 A good handling technique is no substitute for other risk reduction steps such as improvements to the task, load or working environment. In addition, moving the load by rocking, pivoting, rolling or sliding is preferable to lifting it in situations where scope

for risk reduction is limited. However, good handling technique forms a very valuable adjunct to other risk control measures. It requires both training and practice. The training should be carried out in conditions that are as realistic as possible, emphasising its relevance to everyday handling operations.

174 There is no single correct way to lift and many different approaches are put forward. Each has merits and advantages in particular situations or individual circumstances. The content of training in good handling technique, therefore, should be tailored to the particular handling operations likely to be undertaken. It should begin with relatively simple examples and progress to more specialised handling operations as appropriate. The following list illustrates some important points, using a basic lifting operation by way of example:

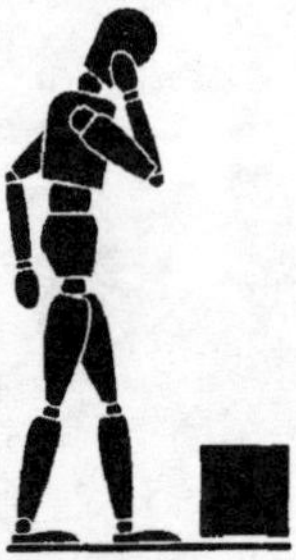

(a) *Stop and think*. Plan the lift. Where is the load going to be placed? Use appropriate handling aids if possible. Do you need help with the load? Remove obstructions such as discarded wrapping materials. For a long lift—such as floor to shoulder height —consider resting the load mid-way on a table or bench to change grip.

(b) *Place the feet*. Have the feet apart, giving a balanced and stable base for lifting (tight skirts and unsuitable footwear make this difficult). Have the leading leg as far forward as is comfortable.

(c) *Adopt a good posture*. Bend the knees so that the hands when grasping the load are as nearly level with the waist as possible. But do not kneel or overflex the knees. Keep the back straight, maintaining its natural curve (tucking in the chin while gripping the load helps). Lean forward a little over the load if necessary to get a good grip. Keep shoulders level and facing in the same direction as the hips.

(d) *Get a firm grip*. Try to keep the arms within the boundary formed by the legs. The optimum position and nature of the grip depends on the circumstances and individual preference, but it must be secure. A hook grip is less fatiguing than keeping the fingers straight. If it is necessary to vary the grip as the lift proceeds, do this as smoothly as possible.

(e) *Don't jerk*. Carry out the lifting movement smoothly, raising the chin as the lift begins, keeping control of the load.

(f) *Move the feet*. Don't twist the trunk when turning to the side.

(g) *Keep close to the load*. Keep the load close to the trunk for as long as possible. Keep the heaviest side of the load next to the trunk. If a close approach to the load is not possible try sliding it towards you before attempting to lift it.

(h) *Put down*, then *adjust*. If precise positioning of the load is necessary, put it down first, then slide it into the desired position.

Figure 21 Basic lifting operation

Vocational qualifications

175 The development of specific statements of what needs to be done, how well and by whom (ie statements of competence) will help to determine the extent of any shortfall in training. Such statements may be embodied in qualifications accredited by the National Council for Vocational Qualifications (NCVQ) and SCOTVEC (the Scottish Vocational Education Council).

The load—providing additional information

176 Regulation 4(1)(b)(iii) can be satisfied in a variety of ways, depending upon the circumstances.

177 The requirement to provide 'general indications' of the weight and nature of the loads to be handled should be addressed during basic training, so that employees are suitably prepared for the operations they are likely to undertake.

178 In addition, where it is reasonably practicable to do so, employers should give more precise information. Employers whose businesses originate loads may find that this information is best given by marking it on the loads.

179 The Regulations impose duties on employers whose employees carry out manual handling. However, those who originate loads—manufacturers, packers, etc—that are likely to undergo manual handling may also have relevant duties, for example under sections 3 or 6 of the HSW Act, for the health and safety of other people at work. They should give particular consideration to making loads easy to grasp and handle and to marking loads clearly with their weight and, where appropriate, an indication of their heaviest side (see paragraphs 141–142).

Reviewing the assessment

180 The assessment should be kept up to date. It should be reviewed if new information comes to light or if there has been a change in the manual handling operations which, in either case, could materially have affected the conclusion reached previously. The assessment should also be reviewed if a reportable injury occurs. It should be corrected or modified where this is found to be necessary.

Regulation 5

Duty of employees

181 Duties are already placed on employees by section 7 of the HSW Act, under which they must:

(a) take reasonable care for their own health and safety and that of others who may be affected by their activities; and

(b) co-operate with their employers to enable them to comply with their health and safety duties.

182 In addition, regulation 12 of the Management of Health and Safety at Work Regulations 1992 requires employees generally to make use of appropriate equipment provided for them, in accordance with their training and the instructions their employer has given them. Such equipment will include machinery and other aids provided for the safe handling of loads.

183 Regulation 5 of the Manual Handling Operations Regulations supplements these general duties in the case of manual handling. It requires employees to follow appropriate systems of work laid down by their employer to promote safety during the handling of loads.

Emergency action

184 These provisions do not preclude well-intentioned improvisation in an emergency, for example during efforts to rescue a casualty, fight a fire or contain a dangerous spillage.

Regulation 7

Extension outside Great Britain

185 The Regulations apply to offshore activities covered by the 1989 Order on or associated with oil and gas installations, including mobile installations, diving support vessels, heavy lift barges and pipe-lay barges.

Regulation 8

Repeals and revocations

186 The Regulations, like the European Directive on manual handling, apply a modern ergonomic approach to the prevention

of injury. They take account of a wide range of relevant factors, including the nature of the task, the load, the working environment and individual capability. The Regulations have, therefore, replaced a number of outdated provisions which concentrated on the weight of the load being handled. The provisions are listed in Schedule 2 to the Regulations.

APPENDIX 1

MANUAL HANDLING RISK ASSESSMENT DETAILED ASSESSMENT GUIDELINES FILTER

Introduction

1 The Manual Handling Regulations set no specific requirements such as weight limits. Instead, they focus on the needs of the individual and set out a hierarchy of measures for safety during manual handling operations:

(a) avoid hazardous manual handling operations so far as is reasonably practicable;

(b) make a suitable and sufficient assessment of any hazardous manual handling operations that cannot be avoided; and

(c) reduce the risk of injury from those operations so far as is reasonably practicable.

Risk assessment filter

2 Where manual handling operations cannot be avoided, employers have a duty to make a suitable and sufficient assessment of the risks to health. This assessment must take into account the range of relevant factors listed in Schedule 1 to the Regulations. A detailed assessment of every manual handling operation, however, could be a major undertaking and might involve wasted effort. Many handling operations, for example lifting a tea cup, will involve negligible handling risk. To help identify situations where a more detailed risk assessment is necessary, HSE has developed a filter to screen out straightforward cases.

3 The filter is based on a set of numerical guidelines developed from data in published scientific literature and on practical

experience of assessing risks from manual handling. They are pragmatic, tried and tested; they are not based on any precise scientific formulae. The intention is to set out an approximate boundary within which the load is unlikely to create a risk of injury sufficient to warrant a detailed assessment.

4 The application of the guidelines will provide a reasonable level of protection to around 95% of working men and women. However, the guidelines should not be regarded as safe weight limits for lifting. There is no threshold below which manual handling operations may be regarded as 'safe'. Even operations lying within the boundary mapped out by the guidelines should be avoided or made less demanding wherever it is reasonably practicable to do so.

5 It is important to remember that the purpose of the guidelines is to avoid wasted time and effort. The use of the filter will only be worthwhile, therefore, where the relevance of the guideline figures can be determined quickly, say within 10 minutes. If it is not clear from the outset that this can be done, it is better to opt immediately for the more detailed risk assessment.

Guidelines for lifting and lowering

6 The guidelines for lifting and lowering operations assume that the load is easy to grasp with both hands and that the operation takes place in reasonable working conditions with the handler in a stable body position. They take into consideration the vertical and horizontal position of the hands as they move the load during the handling operation, as well as the height and reach of the individual handler. For example, if a load is held at arm's length or the hands pass above shoulder height, the capability to lift or lower is reduced significantly.

7 The basic guideline figures for identifying when manual lifting and lowering operations may not need a detailed assessment are set out in Figure 22. If the handler's hands enter more than one of the box zones during the operation, the smallest weight figures apply. It is important to remember, however, that the transition from one box zone to another is not abrupt; an intermediate figure may be chosen where the handler's hands are close to a boundary. Where lifting or lowering with the hands beyond the box zones is unavoidable, a more detailed assessment should always be made.

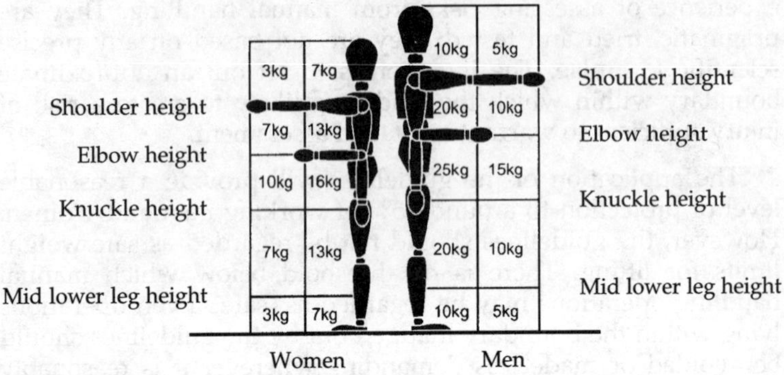

Figure 22 Lifting and lowering

8 These basic guideline figures for lifting and lowering are for relatively infrequent operations—up to approximately 30 operations per hour. The guideline figures will have to be reduced if the operation is repeated more often. As a rough guide, the figures should be reduced by 30% where the operation is repeated once or twice per minute, by 50% where the operation is repeated around five to eight times per minute and by 80% where the operation is repeated more than about 12 times per minute.

9 Even if the above conditions are satisfied, a more detailed risk assessment should be made where:

(a) the worker does not control the pace of work;

(b) pauses for rest are inadequate or there is no change of activity which provides an opportunity to use different muscles;

(c) the handler must support the load for any length of time.

Guidelines for carrying

10 Similar guideline figures apply to carrying operations where the load is held against the body and is carried no further than about 10 m without resting. If the load is carried over a longer distance without resting or the hands are below knuckle height then a more detailed risk assessment should be made.

11 Where the load can be carried securely on the shoulder without first having to be lifted (as for example when unloading sacks from a lorry) the guideline figures can be applied to carrying distances in excess of 10 m.

Guidelines for pushing and pulling

12 For pushing and pulling operations (whether the load is slid, rolled or supported on wheels) the guideline figures assume the force is applied with the hands between knuckle and shoulder height. The guideline figure for starting or stopping the load is a force of about 25 kg (ie about 250 Newtons) for men and about 16 kg (ie about 160 Newtons) for women. The guideline figure for keeping the load in motion is a force of about 10 kg (ie about 100 Newtons) for men and about 7 kg (ie about 70 Newtons) for women.

13 There is no specific limit to the distance over which the load is pushed or pulled provided there are adequate opportunities for rest or recovery.

Guidelines for handling while seated

14 The basic guideline figure for handling operations carried out while seated, shown in Figure 23, is 5 kg for men and 3 kg for women. These guidelines only apply when the hands are within the box zone indicated. If handling beyond the box zone is unavoidable, a more detailed assessment should be made.

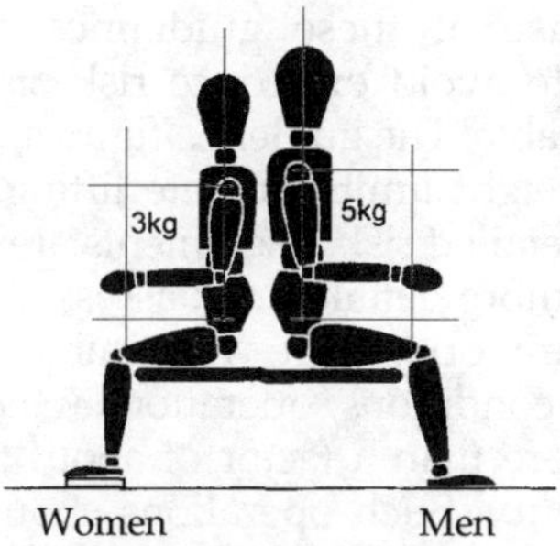

Figure 23 Handling while seated

Other considerations: Twisting

15 In many cases, manual handling operations will involve some twisting (see Figure 24) and this will increase the risk of injury. Where the handling task involves twisting and turning, therefore, a detailed risk assessment should normally be made. However, if the operation is relatively infrequent (see paragraph 8) and there

are no other posture problems then the filter can be used. In such cases, the basic guideline figures shown above should be reduced if the handler twists to the side during the operation. As a rough guide, the figures should be reduced by about 10% where the handler twists through 45° and by about 20% where the handler twists through 90°.

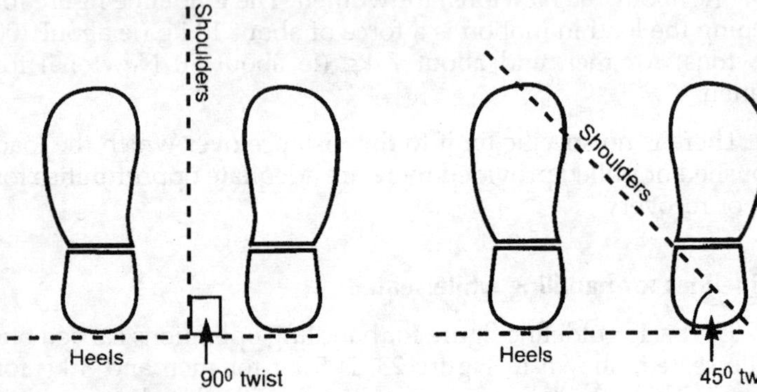

Figure 24 Assessing twist

Remember: The use of these guidelines does not affect the employer's duty to avoid or reduce risk of injury where this is reasonably practicable. The guideline figures, therefore, should not be regarded as weight limits for safe lifting. They are an aid to highlight where detailed risk assessments are most needed. Where doubt remains, a more detailed risk assessment should always be made. Even for a minority of fit, well-trained individuals working under favourable conditions, operations which exceed the guideline figures by more than a factor of about two may represent a serious risk of injury. Such operations should come under very close scrutiny.

APPENDIX 2

EXAMPLE OF AN ASSESSMENT CHECKLIST

1 A suitable and sufficient risk assessment is required when hazardous manual handling is unavoidable. The assessment should identify where the risk lies and suggest an appropriate

range of ideas for reducing the potential for injury. A checklist can help with this process by ensuring a systematic examination of all the potential risk elements.

2 An example of a basic checklist is provided on pages 641–2. Its use will help to highlight the overall level of risk involved and identify how the job may be modified to reduce the risk of injury and make it easier to do. It will also be useful in helping to prioritise the remedial actions needed. The checklist may be copied freely or may be used to help design your own assessment checklist.

3 The following notes are intended to assist in completing the checklist.

(a) *Section A: Describe* the job. There is space available for a diagram to be drawn to summarise the job in a picture, as well as for a written description.

(b) *Section B: Tick* the level of risk you believe to be associated with each of the items on the list. Space is provided for noting the precise nature of the problem and for suggestions about the remedial action that may be taken. It may also be useful to write down the names of the relevant people or groups in your organisation who you will wish to consult about implementing the remedial steps, for example managers, workforce trainers, maintenance personnel or engineers.

Some tasks may involve more than one operator, each with a different level of risk, depending on the exact nature of their duties. If you wish to use the same checklist for all of the operators involved, you can allocate a number (or other identifying mark) to each and use that against each tick (eg ✓^{1}; ✓$^{1/2}$; ✓$^{1/2/3}$; etc) or comment on the checklist form that relates to each particular operator.

(c) *Section C: Decide* whether the overall risk of injury is low, medium or high. This section will help to prioritise remedial action if you have a large number of risk assessments to carry out.

(d) *Section D: Summarise* the remedial steps that should be taken, in order of priority. You may also wish to write in *(I)*, *(M)* or *(L)* alongside each entry to denote whether the action can be taken *(I)mmediately* or is a more *(M)edium-term* or *(L)ong-term* objective. The assessor's name and the date by which the agreed actions should be carried out should be recorded. It may also be useful to enter the target date for reassessment if this is appropriate.

4 When all the manual handling tasks have been assessed, the completed checklists can be compared to help prioritise the most urgent actions. However, there are likely to be several ways to reduce the risks identified and some will be more effective than others. Action on those that can be implemented easily and quickly should not be delayed simply because they may be less effective than others.

5 A check should be carried out at a later date to ensure that the remedial action to remove or reduce the risk of injury has been effective.

6 A worked example of a risk assessment made using the checklist is given on pages 643–4 to show how the checklist might be used in practice.

7 The purpose of the checklist is to help bring out a range of ideas on how the risks identified can be avoided or reduced by making modifications to the load, the task, and the working environment. There are a number of people who may be able to help with suggestions, for example safety representatives, the quality management team within the organisation, and relevant trade associations. There is also a great deal of published information about risk reduction methods. *Solutions you can handle* and *A pain in your workplace*, both published by HSE, give examples that are relevant to situations across many sectors of industry. Trade journals, too, often contain information about products that can be used to help reduce the risk of injury from the manual handling of loads.

APPENDIX 3

REFERENCES AND FURTHER INFORMATION

This Appendix is not reproduced in this book.

MANUAL HANDLING OF LOADS: ASSESSMENT CHECKLIST

Section A - Preliminary: *Circle as appropriate

Job description: Factors beyond the limits of the guidelines?	Is an assessment needed? (ie is there a potential risk for injury, and are the factors beyond the limits of the guidelines?) Yes/No*

If 'Yes' continue. If 'No' the assessment need go no further.

Operations covered by this assessment (detailed description): Locations: Personnel involved: Date of assessment:	Diagrams (other information):

Section B - See over for detailed analysis

Section C - Overall assessment of the risk of injury? Low/ Med/ High*

Section D - Remedial action to be taken:

Remedial steps that should be taken, in order of priority: 1 2 3 4 5 6 7 8	
Date by which action should be taken:	
Date for reassessment:	
Assessor's name:	Signature:

TAKE ACTION ... AND CHECK THAT IT HAS THE DES!RED EFFECT

Section B - More detailed assessment, where necessary:

Questions to consider:	If yes, tick appropriate level of risk			Problems occurring from the task (Make rough notes in this column in preparation for the possible remedial action to be taken)	Possible remedial action (Possible changes to be made to system/task, load, workplace/space, environment. Communication that is needed)
	Low	Med	High		
The tasks - do they involve: ● holding loads away from trunk? ● twisting? ● stooping? ● reaching upwards? ● large vertical movement? ● long carrying distances? ● strenuous pushing or pulling? ● unpredictable movement of loads? ● repetitive handling? ● insufficient rest or recovery? ● a work rate imposed by a process?					
The loads - are they: ● heavy? ● bulky/unwieldy? ● difficult to grasp? ● unstable/unpredictable? ● intrinsically harmful (eg sharp/hot)?					
The working environment - are there: ● constraints on posture? ● poor floors? ● variations in levels? ● hot/cold/humid conditions? ● strong air movements? ● poor lighting conditions?					
Individual capability - does the job: ● require unusual capability? ● hazard those with a health problem? ● hazard those who are pregnant? ● call for special information/training?					
Other factors: Is movement or posture hindered by clothing or personal protective equipment?	Yes/No				

MANUAL HANDLING OF LOADS: ASSESSMENT CHECKLIST WORKED EXAMPLE

Section A - Preliminary: *circle as appropriate

Job description: **Pallet loading : boxes containing coiled wire**	Is an assessment needed? (ie is there a potential risk for injury, and are the factors beyond the limits of the guidelines?) (Yes)/No*

If 'Yes' continue. If 'No' the assessment need go no further.

Operations covered by this assessment (detailed description): **Operator lifts box, with hook grip, from conveyor, which is 20 inches above the ground, turns, walks 3 metres and lowers box onto a pallet on the ground. Boxes are piled six high on pallet.** Locations: **Wire factory only** Personnel involved: **One operator** Date of assessment: **xx June 19xx**	Diagrams (other information): **a) Worker; b) Conveyor; c) 48 kg boxes of wire; d) Pallet.** 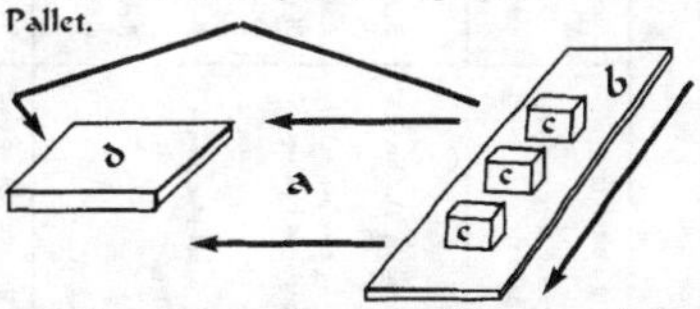 **Arrows show direction of conveyor belt and worker movements between conveyor and pallet**

Section B - See over for detailed analysis

Section C - Overall assessment of the risk of injury? Low/ Med/(High)*

Section D - Remedial action to be taken:

Remedial steps that should be taken, in order of priority:

1 **Review product design to reduce weight of load and improve grip.**

2 **Review process in light of changes agreed in (1), particularly on customer requirements and transportation.**

3 **Seek funding for magnetic lifting aid to help with transfer from conveyor to pallet.**

4 **Seek funding for pallet rotating/height adjustment equipment.**

5 **Operator to attend manual handling training.**

6 **Raise conveyor height by 15 inches.**

7 **Ensure full pallets are removed by pallet truck promptly.**

8 **Operations manager to ensure no rushing on this job.**

Date by which action should be taken: **xx December 19xx**

Date for reassessment: **xx December 20xx**

Assessor's name: **A N Onymous** Signature: **A N Onymous**

Section B - More detailed assessment, where necessary:					
Questions to consider:	If yes, tick appropriate level of risk			Problems occurring from the task (Make rough notes in this column in preparation for the possible remedial action to be taken)	Possible remedial action (Possible changes to be made to system/task, load, workplace/space, environment. Communication that is needed)
	Low	Med	High		
The tasks - do they involve:					
● holding loads away from trunk?			✔	1 Twisting when picking up the box	Remind operator of need to move feet (I).
● twisting?		✔			
● stooping?			✔	2 Stooping when placing box on pallet and stooping when picking box up from the conveyor	Adjust pallet height - Review availability of rotating, height adjusting equipment (I) and raise height of conveyor (M).
● reaching upwards?	✔				
● large vertical movement?	✔				
● long carrying distances?	✔				
● strenuous pushing or pulling?	✔			3 Sometimes extended reaching when placing boxes on pallet.	Provide better information and instruction (I).
● unpredictable movement of loads?	✔				
● repetitive handling?	✔				Review mechanical handling equipment to eliminate manual lifting (I).
● insufficient rest or recovery?	✔				
● a workrate imposed by a process?	✔				
The loads - are they:					
● heavy?			✔	4 Load too heavy. Is the weight of the load a problem for customers too?	Review product and customer needs with a view to improving product design (L).
● bulky/unwieldy?	✔				
● difficult to grasp?		✔		5 Smooth cardboard boxes are difficult to grasp.	Provide boxes with hand grips (M).
● unstable/unpredictable?	✔				
● intrinsically harmful (eg sharp/hot)?	✔				
The working environment - are there:					
● constraints on posture?		✔		6 Bad postures encouraged by obstructions when full pallets are not removed.	Introduce system to ensure full pallets removed promptly - Speak to Operations Manager ***(I)***.
● poor floors?	✔				
● variations in levels?	✔				
● hot/cold/humid conditions?	✔				
● strong air movements?	✔				
● poor lighting conditions?	✔				
Individual capability - does the job:					
● require unusual capability?			✔	7 Operator has no history of back pain problems but clear signs of sweating and straining.	Consider job enlargement to introduce variety and allow for recovery time (M).
● hazard those with a health problem?			✔		Monitor to ensure no rushing (I).
● hazard those who are pregnant?			✔		Speak to trainer about manual handling course (I).
● call for special information/training?		✔			
Other factors: Is movement or posture hindered by clothing or personal protective equipment?	Yes/No				

Appendix 6

Health and Safety (Display Screen Equipment) Regulations 1992

THE REGULATIONS (SI 1992/2792)

Made 5 November 1992

The Secretary of State, in exercise of the powers conferred on her by sections 15(1), (2), (5)(b) and (9) and 82(3)(a) of, and paragraphs 1(1)(a) and (c) and (2), 7, 8(1), 9 and 14 of Schedule 3 to, the Health and Safety at Work etc Act 1974 and of all other powers enabling her in that behalf and for the purpose of giving effect without modifications to proposals submitted to her by the Health and Safety Commission under section 11(2)(d) of the said Act after the carrying out by the said Commission of consultations in accordance with section 50(3) of that Act, hereby makes the following Regulations:

Regulation 1 *(See Guidance Notes, pp 657–665)*

Citation, commencement, interpretation and application

(1) These Regulations may be cited as the Health and Safety (Display Screen Equipment) Regulations 1992 and shall come into force on 1 January 1993.

(2) In these Regulations—

(a) 'display screen equipment' means any alphanumeric or graphic display screen, regardless of the display process involved;

(b) 'operator' means a self-employed person who habitually uses display screen equipment as a significant part of his normal work;

(c) 'use' means use for or in connection with work;

(d) 'user' means an employee who habitually uses display screen equipment as a significant part of his normal work; and

(e) 'workstation' means an assembly comprising—

(i) display screen equipment (whether provided with software determining the interface between the equipment and its operator or user, a keyboard or any other input device),

(ii) any optional accessories to the display screen equipment,

(iii) any disk drive, telephone, modem, printer, document holder, work chair, work desk, work surface or other item peripheral to the display screen equipment, and

(iv) the immediate work environment around the display screen equipment.

(3) Any reference in these Regulations to—

(a) a numbered regulation is a reference to the regulation in these Regulations so numbered; or

(b) a numbered paragraph is a reference to the paragraph so numbered in the regulation in which the reference appears.

(4) Nothing in these Regulations shall apply to or in relation to—

(a) drivers' cabs or control cabs for vehicles or machinery;

(b) display screen equipment on board a means of transport;

(c) display screen equipment mainly intended for public operation;

(d) portable systems not in prolonged use;

(e) calculators, cash registers or any equipment having a small data or measurement display required for direct use of the equipment; or

(f) window typewriters.

Regulation 2 *(See Guidance Notes, pp 665–671)*

Analysis of workstations to assess and reduce risks

(1) Every employer shall perform a suitable and sufficient analysis of those workstations which—

(a) (regardless of who has provided them) are used for the purposes of his undertaking by users; or

(b) have been provided by him and are used for the purposes of his undertaking by operators,

for the purpose of assessing the health and safety risks to which those persons are exposed in consequence of that use.

(2) Any assessment made by an employer in pursuance of paragraph (1) shall be reviewed by him if—

(a) there is reason to suspect that it is no longer valid; or

(b) there has been a significant change in the matters to which it relates;

and where as a result of any such review changes to an assessment are required, the employer concerned shall make them.

(3) The employer shall reduce the risks identified in consequence of an assessment to the lowest extent reasonably practicable.

(4) The reference in paragraph (3) to 'an assessment' is a reference to an assessment made by the employer concerned in pursuance of paragraph (1) and changed by him where necessary in pursuance of paragraph (2).

Regulation 3 *(See Guidance Notes, pp 671–675)*

Requirements for workstations

(1) Every employer shall ensure that any workstation first put into service on or after 1 January 1993 which—

(a) (regardless of who has provided it) may be used for the purposes of his undertaking by users; or

(b) has been provided by him and may be used for the purposes of his undertaking by operators,

meets the requirements laid down in the Schedule to these Regulations to the extent specified in paragraph 1 thereof.

(2) Every employer shall ensure that any workstation first put into service on or before 31 December 1992 which—

(a) (regardless of who provided it) may be used for the purposes of his undertaking by users; or

(b) was provided by him and may be used for the purposes of his undertaking by operators,

meets the requirements laid down in the Schedule to these Regulations to the extent specified in paragraph 1 thereof not later than 31 December 1996.

Regulation 4 *(See Guidance Notes, pp 675–677)*

Daily work routine of users

Every employer shall so plan the activities of users at work in his undertaking that their daily work on display screen equipment is periodically interrupted by such breaks or changes of activity as reduce their workload at that equipment.

Regulation 5 *(See Guidance Notes, pp 677–681)*

Eyes and eyesight

(1) Where a person—

(a) is already a user on the date of coming into force of these Regulations; or

(b) is an employee who does not habitually use display screen equipment as a significant part of his normal work but is to become a user in the undertaking in which he is already employed,

his employer shall ensure that he is provided at his request with an appropriate eye and eyesight test, any such test to be carried out by a competent person.

(2) Any eye and eyesight test provided in accordance with paragraph (1) shall—

(a) in any case to which sub-paragraph (a) of that paragraph applies, be carried out as soon as practicable after being requested by the user concerned; and

(b) in any case to which sub-paragraph (b) of that paragraph applies, be carried out before the employee concerned becomes a user.

(3) At regular intervals after an employee has been provided with an eye and eyesight test in accordance with paragraphs (1)

and (2), his employer shall, subject to paragraph (6), ensure that he is provided with a further eye and eyesight test of an appropriate nature, any such test to be carried out by a competent person.

(4) Where a user experiences visual difficulties which may reasonably be considered to be caused by work on display screen equipment, his employer shall ensure that he is provided at his request with an appropriate eye and eyesight test, any such test to be carried out by a competent person as soon as practicable after being requested as aforesaid.

(5) Every employer shall ensure that each user employed by him is provided with special corrective appliances appropriate for the work being done by the user concerned where—

(a) normal corrective appliances cannot be used; and

(b) the result of any eye and eyesight test which the user has been given in accordance with this regulation shows such provision to be necessary.

(6) Nothing in paragraph (3) shall require an employer to provide any employee with an eye and eyesight test against that employee's will.

Regulation 6 *(See Guidance Notes, pp 681–683)*

Provision of training

(1) Where a person—

(a) is already a user on the date of coming into force of these Regulations; or

(b) is an employee who does not habitually use display screen equipment as a significant part of his normal work but is to become a user in the undertaking in which he is already employed,

his employer shall ensure that he is provided with adequate health and safety training in the use of any workstation upon which he may be required to work.

(2) Every employer shall ensure that each user at work in his undertaking is provided with adequate health and safety training whenever the organisation of any workstation in that undertaking upon which he may be required to work is substantially modified.

Regulation 7 *(See Guidance Notes, pp 683–684)*

Provision of information

(1) Every employer shall ensure that operators and users at work in his undertaking are provided with adequate information about—

(a) all aspects of health and safety relating to their workstations; and

(b) such measures taken by him in compliance with his duties under regulations 2 and 3 as relate to them and their work.

(2) Every employer shall ensure that users at work in his undertaking are provided with adequate information about such measures taken by him in compliance with his duties under regulations 4 and 6(2) as relate to them and their work.

(3) Every employer shall ensure that users employed by him are provided with adequate information about such measures taken by him in compliance with his duties under regulations 5 and 6(1) as relate to them and their work.

Regulation 8

Exemption certificates

(1) The Secretary of State for Defence may, in the interests of national security, exempt any of the home forces, any visiting force or any headquarters from any of the requirements imposed by these Regulations.

(2) Any exemption such as is specified in paragraph (1) may be granted subject to conditions and to a limit of time and may be revoked by the Secretary of State for Defence by a further certificate in writing at any time.

(3) In this regulation—

(a) 'the home forces' has the same meaning as in section 12(1) of the Visiting Forces Act 1952;

(b) 'headquarters' has the same meaning as in article 3(2) of the Visiting Forces and International Headquarters (Application of Law) Order 1965; and

(c) 'visiting force' has the same meaning as it does for the purposes of any provision of Part I of the Visiting Forces Act 1952.

Regulation 9

Extension outside Great Britain

These Regulations shall, subject to regulation 1(4), apply to and in relation to the premises and activities outside Great Britain to which sections 1 to 59 and 80 to 82 of the Health and Safety at Work etc Act 1974 apply by virtue of the Health and Safety at Work etc Act 1974 (Application Outside Great Britain) Order 1989 as they apply within Great Britain.

SCHEDULE
(WHICH SETS OUT THE MINIMUM REQUIREMENTS FOR WORKSTATIONS WHICH ARE CONTAINED IN THE ANNEX TO COUNCIL DIRECTIVE 90/270/EEC ON THE MINIMUM SAFETY AND HEALTH REQUIREMENTS FOR WORK WITH DISPLAY SCREEN EQUIPMENT)

Regulation 3

1 Extent to which employers must ensure that workstations meet the requirements laid down in this Schedule

An employer shall ensure that a workstation meets the requirements laid down in this Schedule to the extent that—

(a) those requirements relate to a component which is present in the workstation concerned;

(b) those requirements have effect with a view to securing the health, safety and welfare of persons at work; and

(c) the inherent characteristics of a given task make compliance with those requirements appropriate as respects the workstation concerned.

2 Equipment

(a) General comment

The use as such of the equipment must not be a source of risk for operators or users.

(b) Display screen

The characters on the screen shall be well-defined and clearly formed, of adequate size and with adequate spacing between the characters and lines.

The image on the screen should be stable, with no flickering or other forms of instability.

The brightness and the contrast between the characters and the background shall be easily adjustable by the operator or user, and also be easily adjustable to ambient conditions.

The screen must swivel and tilt easily and freely to suit the needs of the operator or user.

It shall be possible to use a separate base for the screen or an adjustable table.

The screen shall be free of reflective glare and reflections liable to cause discomfort to the operator or user.

(c) Keyboard

The keyboard shall be tiltable and separate from the screen so as to allow the operator or user to find a comfortable working position avoiding fatigue in the arms or hands.

The space in front of the keyboard shall be sufficient to provide support for the hands and arms of the operator or user.

The keyboard shall have a matt surface to avoid reflective glare.

The arrangement of the keyboard and the characteristics of the keys shall be such as to facilitate the use of the keyboard.

The symbols on the keys shall be adequately contrasted and legible from the design working position.

(d) Work desk or work surface

The work desk or work surface shall have a sufficiently large, low-reflectance surface and allow a flexible arrangement of the screen, keyboard, documents and related equipment.

The document holder shall be stable and adjustable and shall be positioned so as to minimise the need for uncomfortable head and eye movements.

There shall be adequate space for operators or users to find a comfortable position.

(e) Work chair

The work chair shall be stable and allow the operator or user easy freedom of movement and a comfortable position.

The seat shall be adjustable in height.

The seat back shall be adjustable in both height and tilt.

A footrest shall be made available to any operator or user who wishes one.

3 Environment

(a) Space requirements

The workstation shall be dimensioned and designed so as to provide sufficient space for the operator or user to change position and vary movements.

(b) Lighting

Any room lighting or task lighting provided shall ensure satisfactory lighting conditions and an appropriate contrast between the screen and the background environment, taking into account the type of work and the vision requirements of the operator or user.

Possible disturbing glare and reflections on the screen or other equipment shall be prevented by co-ordinating workplace and workstation layout with the positioning and technical characteristics of the artificial light sources.

(c) Reflections and glare

Workstations shall be so designed that sources of light, such as windows and other openings, transparent or translucid walls, and brightly coloured fixtures or walls cause no direct glare and no distracting reflections on the screen.

Windows shall be fitted with a suitable system of adjustable covering to attenuate the daylight that falls on the workstation.

(d) Noise

Noise emitted by equipment belonging to any workstation shall be taken into account when a workstation is being equipped, with a view in particular to ensuring that attention is not distracted and speech is not disturbed.

(e) Heat

Equipment belonging to any workstation shall not produce excess heat which could cause discomfort to operators or users.

(f) Radiation

All radiation with the exception of the visible part of the electromagnetic spectrum shall be reduced to negligible levels from the point of view of the protection of operators' or users' health and safety.

(g) Humidity

An adequate level of humidity shall be established and maintained.

4 Interface between computer and operator/user

In designing, selecting, commissioning and modifying software, and in designing tasks using display screen equipment, the employer shall take into account the following principles:

(a) software must be suitable for the task;

(b) software must be easy to use and, where appropriate, adaptable to the level of knowledge or experience of the operator or user; no quantitative or qualitative checking facility may be used without the knowledge of the operators or users;

(c) systems must provide feedback to operators or users on the performance of those systems;

(d) systems must display information in a format and at a pace which are adapted to operators or users;

(e) the principles of software ergonomics must be applied, in particular to human data processing.

GUIDANCE NOTES

INTRODUCTION

1 This booklet gives guidance on the Health and Safety (Display Screen Equipment) Regulations 1992, which come into force on 1 January 1993. The Regulations implement a European directive, No 90/270/EEC of 29 May 1990, on minimum safety and health requirements for work with display screen equipment.

2 The guidance covers these Regulations only but employers should ensure that they also comply with general duties placed on them by other health and safety legislation, particularly their general obligations under the Health and Safety at Work etc Act 1974 (the HSW Act) and associated legislation. This includes three new sets of Regulations relevant to all or most workplaces—the Management of Health and Safety at Work Regulations 1992, the Workplace (Health, Safety and Welfare) Regulations 1992, and the Provision and Use of Work Equipment Regulations 1992. These are described in the box below.

NEW GENERAL REGULATIONS

Management of Health and Safety at Work Regulations

These Regulations set out broad general duties which apply to almost all kinds of work. They will require employers to:

- assess the risk to the health and safety of their employees and to anyone else who may be affected by their activity, so that the necessary preventive and protective measures can be identified;
- make arrangements for putting into practice the health and safety measures that follow from the risk assessment, covering planning, organisation, control, monitoring and review, in other words, the management of health and safety;
- provide appropriate health surveillance of employees where necessary;
- appoint competent people to help devise and apply the measures needed to comply with employers' duties under health and safety law;

- set up emergency procedures;
- give employees information about health and safety matters;
- co-operate with any other employers who share a work site;
- provide information to people working in their undertaking who are not their employees;
- make sure that employees have adequate health and safety training and are capable enough at their jobs to avoid risk; and
- give some particular health and safety information to temporary workers, to meet their special needs.

The Regulations will also:

- place duties on employees to follow health and safety instructions and report danger; and
- extend the current law which requires employers to consult employees' safety representatives and provide facilities for them. Consultation must now take place on such matters as the introduction of measures that may substantially affect health and safety; the arrangements for appointing competent persons; health and safety information and training required by law; and health and safety aspects of new technology being introduced to the workplace.

Provision and Use of Work Equipment Regulations

These Regulations are designed to pull together and tidy up the laws governing equipment used at work. Instead of piecemeal legislation covering particular kinds of equipment in different industries, they place general duties on employers and list minimum requirements for work equipment to deal with selected hazards whatever the industry.

'Work equipment' is broadly defined to include everything from a hand tool, through machines of all kinds, to a complete plant such as refinery. It therefore includes display screen equipment. 'Use' includes starting, stopping, programming, setting, transporting, repairing, modifying, maintaining, servicing and cleaning.

The general duties will require employers to:

- take into account the working conditions and risks in the workplace when selecting equipment;
- make sure that equipment is suitable for the use that will be made of it and that it is properly maintained; and
- give adequate information, instruction and training.

Specific requirements cover:

- protection from dangerous parts of machinery (replacing the current law on this);
- maintenance operations;
- danger caused by other specified hazards;
- parts and materials at high or very low temperatures;
- control systems and controls;
- isolation of equipment from power sources;
- stability of equipment;
- lighting; and
- warnings and markings.

Workplace (Health, Safety and Welfare) Regulations

These Regulations will replace a total of 38 pieces of older law, including parts of the Factories Act 1961 and the Offices, Shops and Railway Premises Act 1963. They cover many aspects of health, safety and welfare in the workplace and will apply to all places of work except:

- means of transport;
- construction sites;
- sites where extraction of mineral resources or exploration for them is carried out; and
- fishing boats.

Workplaces on agricultural or forestry land away from main buildings are also exempted from most requirements.

The Regulations set general requirements in four broad areas:

- *Working environment*, including: temperature; ventilation; lighting including emergency lighting; room dimensions; suitability of workstations and seating; and outdoor workstations (eg weather protection).
- *Safety*, including: safe passage of pedestrians and vehicles; windows and skylights (safe opening, closing and cleaning); glazed doors and partitions (use of safe material and marking); doors, gates and escalators (safety devices); floors (their construction, and obstructions and slipping and tripping hazards); falls from heights and into dangerous substances; and falling objects.
- *Facilities*, including: toilets; washing, eating and changing facilities; clothing storage; seating; rest areas (and arrangements in them for non-smokers); and rest facilities for pregnant women and nursing mothers.
- *Housekeeping*, including: maintenance of workplace, equipment and facilities; cleanliness; and removal of waste materials.

3 There are some overlaps between general and specific legislation. Where broadly applicable legislation such as the Regulations described in the box imposes a general duty similar to a more specific one in the Display Screen Equipment Regulations—for example the risk assessments required by the Management of Health and Safety at Work Regulations (MHSWR)—the legal requirement is to comply with *both* the more specific and the general duty. However, this should not give rise to any difficulty in practice. For example, in display screen work:

(a) carrying out the suitable and sufficient analysis of workstations and risk assessment required by Regulation 2 (see paragraphs 19–35 below) will also satisfy the MHSWR requirement for risk assessment as far as those workstations are concerned;

(b) ensuring that the requirements for lighting, reflections and glare in the schedule to the Display Screen Regulations are met (see Annex A, paragraphs 20–24) will also satisfy the requirements for suitable and sufficient lighting in the Provision and Use of Work Equipment Regulations and the Workplace (Health,

Safety and Welfare) Regulations, as far as the display screen workstations are concerned.

In these examples, as in other matters, the employer would still have to take other appropriate steps to ensure that the general duties (to carry out risk assessments, ensure suitable lighting etc) were complied with in any other parts of his undertaking where display screen work is not carried out.

STRUCTURE OF THIS BOOKLET

4 The guidance below gives information, explanation or advice in relation to specific requirements of the Regulations. Where the Regulations are self-explanatory no comment is offered.

Regulation 1

Citation, commencement, interpretation and application

5 The definitions of 'display screen equipment', 'workstation', 'user' and 'operator' determine whether or not the Regulations apply in a particular situation.

Which display screen equipment is covered?

6 With a few exceptions (see paragraphs 14–18), the definition of display screen equipment at Regulation 1(2)(a) covers both conventional (cathode ray tube) display screens and other display processes such as liquid crystal displays, and other emerging technologies. Display screens mainly used to display line drawings, graphs, charts or computer generated graphics are included, but screens whose *main* use is to show television or film pictures are not. Judgements about mixed media workstations will be needed to establish the main use of the screen; if this is to display text, numbers and/or graphics, it is within the scope of the Regulations. The definition is not limited to typical office visual display terminals but covers, for example, non-electronic display systems such as microfiche. Process control screens are also covered in principle (where there are 'users') although certain requirements may not apply (see paragraphs 38–40).

7 The use of display screen equipment not covered by these Regulations is still subject to other, general health and safety legislation; see paragraphs 2 and 3. For example, there are

requirements for suitable and sufficient lighting in the Provision and Use of Work Equipment Regulations 1992; and there are general requirements for risk assessment and provision of training and information in the Management of Health and Safety at Work Regulations 1992. Where a display screen is in use but the Display Screen Equipment Regulations do not apply, the assessment of risks and measures taken to control them should take account of ergonomic factors applicable to display screen work. This is also true where these Regulations do not apply because the display screen is not used by a 'user'.

Who is a display screen user or operator?

8 The Regulations are for the protection of people—employees and self-employed—who habitually use display screen equipment for the purposes of an employer's undertaking as a significant part of their normal work.

9 Regulation 1(2)(d) defines the *employees* who are covered as 'users' and all the Regulations apply to protect them, as specified, whether they are required to work:

— at their own employer's workstation;

— at a workstation at home;

— at another employer's workstation. In this case that other employer must comply with Regulations 2, 3, 4, 6(2) and 7, and their own employer with Regulations 5 and 6(1), as is specified in the Regulations concerned.

Regulations 2, 3 and 7 apply, as specified, to protect self-employed people who work at the client employer's workstation and whose use of display screen equipment is such that they would be users if employed. They are defined in Regulation 1(2)(b) as 'operators' for the purposes of the Regulations.

10 Employers must therefore decide which of their employees are display screen users and whether they also make use of other users (employed by other employers) or of operators. Workers who do not input or extract information by means of display screen equipment need not be regarded as users or operators in this context—for example many of those engaged in manufacture, sales, maintenance or the cleaning of display screen equipment. Whether or not those involved in display screen work are users or operators depends on the nature and extent of their use of the equipment.

11 The need for such a definition stems from the fact that possible hazards associated with display screen use are mainly those leading to musculoskeletal problems, visual fatigue and stress (see paragraph 19 and Annex B). The likelihood of experiencing these is related mainly to the frequency, duration, intensity and pace of spells of continuous use of the display screen equipment, allied to other factors, such as the amount of discretion the person has over the extent and methods of display screen use. The combination of factors which give rise to risks makes it impossible to lay down hard and fast rules (eg based on set hours' usage per day or week) about who should be classified as a user or operator.

12 In some cases it will be clear that use of display screen equipment is more or less continuous on most days and the individuals concerned should be regarded as users or operators. This will include the majority of those whose job mainly involves, for example, display screen based data input or sales and order processing. Where use is less continuous or frequent, other factors connected with the job must be assessed. It will generally be appropriate to classify the person concerned as a user or operator if most or all of the following criteria apply:

(a) the individual depends on the use of display screen equipment to do the job, as alternative means are not readily available for achieving the same results;

(b) the individual has no discretion as to use or non-use of the display screen equipment;

(c) the individual needs significant training and/or particular skills in the use of display screen equipment to do the job;

(d) the individual normally uses display screen equipment for continuous spells of an hour or more at a time;

(e) the individual uses display screen equipment in this way more or less daily;

(f) fast transfer of information between the user and screen is an important requirement of the job;

(g) the performance requirements of the system demand high levels of attention and concentration by the user, for example, where the consequences of error may be critical.

Some examples to illustrate these factors are included in the box. This is *not* an exhaustive list of display screen jobs, but a list of examples chosen to illuminate the above criteria.

WHO IS A DISPLAY SCREEN USER?

Some examples

Definite display screen users

Word processing pool worker employed on full-time text input using dedicated display screen equipment. A mix of checking from screen, keyboard input and formatting. Some change of posture involved in collecting work, operating printer etc. Often five hours in total on the work itself with a lunch break and at least two breaks morning and afternoon. Part-time workers, required to work fewer hours but spending all or most of their working time on this kind of work would also be included.

Secretary or typist who uses a dedicated word processing system and laser printer. Word processing of reports, memos, letters from manuscript and dictation, combined with electronic mail. Some variation in workload with a concomitant degree of control over scheduling throughout the day. Typically around two or three hours daily.

Data Input Operator employed full-time on continuous processing of invoices. Predominantly numeric input using numeric key pad. Other keystroke monitoring with associated bonus system. Part-timers, or other staff temporarily assigned to this work to deal with peak workloads, would be definite 'users' while spending all or most of their working time on these duties.

News sub-editor making use of display screen equipment more or less continuously with peak workloads. Some text input to abridge/precis stories, but mainly scanning copy for fact, punctuation, grammar and size.

Journalist whose pattern of work may be variable but includes substantial use of display screen equipment. Information collected by field or telephone interviews (which may involve use of a portable computer) followed by, typically, several hours text input while working on a story. Work likely to be characterised by deadlines and interruptions. Some days may contain periods of less intense work but with more prolonged keyboard text entry and composition.

Tele-Sales/customer complaints/accounts enquiry/directory enquiry operator employed on mainly full-time display screen use while taking telephone enquiries from customers/public.

Air traffic controller whose main task is monitoring of purpose designed screens for air traffic movements combined with communication with air crew on navigation etc. High visual and mental workload. Shift work.

Financial dealer using a dedicated workstation typically with multiple display screens. Variable/unpredictable workload. Often highly stressful situations with information overload. Often long hours.

Graphic designer working on multimedia applications. Intensive scrutiny of images at high resolution. Large screens. Page make-up. Multiple input devices. Colour systems critical.

Librarian carrying out intensive text input on dedicated equipment to add to information held on databases; accessing and checking on records held on databases, eg bibliographic and lending references; creating summaries and reports, combining data held on the equipment and new copy inputted into the system. Display screen work either intensive throughout the day on most days, or more intermittent but still forming at least half of the librarian's total working time.

Possible display screen users—depending on the circumstances

The following are examples of jobs whose occupants may or may not be designated as display screen users, depending on circumstances. In reaching a decision, employers will need to judge the relative importance of different aspects of the work, weighing these against the factors discussed in paragraph 12 and bearing in mind the risks to which the job-holder may be exposed. If there is doubt over whether an individual is a display screen worker, carrying out a risk assessment (see Regulation 2) should help in reaching a decision.

Scientist/technical adviser having use of dedicated display screen equipment. Word processing of a few letters/memos per day. Monitoring of electronic mail for a short period, average 10 minutes, on most days. At irregular intervals, uses display screen equipment intensively for data analysis of research results.

Discussion: This scientist's daily use of display screen equipment is relatively brief, non-intense and he or she would have a good deal of discretion over when and how the equipment was used. Judged against this daily use, he or she would not be a 'user'. However, this decision might be reversed if the periods of use for analysis of research results were at all frequent, of long duration and intensive.

Client manager in a large management accounting consultancy. Dedicated display screen equipment on desk. Daily scanning and transmitting of electronic mail. Typically 1½–2 hours daily.

Discussion: Whether or not this manager is a user will depend on the extent and nature of his or her use of electronic mail. For example, how continuous is use of the screen and/or keyboard during each period of use; is there discretion as to the extent of use of electronic mail; how long is the total daily use?

Building society customer support officer with shared use of office, desk and display screen workstation. Display screen equipment used during interviews with clients to interrogate HQ database to obtain customer details, transactions etc.

Discussion: Decision will be influenced by what proportion of the individual worker's time is spent using the display screen equipment; are there any prolonged and/or intensive periods of use; and what are the consequences of errors (this factor may be relevant if the job involves inputting financial data as well as searching a database).

Airline check-in clerk whose workload in job as a whole varies during day, with occasional peaks of intensive work as flight times get near. Use of display screen equipment follows a predictable pattern; typically, used as part of most transactions but may not be a significant proportion of total working time.

Discussion: There needs to be consideration of how equipment is used and for what purpose. Is the display screen used during most parts of the check-in process or only a few of them? Is the workload of transactions high? What proportion of each transaction involves viewing the screen or keying in data? Is interaction with the screen rushed and intensive? What are the consequences of errors?

Community care worker using a portable computer to make notes during and/or following interviews or visits in the field.

Discussion: Decisions on whether or not those using laptops are 'users' need to be made on the same basis as if they were using non-portable equipment. While some of the specific minimum requirements in the Schedule may not be applicable to portables in prolonged use, as the inherent characteristics of the task may rule them out, it is important that such work is properly assessed, that users are trained, and that measures are taken to control risks.

Receptionist whose job involves frequent use of display screen equipment, for example to check or enter details of each visitor and/or provide them with information.

Discussion: The nature, frequency and duration of periods of display screen work need to be assessed. Some, perhaps most, receptionists would not be users, if most of their work consists of face to face contact and/or phone calls, with a display screen only being used occasionally.

Definitely not display screen users

Senior manager in a large organisation using display screen for occasional monitoring of state of markets or other data, or more frequent but brief enquiries. Low dependency, high control.

Senior manager using display screen equipment at month end for generation/manipulation of financial statistics for board presentation.

Receptionist if work is mainly concerned with customer/public interaction, with the possibilities of interrogating display screen occasionally for limited purposes such as obtaining details of the organisation (telephone numbers, location etc).

13 The table on page 664 shows how the criteria in paragraph 12 relate to the job examples in the box.

Application

14 Where any of the exclusions in Regulation 1(4) are operative, none of the specific duties in the Regulations apply to or in connection with the use of the equipment concerned. However, the proviso at paragraph 7 applies here too. Employers should still

APPLICATION OF 'USER' CRITERIA TO PARTICULAR JOBS

CATEGORY	*Job*	*(a) Dependency*	*(b) Discretion*	*(c) Significant training*	*(d) Prolonged spells > 1 hr*	*(e) Daily use*	*(f) Fast information transfer*	*(g) Criticality of errors*
DEFINITELY USERS	Word processing	H	L	Yes	Frequent	Yes	Yes	M
	Secretary	M–H	M	Yes	Frequent	Yes	Yes	M
	Data input	H	L	Yes	Frequent	Yes	Yes	M
	Sub editor	H	L	Yes	Frequent	Yes	Yes	M
	Journalist	M–H	M	Yes	Frequent	Yes	Yes	L
	Tele-sales etc	H	L	Yes	Frequent	Yes	Maybe (paced by incoming calls)	M
	Air traffic control	H	L	Yes	Frequent	Yes	Yes	H
	Financial dealer	H	M	Yes	Frequent	Yes	Yes	H
	Graphic designer	M–H	L	Yes	Frequent	Yes	Maybe	M
	Librarian	H	L	Yes	Variable	Yes	Yes	M
MAY BE USERS	Client manager	M–H	H	No	1 per day	Yes	No	M
	Scientist	L–H	H	No	1 per day	Yes	No	M
	Building society officer	M	M–H	No	Occasional	Yes	No	M
	Airline check in	M–H	L	Yes	Infrequent	Yes	Variable	M
	Community care	L–H	L–H	No	Infrequent	Maybe	Variable	M
	Receptionist 1	M	M	Yes	Infrequent	Yes	Maybe	M
NOT USERS	Senior manager 1	L–M	H	No	Infrequent	Yes	No	M
	Senior manager 2	L–M	H	No	Infrequent	No	No	M
	Receptionist 2	L–M	H	No	Infrequent	No	No	L

KEY

H = High
M = Medium
L = Low

ensure that, so far as is reasonably practicable, the health and safety of those using the equipment are not put at risk. The general duties on employers and others under the Health and Safety at Work etc Act 1974, and other general health and safety legislation (see paragraphs 2 and 3), are still applicable and particular attention should be paid to ergonomics in this context.

15 The exclusion in Regulation 1(4)(c) is for display screen equipment mainly provided for short-term operation by the general public, such as cashpoint machines at banks and microfiche readers and computer terminals in public libraries. It does not extend to display screen equipment available for operation by the public but mainly provided for use by users.

16 Portable display screen equipment (such as laptop computers) comes under the exclusion in Regulation 1(4)(d) only if it is not in prolonged use. While there are no hard and fast rules on what constitutes 'prolonged' use, portable equipment that is habitually in use by a display screen user for a significant part of his or her normal work, as explained in paragraphs 11–13, should be regarded as covered by the Regulations. While some of the specific minimum requirements in the Schedule may not be applicable to portables in prolonged use, employers should still ensure that such work is assessed and measures taken to control risks.

17 The exclusion in Regulation 1(4)(e) for small data or measurement displays is there because such displays are usually not intensively monitored by workers for long continuous spells. This exclusion covers, for example, much scientific and medical equipment, such as cardiac monitors, oscilloscopes, and instruments with small displays showing a series of digits.

18 The exclusion in Regulation 1(4)(f) is for window typewriters having a small display showing no more than a few lines of text.

Regulation 2

Analysis of workstations to assess and reduce risks

19 Possible risks which have been associated with display screen equipment work are summarised at Annex B. The principal risks relate to physical (musculoskeletal) problems, visual fatigue and mental stress. These are not unique to display screen work nor an inevitable consequence of it, and indeed research shows that the

risk to the individual user from typical display screen work is low. However, in display screen work as in other types of work, ill health can result from poor work organisation, working environment, job design and posture, and from inappropriate working methods. As discussed in Annex B, some types of display screen work have been associated with chronic musculoskeletal disorders. While surveys indicate that only a very small proportion of display screen workers are likely to be involved, the number of cases may still be significant as display screen workers are so numerous. All the known health problems that may be associated with display screen work can be prevented altogether by good design of the workplace and the job, and by worker training and consultation.

20 Employers will need to assess the extent to which any of the above risks arise for display screen workers using their work-stations who are:

— users employed by them;

— users employed by others (eg agency employed 'temps');

— operators, ie self-employed contractors who would be classified as users if they were employees (eg self-employed agency 'temps', self-employed journalists).

Individual workstations used by any of these people will need to be analysed and risks assessed. If employers require their employees to use workstations at home, these too will need to be assessed (see paragraph 26).

If there is doubt whether any individual is a user or operator, carrying out a risk assessment should help in reaching a decision.

Suitable and sufficient analysis and risk assessment

21 Risk assessment should first identify any hazards and then evaluate risks and their extent. A *hazard* is something with the potential to cause harm; *risk* expresses the likelihood that the harm from a particular hazard is realised. The *extent of the risk* takes into account the number of people who might be exposed to a risk and the consequences for them. Analysis of display screen work-stations should include a check for the presence of desirable features as well as making sure that bad points have been eliminated. In general, the risks outlined above will arise when the work, workplace and work environment do not take account of worker requirements. Since any risks to health may arise from a

combination of risk factors, a suitable and sufficient analysis should:

(a) be systematic—including investigation of non-obvious causes of problems. For example, poor posture may be a response to screen reflections or glare, rather than poor furniture;

(b) be appropriate to the likely degree of risk. This will largely depend on the duration, intensity or difficulty of the work undertaken, for example the need for prolonged high concentration because of particular performance requirements;

(c) be comprehensive, covering organisational, job, workplace and individual factors;

(d) incorporate information provided by both employer and worker.

The form of the assessment

22 In the simplest and most obvious cases which can be easily repeated and explained at any time an assessment need not be recorded. This might be the case, for example, if no significant risks are indicated and no individual user or operator is identified as being especially at risk. Assessments of short-term or temporary workstations may also not need to be recorded, unless risks are significant. However, in most other cases assessments need to be recorded and kept readily accessible to ensure continuity and accuracy of knowledge among those who may need to know the results (eg where risk reduction measures have yet to be completed). Recorded assessments need not necessarily be a paper and pencil record but could be stored electronically.

23 Information provided by users is an essential part of an assessment. A useful way of obtaining this can be through an ergonomic checklist, which should preferably be completed by users. Other approaches are also possible. For example, more objective elements of the analysis (eg nature of work, chair adjustability, keyboard characteristics etc) could be assessed generically in respect of particular types of equipment or groups of worker. Other aspects of workstations would still need to be assessed individually through information collected from users, but this could then be restricted to subjective factors (eg relating to comfort). *Whatever type of checklist is used, employers should ensure workers have received the necessary training before being asked to complete one.*

24 The form of the assessment needs to be appropriate to the nature of the tasks undertaken and the complexity of the workstation. For many office tasks the assessment can be a judgement based on responses to the checklist. Where particular risks are apparent, however, and for complex situations, eg where safety of others is a critical factor, a more detailed assessment may be appropriate. This could include, for example, a task analysis where particular job stresses had been identified, recording of posture, physical measurement of workstations; or quantitative surveys of lighting and glare.

Shared workstations

25 Where one workstation is used by more than one worker, whether simultaneously or in shifts, it should be analysed and assessed in relation to all those covered by the Regulations (see paragraph 9).

Assessment of risks to homeworkers

26 If a display screen user (ie an employee) is required by his or her employer to work at home, whether or not the workstation is provided in whole or part by the employer, the risks must be assessed. An ergonomic checklist which the homeworker completes and submits to the employer for assessment is the most practicable means. The assessment will need to cover any need for extra or special training and information provision for homeworkers to compensate for the absence of direct day to day employer oversight and control of their working methods.

Who should do assessments?

27 Those responsible for the assessment should be familiar with the main requirements of the Regulations and have the ability to:

(a) assess risks from the workstation and the kind of display screen work being done, for example, from a checklist completed by them or others;

(b) draw upon additional sources of information on risk as appropriate;

(c) based upon the assessment of risk, draw valid and reliable conclusions;

(d) make a clear record of the assessment and communicate the findings to those who need to take appropriate action;

(e) recognise their own limitations as to assessment so that further expertise can be called on if necessary.

28 Assessments can be made by health and safety personnel, or line managers with, or trained for, these abilities. It may be necessary to call in outside expertise where, for example, display screen equipment or associated components are faulty in design or use, where workstation design is complex, or where critical tasks are being performed.

29 The views of individual users about their workstations are an essential part of the assessment, as noted in paragraph 23. Employees' safety representatives should also be encouraged to play a full part in the assessment process. In particular, they should be encouraged to report any problems in display screen work that come to their attention.

Review of assessment

30 The assessment or relevant parts of it should be reviewed in the light of changes to the display screen worker population, or changes in individual capability and where there has been some significant change to the workstation such as:

(a) a major change to software used;

(b) a major change to the hardware (screen, keyboard, input devices etc);

(c) a major change in workstation furniture;

(d) a substantial increase in the amount of time required to be spent using display screen equipment;

(e) a substantial change in other task requirements (eg more speed or accuracy);

(f) if the workstation is relocated;

(g) if the lighting is significantly modified.

Assessments would also need to be reviewed if research findings indicated a significant new risk, or showed that a recognised hazard should be re-evaluated.

31 Because of the varying nature and novelty of some display

screen tasks, and because there is incomplete understanding of the development of chronic ill-health problems (particularly musculo-skeletal ones), prediction of the nature and likelihood of problems based upon a purely objective evaluation of equipment may be difficult. It is therefore most important that employers should encourage early reporting by users of any symptoms which may be related to display screen work. The need to report and the organisational arrangements for making a report should be covered in training.

Reducing risks

32 The assessment will highlight any particular areas which may give rise for concern, and these will require further evaluation and corrective action as appropriate. The four year lead-in period for the 'minimum requirements' for workstations which are not new (see paragraph 41) does *not* apply to the requirement to reduce the risk. Risks identified in the assessment must be remedied as quickly as possible. For typical applications of display screens, such as VDUs in offices, remedial action is often straightforward, for example:

(a) *postural problems* may be overcome by simple adjustments to the workstation such as repositioning equipment or adjusting the chair. Postural problems can also indicate a need to provide reinforced training of the user (for example on correct hand position, posture, how to adjust equipment). New equipment such as a footrest or document holder may be required in some cases;

(b) *visual problems* may also be tackled by straightforward means such as repositioning the screen or using blinds to avoid glare, placing the screen at a more comfortable viewing distance from the user, or by ensuring the screen is kept clean. In some cases, new equipment such as window blinds or more appropriate lighting may be needed;

(c) *fatigue and stress* may be alleviated by correcting obvious defects in the workstation as indicated above. In addition, as in other kinds of work, good design of the task will be important. Wherever possible the task should provide users with a degree of personal control over the pace and nature of their tasks. Proper provision must be made for training, advice and information, not only on health and safety risks but also on the use of software. Further advice is given at paragraphs 31–34 of Annex A.

33 It is important to take a systematic approach to risk reduction and recognise the limitations of the basic assessment. Observed problems may reflect the interaction of several factors or may have causes that are not obvious. For example, backache may turn out to have been caused by the worker sitting in an abnormal position in order to minimise the effects of reflections on the screen. If the factors underlying a problem appear to be complex, or if simple remedial measures do not have the desired effect, it will generally be necessary to obtain expert advice on corrective action.

Sources of information and advice

34 Annex C contains a list of relevant HSE guidance documents, for example on lighting and seating, and other publications (not reproduced in this book). Further advice on health problems that may be connected with display screen work could be obtained from in-house safety or occupational health departments where applicable or, if necessary, from Employment Medical Advisory Service staff in HSE (listed in your local telephone directory under 'Health and Safety Executive'). Expert advice may be obtained from independent specialists in relevant professional disciplines such as ergonomics, or lighting design.

Standards

35 Ergonomic specifications for use of display screen equipment are contained in various international, European and British standards. Further information is given at Annex A. Compliance with relevant parts of these standards will generally not only satisfy, but go beyond the requirements of the Regulations, because such standards aim to enhance performance as well as health and safety.

Regulation 3

Requirements for workstations

36 Regulation 3 refers to the Schedule to the Regulations which sets out minimum requirements for display screen workstations, covering the equipment, the working environment, and the inter-

face between computer and user/operator. Figure 1 (on page 684) summarises the main requirements. Annex A contains more information on those requirements of the Schedule which call for some interpretation.

37 Regulation 3 and the Schedule must be complied with in respect of all workstations that may be used by a display screen user or operator. Where an employer decides that a particular workstation is not used by a display screen user or operator and is unlikely to be used by one in future, there is no legal need for that workstation to comply with Regulation 3 or the Schedule, though, where it is applicable, compliance will in most cases enhance performance and efficiency. Where employers have workstations that do not comply they should take steps to ensure that display screen users or operators do not use them.

They should also bear in mind their general responsibilities under the Health and Safety at Work Act to ensure health and safety of all those at work—see paragraph 42.

Application of the Schedule

38 By virtue of paragraph 1 of the Schedule, the requirements apply only in so far as:

(a) *the components concerned (eg document holder, chair or desk) are present at the workstation*. Where a particular item is mentioned in the Schedule, this should not be interpreted as a requirement that all workstations should have one, unless risk assessment under Regulation 2 suggests the item is necessary;

(b) *they relate to worker health, safety and welfare*. For the purposes of these Regulations, it is only necessary to comply with the detailed requirements in paragraphs 2, 3 and 4 of the Schedule if this would actively secure the health, safety or welfare of persons at work. The requirements in the Schedule do not extend to the efficiency of use of display screen equipment, workstations or software. However, these matters are covered, in addition to worker health and safety, in BS 7179 and other standards, and in international standards in preparation (see Annex A). Compliance with such standards, where they are appropriate, should enhance efficiency as well as ensuring that relevant health and safety requirements of the Schedule are also satisfied;

(c) *the inherent requirements or characteristics of the task make compliance appropriate*; for example, where the task could not be carried out successfully if all the requirements in the Schedule were complied with. (Note that it is the demands of the task, rather than the capabilities of any particular equipment, that are the deciding factor here.)

39 In practice, the detailed requirements in paragraphs 2 to 4 of the Schedule are most likely to be fully applicable in typical office situations, for example where a VDU is used for tasks such as data entry or word processing. In more specialised applications of display screens compliance with particular requirements in the Schedule may be inappropriate where there would be no benefit to, or even adverse effects on, health and safety. Where display screen equipment is used to control machinery, processes or vehicle traffic, it is clearly essential to consider the implications of any design changes for the rest of the workforce and the public, as well as the health and safety of the screen user.

40 The following examples illustrate how these factors can operate in practice. They each include a reference to the relevant part of paragraph 1 of the Schedule:

(a) where, as in some control-room applications, a screen is used from a standing position and without reference to documents, a work surface and chair may be unnecessary (Schedule 1(a));

(b) some individuals who suffer from certain back complaints may benefit from a chair with a fixed back rest or a special chair without a back rest (Schedule 1(b));

(c) wheelchair users work from a 'chair' that may not comply with the requirements in paragraph 2(e) of the Schedule. They may have special requirements for work surface (eg height); in practice some wheelchair users may need a purpose-built workstation but others may prefer to use existing work surfaces. Clearly the needs of the individual here should have priority over rigid compliance with paragraph 2 of the Schedule (Schedule 1(b));

(d) where a user may need to rapidly locate and operate emergency controls, placing them on a detachable keyboard may be inappropriate (Schedule 1(b) and (c));

(e) where there are banks of screens as in process control or air traffic control for example, individually tilting and swivelling screens may be undesirable as the screens may need to be aligned with one another and/or be aligned for easy viewing from the operator's seat. Detachable keyboards may also be undesirable if a particular keyboard needs to be associated with a particular screen and/or instrumentation in a multi-screen array (Schedule 1(c) and (b));

(f) a brightness control would be inappropriate for process control screens used to display alarm signals—turning down the brightness could cause an alarm to be missed (Schedule 1(b) and (c));

(g) screens that are necessarily close to other work equipment (for example, in a fixed assembly such as a control room panel) that needs to be well-illuminated will need carefully positioned local lighting—it may then be inappropriate for the screen to tilt and swivel as this could give rise to strong reflections on the screen (Schedule 1(b));

(h) where microfiche is used to keep records of original documents, screen characters may not be well-defined or clearly formed if the original was in poor condition or was badly photographed (Schedule 1(c));

(i) radar screens used in air traffic control have characters which have blurred 'tails' and hence might be considered to be not well-defined and clearly formed; however, long-persistence phosphors are deliberately used in these screens in order to indicate the direction of movement of the aircraft (Schedule 1(c));

(j) screens forming part of a simulator for ship or aircraft crew training may have special features that do not comply with the Schedule but are necessary if the simulator is to accurately mimic the features of the exempt display screen equipment on the ship or aircraft (Schedule 1(c)).

Transitional period for existing equipment

41 Employers are required to ensure that workstations, whether or not they are new, which are put into service in their undertakings on or after the coming into force of these Regulations comply with the Schedule where it is relevant. Workstations already in service should comply by 31 December 1996. If new display screen equipment is put into service at an existing workstation, the whole workstation concerned should be regarded as new and brought into compliance with the Schedule straight away. However, if any other

part of an existing workstation is changed, only the new component need comply with the Schedule at once; the remainder of the workstation need not comply until 31 December 1996.

42 Where the Schedule does not apply, either because its requirements are not applicable (under paragraph 1) or the workstation is not new, employers must still comply with other provisions of these Regulations as well as with the Health and Safety at Work Act to ensure that risks to users and operators are reduced to the lowest extent reasonably practicable. Thus:

(a) if assessment of an existing workstation shows there is a risk to users or operators, the employer should take immediate steps to reduce the risk; or

(b) where paragraph 1(a) or (c) of the Schedule is applicable and the minimum requirements in paragraphs 2, 3 and 4 of the Schedule are therefore not being followed, the employer must ensure that the health and safety of users and operators are adequately safeguarded by whatever other means are appropriate, reasonably practicable and necessary.

Regulation 4

Daily work routine of users

43 In most tasks, natural breaks or pauses occur as a consequence of the inherent organisation of the work. Whenever possible, jobs at display screens should be designed to consist of a mix of screen-based and non screen-based work to prevent fatigue and to vary visual and mental demands. Where the job unavoidably contains spells of intensive display screen work (whether using the keyboard or input device, reading the screen, or a mixture of the two), these should be broken up by periods of non-intensive, non-display screen work. Where work cannot be so organised, eg in jobs requiring only data or text entry requiring sustained attention and concentration, deliberate breaks or pauses must be introduced.

Nature and timing of breaks or changes of activity

44 Where the display screen work involves intensive use of the keyboard, any activity that would demand broadly similar use of the arms or hands should be avoided during breaks. Similarly, if the display screen work is visually demanding any activities during breaks should be of a different visual character. Breaks must also

allow users to vary their posture. Exercise routines which include blinking, stretching and focussing eyes on distant objects can be helpful and could be covered in training programmes.

45 It is not appropriate to lay down requirements for breaks which apply to all types of work; it is the nature and mix of demands made by the job which determine the length of break necessary to prevent fatigue. But some general guidance can be given:

(a) breaks should be taken before the onset of fatigue, not in order to recuperate and when performance is at a maximum, before productivity reduces. The timing of the break is more important than its length;

(b) breaks or changes of activity should be included in working time. They should reduce the workload at the screen, ie should not result in a higher pace or intensity of work on account of their introduction;

(c) short, frequent breaks are more satisfactory than occasional, longer breaks: eg a 5–10 minute break after 50–60 minutes continuous screen and/or keyboard work is likely to be better than a 15 minute break every 2 hours;

(d) if possible, breaks should be taken away from the screen;

(e) informal breaks, that is time spent not viewing the screen (eg on other tasks), appear from study evidence to be more effective in relieving visual fatigue than formal rest breaks;

(f) wherever practicable, users should be allowed some discretion as to how they carry out tasks; individual control over the nature and pace of work allows optimal distribution of effort over the working day.

The employer's duty to plan activities

46 The employer's duty under Regulation 4 to plan the activities of users can be satisfied by arranging things so that users are able to benefit from breaks or changes of activity, and encouraging them to do so. The duty to plan does not imply a need for the employer to draw up a precise and detailed timetable for periods of DSE work and breaks.

47 It is generally best for users to be given some discretion over when to take breaks. In such cases the employer's duty to plan activities may be satisfied by allowing an adequate degree of

flexibility for the user to organise their own work. However, users given total discretion may forego breaks in favour of a shorter working day, and thus may suffer fatigue. Employers should ensure that users are given adequate information and training on the need for breaks (see paragraphs 64 and 66). Where users forego breaks despite this, it may be necessary for employers to lay down minimum requirements for the frequency of breaks while still allowing users some flexibility.

48 The employer's duty is to plan activities so that breaks or changes of activity are taken by users during their normal work. There are a few situations, for example where users working in a control room are handling an unforeseen emergency, where other health and safety considerations may occasionally dictate that normal breaks are not taken.

Regulation 5

Eyes and eyesight

49 There is no reliable evidence that work with display screen equipment causes any permanent damage to eyes or eyesight, but it may make users with pre-existing vision defects more aware of them. This (and/or poor working conditions) may give some users temporary visual fatigue or headaches. Uncorrected vision defects can make work at display screens more tiring or stressful than it should be, and correcting defects can improve comfort, job satisfaction and performance. (Note that some display screen work may also require specific visual capabilities such as colour discrimination).

Eye and eyesight test

50 Regulations 5(1) and 5(2) require employers to provide users who so request it with an appropriate eye and eyesight test. In Great Britain an 'appropriate eye and eyesight test' means a 'sight test' as defined in the Opticians Act legislation.[1] The test includes a test of

1 The Opticians Act 1989, s 36(2), defines testing sight as 'determining whether there is any and, if so, what defect of sight and of correcting, remedying or relieving any such defect of an anatomical or physiological nature by means of an optical appliance prescribed on the basis of the determination'. The test is defined in further detail in the Sight Testing Examination and Prescription (No 2) Regulations 1989.

vision and an examination of the eye. For the purpose of the Display Screen Equipment Regulations, the test should take account of the nature of the user's work, including the distance at which the screen is viewed. Display screen users are not obliged to have such tests performed but where they choose to exercise their entitlement, employers should offer an examination by a registered ophthalmic optician, or a registered medical practitioner with suitable qualifications ('optometrist' and 'doctor' respectively in the paragraphs below). (All registered medical practitioners, including those in company occupational health departments, are entitled to carry out sight tests but normally only those with an ophthalmic qualification do so).

51 Regulation 5(1) gives employers a duty to ensure the provision of appropriate eye and eyesight tests on request:

(a) to their employees who are already users when the Regulations come into force;

(b) and (thereafter) to any of their non-user employees who are to become users.

The Regulations do not give employers any duty to offer eye and eyesight tests to persons not in their employment, such as applicants for jobs. However, where somebody has been recruited and is to work with display screen equipment to the extent that they will become a user, Regulation 5(1)(b) becomes applicable. Hence where a newly recruited employee of this kind—whether or not they have been a user in any previous employment in a different undertaking—requests one, an appropriate eye and eyesight test should be arranged by their new employer. The test should be carried out before the newly recruited employee becomes a user, as required by Regulation 5(2)(b). This does not mean that new recruits must be given a test before doing any display screen work, but they would have to be given a test (if they requested one) before doing sufficient display screen work for this to be regarded as a significant part of their normal work. For guidance on what this means in practice, see paragraphs 10–13 on the definition of a user.

52 The British College of Optometrists has produced a statement of good practice for optometrists, obtainable from them. Among other things, it makes clear that the purpose of the eye test by an optometrist or doctor under Regulation 5 is to decide whether the user has any defect of sight which requires correction when working with a display screen. It follows that users need to be able to describe their display screen and working environment when

they have the eye test. As the College points out, the optometrist will need to make a report to the employer, copied to the employee, stating clearly whether or not a corrective appliance is needed specifically for display screen work and when re-examination should take place. Any prescription, or other confidential clinical information from the eye test, can only be provided to the employer with the employee's consent.

Vision screening tests

53 Vision screening tests are a means of identifying individuals with defective vision who need a full sight test (see paragraph 50). These tests are not designed to screen for eye defects, such as injury or disease, that may not at first affect vision. Where companies offer vision screening facilities, some users may opt for a vision screening test to check their need for a full sight test. Other users, however, may choose at the outset to exercise their entitlement to a full sight test, and in such cases the employer must arrange for the test specified in paragraph 50 to be provided.

54 Where the user opts for vision screening, the screening instrument or other test method used should be capable of testing vision at the distances appropriate to the user's display screen work, including the intermediate distance at which screens are viewed (normally 50–60 cm). Where test results indicate that vision is defective at the relevant distances, the user should be informed and referred to an optometrist or doctor for a full sight test.

55 Those conducting eyesight screening tests should have basic knowledge of the eye and its function and be competent in operation of the instrument and/or tests. Both the test results and the need for further referral should be assessed by those with medical, ophthalmic, nursing or paramedical skills.

Regularity of provision of eye and eyesight tests

56 Regulation 5 requires that eye and eyesight tests are provided:

(a) as soon as practicable after display screen users have made a request;

(b) for employees who are to become users, and have made a request. In such cases the test must be carried out before the employee becomes a user;

(c) for users at regular intervals thereafter to check the need for special corrective appliances for display screen work, provided that they want the tests. Employers should be guided by the clinical judgement of the optometrist or doctor on the frequency of repeat testing. The frequency of repeat testing needed will vary between individuals, according to factors such as age. However, employers are not responsible for any corrections for vision defects or examinations for eye complaints which are not related to display screen work which may become necessary within the period. These are the responsibility of the individual concerned;

(d) for users experiencing visual difficulties which may reasonably be considered to be related to the display screen work, for example visual symptoms such as eyestrain or focussing difficulties.

57 Where an eye test by an optometrist suggests that a user is suffering eye injury or disease, the user will be referred to his or her registered medical practitioner for further examination. This examination is free of charge under the National Health Service.

Corrective appliances

58 'Special' corrective appliances (normally spectacles) provided to meet the requirements of the Regulations will be those appliances prescribed to correct vision defects at the viewing distance or distances used specifically for the display screen work concerned. 'Normal' corrective appliances are spectacles prescribed for any other purpose. It should be noted that experience has shown that in most working populations only a minority (usually less than 10%) will need special corrective appliances for display screen work. Those who need special corrective appliances may include users who already wear spectacles or contact lenses, or others who have uncorrected vision defects.

59 Anti-glare screens, and so-called 'VDU spectacles' and other devices that purport to protect against radiation, are not special corrective appliances (see paragraphs 27–30 of Annex A for advice on radiation).

Employers' liability for costs

60 The provision of eye and eyesight tests and of special corrective appliances under the Regulations is at the expense of the *user's employer*. This is the case even if the user works on other employers'

workstations. Employers are free to specify that users' tests and correction are provided by a particular company or professional. 'Normal' corrective appliances are at the user's own expense.

61 Users needing special corrective appliances may be prescribed a special pair of spectacles for display screen work. Employers' liability for costs is restricted to payment of the cost of a basic appliance, ie of a type and quality adequate for its function. If users wish to choose more costly appliances (eg with designer frames; or lenses with optional treatments not necessary for the work), the employer is not obliged to pay for these. In these circumstances employers may either provide a basic appliance as above, or may opt to contribute a portion of the total cost of a luxury appliance equal to the cost of a basic appliance.

62 If users are permitted by their employers to choose spectacles to correct eye or vision defects for purposes which include display screen work but go wider than that, employers need contribute only the costs attributable to the requirements of the display screen work involved.

Regulation 6

Provision of training

63 In accordance with this Regulation, employers should ensure that all users who make use of their workstations or are required to use other workstations have been provided with *health and safety* training, in addition to the training received in order to do the work itself. In practice, there may be considerable overlap between general training requirements and specific health and safety ones (for example the development of keyboard skills) and they are best done together. They will then reinforce each other and facilitate efficient and effective use of the equipment as well as avoidance of risk. The purpose of training is to increase the user's competence to use workstation equipment safely and reduce the risk to their or anyone else's health. In considering the extent of any training which will be necessary in a particular case, the employer needs to make up any shortfalls between the user's existing competence and that necessary to use the equipment in a safe and healthy way. The development of specific statements of what the user needs to do and how well (ie statements of competence) will assist the employer to determine the extent of any shortfall.

64 Training will need to be adapted to the requirements of the particular display screen tasks, be adapted to users' skills and

capabilities and be refreshed or updated as the hardware, software, workstation, environment or job are modified. (A workstation should be regarded as having been 'substantially modified' for the purposes of Regulation 6(2) if there has been a significant change to it, as set out in paragraph 30). Special training or retraining needs may need to be considered for rehabilitation of people absent for long periods, particularly if ill-health problems are related to the visual, musculoskeletal or stress-related risks referred to earlier. Organisations should develop systems for identifying the occasions when any of these needs for training arise.

65 The health and safety training should be aimed at reducing or minimising the three risk areas outlined at paragraph 19 (page 665) and in Annex B, with reference to the part played by the individual user. To do this, six inter-related aspects of training should be covered:

(a) The user's role in correct and timely detection and recognition of hazards and risks. This should cover both the absence of desirable features (chair comfort) and the presence of undesirable ones (screen reflections and glare) together with information on health risks and how problems may be manifested.

(b) A simple explanation of the causes of risk and the mechanisms by which harm may be brought about, for example poor posture leading to static loading on the musculoskeletal system and eventual fatigue and pain.

(c) User initiated actions and procedures which will bring risks under control and to acceptable levels. Training should cover the following:

- the desirability of comfortable posture and the importance of postural change;
- the use of adjustment mechanisms on equipment, particularly furniture, so that stress and fatigue can be minimised;
- the use and arrangement of workstation components to facilitate good posture, prevent over-reaching and avoid glare and reflections on the screen;
- the need for regular cleaning (or inspection) of screens and other equipment for maintenance;

— the need to take advantage of breaks and changes of activity.

(d) Organisational arrangements by which symptoms or problems with the workstation can be communicated to management.

(e) Information on these Regulations, particularly as regards eyesight, rest pauses and the contents of Annex A.

(f) The user's contribution to assessments.

66 New users could be given such training at the same time as they are trained on how to use the equipment. The information required to be provided under Regulation 7 will reinforce the training and could usefully be in the form of posters or cards with pictorial reminders of some of the essential points. Figure 2 (on page 685) provides an example.

Regulation 7

Provision of information

67 There is a general requirement under the Management of Health and Safety at Work Regulations 1992 for employers to provide information on risks to health and safety to all their own employees as well as to employees of other employers on site, to visiting employees, and to the self-employed. Under Regulation 7 of the Display Screen Regulations specific information should be provided as follows:

		Information on:					
		Risks from display screen equipment and workstations	*Risk assessment and measures to reduce the risks (Regs 2 and 3)*	*Breaks and activity changes (Reg 4)*	*Eye and eyesight tests (Reg 5)*	*Initial training (Reg 6(1))*	*Training when workstation modified (Reg 6(2))*
DOES EMPLOYER HAVE TO PROVIDE INFORMATION TO DISPLAY SCREEN WORKERS WHO ARE:	Users employed by the undertaking	Yes	Yes	Yes	Yes	Yes	Yes
	Users employed by other employer	Yes	Yes	Yes	No	No	Yes
	Operators in the undertaking	Yes	Yes	No	No	No	No

68 The information should among other things include reminders of the measures taken to reduce the risks such as the system for reporting problems, the availability of adjustable window covering and furniture, *and of how to make use of them*. It will thus reinforce any training provided by the employer and be a useful reminder to those trained already.

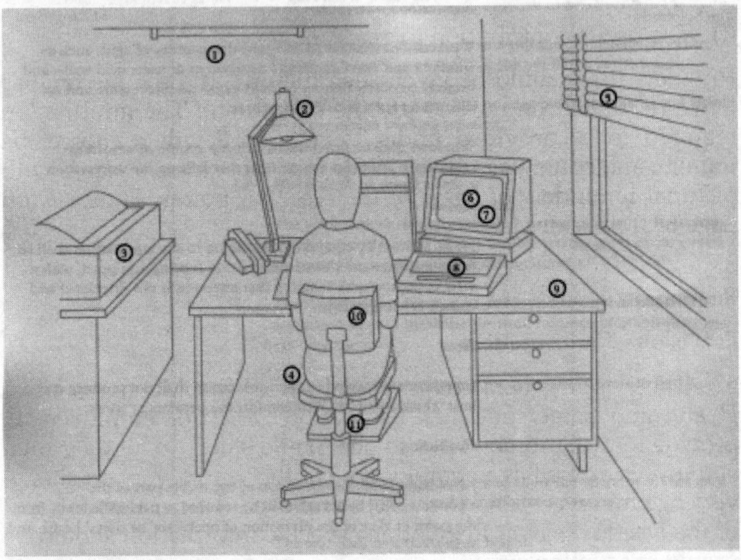

Figure 1

SUBJECTS DEALT WITH IN THE SCHEDULE

① ADEQUATE LIGHTING

② ADEQUATE CONTRAST, NO GLARE OR DISTRACTING REFLECTIONS

③ DISTRACTING NOISE MINIMISED

④ LEG ROOM AND CLEARANCES TO ALLOW POSTURAL CHANGES

⑤ WINDOW COVERING

⑥ SOFTWARE APPROPRIATE TO TASK, ADAPTED TO USER, PROVIDES FEEDBACK ON SYSTEM STATUS, NO UNDISCLOSED MONITORING

⑦ SCREEN: STABLE IMAGE, ADJUSTABLE, READABLE, GLARE/ REFLECTION FREE

⑧ KEYBOARD: USABLE, ADJUSTABLE, DETACHABLE, LEGIBLE

⑨ WORK SURFACE: ALLOW FLEXIBLE ARRANGEMENTS, SPACIOUS, GLARE FREE

⑩ WORK CHAIR: ADJUSTABLE

⑪ FOOTREST

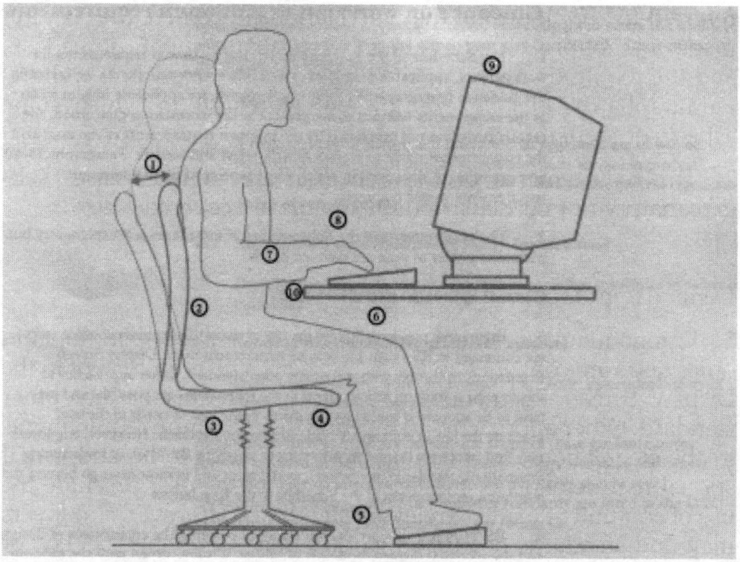

Figure 2

SEATING AND POSTURE FOR TYPICAL OFFICE TASKS

1. SEAT BACK ADJUSTABILITY
2. GOOD LUMBAR SUPPORT
3. SEAT HEIGHT ADJUSTABILITY
4. NO EXCESS PRESSURE ON UNDERSIDE OF THIGHS AND BACKS OF KNEES
5. FOOT SUPPORT IF NEEDED
6. SPACE FOR POSTURAL CHANGE, NO OBSTACLES UNDER DESK
7. FOREARMS APPROXIMATELY HORIZONTAL
8. MINIMAL EXTENSION, FLEXION OR DEVIATION OF WRISTS
9. SCREEN HEIGHT AND ANGLE SHOULD ALLOW COMFORTABLE HEAD POSITION
10. SPACE IN FRONT OF KEYBOARD TO SUPPORT HANDS/WRISTS DURING PAUSES IN KEYING

ANNEX A
GUIDANCE ON WORKSTATION MINIMUM REQUIREMENTS

1 The Schedule to the Regulations sets out minimum requirements for workstations, applicable mainly to typical office workstations. As explained in the guidance (paragraph 38) these requirements are applicable only in so far as the components

referred to are present at the workstation concerned, the requirements are not precluded by the inherent requirements of the task, and the requirements relate to worker health, safety and welfare. Paragraphs 39–40 give examples of situations in which some aspects of these minimum requirements would not apply.

2 The requirements of the Schedule are in most cases self-explanatory but particular points to note are covered below.

General approach: use of standards

3 Ergonomic requirements for the use of visual display units in office tasks are contained in BS 7179. There is no requirement in the Display Screen Regulations to comply with this or any other standard. Other approaches to meeting the minimum requirements in the Regulations are possible, and may have to be adopted if special requirements of the task or needs of the user preclude the use of equipment made to relevant standards. However, employers may find standards helpful as workstations satisfying BS 7179, or forthcoming international standards, would meet and in most cases go beyond the minimum requirements in the Schedule to the Regulations.

4 BS 7179 is a six-part interim standard covering the ergonomics of design and use of visual display terminals in offices; it is concerned with the efficient use of VDUs as well as with user health, safety and comfort. BS 7179 has been issued by the British Standards Institution in recognition of industry's immediate need for guidance and is intended for the managers and supervisors of VDU users as well as for equipment manufacturers. While originally confined to office VDU tasks, many of the general ergonomic recommendations in BS 7179 will be relevant to some non-office situations.

5 International standards are in preparation that will cover the same subject in an expanded form. BS 7179 will be withdrawn when the European standards organisation CEN (Comité Européen de Normalisation) issues its multipart standard (EN 29241) concerned with the ergonomics of design and use of visual display terminals for office tasks. This CEN Standard will in turn be based on an ISO Standard (ISO 9241) that is currently being developed. The eventual ISO and CEN standards will cover screen and keyboard design and evaluation, workstation design and environmental requirements, non-keyboard input devices

and ergonomic requirements for software design and usability. While the CEN standard is not formally linked to the Display Screen Equipment directive, one of its aims is to establish appropriate levels of user health, safety and comfort. Technical data in the various parts of the CEN standard (and currently BS 7179) may therefore help employers to meet the requirements laid down in the Schedule to the Regulations.

6 There are other standards that deal with requirements for furniture, some of which are cross-referenced by BS 7179. These include BS 3044, which is a guide to ergonomic principles in the design and selection of office furniture generally. There is also now a separate standardisation initiative within CEN concerned with the performance requirements for office furniture, including dimensioning appropriate for European user populations. Details of relevant British, European and international standards can be obtained from the Department of Trade and Industry.

7 Other more detailed and stringent standards are relevant to certain specialised applications of display screens, especially those where the health or safety of persons other than the screen user may be affected. Some examples in particular subject areas are:

(1) *Process control*

A large number of British and international standards are or will be relevant to the design of display screen interfaces for use in process control—such as the draft Standard ISO 11064 on the general ergonomic design of control rooms.

(2) *Applications with machinery safety implications*

Draft Standard prEN 614 Pt 1—Ergonomic design principles in safety of machinery.

(3) *Safety of programmable electronic systems*

Draft document IEC 65A (Secretariat) 122 Draft: Functional safety of electrical/electronic programmable systems.

Applications such as these are outside the scope of these Guidance Notes. Anyone involved in the design of such display screen interfaces and others where there may be safety considerations for non-users should seek appropriate specialist advice. Many relevant standards are listed in the DTI publication *Directory of HCI Standards*.

Equipment

Display screen

8 Choice of display screen should be considered in relation to other elements of the work system, such as the type and amount of information required for the task, and environmental factors. A satisfactory display can be achieved by custom design for a specific task or environment, or by appropriate adjustments to adapt the display to suit changing requirements or environmental conditions.

Display stability

9 Individual perceptions of screen flicker vary and a screen which is flicker-free to 90% of users should be regarded as satisfying the minimum requirement. (It is not technically feasible to eliminate flicker for all users). A change to a different display can resolve individual problems with flicker. Persistent display instabilities—flicker, jump, jitter or swim—may indicate basic design problems and assistance should be sought from suppliers.

Brightness and contrast

10 Negative or positive image polarity (light characters on a dark background, dark characters on a light background respectively) is acceptable, and each has different advantages. With negative polarity flicker is less perceptible, legibility is better for those with low acuity vision, and characters may be perceived as larger than they are; with positive polarity, reflections are less perceptible, edges appear sharper and luminance balance is easier to achieve.

11 It is important for the brightness and contrast of the display to be appropriate for ambient lighting conditions; trade-offs between character brightness and sharpness may be needed to achieve an acceptable balance. In many kinds of equipment this is achieved by providing a control or controls which allow the user to make adjustments.

Screen adjustability

12 Adjustment mechanisms allow the screen to be tilted or swivelled to avoid glare and reflections and enable the worker to maintain a natural and relaxed posture. They may be built into the screen, form part of the workstation furniture or be provided by

separate screen support devices; they should be simple and easy to operate. Screen height adjustment devices, although not essential, may be a useful means of adjusting the screen to the correct height for the worker. (The reference in the Schedule to adjustable tables does not mean these have to be provided).

Glare and reflections

13 Screens are generally manufactured without highly reflective surface finishes but in adverse lighting conditions, reflection and glare may be a problem. Advice on lighting is in paragraphs 20–24.

Keyboard

14 Keyboard design should allow workers to locate and activate keys quickly, accurately and without discomfort. The choice of keyboard will be dictated by the nature of the task and determined in relation to other elements of the work system. Hand support may be incorporated into the keyboard for support while keying or at rest depending on what the worker finds comfortable, may be provided in the form of a space between the keyboard and front edge of the desk, or may be given by a separate hand/wrist support attached to the work surface.

Work desk or work surface

15 Work surface dimensions may need to be larger than for conventional non-screen office work, to take adequate account of:

(a) the range of tasks performed (eg screen viewing, keyboard input, use of other input devices, writing on paper etc);

(b) position and use of hands for each task;

(c) use and storage of working materials and equipment (documents, telephones etc).

16 Document holders are useful for work with hard copy, particularly for workers who have difficulty in refocusing. They should position working documents at a height, visual plane and, where appropriate, viewing distance similar to those of the screen; be of low reflectance; be stable; and not reduce the readability of source documents.

Work chair

17 The primary requirement here is that the work chair should allow the user to achieve a comfortable position. Seat height adjustments should accommodate the needs of users for the tasks performed. The Schedule requires the seat to be adjustable in height (ie relative to the ground) and the seat back to be adjustable in height (also relative to the ground) and tilt. Provided the chair design meets these requirements and allows the user to achieve a comfortable posture, it is not necessary for the height or tilt of the seat back to be adjustable independently of the seat. Automatic backrest adjustments are acceptable if they provide adequate back support. General health and safety advice and specifications for seating are given in the HSE publication *Seating at Work* (HS(G)57). A range of publications with detailed advice covering comfort and performance as well as health and safety is included in Annex C (not reproduced in this book).

18 Footrests may be necessary where individual workers are unable to rest their feet flat on the floor (eg where work surfaces cannot be adjusted to the right height in relation to other components of the workstation). Footrests should not be used when they are not necessary as this can result in poor posture.

Environment

Space requirements

19 Prolonged sitting in a static position can be harmful. It is most important that support surfaces for display screen and other equipment and materials used at the workstation should allow adequate clearance for postural changes. This means adequate clearances for thighs, knees, lower legs and feet under the work surface and between furniture components. The height of the work surface should allow a comfortable position for the arms and wrists, if a keyboard is used.

Lighting, reflections and glare

20 Lighting should be appropriate for all the tasks performed at the workstation, eg reading from the screen, keyboard work, reading printed text, writing on paper etc. General lighting—by artificial or natural light, or a combination—should illuminate the entire room to an adequate standard. Any supplementary

individual lighting provided to cater for personal needs or a particular task should not adversely affect visual conditions at nearby workstations.

Illuminance

21 High illuminances render screen characters less easy to see but improve the ease of reading documents. Where a high illuminance environment is preferred for this or other reasons, the use of positive polarity screens (dark characters on a light background) has advantages as these can be used comfortably at higher illuminances than can negative polarity screens.

Reflections and glare

22 Problems which can lead to visual fatigue and stress can arise for example from unshielded bright lights or bright areas in the worker's field of view; from an imbalance between brightly and dimly lit parts of the environment; and from reflections on the screen or other parts of the workstation.

23 Measures to minimise these problems include: shielding, replacing or repositioning sources of light; rearranging or moving work surfaces, documents or all or parts of workstations; modifying the colour or reflectance of walls, ceilings, furnishings etc near the workstation; altering the intensity of vertical to horizontal illuminance; or a combination of these. Anti-glare screen filters should be considered as a last resort if other measures fail to solve the problem.

24 General guidance on minimum lighting standards necessary to ensure health and safety of workplaces is available in the HSE guidance note *Lighting at Work* (HS(G)38). This does not cover ways of using lighting to maximise task performance or enhance the appearance of the workplace, although it does contain a bibliography listing relevant publications in this area. Specific and detailed guidance is given in the CIBSE Lighting Guide 3 *Lighting for visual display terminals*.

Noise

25 Noise from equipment such as printers at display screen workstations should be kept to levels which do not impair

concentration or prevent normal conversation (unless the noise is designed to attract attention, eg to warn of a malfunction). Noise can be reduced by replacement, sound-proofing or repositioning of the equipment; sound insulating partitions between noisy equipment and the rest of the workstation are an alternative.

Heat and humidity

26 Electronic equipment can be a source of dry heat which can modify the thermal environment at the workstation. Ventilation and humidity should be maintained at levels which prevent discomfort and problems of sore eyes.

Radiation

27 The Schedule requires radiation with the exception of the visible part of the electromagnetic spectrum (ie visible light) to be reduced to negligible levels from the point of view of the protection of users' health and safety. In fact so little radiation is emitted from current designs of display screen equipment that no special action is necessary to meet this requirement (see also Annex B, paragraphs 8–10).

28 Taking cathode ray tube displays as an example, ionising radiation is emitted only in exceedingly small quantities, so small as to be generally much less than the natural background level to which everyone is exposed. Emissions of ultraviolet, visible and infrared radiation are also very small, and workers will receive much less than the maximum exposures generally recommended by national and international advisory bodies.

29 For radio frequencies, the exposures will also be well below the maximum values generally recommended by national and international advisory bodies for health protection purposes. The levels of electric and magnetic fields are similar to those from common domestic electrical devices. Although much research has been carried out on possible health effects from exposure to electromagnetic radiation, no adverse health effects have been shown to result from the emissions from display screen equipment.

30 Thus it is not necessary, from the standpoint of limiting risk to human health, for employers or workers to take any action to reduce radiation levels or to attempt to measure emissions; in fact the latter is not recommended as meaningful interpretation of the

data is very difficult. There is no need for users to be given protective devices such as anti-radiation screens.

Task design and software

Principles of task design

31 Inappropriate task design can be among the causes of stress at work. Stress jeopardises employee motivation, effectiveness and efficiency and in some cases it can lead to significant health problems. The Regulations are only applicable where health and safety rather than productivity is being put at risk; but employers may find it useful to consider both aspects together as task design changes put into effect for productivity reasons may also benefit health, and vice versa.

32 In display screen work, good design of the task can be as important as the correct choice of equipment, furniture and working environment. It is advantageous to:

(a) design jobs in a way that offers users variety, opportunities to exercise discretion, opportunities for learning, and appropriate feedback, in preference to simple repetitive tasks whenever possible. (For example, the work of a typist can be made less repetitive and stressful if an element of clerical work is added);

(b) match staffing levels to volumes of work, so that individual users are not subject to stress through being either overworked or underworked;

(c) allow users to participate in the planning, design and implementation of work tasks whenever possible.

Principles of software ergonomics

33 In most display screen work the software controls both the presentation of information on the screen and the ways in which the worker can manipulate the information. Thus software design can be an important element of task design. Software that is badly designed or inappropriate for the task will impede the efficient completion of the work and in some cases may cause sufficient stress to affect the health of a user. Involving a sample of users in the purchase or design of software can help to avoid problems.

34 Detailed ergonomic standards for software are likely to be developed in future as part of the ISO 9241 standard; for the

moment, the Schedule lists a few general principles which employers should take into account. Requirements of the organisation and of display screen workers should be established as the basis for designing, selecting, and modifying software. In many (though not all) applications the main points are:

Suitability for the task

— Software should enable workers to complete the task efficiently, without presenting unnecessary problems or obstacles.

Ease of use and adaptability

— Workers should be able to feel that they can master the system and use it effectively following appropriate training;

— The dialogue between the system and the worker should be appropriate for the worker's ability;

— Where appropriate, software should enable workers to adapt the user interface to suit their ability level and preferences;

— The software should protect workers from the consequences of errors, for example by providing appropriate warnings and information and by enabling lost data to be recovered wherever practicable.

Feedback on system performance

— The system should provide appropriate feedback, which may include error messages; suitable assistance (help) to workers on request; and messages about changes in the system such as malfunctions or overloading;

— Feedback messages should be presented at the right time and in an appropriate style and format. They should not contain unnecessary information.

Format and pace

— Speed of response to commands and instructions should be appropriate to the task and to workers' abilities;

— Characters, cursor movements and position changes should where possible be shown on the screen as soon as they are input.

Performance monitoring facilities

— Quantitative or qualitative checking facilities built into the software can lead to stress if they have adverse results such as an over-emphasis on output speed;

— It is possible to design monitoring systems that avoid these drawbacks and provide information that is helpful to workers as well as managers. However, in all cases workers should be kept informed about the introduction and operation of such systems.

ANNEX B
DISPLAY SCREEN EQUIPMENT: POSSIBLE EFFECTS ON HEALTH

The main hazards

1 The introduction of VDUs and other display screen equipment has been associated with a range of symptoms related to the visual system and working posture. These often reflect bodily fatigue. They can readily be prevented by applying ergonomic principles to the design, selection and installation of display screen equipment, the design of the workplace, and the organisation of the task.

Upper limb pains and discomfort

2 A range of conditions of the arm, hand and shoulder areas linked to work activities are now described as work related upper limb disorders. These range from temporary fatigue or soreness in the limb to chronic soft tissue disorders like peritendinitis or carpal tunnel syndrome. Some keyboard operators have suffered occupational cramp.

3 The contribution to the onset of any disorder of individual risk factors (eg keying rates) is not clear. It is likely that a combination of factors are concerned. Prolonged static posture of the back, neck and head are known to cause musculoskeletal problems. Awkward positioning of the hands and wrist (eg as a result of poor working technique or inappropriate work height) are further likely factors. Outbreaks of soft tissue disorders among keyboard workers have often been associated with high workloads

combined with tight deadlines. This variety of factors contributing to display screen work risk requires a risk reduction strategy which embraces proper equipment, furniture, training, job design and work planning.

Eye and eyesight effects

4 Medical evidence shows that using display screen equipment is not associated with damage to eyes or eyesight; nor does it make existing defects worse. But some workers may experience *temporary* visual fatigue, leading to a range of symptoms such as impaired visual performance, red or sore eyes and headaches, or the adoption of awkward posture which can cause further discomfort in the limb. These may be caused by:

(a) staying in the same position and concentrating for a long time;

(b) poor positioning of the display screen equipment;

(c) poor legibility of the screen or source documents;

(d) poor lighting, including glare and reflections;

(e) a drifting, flickering or jittering image on the screen.

Like other visually demanding tasks, VDU work does not cause eye damage but it may make workers with pre-existing vision defects more aware of them. Such uncorrected defects can make work with a display screen more tiring or stressful than would otherwise be the case.

Fatigue and stress

5 Many symptoms described by display screen workers reflect stresses arising from their task. They may be secondary to upper limb or visual problems but they are more likely to be caused by poor job design or work organisation, particularly lack of sufficient control of the work by the user, under-utilisation of skills, high-speed repetitive working or social isolation. All these have been linked with stress in display screen work, although clearly they are not unique to it; but attributing individual symptoms to particular aspects of a job or workplace can be difficult. The risks of display screen workers experiencing physical fatigue and stress can be minimised, however, by following the principles underlying the Display Screen Equipment Regulations 1992 and guidance, ie by

careful design, selection and disposition of display screen equipment; good design of the user's workplace, environment and task; and training, consultation and involvement of the user.

Other concerns

Epilepsy

6 Display screen equipment has not been known to induce epileptic seizures. People suffering from the very rare (1 in 10,000 population) photosensitive epilepsy who react adversely to flickering lights and patterns also find they can safely work with display screens. People with epilepsy who are concerned about display screen work can seek further advice from local offices of the Employment Medical Advisory Service.

Facial dermatitis

7 Some VDU users have reported facial skin complaints such as occasional itching or reddened skin on the face and/or neck. These complaints are relatively rare and the limited evidence available suggests they may be associated with environmental factors, such as low relative humidity or static electricity near the VDU.

Electromagnetic radiation

8 Anxiety about radiation emissions from display screen equipment and possible effects on pregnant women has been widespread. However, there is substantial evidence that these concerns are unfounded. The Health and Safety Executive has consulted the National Radiological Protection Board, which has the statutory function of providing information and advice on all radiation matters to Government Departments, and the advice below summarises scientific understanding.

9 The levels of ionising and non-ionising electromagnetic radiation which are likely to be generated by display screen equipment are well below those set out in international recommendations for limiting risk to human health created by such emissions and the National Radiological Protection Board does not consider such levels to pose a significant risk to health. No special protective measures are therefore needed to protect the health of people from this radiation.

Effects on pregnant women

10 There has been considerable public concern about reports of higher levels of miscarriage and birth defects among some groups of visual display unit (VDU) workers in particular due to electromagnetic radiation. Many scientific studies have been carried out, but taken as a whole their results do not show any link between miscarriages or birth defects and working with VDUs. Research and reviews of the scientific evidence will continue to be undertaken.

11 In the light of the scientific evidence pregnant women do not need to stop work with VDUs. However, to avoid problems caused by stress and anxiety, women who are pregnant or planning children and worried about working with VDUs should be given the opportunity to discuss their concerns with someone adequately informed of current authoritative scientific information and advice.

ANNEX C
FURTHER SOURCES OF INFORMATION

This Annex is not reproduced in this book.

Appendix 7

Health and Safety (Consultation with Employees) Regulations 1996

THE REGULATIONS (SI 1996/1513)

Made 10 June 1996

The Secretary of State, being a Minister designated for the purposes of section 2(2) of the European Communities Act 1972 in relation to measures relating to consultation with employees on health and safety at work, in exercise of the powers conferred on him by the said section 2(2) and of all other powers enabling him in that behalf, hereby makes the following Regulations:

Regulation 1

Citation, extent and commencement

These Regulations, which extend to Great Britain, may be cited as the Health and Safety (Consultation with Employees) Regulations 1996 and shall come into force on 1 October 1996.

Regulation 2

Interpretation

(1) In these Regulations, unless the context otherwise requires—

— 'the 1974 Act' means the Health and Safety at Work etc Act 1974;

— 'the 1977 Regulations' means the Safety Representatives and Safety Committees Regulations 1977;

— 'employee' has the meaning assigned to it by section 53(1) of the 1974 Act but shall not include a person employed as a domestic servant in a private household; and 'employer' shall be construed accordingly;

— 'the relevant statutory provisions' has the meaning assigned to it by section 53(1) of the 1974 Act;

— 'representatives of employee safety' shall be construed in accordance with regulation 4(1)(b);

— 'safety representative' has the meaning assigned to it by regulation 2(1) of the 1977 Regulations;

— 'workplace' means, in relation to an employee, any place or places where that employee is likely to work or which he is likely to frequent in the course of his employment or incidentally to it and, in relation to a representative of employee safety, any place or places where the employees he represents are likely so to work or frequent.

(2) Any reference in these Regulations to consulting employees directly or consulting representatives of employee safety is a reference to consulting them pursuant to regulation 3 and regulation 4(1)(a) or (b), as the case may be.

(3) Unless the context otherwise requires, any reference in these Regulations to—

(a) a numbered regulation or schedule is a reference to the regulation or schedule in these Regulations so numbered; and

(b) a numbered paragraph is a reference to the paragraph so numbered in the regulation or schedule in which the reference appears.

Regulation 3 *(See Guidance Notes, pp 709–711)*

Duty of employer to consult

Where there are employees who are not represented by safety representatives under the 1977 Regulations, the employer shall consult those employees in good time on matters relating to their health and safety at work and, in particular, with regard to—

(a) the introduction of any measure at the workplace which may substantially affect the health and safety of those employees;

(b) his arrangements for appointing or, as the case may be, nominating persons in accordance with [regulations 7(1) and

8(1)(b) of the Management of Health and Safety at Work Regulations 1999];[1]

(c) any health and safety information he is required to provide to those employees by or under the relevant statutory provisions;

(d) the planning and organisation of any health and safety training he is required to provide to those employees by or under the relevant statutory provisions; and

(e) the health and safety consequences for those employees of the introduction (including the planning thereof) of new technologies into the workplace.

Regulation 4 *(See Guidance Notes, pp 712–715)*

Persons to be consulted

(1) The consultation required by regulation 3 is consultation with either—

(a) the employees directly; or

(b) in respect of any group of employees, one or more persons in that group who were elected, by the employees in that group at the time of the election, to represent that group for the purposes of such consultation (and any such persons are in these Regulations referred to as 'representatives of employee safety').

(2) Where an employer consults representatives of employee safety he shall inform the employees represented by those representatives of—

(a) the names of those representatives; and

(b) the group of employees represented by those representatives.

(3) An employer shall not consult a person as a representative of employee safety if—

(a) that person has notified the employer that he does not intend to represent the group of employees for the purposes of such consultation;

1 Words from 'regulations 7(1) and' to 'Work Regulations 1999' in square brackets substituted by SI 1999/3242, reg 29(2), Sch 2. Date in force: 29 December 1999: see SI 1999/3242, reg 1(1).

(b) that person has ceased to be employed in the group of employees which he represents;

(c) the period for which that person was elected has expired without that person being re-elected; or

(d) that person has become incapacitated from carrying out his functions under these Regulations;

and where pursuant to this paragraph an employer discontinues consultation with that person he shall inform the employees in the group concerned of that fact.

(4) Where an employer who has been consulting representatives of employee safety decides to consult employees directly he shall inform the employees and the representatives of that fact.

Regulation 5 *(See Guidance Notes, p 715)*

Duty of employer to provide information

(1) Where an employer consults employees directly he shall, subject to paragraph (3), make available to those employees such information, within the employer's knowledge, as is necessary to enable them to participate fully and effectively in the consultation.

(2) Where an employer consults representatives of employee safety he shall, subject to paragraph (3), make available to those representatives such information, within the employer's knowledge, as is—

(a) necessary to enable them to participate fully and effectively in the consultation and in the carrying out of their functions under these Regulations;

(b) contained in any record which he is required to keep by regulation 7 of the Reporting of Injuries, Diseases and Dangerous Occurrences Regulations 1995 and which relates to the workplace or the group of employees represented by those representatives.

(3) Nothing in paragraph (1) or (2) shall require an employer to make available any information—

(a) the disclosure of which would be against the interests of national security;

(b) which he could not disclose without contravening a prohibition imposed by or under any enactment;

(c) relating specifically to an individual, unless he has consented to its being disclosed;

(d) the disclosure of which would, for reasons other than its effect on health or safety, cause substantial injury to the employer's undertaking or, where the information was supplied to him by some other person, to the undertaking of that other person; or

(e) obtained by the employer for the purpose of bringing, prosecuting or defending any legal proceedings;

or to provide or allow the inspection of any document or part of a document which is not related to health or safety.

Regulation 6 *(See Guidance Notes, p 715)*

Functions of representatives of employee safety

Where an employer consults representatives of employee safety each of those representatives shall, for the period for which that representative is so consulted, have the following functions—

(a) to make representations to the employer on potential hazards and dangerous occurrences at the workplace which affect, or could affect, the group of employees he represents;

(b) to make representations to the employer on general matters affecting the health and safety at work of the group of employees he represents and, in particular, on such matters as he is consulted about by the employer under regulation 3; and

(c) to represent the group of employees he represents in consultations at the workplace with inspectors appointed under section 19(1) of the 1974 Act.

Regulation 7 *(See Guidance Notes, pp 716–720)*

Training, time off and facilities for representatives of employee safety and time off for candidates

(1) Where an employer consults representatives of employee safety, he shall—

(a) ensure that each of those representatives is provided with such training in respect of that representative's functions under these Regulations as is reasonable in all the circumstances and the employer shall meet any reasonable costs associated with such training including travel and subsistence costs; and

(b) permit each of those representatives to take such time off with pay during that representative's working hours as shall be necessary for the purpose of that representative performing his functions under these Regulations or undergoing any training pursuant to paragraph (1)(a).

(2) An employer shall permit a candidate standing for election as a representative of employee safety reasonable time off with pay during that person's working hours in order to perform his functions as such a candidate.

(3) Schedule 1 (pay for time off) and Schedule 2 (provisions as to [employment tribunals])[2] shall have effect.

(4) An employer shall provide such other facilities and assistance as a representative of employee safety may reasonably require for the purpose of carrying out his functions under these Regulations.

Regulation 8 *(See Guidance Notes, p 721)*

Amendment of the Employment Rights Act 1996

In sections 44(1) (health and safety cases: right not to suffer detriment) and 100(1) (health and safety cases: unfair dismissal) of the Employment Rights Act 1996,[3] after paragraph (b) insert—

'(ba) the employee took part (or proposed to take part) in consultation with the employer pursuant to the Health and Safety (Consultation with Employees) Regulations 1996 or in an election of representatives of employee safety within the meaning of those Regulations (whether as a candidate or otherwise),'.

Regulation 9

Exclusion of civil liability

Breach of a duty imposed by these Regulations shall, subject to regulation 7(3) and Schedule 2, not confer any right of action in any civil proceedings.

2 Words 'employment tribunals' in square brackets substituted by the Employment Rights (Dispute Resolution) Act 1998, s 1(2)(b). Date in force: 1 August 1998: see SI 1998/1658, art 2(1), Sch 1.

3 1996 c 18.

Regulation 10 *(See Guidance Notes, p 721)*

Application of health and safety legislation

Sections 16 to 21, 23, 24, 26, 28, 33, 34, 36 to 39, 42(1) to (3) and 46 of the 1974 Act, the Health and Safety (Enforcing Authority) Regulations 1989 and the Health and Safety (Training for Employment) Regulations 1990 shall apply as if any references therein to health and safety regulations or to the relevant statutory provisions included references to these Regulations.

Regulation 11

Application to the Crown and armed forces

(1) Section 48 of the 1974 Act shall, subject to paragraph (2), apply in respect of these Regulations as it applies in respect of regulations made under Part I of that Act.

(2) These Regulations shall apply in respect of members of the armed forces of the Crown subject to the following—

(a) references to 'representatives of employee safety' (in regulation 4(1)(b) and elsewhere) shall, in respect of any group of employees, be references to one or more persons in that group who were appointed by the employer to represent that group for the purposes of such consultation;

(b) references to 'elected' and 're-elected' in regulation 4(3)(c) shall be, respectively, references to 'appointed' and 're-appointed'; and

(c) regulation 7(1)(b), (2) and (3) shall not apply.

Regulation 12

Disapplication to sea-going ships

These Regulations shall not apply to or in relation to the master or crew of a sea-going ship or to the employer of such persons in respect of the normal ship-board activities of a ship's crew under the direction of the master.

Regulation 13

Amendment of the 1977 Regulations

In regulation 3(1) of the 1977 Regulations the words, 'except in the case of employees employed in a mine within the meaning of section 180 of the Mines and Quarries Act 1954 which is a coal mine' shall be omitted.

SCHEDULE 1
PAY FOR TIME OFF

Regulation 7(3)

1

Subject to paragraph 3 below, where a person is permitted to take time off in accordance with regulation 7(1)(b) or 7(2), his employer shall pay him—

(a) where the person's remuneration for the work he would ordinarily have been doing during that time does not vary with the amount of work done, as if he had worked at that work for the whole of that time;

(b) where the person's remuneration for that work varies with the amount of work done, an amount calculated by reference to the average hourly earnings for that work (ascertained in accordance with paragraph 2).

2

The average hourly earnings referred to in paragraph 1(b) are the average hourly earnings of the person concerned or, if no fair estimate can be made of those earnings, the average hourly earnings for work of that description of persons in comparable employment with the same employer or, if there are no such persons, a figure of average hourly earnings which is reasonable in all the circumstances.

3

Any payment to a person by an employer in respect of a period of time off—

(a) if it is a payment which discharges any liability which the employer may have under sections 168 or 169 of the Trade Union and Labour Relations (Consolidation) Act 1992, in respect of that period, shall also discharge his liability in respect of the same period under regulation 7(1)(b) or 7(2);

(b) if it is a payment under any contractual obligation, shall go towards discharging the employer's liability in respect of the same period under regulation 7(1)(b) or 7(2);

(c) if it is a payment under regulation 7(1)(b) or 7(2), shall go towards discharging any liability of the employer to pay contractual remuneration in respect of the same period.

SCHEDULE 2
PROVISIONS AS TO [EMPLOYMENT TRIBUNALS][4]

Regulation 7(3)

1

An [employment tribunal][5] shall have jurisdiction to determine complaints in accordance with the following provisions of this Schedule.

2

A person (referred to in this Schedule as the 'complainant') may present a complaint to an [employment tribunal][6] that—

(a) his employer has failed to permit him to take time off in accordance with regulation 7(1)(b) or 7(2); or

(b) his employer has failed to pay him in accordance with regulation 7(1)(b) or 7(2) and Schedule 1.

3

An [employment tribunal][7] shall not consider a complaint under paragraph 2 unless it is presented within three months of the date

4 Words 'Employment Tribunals' in square brackets substituted by the Employment Rights (Dispute Resolution) Act 1998, s 1(2)(b). Date in force: 1 August 1998: see SI 1998/1658, art 2(1), Sch 1.

5 Words 'employment tribunal' in square brackets substituted by the Employment Rights (Dispute Resolution) Act 1998, s 1(2)(a). Date in force: 1 August 1998: see SI 1998/1658, art 2(1), Sch 1.

6 Words 'employment tribunal' in square brackets substituted by the Employment Rights (Dispute Resolution) Act 1998, s 1(2)(a). Date in force: 1 August 1998: see SI 1998/1658, art 2(1), Sch 1.

7 Words 'employment tribunal' in square brackets substituted by the Employment Rights (Dispute Resolution) Act 1998, s 1(2)(a). Date in force: 1 August 1998: see SI 1998/1658, art 2(1), Sch 1.

when the failure occurred or within such further period as the tribunal considers reasonable in a case where it is satisfied that it was not reasonably practicable for the complaint to be presented within the period of three months.

4

Where an [employment tribunal][8] finds a complaint under paragraph 2(a) well-founded the tribunal shall make a declaration to that effect and may make an award of compensation to be paid by the employer to the complainant which shall be of such amount as the tribunal considers just and equitable in all the circumstances having regard to the employer's default in failing to permit time off to be taken by the complainant and to any loss sustained by the complainant which is attributable to the matters complained of.

5

Where on a complaint under paragraph 2(b) an [employment tribunal][9] finds that the employer has failed to pay the complainant the whole or part of the amount required to be paid in accordance with regulation 7(1)(b) or 7(2) and Schedule 1, the tribunal shall order the employer to pay the complainant the amount which it finds due to him.

GUIDANCE NOTES

INTRODUCTION

1 The law requires you to consult your employees on matters that affect their health and safety. This guide tells you about Regulations which will apply to you if you have any employees who are not covered by representatives appointed by recognised trade unions. It:

8 Words 'employment tribunal' in square brackets substituted by the Employment Rights (Dispute Resolution) Act 1998, s 1(2)(a). Date in force: 1 August 1998: see SI 1998/1658, art 2(1), Sch 1.

9 Words 'employment tribunal' in square brackets substituted by the Employment Rights (Dispute Resolution) Act 1998, s 1(2)(a). Date in force: 1 August 1998: see SI 1998/1658, art 2(1), Sch 1.

(a) describes what you must consult those employees about;

(b) explains that you can choose to consult those employees:

(i) directly, or

(ii) through elected representatives;

and

(c) sets out what is involved if you choose to consult through elected representatives.

2 Proper consultation with your employees on health and safety matters can make a significant contribution to creating and maintaining an effective 'health and safety culture' within your business. The effect on employees' motivation, and awareness of the importance of health and safety, should be entirely positive with the potential for greater efficiency and a reduction in accidents or incidents of work-related ill-health.

Regulation 3

Duty of employer to consult

To whom does this apply?

3 These Health and Safety (Consultation with Employees) Regulations (HSCER) 1996 apply to all employers and employees in Great Britain, other than:

(a) those whose employees are all covered by safety representative(s) appointed by recognised trade union(s) under the Safety Representatives and Safety Committees Regulations (SRSCR) 1977;

(b) domestic staff employed in private households; or

(c) the master or crew of a sea-going ship or to their employer in respect of the normal ship-board activities of a ship's crew under the direction of the master.

4 The SRSCR 1977 provide for the appointment of safety representatives by recognised trade unions. Those Regulations specify the functions of such safety representatives and set out the obligations of employers towards them. They remain in force in their entirety. The Offshore Installations (Safety Representatives and Safety Committees) Regulations 1989 apply offshore. The

Marine Safety Agency are preparing regulations to apply in the maritime sector.[10]

The duty to consult: what must I consult my employees about?

5 The HSCER 1996 require you to consult your employees on:

(a) any measure at the workplace which may substantially affect their health and safety;

(b) your arrangements for getting a competent person[11] or persons to help you comply with health and safety requirements and to implement evacuation procedures. The Management of Health and Safety at Work Regulations 1992 ('the Management Regulations') require you to make such an appointment unless you are competent to deal with these matters yourself; under the HSCER 1996 you will need to consult your employees on how you plan to go about this;

(c) the information you must give them on risks to health and safety, and preventive measures. This will include the information you are already required by other Regulations to give your employees. Appendix A sets out the details. Under the Management Regulations, for example, among other things,[12] you have to tell your employees about the risks identified by the risk assessment you must carry out and your preventive and protective measures. You must also tell them about your emergency procedures. Now, under the HSCER 1996, you will have to consult your employees about these matters before you tell them what has been decided, and before you make changes;

10 New regulations were due to be made by the Marine Safety Agency in 1996 to come into force on 1 January 1997.

11 The 'competent person' as specified under the Management Regulations is distinct from the elected 'representative of employee safety' defined under the HSCER 1996: in brief, the 'competent person' advises the employer; the 'representative of employee safety' represents the views of the employees.

12 Requirements to provide information to your employees under the Management Regulations include information on: the risks to their health and safety identified by your risk assessment; the preventive and protective measures designed to ensure their health and safety; the procedures to be followed in the event of an emergency in your undertaking; the identity of any 'competent person' or persons nominated by you to help with the implementation of those procedures; and risks notified to you by another employer with whom you share a workplace, arising out of, or in connection with, the conduct of the second employer's undertaking.

(d) the planning and organising of any health and safety training you must provide to employees under health and safety law, for example when your employees are first recruited and when they are to be exposed to new or increased risks (as required by the Management Regulations). Other Regulations are relevant where your employees are exposed to particular risks or hazards. Appendix B sets out what applies. In addition, you also have to ensure that any representatives of employee safety elected under the HSCER 1996 are provided with relevant training (see paragraphs 23–24); and

(e) the health and safety consequences for them of new technology that you plan to bring into the workplace. This will cover the introduction of any new technology if there could be implications for their health and safety, and for the risks and hazards to which they are exposed.

When do I have to consult?

6 The HSCER 1996 require that you must consult your employees 'in good time'. That means that, wherever a decision involving work equipment, processes or organisation could have health and safety consequences for employees, before making that decision you must allow time:

(a) to provide the employees, or their elected representatives, with information about what you propose to do;

(b) to give the employees, or their elected representatives, an opportunity to express their views about the matter in the light of that information;

and then

(c) for you to take account of any response from employees or their elected representatives.

The difference between informing and consulting

7 You are already required by the Management Regulations to provide your employees with comprehensible and relevant information on the matters set out previously in footnote 12. The difference between providing information to your employees and consulting them, which the HSCER 1996 require, is that consultation involves listening to their views and taking account of what they say before any decision is taken.

Regulation 4

Persons to be consulted

How to consult: what if I recognise a trade union?

8 Some or all of your employees may be represented by one or more trade unions which you recognise. If so, you are already required to consult health and safety representatives appointed by each union under the SRSCR 1977.[13] When appointing such representatives, the union must notify you in writing of the group or groups of employees they represent. They may, for example, tell you that they are representing all the workers in a particular category, or, alternatively, only their own members. You may continue to implement existing arrangements that comply with the SRSCR 1977 in respect of these employees. The HSCER 1996 apply only to any employees not covered by such arrangements including where recognised trade unions have not appointed representatives, or they are not about to, and mean that you must now also consult any such employees. They therefore 'top up' the consultation requirement of the SRSCR 1977. You need to check whether the trade union safety representative(s) do indeed cover all of the employees or not.

What about my other employees?

9 If some employees are not members of a trade union which you recognise, but are members of a group of employees for which a union is recognised, they may be covered by the consultation arrangements with the union safety representative and paragraph 8 applies. If they are not members of such a group, or if the union safety representative(s) cover only their own members, then you must make arrangements to consult them under the HSCER 1996. If there is no recognised trade union, you must make arrangements to consult all your employees under the HSCER 1996.

10 You are not obliged to keep written records of your consultation. You may, however, need to convince health and safety inspectors that you have complied with the HSCER 1996; and they will also talk to safety representatives and/or employees.

13 In 1992 specific provisions about consultation rights for such union-appointed representatives were added to the SRSCR 1977. Guidance on Safety Representatives and Safety Committees is available from HSE Books, P0 Box 1999, Sudbury, Suffolk, CO10 6FS: ISBN 0 7176 1220 1.

The self-employed

11 The HSCER 1996 do not apply to the self-employed. However, you should bear in mind that case law has been established that workers who are categorised as self-employed for tax or other purposes may in fact be employees in respect of health and safety law. This depends on individual circumstances (eg the worker's degree of independence, whether they are using their own tools etc).[14]

Non-employees

12 Seconded staff, long-term agency staff and other workers on your premises who are not your employees do not have to be consulted by you, though you must provide them with the information they need in order to work safely, under the Management Regulations (see paragraph 5(c) and footnote 12). You may choose, voluntarily, to include them in your arrangements for consulting your employees. This may be worthwhile, particularly if they have been working for you for a period of weeks or months or are exposed to particular risks and hazards at work.

Non-employed trainees

13 However, non-employed trainees such as student nurses and others participating in work experience schemes are covered by the HSCER 1996 as they are treated as employees under Health and Safety law.

Construction industry

14 One special case is that of construction, where firstly the employer may not be in control of all the risks and secondly there are self-employed people who are exposed to the risks. The Construction (Design and Management) Regulations 1994 address this situation by requiring the principal contractor to ensure all employees and self-employed people carrying out the construction work are able to discuss issues which affect their health and safety and to offer him/her advice. The principal contractor

14 For a detailed review see the IRS Industrial Relations Law Bulletin, parts 533 (November 1995) and 534 (December 1995).

must ensure that there are arrangements in place for the co-ordination of site-specific consultation undertaken by employers.[15]

Choose how to consult

15 When you consult employees under the HSCER 1996 you have a choice between consultation:

(a) through one or more elected representatives (and it is important that they are independent of you in conveying the views of their colleagues); or

(b) directly with each employee.

You can choose whatever means suit you and your employees best.

16 For example:

(a) you may have briefing meetings allowing feedback of representatives' or employees' views up the management chain. Such meetings might deal with health and safety matters alone, or might easily be adapted to include them as agenda items among other issues. You may have employee quality circles which could be adapted to deal with consultation. Similarly staff councils, notice boards, newsletters, electronic mail and surveys may also be used;

(b) if you run a very small firm, you may talk to your employees on a regular basis, and take account of what they say. If so, you may need to do no more if such informal arrangements are adequate for consultation on health and safety matters affecting the employees.

17 Where elected representatives exist, there would be nothing to prevent you from consulting employees directly on particular matters. Similarly, you might consult on any particular issue both directly and indirectly.

18 What matters is that your employees, or their representatives, are made aware of:

(a) when their views are being sought about health and safety;

15 For further information see 26 L54 *Managing construction for health and safety—Construction (Design and Management) Regulations 1994. Approved Code of Practice* HSE Books 1995 ISBN 0 7176 0792 5; CONIAC guidance: *A guide to managing health and safety in construction* HSE Books ISBN 0 7176 0755 0.

(b) how they can give their views to you as their employer; and

(c) their right to take part in discussions on all questions relating to their health and safety at work.

19 If you decide to consult via elected representatives but no candidates come forward for election, then you will have to consult directly. If only one candidate is nominated for a vacancy then there would be no need to hold a formal election.

Regulation 5

Duty of employer to provide information

What sort of information do I have to provide?

20 You must provide enough information to allow your employees to understand:

(a) what the likely risks and hazards arising from their work, or changes to their work, may be (including details of injuries, diseases and dangerous occurrences that have been reported);

(b) the measures in place, or which will be introduced, to eliminate or reduce them; and

(c) what employees ought to do when encountering risks and hazards.

But you do not have to provide information for consultation if the exemptions listed in Regulation 5(3) apply to it.

Regulations 6 and 7, Schedules 1 and 2 and guidance paragraphs 21 to 28 below relate specifically to consultation via elected representatives.

Regulation 6

Functions of representatives of employee safety

Statutory functions of elected representatives

21 As long as a person holds office as an elected representative of employee safety and is being consulted by the employer, the law gives them the functions described in Regulation 6. They can make representations to you on their own initiative; they are not restricted to the occasions when you consult them; and the HSCER 1996 do not impose any legal duties upon them.

Regulation 7

Training, time off and facilities for representatives of employee safety and time off for candidates

Special facilities for elected representatives

22 So that elected representatives can carry out their functions properly, you must:

(a) provide them with facilities and assistance which they may reasonably require for the purpose of carrying out their functions under the HSCER 1996. What is needed in particular circumstances will vary widely. It could involve access to:

(i) lists showing the names and workplaces of employees in a representative's constituency;

(ii) communications, distribution and photocopying facilities for the representative to communicate with represented employees;

(iii) you or your senior management to discuss health and safety issues whenever the representative(s) might reasonably wish to do so;

(b) give them reasonable time off with pay,[16] during working hours, as necessary to perform the representative's functions under the HSCER 1996, and for reasonable training in respect of those functions. In practice, this means they should carry out their duties as part of their normal job, and you will need to take account of this in their work load;

(c) ensure that they are provided with adequate training (see paragraphs 23–24); and

(d) ensure that they are protected against detriment (see paragraph 30).

Training for elected representatives

23 Individuals' abilities and needs will vary widely. Some may need to undertake a structured programme that will equip them

16 When elected representatives are entitled to take time off under the HSCER 1996, you must give them the pay they would ordinarily receive for the work they would otherwise have been doing during that time. Schedule 1 to the Regulations sets out the requirements in detail.

with the necessary skills and knowledge. Others may already be sufficiently competent in respect of both representational skills and knowledge of health and safety issues and current legislation[17] so that initially little or no special training will be required. Further training may be needed in the light of legislative changes, different working conditions, new hazards etc. You must pay any reasonable costs associated with such training, including any travel and subsistence costs.

24 Once training needs are identified and agreed, you will have to identify sources. There may be various possibilities, including:

(a) participation in health and safety training courses which trade unions run for the representatives they appoint under the SRSCR 1977[18]—and particularly apt if the elected representative happens to be a member of a union which offers or has access to such courses;

(b) health and safety courses which may be offered by your trade association, specialist personnel organisations, the TUC, individual trade unions, local Chambers of Commerce, TECs, colleges and similar institutions; and

(c) distance learning materials, including (for example) HSE's booklet and audio package *You can do it*.[19]

Preliminary arrangements for the election

25 When making the arrangements for electing representatives of employee safety, it may be useful to consider the following:

17 The Management Regulations already require that all employees are provided with adequate health and safety training when they are recruited and if they are to be exposed to new or increased risks. The need for further training must be considered when employees transfer or take on new responsibilities, and when there is a change in the work equipment or systems of work in use. Other regulations have more specific requirements.

18 There are some differences in the legal functions of trade union safety representatives as compared with elected representatives of employee safety (the SRSCR 1977 give trade union safety representatives the additional functions of undertaking inspections and investigations). This means the training that will be needed so that they can carry out their statutory functions is not precisely the same.

19 *You can do it* ISBN 0 7176 0726 7 is available from HSE Books, P0 Box 1999, Sudbury, Suffolk CO10 6FS.

(a) Constituencies

The HSCER 1996 require that any elected representative of employee safety must be employed in the group ('the constituency') which elects them. There is no legal definition of the group. Constituencies may, for example, cover multiple sites if this is appropriate to your particular business. Elected representatives may well have to travel, for example, to consult dispersed or mobile constituents. The constituencies can be specified in the light of the circumstances that apply to your business—see paragraph (d) for some factors you may want to bear in mind. There are no set rules on how the election must be organised, or which employees must be part of any particular constituency.

(b) How often to hold elections

In deciding what interval to allow between elections, account may be taken of the benefits that may follow from giving any elected representative time to develop experience so as better to discharge the functions and responsibilities of the post. On the other hand, where your employees change quite frequently, someone elected a couple of years ago may no longer be the choice of the current workforce.[20]

(c) Optional functions for elected representatives

Elected representatives do not have to be confined to consultation on matters set out in these particular Regulations. For example, you might find it convenient to consult these representatives on matters other than health and safety. Alternatively, some employers have non-union safety representatives who carry out investigations of accidents and complaints and sit on safety committees, in the same way as representatives appointed under the SRSCR 1977. It would need to be clear to all those taking part in the election that the elected representative(s) would include the particular functions and consultation rights provided for by these Regulations, as well as any others you wished to add.

20 Under the Offshore Installations (Safety Representatives etc) Regulations 1989, there must be at least two constituencies for every installation; limits are set for the size of constituencies (no fewer than three, no more than 40); and elections must be held at least every two years. But this model may not meet the health and safety needs of any other situation.

(d) Number of representatives

When deciding how many representatives should be elected from a constituency, it may be useful to take into account:

(i) the total numbers to be represented;

(ii) the variety of different groupings (eg by occupation, location, type of work or shift patterns) into which the employees might be divided; and

(iii) the nature of the work activities they undertake and the degree and character of the health and safety risks to which some or all of them may be exposed.

Before the election

26 It is important that the election is properly conducted so that the result reflects the wishes of those in the constituency. In advance of the election, relevant employees should be aware of:

(a) which other employees or groups of employees are part of their constituency;

(b) the length of time the elected representative(s) will hold office; and

(c) the nature of the health and safety functions to be undertaken by the elected representative(s).

How to conduct an election

27 Some general principles which may be followed, in order to make sure the election is fair and the representative(s) are independent of you, are that:

(a) each of the employees to be represented must be able, if they wish, to stand as a candidate on an equal basis to any other, and without any cost or disadvantage to themselves;

(b) a candidate standing for election as a representative of employee safety must also be permitted reasonable time off with pay during working hours in order to perform his/her functions as a candidate;

(c) if the election is contested:

(i) no-one eligible to vote in the ballot is subjected to intimidation or interference from any source intended to influence the way in which they cast their vote;

(ii) steps are taken to prevent the ballot papers from being tampered with at any stage (giving each one a serial number can be a useful further protection);

(iii) as far as possible, each of the employees to be represented has an equal opportunity to cast their vote, without any cost or disadvantage to themselves, for the candidate or candidates of their choice;

(iv) votes can be cast in secret;

(v) the result is determined by the number of votes cast for each candidate; and

(d) the election result is communicated, with sufficient information (eg on the number of votes cast for each candidate in a contested election), as soon as practicable, to all the candidates who stood and to the employees to be represented by the elected representative, eg by means of notices at relevant workplaces and/or through company newsletters etc.

28 To make sure that the election is properly carried out, you might want to get help from an outside body. There are some organisations that help to organise workplace elections and provide independent scrutiny.

Other help

29 Consulting your employees, either directly or through representatives, should bring benefits for health and safety in your undertaking. If you run into problems, HSE or your local authority can provide advice on health and safety issues. They will not get involved in industrial relations matters: you should seek to resolve any disputes through your normal procedures, and you can get further help from ACAS. (The telephone number of ACAS can be found in any telephone directory.) Independent external information and advice is also available from public libraries, Citizens Advice Bureaux, the Fire Service and safety bodies such as the Royal Society for the Prevention of Accidents.

Regulation 8

Amendment of the Employment Rights Act 1996

Protection against detriment

30 None of your employees may suffer any detriment because of anything reasonable they do, or propose to do, in connection with consultation on health and safety matters. If you penalise an employee (for example, by denying them promotion or opportunities for extra earnings) or dismiss them for such a reason, they could complain to an employment tribunal.[21] The same applies if you penalise an employee because of participation in an election for a representative of employee safety; because they are or have been a candidate in such an election; or because of what they do when they are acting as such a representative.

Regulation 10

Application of health and safety legislation

31 This means that the HSCER 1996 are, in respect of such matters as who enforces them and powers of inspection, the same as Regulations made under the Health and Safety at Work Act. In particular, they apply to non-employed trainees, see paragraph 13.

APPENDIX A
REQUIREMENTS FOR INFORMATION FOR EMPLOYEES IN EXISTING LEGISLATION

The following health and safety legislation requires employers to give information to their employees. Each entry contains a brief summary of what is required but you will need to find out your precise duties from the publications referred to in footnotes 22–47.

21 These protections apply to any employee, regardless of the length of time they have been working for you.

General health and safety

Management of Health and Safety at Work Regulations 1992 (MHSW)[22]

Information on:

(a) risks to health and safety;

(b) preventive and protective measures;

(c) emergency procedures (including who is responsible for evacuation); and

(d) temporary employees on fixed contracts: any special occupational skills or qualifications needed for the work and any requirements for health surveillance.

Health and Safety (First Aid) Regulations 1981[23]

First aid arrangements: including facilities, responsible personnel and where first aid equipment is kept.

Health and Safety (Safety Signs and Signals) Regulations 1995[24]

Each employee must be given clear and relevant information on the measures to be taken in connection with safety signs.

Health and Safety Information for Employees Regulations 1989[25]

Information about employees' health and safety welfare in the form of:

(a) an approved poster to be displayed where it can be easily read as soon as is reasonably practicable after any employees are taken on; or

22 L21 *Management of Health and Safety at Work Regulations 1992. Approved Code of Practice* HSE Books 1992 ISBN 0 7176 0412 8.

23 COP42 *First aid at work. Health and Safety (First Aid) Regulations 1981. Approved Code of Practice and guidance* HSE Books ISBN 0 7176 0426 8.

24 L64 *Safety Signs and Signals: Guidance on Regulations* HSE Books ISBN 0 7176 0870 0.

25 Poster—Health and Safety Law: What You Should Know HMSO ISBN 0 1170 1424 9.

(b) an approved leaflet to be given to employees as soon as practicable after they start.

Health and Safety (Consultation with Employees) Regulations 1996[26]

Necessary information to enable your employees to fully take part in consultation and to understand:

(a) what the likely risks and hazards arising from their work, or changes to their work, might be;

(b) the measures in place, or to be introduced, to eliminate or reduce them; and

(c) what employees ought to do when encountering risks and hazards.

See paragraph 5 of these Guidance Notes.

Safety Representatives and Safety Committees Regulations 1977[27]

Necessary information to assist the work of safety representatives nominated in writing by a recognised trade union. (See also Railways (Safety Case) Regulations 1994.)

Health hazards

Control of Substances Hazardous to Health Regulations 1994[28]

Information on:

(a) risks to health created by exposure to substances hazardous to health (including eg high hazard biological agents);

(b) precautions;

(c) results of any required exposure monitoring; and

(d) collective results of any required health surveillance.

26 *A guide to the Health and Safety (Consultation with Employees) Regulations 1996* HSE Books 1996 ISBN 0 7176 1234 1.

27 COP 1 *Safety representatives and safety committees* Rev ISBN 0 7176 1220 1.

28 L5 *General COSHH ACOP (Control of substances hazardous to health), Carcinogens ACOP (Control of carcinogenic substances) and Biological agents ACOP (Control of biological agents). Control of Substances Hazardous to Health. Approved Code of Practice 1995* HSE Books ISBN 0 7176 0819 0.

Health and Safety (Consultation with Employees) Regulations 1996

Chemicals (Hazards Information and Packaging) Regulations 1995[29]

Safety data sheets or the information they contain to be made available to employees (or to their appointed representatives).

Manual Handling Operations Regulations 1992[30]

Information on:

(a) the weight of loads for employees undertaking manual handling; and

(b) the heaviest side of any load whose centre of gravity is not positioned centrally.

Health and Safety (Display Screen Equipment) Regulations 1992[31]

Health and safety information about display screen work for both operators and users. (The Regulations define who is an operator and who is a user.)

Noise at Work Regulations 1989[32]

Information for employees likely to be exposed to daily personal noise levels at 85dB(A) or above:

29 L62 *Safety data sheets for substances and preparations dangerous for supply (second edition). Guidance on regulation 6 of the CHIP Regulations 1994. Approved Code of Practice* HSE Books 1995 ISBN 0 7176 0859 X. L63 *Approved guide to the classification and labelling of substances dangerous for supply—CHIP* 2. HSE Books 1995 ISBN 0 7176 0860 3.

30 L23 *Manual Handling. Manual Handling Operations Regulations 1992. Guidance on regulations* HSE Books 1992 ISBN 0 7176 0411 X. IND(G) 143L leaflet: *Getting to grips with manual handling* HSE Books ISBN 0 7176 0966 9.

31 L26 *Display screen equipment work. Health and Safety (Display Screen Equipment) Regulations 1992. Guidance on Regulations* HSE Books 1992 ISBN 0 7176 0410 1.

32 IND(G)75L(Rev) leaflet: *Noise at work: introducing the Noise at Work Regulations 1989* HSE Books. IND(G)99L leaflet: *Noise at work: advice for employees* HSE Books 1995 ISBN 0 7176 0962 6. IND(G)193L leaflet: *Health surveillance in noisy industries* HSE Books 1995 ISBN 0 7176 0933 2. IND(G)200L leaflet: *Ear protection in noisy firms—employers' duties explained* HSE Books 1995 ISBN 0 7176 0924 3.
Noise at work. Noise guide No 1: Legal duties of employers to prevent damage to hearing. Noise guide No 2: Legal duties of designers, manufacturers, importers and suppliers to prevent damage to hearing. The Noise at Work Regulations 1989 (one volume) HSE Books 1989 ISBN 0 7176 0454 3.

(a) noise exposure: level, risk of damage to hearing and action employees can take to minimise that risk;

(b) personal ear protectors (to be provided by employers): how to get them, where and when they should be worn, how to look after them and how to report defective ear protectors/noise control equipment;

(c) when to seek medical advice on loss of hearing; and

(d) employees' duties under the Regulations.

In addition, ear protection zones to be marked for employees likely to be exposed to daily personal noise levels at 90dB(A) or above (as far as is reasonably practicable).

Control of Asbestos at Work Regulations 1987[33]

Information about risks and precautions for:

(a) employees liable to be exposed to asbestos; and

(b) employees who carry out any work connected with your duties under these Regulations.

Control of Lead at Work Regulations 1980[34]

Information about risks and precautions for:

(a) employees liable to be exposed to lead; and

(b) employees who carry out any work connected with your duties under these Regulations.

Ionising Radiations Regulations 1985[35]

Information:

(a) to enable employees working with ionising radiations to meet the requirements of the Regulations;

33 L27 *The Control of Asbestos at Work. Control of Asbestos at Work Regulations. 1987. Approved Code of Practice* (2nd edn) HSE Books 1993 ISBN 0 11 882037 0. This legislation is being reviewed as part of the Commission's programme to reform health and safety legislation.

34 COP2 *Control of lead at work. Approved code of practice* HSE Books revised June 1985 ISBN 0 7176 1046 2. This legislation is being reviewed as part of the Commission's programme to reform health and safety legislation.

35 Approved Codes Of Practice:
Parts 1 and 2 (L58): *The protection of persons against ionising radiations arising*

(b) on health hazards for particular employees classified in the Regulations, the precautions to be taken and the importance of complying with medical and technical requirements; and

(c) for female employees on the possible hazard to the unborn child and the importance of telling the employer as soon as they find out they are pregnant.

Control of Pesticides Regulations 1986[36]

Information on risks to health from exposure to pesticides and precautions.

Safety hazards

Provision and Use of Work Equipment Regulations 1992[37]

Information on:

(a) conditions and methods of use of work equipment (including hand tools); and

(b) foreseeable abnormal situations: what to do and lessons learned from previous experience.

Personal Protective Equipment at Work Regulations 1992[38]

Information on:

(a) risk(s) that the personal protective equipment (PPE) will avoid or limit;

(b) the PPE's purpose and the way it must be used; and

from any work activity HSE Books 1994 ISBN 0 7176 0508 6.
Part 3 (COP 23): *Exposure to radon: the Ionising Radiations Regulations 1985. Approved Code of Practice* HSE Books 1988 ISBN 0 11 883978 0.
Part 4 (L7): *Dose limitation—restriction of exposure: additional guidance on regulation 6 of the Ionising Radiations Regulations 1985. Approved Code of Practice* HSE Books 1991 ISBN 0 11 885605 7.

36 L9(rev) *The safe use of pesticides for non-agricultural purposes. Control of. Substances Hazardous to Health Regulations 1994. Approved Code of Practice* HSE Books ISBN 0 7176 0542 6.

37 L22 *Work equipment. Provision and Use of Work Equipment Regulations 1992. Guidance on Regulations* HSE Books 1992 ISBN 0 7176 0414 4.

38 L25 *Personal Protective Equipment at Work Regulations 1992. Guidance on Regulations* HSE Books ISBN 0 7176 0415 2.

(c) what your employee needs to do to keep the PPE in working order and good repair.

Special hazards

Control of Industrial Major Accident Hazards Regulations 1984[39]

Information necessary to ensure employees' safety.

Railways (Safety Case) Regulations 1994[40]

Safety cases need to demonstrate that there are adequate arrangements for giving health and safety information to everyone concerned with the operation.

Safety representatives must be consulted on the preparation or revision of safety cases.

Road Traffic (Carriage of Dangerous Substances in Road Tankers and Tank Containers) Regulations 1992[41] *and*
Road Traffic (Carriage of Dangerous Substances in Packages etc) Regulations 1992[42]

Information in writing (required by both sets of Regulations) for drivers on:

(a) specified information about the load, eg what the substance is and the amount being carried; and

(b) hazards created by the substance and how to deal with emergencies.

(NB Both sets of Regulations are to be replaced by the Carriage of Dangerous Goods by Road Regulations on 1 January 1997 but the requirements will remain largely the same.)

39 HS(R)21 *A guide to the Control of Industrial Major Accident Hazards Regulations 1984* HSE Books 1990 ISBN 0 11 885579 4. HS(G)25 *Control of Industrial Major Accident Hazard Regulations 1984 (CIMAH). Further guidance on emergency plans* HSE Books 1985 ISBN 0 11 883831 8.

40 L52 *Railways (Safety Case) Regulations 1994* HSE Books ISBN 0 7176 0699 6.

41 Road Traffic (Carriage of Dangerous Substances in Road Tankers and Tank Containers) Regulations 1992: SI 1992 (743) ISBN 0 11 023743 9.

42 Road Traffic (Carriage of Dangerous Substances in Packages etc) Regulations 1992: SI 1992 (742) ISBN 0 11 023742 0.

Pressure Systems and Transportable Gas Containers Regulations 1989[43]

Operators (employees) of installed or mobile pressure systems to be informed about:

(a) safe operation of the system; and

(b) action to be taken in the case of an emergency.

Dangerous Substances in Harbour Areas Regulations 1987[44]

Information for employees handling dangerous substances to ensure their own health and safety (and that of others).

Management and Administration of Safety and Health at Mines Regulations 1993[45]

Information on:

(a) risks to health and safety; and

(b) preventive and protective measures.

Nuclear Installations Act 1965 (as amended)[46]

Employers who hold site licences must give information on:

(a) safety; and

(b) effective implementation of emergency arrangements.

43 HS(R)30 *A guide to the Pressure System and Transportable Gas Containers Regulations 1989* HSE Books 1990 ISBN 0 7176 0489 6. *An open learning course on the Pressure System and Transportable Gas Containers Regulations 1989* HSE Books ISBN 0 7176 0687 2.

44 COP 18 *Dangerous substances in harbour areas. The Dangerous Substances in Harbour Areas Regulations 1987. Approved Code of Practice* HSE Books 1987 ISBN 0 11 883857 1. HS(R)27 *A guide to the Dangerous Substances in Harbour Areas Regulations 1987* HSE Books 1988 ISBN 0 11 883991 8.

45 L44 *The management and administration of safety and health at mines. Management and Administration of Safety and Health at Mines Regulations 1993. Approved Code of Practice* HSE Books 1993 ISBN 0 7176 0618 X.

46 HS(G)120 *Nuclear site licences: notes for applicants* HSE Books 1994 ISBN 0 7176 0795 X.

Construction (Design and Management) Regulations 1994[47]

Contractors must provide employees who are engaged in construction work with information:

(a) on the risks to their health and safety; and

(b) as required under the Management of Health and Safety Regulations.

Principal contractors are responsible for ensuring that this happens. They should also display all notifiable project information on a notice and bring this and any site rules to the attention of everyone who may be affected by them.

APPENDIX B
REQUIREMENTS FOR INSTRUCTIONS AND TRAINING FOR EMPLOYEES IN EXISTING LEGISLATION

The following health and safety legislation requires employers to instruct and train their employees. Each entry contains a brief summary of what is required but you will need to find out your precise duties from the publications referred to in footnotes 48–78.

General health and safety

Management of Health and Safety at Work Regulations 1992[48]

Health and safety training:

(a) on recruitment;

(b) on being exposed to new or increased risks; and

(c) repeated as appropriate.

47 L54 *Managing construction for health and safety—Construction (Design and Management)Regulations 1994. Approved Code of Practice* HSE Books 1995 ISBN 0 7176 0792 5. CONIAC guidance: *A guide to managing health and safety in construction* HSE Books ISBN 0 7176 0755 0.

48 L21 *Management of Health and Safety at Work Regulations 1992. Approved Code of Practice* HSE Books 1992 ISBN 0 7176 0412 8.

Health and Safety (First Aid) Regulations 1981[49]

First-aiders provided under the Regulations must have received training approved by HSE.

Health and Safety (Safety Signs and Signals) Regulations 1995[50]

Each employee must be given instruction and training on:

(a) the meaning of safety signs; and

(b) measures to be taken in connection with safety signs.

Health and Safety (Consultation with Employees) Regulations 1996[51]

Training for employee representatives in their functions as representatives (as far as is reasonable). You are required to meet the costs of this training, including travel and subsistence and giving time off with pay for training. See paragraphs 23 and 24 of these Guidance Notes.

Safety Representatives and Safety Committees Regulations 1977[52]

Sufficient time off with pay for safety representatives to receive adequate training in their functions as a safety representative.

Health hazards

Control of Substances Hazardous to Health Regulations 1994[53]

Instruction and training in:

(a) risks created by exposure to substances hazardous to health (eg high hazard biological agents) and precautions;

49 COP42 *First aid at work. Health and Safety (First Aid) Regulations 1981. Approved Code of Practice and guidance* HSE Books ISBN 0 7176 0426 8.

50 L64 *Safety Signs and Signals: Guidance on Regulations* HSE Books ISBN 0 7176 0870 0.

51 *A guide to the Health and Safety (Consultation with Employees) Regulations 1996* HSE Books 1996 ISBN 0 7176 1234 1.

52 COP 1 *Safety representatives and safety committees* Rev ISBN 0 7176 1220 1.

53 L5 *General COSHH ACOP (Control of substances hazardous to health). Carcinogens ACOP (Control of carcinogenic substances) and Biological agents ACOP (Control of biological agents). Control of Substances Hazardous to Health. Approved Code of Practice 1995* HSE Books ISBN 0 7176 0819 0.

(b) results of any required exposure monitoring; and

(c) collective results of any required health surveillance.

Genetically Modified Organisms (Contained Use) Regulations 1992[54]

Training and local rules for the safety of employees whose work involves genetically modified organisms.

Health and Safety (Display Screen Equipment) Regulations 1992[55]

Adequate health and safety training in the use of any workstation to be used.

Noise at Work Regulations 1989[56]

Instruction and training for employees likely to be exposed to daily personal noise levels at 85dB(A) or above:

(a) noise exposure: level, risk of damage to hearing and action employees can take to minimise that risk;

(b) personal ear protectors (to be provided by employers): how to get them, where and when they should be worn, how to look after them and how to report defective ear protectors/noise control equipment;

(c) when to seek medical advice on loss of hearing; and

(d) employees' duties under the Regulations.

54 Genetically Modified Organisms (Contained Use) Regulations 1992 SI 1992 (3217) ISBN 0 11 025332 9.

55 L26 *Display screen equipment work. Health and Safety (Display Screen Equipment) Regulations 1992. Guidance on Regulations* HSE Books 1992 ISBN 0 7176 0410 1.

56 IND(G)75L(Rev) leaflet: *Noise at work: introducing the Noise at Work Regulations 1989* HSE Books. IND(G)99L leaflet: *Noise at work: advice for employees* HSE Books 1995 ISBN 0 7176 0962 6. IND(G)193L leaflet: *Health surveillance in noisy industries* HSE Books 1995 ISBN 0 7176 0933 2. IND(G)200L leaflet: *Ear protection in noisy firms—employers' duties explained* HSE Books 1995 ISBN 0 7176 0924 3.
Noise at work. Noise guide No 1: Legal duties of employers to prevent damage to hearing. Noise guide No 2: Legal duties of designers, manufacturers, importers and suppliers to prevent damage to hearing. The Noise at Work Regulations 1989 (one volume) HSE Books 1989 ISBN 0 7176 0454 3.

Health and Safety (Consultation with Employees) Regulations 1996

Control of Asbestos at Work Regulations 1987[57]

Instruction and training about risks and precautions for:

(a) employees liable to be exposed to asbestos; and

(b) employees who carry out any work connected with your duties under these Regulations.

Control of Lead at Work Regulations 1980[58]

Instruction and training about risks and precautions for:

(a) employees liable to be exposed to lead; and

(b) employees who carry out any work connected with your duties under these Regulations.

Ionising Radiations Regulations 1985[59]

Instruction and training to enable employees working with ionising radiations to meet the requirements of the Regulations, eg in radiation protection for particular groups of employees classified in the Regulations.

Safety hazards

Provision and Use of Work Equipment Regulations 1992[60]

Employees who use work equipment (including hand tools) and

57 L27 *The Control of Asbestos at Work. Control of Asbestos at Work Regulations. 1987. Approved Code of Practice* (2nd edn) HSE Books 1993 ISBN 0 11 882037 0. This legislation is being reviewed as part of the Commission's programme to reform health and safety legislation.

58 COP2 *Control of lead at work. Approved code of practice* HSE Books (revised) June 1985 ISBN 0 7176 1046 2. This legislation is being reviewed as part of the Commission's programme to reform health and safety legislation.

59 Approved Codes Of Practice:
Parts 1 and 2 (L58): *The protection of persons against ionising radiations arising from any work activity* HSE Books 1994 ISBN 0 7176 0508 6.
Part 3 (COP 23): *Exposure to radon: the Ionising Radiations Regulations 1985. Approved Code of Practice* HSE Books 1988 ISBN 0 11 883978 0.
Part 4 (L7): *Dose limitation—restriction of exposure: additional guidance on regulation 6 of the Ionising Radiations Regulations 1985. Approved Code of Practice* HSE Books 1991 ISBN 0 11 885605 7.

60 L22 *Work equipment. Provision and Use of Work Equipment Regulations 1992. Guidance on Regulations* HSE Books 1992 ISBN 0 7176 0414 4.

those who manage or supervise the use of work equipment need health and safety training in:

(a) methods which must be used; and

(b) any risks from use and precautions.

Personal Protective Equipment at Work Regulations 1992[61]

Employees who must be provided with personal protective equipment (PPE) need instruction and training in:

(a) risk(s) the PPE will avoid or limit;

(b) the PPE's purpose and the way it must be used; and

(c) how to keep the PPE in working order and good repair.

Office, Shops and Railway Premises Act 1963, section 19 (Training and supervision of persons working at dangerous machines)[62]

Training and supervision of people working at machines.

Factories Act 1961, section 21 (Training and supervision of young persons working at dangerous machines)[63]

Training, instruction and supervision of young people working on prescribed machinery.

Woodworking Machines Regulations 1974[64]

Prescribed training, instruction and supervision of employees on woodworking machines.

61 L25 *Personal Protective Equipment at Work Regulations 1992. Guidance on Regulations* HSE Books ISBN 0 7176 0415 2.

62 Office, Shops and Railway Premises Act 1963 Ch 41 ISBN 0 1085 0111 6. This legislation is being reviewed as part of the Commission's programme to reform health and safety legislation.

63 This legislation is being reviewed as part of the Commission's programme to reform health and safety legislation.

64 L4 *Woodworking Machines Regulations 1974. Guidance on Regulations* HSE Books 1991 ISBN 0 11 885592 1 (Due to be revised in January 1997). This legislation is being reviewed as part of the Commission's programme to reform health and safety legislation.

Health and Safety (Consultation with Employees) Regulations 1996

Power Presses Regulations 1965[65]

Prescribed training and competence for appointed people preparing, installing and adjusting tools or safety devices on power presses.

The Abrasive Wheels Regulations 1970[66]

Prescribed training and competence for appointed people mounting abrasive wheels.

Special hazards

Control of Industrial Major Accident Hazards Regulations 1984[67]

Training for employees necessary to ensure their safety under the Regulations.

Road Traffic (Carriage of Dangerous Substances in Road Tankers and Tank Containers) Regulations 1992;[68] *and*
Road Traffic (Carriage of Dangerous Substances in Packages etc) Regulations 1992[69]

> *. . . [The Carriage of Dangerous Goods by Road Regulations 1996 (SI 1996 No 2095), reg 29, revokes both sets of Road Traffic regulations (SI 1992/742, SI 1992/743)]*

Road Traffic (Carriage of Explosives) Regulations 1989[70]

> *. . . [The Carriage of Explosives by Road Regulations 1996 (SI 1996 No 2093), reg 34, revokes the Road Traffic (Carriage of Explosives) Regulations 1989 (SI 1989 No 615)]*

65 L2 *Power presses. The Power Presses Regulations 1965 and 1972* HSE Books 1991 ISBN 0 11 885534 4. This legislation is being reviewed as part of the Commission's programme to reform health and safety legislation.
66 HS(G)17 *Safety in the use of abrasive wheels* HSE Books 1992 ISBN 0 7176 0466 7. This legislation is being reviewed as part of the Commission's programme to reform health and safety legislation.
67 HS(R)21 *A guide to the Control of Industrial Major Accident Hazards Regulations 1984* HSE Books 1990 ISBN 0 11 885579 4. HS(G)25 *Control of Industrial Major Accident Hazard Regulations 1984 (CIMAH). Further guidance on emergency plans* HSE Books 1985 ISBN 0 11 883831 8.
68 Road Traffic (Carriage of Dangerous Substances in Road Tankers and Tank Containers) Regulations 1992: SI 1992 (743) ISBN 0 11 023743 9.
69 Road Traffic (Carriage of Dangerous Substances in Packages etc) Regulations 1992: SI 1992 (742) ISBN 0 11 023742 0.
70 Road Traffic (Carriage of Explosives) Regulations SI 1989 (615) ISBN 0 11 096615 5.

Road Traffic (Training of Drivers of Vehicles Carrying Dangerous Goods) Regulations 1992[71]

> *. . . [The Carriage of Dangerous Goods by Road (Driver Training) Regulations 1996 (SI 1996 No 2094), reg 13, revokes the Road Traffic (Training of Drivers of Vehicles Carrying Dangerous Goods) Regulations 1992 (SI 1992 No 744)]*

Training for employees necessary to ensure their safety under the Regulations.

In some cases, the driver has to attend an approved training course, pass an approved examination and hold a related training certificate.

Pressure Systems and Transportable Gas Containers Regulations 1989[72]

Operators (employees) of installed or mobile pressure systems to be instructed on:

(a) safe operation of the system; and

(b) action to be taken in the case of an emergency.

Dangerous Substances in Harbour Areas Regulations 1987[73]

Employees handling dangerous substances to be instructed and trained to ensure their own health and safety (and that of others).

Management and Administration of Safety and Health at Mines Regulations 1993[74]

Instruction and training in:

(a) risks to health and safety; and

(b) preventive and protective measures.

71 Road Traffic (Training of Drivers of Vehicles Carrying Dangerous Goods) Regulations 1992 SI 1992/744 ISBN 0 11 023744 7.

72 HS(R)30 *A guide to the Pressure System and Transportable Gas Containers Regulations 1989* HSE Books 1990 ISBN 0 7176 0489 6. *An open learning course on the Pressure System and Transportable Gas Containers Regulations 1989* HSE Books ISBN 0 7176 0687 2.

73 COP 18 *Dangerous substances in harbour areas. The Dangerous Substances in Harbour Areas Regulations 1987. Approved Code of Practice* HSE Books 1987 ISBN 0 11 883857 1. HS(R)27 *A guide to the Dangerous Substances in Harbour Areas Regulations 1987* HSE Books 1988 ISBN 0 11 883991 8.

74 L44 *The management and administration of safety and health at mines. Management and Administration of Safety and Health at Mines Regulations 1993. Approved Code of Practice* HSE Books 1993 ISBN 0 7176 0618 X.

Health and Safety (Consultation with Employees) Regulations 1996

Nuclear Installations Act 1965 (as amended)[75]

Employers who hold nuclear site licences must give instruction and training on:

(a) safety; and

(b) effective implementation of emergency arrangements.

Fire Precautions (Sub-surface Railway Stations) Regulations 1989[76]

Instruction and training in fire precautions and emergency procedures.

Agriculture (Circular Saws) Regulations 1959 (remaining parts)[77]

Training in the operation of a circular saw (or similar type).

Construction (Design and Management) Regulations 1994[78]

Principal contractors are responsible for ensuring (as far as is reasonably practicable) that contractors comply with the Management of Health and Safety at Work Act.

APPENDIX C
HELP WITH HOLDING ELECTIONS

This Appendix is not reproduced in this book.

APPENDIX D
REFERENCES

This Appendix has been reproduced in these Guidance Notes as footnotes 22–62 and 64–78.

75 HS(G)120 *Nuclear site licences: notes for applicants* HSE Books 1994 ISBN 0 7176 0795 X.

76 Fire Precautions (Sub-surface Railway Stations) Regulations 1988 SI 1988 (1401).

77 AS 11 *Circular saws* Free leaflet HSE Books. This legislation is being reviewed as part of the Commission's programme to reform health and safety legislation.

78 L54 *Managing construction for health and safety—Construction (Design and Management)Regulations 1994. Approved Code of Practice* HSE Books 1995 ISBN 0 7176 0792 5. CONIAC guidance: *A guide to managing health and safety in construction* HSE Books ISBN 0 7176 0755 0.

Index